HODGE

The Dynamic Earth

The Dynamic Earth

an introduction to physical geology

Brian J. Skinner

Yale University

Stephen C. Porter

University of Washington

WILEY

John Wiley & Sons

New York Chichester Brisbane
Toronto Singapore

Cover Photo: *Eruption of Mt. St. Helen's, from Krafft/*
Explorer/Science Source/Photo Researchers.

Library of Congress Cataloging in Publication Data:

Skinner, Brian J.
  The dynamic earth.
  Bibliography: p.
  Includes index.
  1. Physical geology. I. Porter, Stephen C.
II. Title.
QE28.2.S55 1989   551   88-20725
ISBN 0-471-60618-9

Printed in the United States of America

10 9 8 7 6 5 4 3

Preface

Geology is the science of the Earth. It is a special science because its laboratory is the world in which we live. Only rarely, however, can we carry out experiments in the geological laboratory. Either the scales of space and time needed for such experiments are too large, or the experiments would cause our environment to change in some unfortunate way. Geologists must study the Earth as it exists. From the assembled observations they draw conclusions about the processes that shape the Earth and events that have affected it over the past 4–5 billion years.

Revolutionary advances in the breadth and depth of our knowledge of the Earth have occurred over the past 35 years. At no previous time during human history have so many dramatic discoveries been made within such a short time. The revolutionary discoveries continue unabated. Geology is a field in ferment, a subject laced with challenging excitement; new discoveries, new insights, and new theories heighten the excitement almost every day.

A Geological Revolution

A few short years ago the suggestion that the outer layer of the Earth moves laterally at rates of up to 10 cm/yr was embodied in a theory called plate tectonics. Evidence to support the theory was discovered in rocks on the seafloor, and while the evidence was compelling that movement had happened in the past, it did not prove that such movements were occurring today. In 1986, measurements using satellites and lasers demonstrated, for the first time, that continents really are moving. Plate tectonics is no longer just a theory; it's a fact.

The plate tectonics revolution came about through such seemingly unrelated studies as exploration of the ocean floor, seismic studies of the Earth's core, and long-term measurements of the strength of the magnetic field. However, the studies are not unrelated. The Earth's magnetism arises in the core, and rocks of the seafloor are influenced by the magnetic field in distinctive ways. The realization that many of the Earth's processes, whether large or small, can interact in many unsuspected ways is forcing geologists to reexamine every scrap of evidence, and to rethink every conclusion. The discovery of plate tectonics was one result of that rethinking.

Life and the Earth

Perhaps the most profound geological discovery will be one that is still emerging. It is slowly becoming apparent that the Earth is the way it is because there is life on Earth, and that the converse is also true—all living things on the Earth (microorganisms, plants, and animals) are the way they are because the Earth is the way it is. The compositions of the atmosphere, the ocean, and even the rocks beneath our feet are strongly influenced by the activities of living matter; and living things, in turn, are influenced and controlled by the atmosphere, ocean, rocks, and soils.

Each and every one of us plays a part in the changes that ceaselessly work to maintain the balance of the Earth. But modern civilization is causing changes that threaten to upset the long-maintained balance between life and the Earth. Our individual contributions are tiny, but the sum of all human activities is large. We influence the atmosphere, streams, lakes, and oceans; we affect rates of erosion, and the way deserts expand or contract; we cover the surface of the land with roads and cities, and we redistribute the Earth's materials by digging them up and then using them for the multitude of things we need for living in the complex society we have developed. We human beings have become a vital force in the shaping of our own environment.

Within the solar system our planet is unique. No other planet is known to support life. None has such a friendly atmosphere nor such a comfortable range of climates. None has an ocean of water. So far as we can tell, no other planet can supply the wide range of minerals and fuels we human beings have come to need in order to support our complex society.

The human race is completely dependent on the resources of the Earth. These resources include air, water, and soil, as well as the materials we grow and dig from the ground. We have come to depend on so many things in order to live that we seem to teeter on the brink of a crisis. The crisis is one of balance. Because there are so many human beings, there is a possibility that the friendly balance between them and nature could be upset.

Population experts calculate that the number of human beings will rise from 1988's 5 billion people to about 7 billion by the year 2000, and a century from now to about 12 billion. Can the Earth supply all the food and materials needed for 12 billion people? Can it withstand the changes, the pollution, the environmental stress that 12 billion people will cause?

With a limited view confined to the natural community around us, we sometimes fail to realize the magnitude of the changes that human activities create in the Earth's natural features. A major example is the group of changes being wrought by strip mining of coal in the Appalachian region of the United States. Destruction of terrain and pollution of water are clearly evident in the huge continuous gashes along hillsides. Broad horizontal shelves floored with broken rock waste and backed by sheer rock walls are the most conspicuous feature of many Appalachian landscapes (Fig. P.1). Such shelves now aggregate more than 30,000 km in length, a distance that is more than one half the circumference of the Earth. As stripping proceeds, it destroys forests and soils, reduces the amount and quality of the water in the ground, and pollutes streams by pouring into them quantities of debris and acid chemicals released from the coal. Landslides are created, and this in turn causes erosion that leads to the piling up of debris on valley floors.

FIGURE P.1 Coal mining in the Appalachians, Tremont, Pennsylvania. Benches cut in the steep hillsides expose coal seams and change the shape of the land.

Slopes once forested become dangerous eyesores; entire stream systems can become clogged and lifeless.

The Earth, Small but Intricate

Although small, Planet Earth is made up of different materials, all of them involved in a wonderful array of interlocking movements, some slow, some fast. There is movement in the various layers that form the Earth's interior. The Earth's outermost layer is continually being destroyed and renewed as rocks form soil, then streams and winds transport the soil particles to new resting places. At the surface of the Earth all living things interact with one another and with the Earth's nonliving parts such as the atmosphere, seawater, and soils. It is especially necessary that we all understand and appreciate the repeating, cyclic pattern of activities at the Earth's surface, and how human activities can cause disruptions to the natural cycles.

To you, the reader, the facts of physical geology and related facts about the world of plants and animals offer a graphic explanation of these Earth systems. Your generation has now become responsible for maintenance of the systems. The Earth sciences, of which physical geology is the foundation, are the study of our environment. To the average educated man or woman, regardless of occupation, this knowledge is necessary as never before. A broad view of the Earth's natural activities is an essential beginning to the person who aims at a professional career in any of the many aspects of environmental studies. Just as the recognition of plate tectonics created a scientific revolution, so is the realization of the Earth's finite limits creating a social revolution. However, an understanding of the limits can come only through scientific study of the Earth itself and how it works. This is the domain of **geology.**

BRIAN J. SKINNER
STEPHEN C. PORTER

Acknowledgments

The roots of this book stretch back a long way to a suggestion made by the late Donald Deneck, editor for the Earth Sciences at John Wiley and Sons. The nurturing and fruition of the volume are due to Clifford Mills, the present editor. To both of these men we are exceedingly grateful for their help and encouragement. The need for flexibility in length and level in introductory college texts was recognized by both Mills and Deneck, and at the urging of Clifford Mills we undertook the task of reducing the length, and to some extent the range of material presented in *Physical Geology* (1987). We are greatly indebted to all of the staff with whom we have worked at John Wiley and Sons. Their professional skills, experience, and competence make an author's lot much easier than it might otherwise be.

We have not attempted to avoid difficult or controversial topics—to do so would be to diminish the pleasurable challenge of science. We have included those topics and issues we consider to be most important, and we have attempted to do so in a way that is understandable, rigorous, challenging, and most importantly, fun. We have very much enjoyed preparing this volume; we hope you, the reader, will sense and share our enjoyment.

We are indebted to the many people who provided elegant colored photographs that appear in the book. Most of the photographers are geologists and their discerning eyes can be sensed through the beautiful photographs they took. Their names are listed in the Photo Credits at the back of the book, but being so placed is no reflection on their importance. Through their eyes we believe a generation of students will see the world in a new way.

We especially thank the thoughtful and dedicated teachers who suggested how *Physical Geology* might be condensed while increasing clarity and vitality, and the exceptionally thoughtful comments of those who reviewed the final text: Gary C. Allen, University of New Orleans, New Orleans, Louisiana; J.C. Allen, Bucknell University, Lewisburg, Pennsylvania; N.L. Archbold, Western Illinois University, Macomb, Illinois; Collette D. Burke, Wichita State University, Wichita, Kansas; Robert A. Christman, Western Washington University, Bellingham, Washington; Nicholas K. Coch, Queens College, Flushing, New York; Kristine J. Crossen, Anchorage Community College, Anchorage, Alaska; Grenville Draper, Florida International University, Miami, Florida; M. Ira Dubins, State University College, Oneonta, New York; Ann G. Harris, Youngstown State University, Youngstown, Ohio; Robert L. Hopper, University of Wisconsin, Eau Claire, Wisconsin; Peter L. Kresan, University of Arizona, Tucson, Arizona; Albert M. Kudo, University of New Mexico, Albuquerque, New Mexico; Nancy Lindsley-Griffin, University of Nebraska, Lincoln, Nebraska; William W. Locke, Montana State University, Bozeman, Montana; David N. Lumsden, Memphis State University, Memphis, Tennessee; Gerald Matisoff, Case Western Reserve University, Cleveland, Ohio; Gary Peters, California State University, Long Beach, California; Donald Ringe, Central Washington University, Ellensburg, Washington; Graham Thompson, University of Montana, Missoula, Montana; Charles P. Thornton, Pennsylvania State University, State College, Pennsylvania; James B. Van Alstine, University of Minnesota, Morris, Minnesota; Margaret S. Woyski, California State University, Fullerton, California.

There are certain features in *The Dynamic Earth* that do not appear in *Physical Geology*. The topical essay has been added to each chapter to try and present some of the excitement of current geological research to the reader. For students we have added review questions and lists of important words and phrases to each chapter. Some of the chapters have been updated significantly from the parent volume. This is especially so for some of the chapters dealing with surficial processes; the chapter on mass-wasting is an example. To the best of our ability this volume is as up-to-date as it can be. We have made every effort to introduce topics that are the focus of major research efforts today, but not to downplay the great advances in the earth sciences of the past century.

Above all, we hope that you, the readers, enjoy the volume. If you find deficiencies in what you read, please let us know.

SI UNITS

Regardless of the field of specialization, all scientists use the same units and scales of measurement. They do so to avoid confusion and the possibility that mistakes can creep in when data are converted from one system of units, or one scale, to another. By international agreement the SI units are used by all, and they are the units used in this text. SI is the abbreviation of Système International d'Unités (in English, the International System of Units).

Some of the SI units are likely to be familiar, some unfamiliar. The SI unit of length is the meter (m), of area the square meter (m^2), and of volume the cubic meter (m^3). The SI unit of mass is the kilogram (kg), and of time the second (s). The other SI units used in this book can be defined in terms of these basic units. Three important ones are:

The newton (N), a unit of force defined as that force needed to accelerate a mass of 1 kg by 1 m/s^2; hence 1 N = 1 $kg \cdot m/s^2$. (The period between kg and m indicates multiplication.)

The Joule (J), a unit of energy or work, defined as the work done when a force of 1 newton is displaced a distance of 1 meter; hence 1 J = 1 N·m. One important form of energy so far as the Earth is concerned is heat. The outward flow of the Earth's internal heat is measured in terms of the number of joules flowing outward from each square centimeter each second; thus, the unit of heat flow is $J/cm^2/s$.

The pascal (Pa), a unit of pressure defined as a force of 1 newton applied across an area of 1 square meter; hence 1 Pa = 1 N/m^2. The pascal is a numerically small unit. Atmospheric pressure, for example (15 $lb/in.^2$), is 101,300 Pa. Pressure within the Earth reaches millions or billions of pascals. For convenience, earth scientists sometimes use 1 million pascals (megapascal, or MPa) as a unit.

Temperature is a measure of the internal kinetic energy (expressed as movement) of the atoms and molecules in a body. In the SI system, temperature is measured on the Kelvin scale (K). The temperature intervals on the Kelvin scale are arbitrary, and they are the same as the intervals on the more familiar Celsius scale (°C). The difference between the two scales is that the Celsius scale selects 100°C as the temperature at which water boils at sea level, and 0°C as the freezing temperature of water at sea level. Zero degrees Kelvin, on the other hand, is absolute zero, the temperature at which all atomic and molecular motions cease. Thus, 0°C is equal to 273.15 K, and 100°C is 373.15 K. The temperatures of processes on and within the Earth tend to be at or above 273.15 K. Despite the inconsistency, earth scientists still use the Celsius scale when geological processes are discussed.

Contents

The Dynamic Earth

Thin tongues of recently cooled pahoe-hoe lava cover the flanks of Mauna Ulu, formed above a vent on the east rift zone of Kilauea, Hawaii.

The Earth's Materials

CHAPTER 1

The Earth, Inside and Out

The Earth seen from space. Africa is on the right-hand side, South America on the far left. Europe and the Arabian Peninsula can be seen across the upper-right-hand edge, separated from Africa by the Mediterranean Sea and the Red Sea. The spiral cloud patterns arise from the global wind circulation.

MATERIALS AND PROCESSES

Living on the Earth's solid surface we are very aware of the *materials* that surround us and the *processes* that alter them. The materials are substances such as rock, sand, clay, and organic matter. The processes change the materials and move them around.

Some examples of processes are: the breakdown of materials to form soil; water flowing in a river and carrying mud or sand; surf pounding against a shore and making a beach; and air flowing (wind), carrying sand and dust. These and other processes continually move materials from place to place. The solid particles they are moving are mostly bits of rock.

Rock.   *Rock* is any naturally formed, nonliving, firm, and coherent aggregate mass of solid matter that constitutes part of a planet. Note that the definition specifies a coherent aggregate. A pile of loose sand grains is not a rock because the grains are not coherent. A tree is not a rock, even though it is solid, because it is living matter. But coal, which is a compressed and coherent aggregate of twigs, leaves, and other bits of dead plant matter, is a rock.

Cycles of the Earth

A great number of processes endlessly change the Earth's surface, and they do so repetitively. The water that flows off through gutters, sewers, and rivers comes from rain, which comes from clouds, which in turn form from moisture which has evaporated from the ocean. As rivers empty into the ocean, the traveling water—ocean to land and land to ocean—completes a cycle and is ready to begin anew. This cycle, which is repeated endlessly, is called the hydrologic cycle. It is discussed in greater detail later in this chapter and also in chapters 9 and 10.

The rock particles carried in the flowing water are part of another cycle—the rock cycle—which interlocks with the hydrologic cycle. The particles of rock came originally from the firm rock that forms all continents and that is continually being broken into particles. Like the water itself, the rock particles are on their way to the ocean. There they will be spread out and deposited, and they will eventually form new rock. The rock cycle is also discussed in more detail later in this chapter.

External Processes

The Earth's surface is continually being changed by a complex group of processes called *erosion* ("wearing away"). Through erosion rocks are broken down physically and chemically into small fragments and the fragments are then moved around by wind, water, and ice. All the activities involved in erosion, and also in the moving around of the eroded materials, are together called *external processes* because they operate at or near the Earth's solid surface. The energy needed to drive external processes comes from the Sun's heat.

Internal Processes

It is not possible to see the interior of the Earth, but it is possible, nevertheless, to reason that activities must be occurring there. Evidence is provided by earthquakes, volcanic eruptions, and the thrusting up of mountain ranges. All activities involving changes to rocks in the Earth's interior are called *internal processes.* The energy that drives internal processes comes from the Earth's internal heat.

Heat Energy

Activities and energy are so intimately related that *energy* is defined as the capacity to produce activity. Turn a page of this book—you are using energy. Walking outside, driving a car, or merely turning on a light or sleeping, whatever the activity, you are using energy. Energy is vital for our existence, and vital, also, for the Earth's existence. Without energy, the Earth would be a lifeless planet.

Stored energy is said to be *potential energy.* When potential energy is released it can take many forms, each of which produces characteristic activities. We speak of *kinetic energy,* meaning the ability of a moving body to generate activity. A boulder poised on the edge of a cliff has potential energy that is converted to kinetic energy when it falls. Other forms of energy are heat, electrical, chemical, and atomic. Each of these kinds of energy is important for one or another of the Earth's processes (Table 1.1), but it is heat energy that drives most of these.

Heat and Temperature

Heat is the kinetic energy possessed by the motions of atoms. All atoms move constantly. The term *temperature* refers to hotness, or degree of heat. Temperature is an indication of the average speed of the moving atoms.

The faster atoms move, the higher the temperature and the hotter a body feels. In a solid object atoms move within confined spaces—in a sense

TABLE 1.1 *Activities Produced by Common Forms of Energy*

Energy	Common Activity
Kinetic	Flowing water, wind, waves, landslides
Heat	Volcanoes, hot springs, rainstorms
Chemical	Decaying vegetation, forest fires, rusting, burning coal
Electrical	Lightning, aurora
Radiant	Daylight, sunburn
Atomic	Heating Earth's interior

they rattle around their assigned spaces in the solid structure. If they move fast enough (become hot enough), they break out of their fixed positions and the solid is said to have melted. With faster movement still, which means more heat, atoms move with complete freedom, and the liquid is then said to have vaporized.

Temperature is measured by various scales. A common one is the Celsius scale, in which 100°C is selected as the temperature (or speed of the atoms) just sufficient to boil water at sea level, and 0°C as the freezing point of water.

Transmission of Heat

Heat energy is transmitted in three ways. The first is **conduction,** which occurs when atoms pass on some of their motions to adjacent atoms in a solid. Conduction is the process by which heat is transmitted through the side of a cup of hot coffee; it is also the way by which much of the Earth's internal heat is transmitted through rock and reaches the surface. Conduction, therefore, is the main way by which heat is transmitted through a solid; it is a slow process because the transfer occurs atom by atom.

A second and much faster way by which heat can be transmitted is **convection.** Convection occurs in liquids and gases in which the distribution of heat is uneven. When a liquid or gas is heated it expands and its density (mass per unit volume) decreases. Hot and less dense material floats upward, while colder and more dense material sinks to replace it, thereby setting up a convection cell or **convection current.**

The third way heat is transmitted is by **radiation.** Electromagnetic waves, like light and radio waves, transmit energy and they can pass through empty space. When electromagnetic rays hit a solid surface, or pass through a gas or liquid, some of the energy is absorbed and the temperature of the absorbing body rises. Heat reaches the Earth from the Sun through radiation.

Energy from the Sun

Electromagnetic waves reach the Earth from all the stars in the universe, but the total amount of heat energy they carry is tiny by comparison with the heat energy radiated to the Earth by the Sun. When the Sun's rays reach the Earth's atmosphere, approximately 40 percent is simply reflected back into space. The remaining 60 percent is absorbed, partly by the Earth's atmosphere and partly by the land and the sea. The energy absorbed by the sea warms the water and causes evaporation. The resulting water vapor forms clouds and eventually rain, snow, and other forms of precipitation. The energy absorbed by the land eventually warms the air and causes convection. Winds are the result of convection and as they blow over the sea they create waves. Thus, all external processes that change the Earth's surface—rain, glaciers, streams, winds, and waves—are driven by the Sun's energy.

Energy from the Earth's Interior

Everyone who goes down a mine discovers that rock temperatures increase with depth. Measurements made in deep drillholes and mines show that the rate of temperature increase (the **geothermal gradient**) varies with depth and from place to place in the world. The range is from 15 to 75°C/km. Direct temperature measurements cannot be made beyond the deepest drillholes, which are little more than 10 km deep. But the Earth's radius is 6371 km! Therefore, indirect means must be used to estimate temperatures in the deep interior. Physical properties that vary with temperature, such as the speeds of earthquake waves, can be used for this. The resulting estimate shows that temperatures continue to increase toward the Earth's center and eventually reach values of 5000°C or more at its center. When the Earth's size is considered, it is obvious that a vast amount of energy must be stored as heat within it.

The Earth's internal heat continually flows outward toward the surface. The average value of the **heat flow** that reaches the Earth's surface is about 6.3×10^{-6} J/cm^2/s. To sense how small this amount of heat energy is, imagine that we have an instrument that can catch and use all the heat reaching 1 m^2 of the Earth's surface. We would have to catch the escaping heat for approximately 14 days and nights before enough energy was gathered to raise the temperature of a cup of water to 100°C so that boiling would commence. Nevertheless, because the area of the Earth's surface is enormous, the total amount of internal heat energy that reaches the surface is enormous—it is estimated to be 2700 billion J/s! The heat loss is not constant everywhere. Just as the geothermal gradient varies from place to place, so does the heat flow, which is greatest near young volcanoes and active hot springs. Variations in surface heat flow tell us a great deal about activities below the Earth's surface, as we shall see in later chapters.

THE INTERNAL STRUCTURE OF THE EARTH

Pressure increases with depth due to the weight of overlying rock. As with temperature, internal pressures cannot be measured directly. But they can be estimated by means of indirect measure-

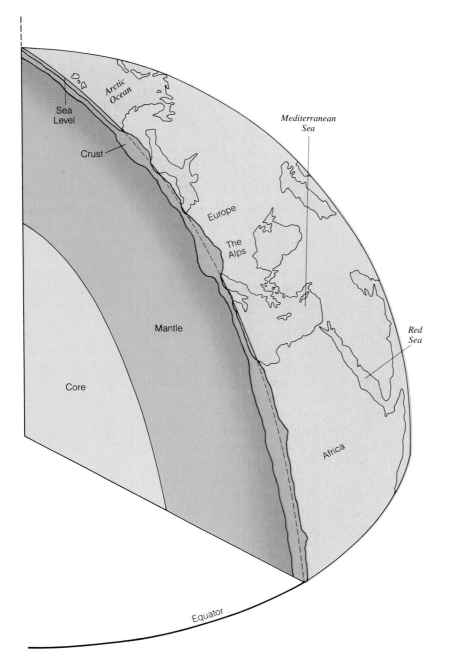

FIGURE 1.1 A slice of the Earth, revealing layers with distinctly different composition. The slice cuts through the North Pole. Note that crust is thicker under the continents and thinner under the ocean floor.

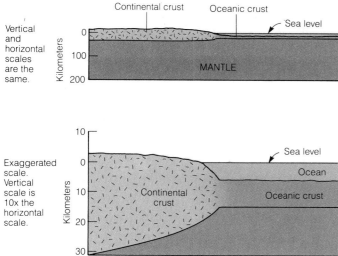

FIGURE 1.2 Section through the crust and upper part of the mantle. The crust is of two different kinds. Continental crust underlies the continents and is 20 to 70 km thick with an average of about 40 km; oceanic crust underlies the oceans and is only about 8 km thick.

ments. One way to do this is to measure how the density of rock changes with depth below the surface; this can also be done by measuring the speeds with which earthquake waves pass through the Earth. They move more quickly through more dense rocks (see chapter 14). From such measurements it can be calculated that density increases with depth. Increasing density is caused by increasing pres-

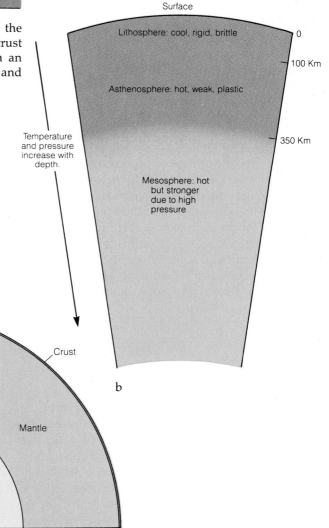

FIGURE 1.3 Layering of physical properties in the Earth. (a) Right side shows the compositional layering of crust, mantle, and core. Left side depicts changing physical properties with depth. Note that compositional changes and physical property changes do not coincide. (b) Expanded view of the upper portion of the left side of part a.

sure. But density does not increase smoothly with depth. There are some depths where abrupt increases occur. Thus we infer that the solid Earth does not have a uniform composition, but must instead consist of distinct layers with different densities and different compositions.

Layers of Differing Composition

Three compositional layers are recognized (Fig. 1.1). At the Earth's center is the most dense of the three layers, the *core.* The core is a spherical mass, largely of metallic iron, with lesser amounts of nickel, sulfur, and silicon.

The thick shell of dense, rocky matter that surrounds the core is called the *mantle.* The mantle is less dense than the core, but it is more dense than the outermost layer. Above the mantle lies the thinnest and outermost layer, the *crust,* which consists of rocky matter that is less dense than the rocks of the mantle below.

The core and the mantle have roughly constant thicknesses. The crust, on the other hand, is far from uniform and differs in thickness from place to place by a factor of nine. The crust beneath the oceans, the *oceanic crust,* has an average thickness of about 8 km, whereas the *continental crust* ranges from 20 to 70 km in thickness (Fig. 1.2).

Slight compositional variations probably exist within the mantle, but we know little about them. The crust is quite varied in its composition but it is very different from the mantle, and the boundary between them is sharp.

Layers of Differing Physical Properties

In addition to compositional layering, a layering of physical properties occurs within the Earth. The places where physical properties change do not coincide exactly with the compositional boundaries between core, mantle, and crust (Fig. 1.3). Within the core an inner region exists where pressures are so great that iron is solid despite its high temperature. Surrounding the inner core is a zone where temperature and pressure are so balanced that the iron is molten and exists as a liquid. Analogous changes in physical properties occur in the upper part of the mantle, except that melting is not involved. Rather, the change involves the strength of rocks.

The Mesosphere

The strength of a rock is controlled by both temperature and pressure. When a solid is heated it loses strength. When it is compressed it gains strength. In the lower part of the mantle rocks are so highly compressed they have considerable strength even though the temperature is high. The region of high temperature and high strength persists from the core–mantle boundary to a depth of about 350 km and is called the *mesosphere* ("intermediate, or middle sphere") (Fig. 1.3).

The Asthenosphere

From a depth of 350 to 100 km, in the upper part of the mantle, is a region called the *asthenosphere* ("weak sphere"), where the balance between temperature and pressure is such that rocks have little strength. Instead of being strong, like rocks in the mesosphere, or the rocks we see at the Earth's surface, rocks in the asthenosphere are weak and easily deformed, like butter or warm tar.

The Lithosphere

Above the asthenosphere and corresponding approximately to the outer 100 km of the solid Earth is a region where rocks are stronger and more rigid than those in the plastic asthenosphere. This hard outer region is called the *lithosphere* ("rock sphere").

The boundary between the lithosphere and the asthenosphere is again one where the balance between temperature and pressure causes the physical properties of rock to change rapidly. Above the boundary, rocks in the lithosphere are strong, and can only be deformed or broken with difficulty.

SURFACE FEATURES OF THE CRUST

Continents and Ocean Basins

The ocean covers some 71 percent of the world's surface, and its average depth is 3.7 km. The depth is very irregular, however; the greatest depth of 11 km is reached near the island of Guam, in the western Pacific.

The remaining 29 percent of the world's surface is occupied by land, which has an average height of 0.8 km above mean sea level. If it were possible briefly to remove all the water from the ocean and then to view the dry Earth from a spaceship, it would be possible to contrast the ocean basins and

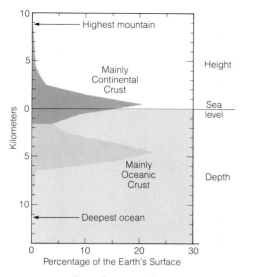

FIGURE 1.4 Distribution of the areas of the Earth's solid surface expressed as a percentage. Note that the surface of the continental crust is considerably higher than the surface of the oceanic crust. (*Source:* After Wyllie, 1976.)

the continents. It would be observed that the continents rise abruptly and stand about 4.5 km above the average depth of the floor of the ocean basin. This is because continental crust is relatively light (it has a density of 2.7 g/cm³), while oceanic crust, by comparison, is relatively heavy (it has a density closer to 3.2 g/cm³). In a sense, the lithosphere is "floating" on the asthenosphere. Lithosphere capped by continental crust stands high because it is light while that capped by oceanic crust is more dense and heavier so it sits lower (Fig. 1.4).

The Shape of Ocean Basins

Continental Shelf, Slope, and Rise

Modern shorelines lie above the continental margin. This is so because some of the ocean water spills out of the ocean basin onto the continent (Fig.

1.5). As a result each continent is surrounded by a flooded margin, of variable width, known as the **continental shelf.** The geological edge of the ocean basin is not the shoreline, but rather it is the **continental slope,** a pronounced slope beyond the seaward margin of the continental shelf. Taking the continental slope as the continental margin, only 60 percent of the Earth's surface is occupied by ocean basins, while 40 percent is occupied by continents. Thus 25 percent of the continental crust is covered by seawater (Fig. 1.6).

The **continental rise** lies at the base of the continental slope. It is a region of gently changing slope where the floor of the ocean basin meets the margin of the continent. The rise is actually part of the floor of the ocean basin, but it is a distinctive part because it is covered by a thick pile of erosional debris shed from the adjacent continent.

Submarine Canyons

The continental slope and rise, like the shelf, generally have a rather smooth surface. However, cut into them are many remarkable valleys, some so deep that they have been called **submarine canyons** (Fig. 1.7). These valleys, which are found on the edge of all continents, are as much as 1 km deeper than the adjacent seafloor and have steep sided slopes. Some are as long as 370 km, but the average probably is closer to 50 km.

Abyssal Plains

Beyond the continental slope and rise lies the strange, rarely seen world of the deep ocean floor. With newly perfected devices for sounding the sea bottom, for making submarine dives of limited duration, and for sampling its sediment, teams of oceanographers and seagoing geologists have explored and mapped the ocean floor so that almost as much is now known about the seafloor as is known about the land surface.

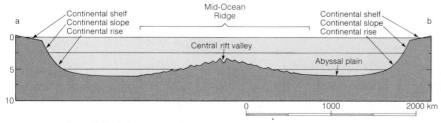

FIGURE 1.5 Simplified diagram of a section across a portion of the Atlantic Ocean showing the major topographic features. The profile is along the line a–b in Figure 1.10.

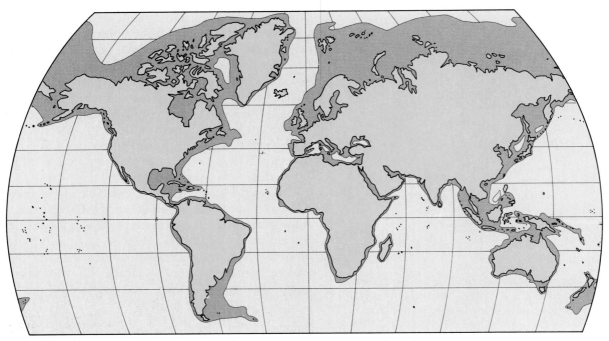

FIGURE 1.6 The continental shelves and slopes (shown in gray), together, form about 15 percent of the continents. The map projection exaggerates the shelves in the arctic region, which are nevertheless extensive.

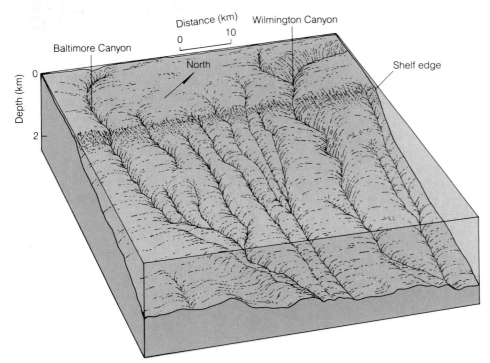

FIGURE 1.7 Submarine canyons cutting the continental shelf of the eastern United States off Maryland. The canyons, which head at or near the shelf edge, are as much as 1 km deep and have tributaries like those of large streams on the land. (*Source:* After B. A. McGregor, 1984.)

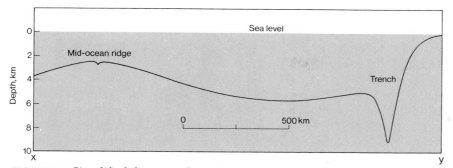

FIGURE 1.8 Simplified diagram of a section across a portion of the Pacific margin of South America. The profile is along the line x–y in Figure 1.10. Note that the side of the trench adjacent to the continent (to the right) is steeper than the oceanic side.

A major topographic feature of the seafloor is the **abyssal plain,** a large flat area of deep seafloor. Abyssal plains lie adjacent to the continental rise (Fig. 1.5). They generally are found at depths of 3000 to 6000 m and range in width from about 200 to 2000 km. Such features are most common in the Atlantic and Indian oceans, which have large sediment-laden rivers entering them and are not bounded by basins that can act as sediment traps. Where an abyssal plain occurs, the original seafloor topography has been completely buried beneath a blanket of fine sediment.

Ridges and Trenches

Two particularly prominent features of the seafloor are (1) **midocean ridges,** which are continuous rocky ridges on the ocean floor, measuring from many hundred to a few thousand kilometers wide, with a relief of more than 0.6 km; and (2) **trenches,** which are long, narrow, deep basins in the seafloor (Fig. 1.8).

The oceanic ridge system is a chain of mountains some 84,000 km long that twists and branches in a complex pattern through the ocean basins. This great mountain chain would be one of the most impressive features we would see if we could view a dry Earth from out in space.

A narrow valley, or rift, runs down the center of an oceanic ridge. It is characterized by unusually high heat flow and intense volcanic activity. At several places around the world the oceanic ridge reaches sea level and forms volcanic islands. The largest of these is Iceland, which lies on the center of the Mid-Atlantic Ridge. As a result we can examine part of the central rift valley on land (Fig. 1.9).

FIGURE 1.9 The central rift valley of the Mid-Atlantic Ridge can be seen in Iceland. A scarp 40 m high borders the western margin of the volcanically active and slowly widening rift.

PLATE TECTONICS AND THE EVER-MOVING LITHOSPHERE

All of the major features on the Earth's surface, whether submerged or on land, arise as a result of internal processes. By far the most important result of internal processes is the lateral motion of the lithosphere over the asthenosphere. Such motions involve complicated events, all of which are embraced by the term *tectonics.*

Tectonics. The word *tectonics* is derived from a Greek word, *tekton,* that means carpenter or builder. **Tectonics** is the study of movement and deformation of the crust on a large scale.

Plate Tectonics. The special branch of tectonics that deals with the processes by which the lithosphere is moved laterally over the asthenosphere is called *plate tectonics.* The strong rocky lithosphere not only moves, but it does so in a series of platelike pieces; it is the movement of the plates that causes continents and ocean basins to be where they are and to have the shapes they do. The plates range from several hundred to several thousand kilometers in width.

The very suggestion that a process such as plate tectonics might occur was first made in the late 1960s. Plate tectonics has now been proven, but recognition of the process is so new many of the details are still being investigated. The discoveries and new understandings that have come from plate tectonics are so profound, however, that the concept has sparked a modern geological revolution.

Continental Drift

Today the lithosphere is broken into six large plates and numerous smaller ones. The plates move at speeds ranging from 1 to 12 cm a year (Fig. 1.10). As a plate moves, everything on the plate moves too. If the cap of a plate is partly oceanic crust and partly continental crust, then both the ocean floor and the continent move with the same speed and in the same direction.

The idea that seafloor might move was first proposed in the early 1960s. But the realization that continents might move is much older and goes back to the early years of the present century. The idea of continental movement was most forcefully proposed by a German scientist, Alfred Wegener. The concept came to be called *continental drift.* When first proposed the idea did not receive widespread support because at the time no adequate explanation could be offered as to how it could happen. Plate tectonics provided the answer.

The original suggestion for continental drift was that continents must somehow slide across the seafloor. It was soon realized that friction would not allow this to happen. Rocks on the floor of the ocean basin are too rigid and strong for continents to slide over them. Eventually, following the discovery that the seafloor also moves and that the asthenosphere is weak and easily deformed, geologists realized that the entire lithosphere was in motion, not just the continents.

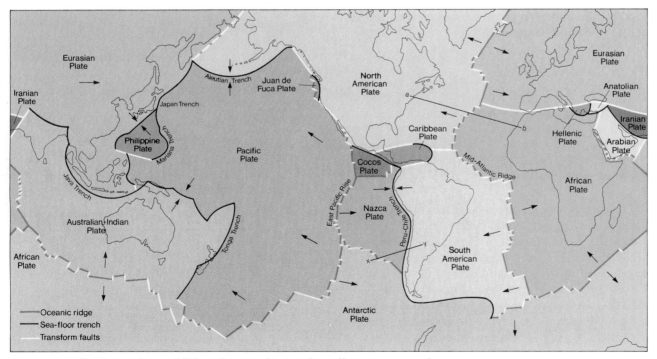

FIGURE 1.10 Six large plates of lithosphere and several smaller ones cover the Earth's surface and move continuously, in the direction shown by arrows. Plates have three kinds of margins: (1) spreading centers delineated by midocean ridges; (2) subduction zones, delineated by seafloor trenches; and (3) transform faults. The profile shown in Figure 1.5 lies along the line a–b, that in Figure 1.8 lies along the line x–y.

Unanswered questions still exist concerning plate tectonics. For example, what were the shapes and sizes of plates from past ages? The evidence is convincing that at times in the past sometimes fewer and at other times more plates were present. Plates change both in size and shape because new, smaller plates can form through the breaking up of larger plates. Also, larger plates can form by the collision and welding together of smaller plates. As discussed in later chapters, past breakups and weldings can be inferred from the geological record, but the evidence is rarely easy to decipher.

Plate Motions and Plate Margins

When we examine how plates move a good analogy is conveyor belts. In a conveyor, the belt continually appears from below, moves along the length, then turns down, and passes temporarily from sight as it completes its circuit. Although broad and irregular rather than long and narrow, a plate of lithosphere acts like the top of a slowly moving conveyor belt. One edge or margin of most plates is a long fracture in the oceanic crust that coincides with the midocean ridge. The plate moves away from the ridge just as if it were a continuous belt rising up the fracture from the mantle below. The analogy is only partly correct, because the plate is not rising as a solid ribbon. It is being created by the formation of new oceanic crust along the midocean ridge.

Formation of Oceanic Crust at a Spreading Center

It is not possible to see into the mantle beneath the midocean ridges, but it is possible to infer what must be happening. Hot, plastic rock in the asthenosphere rises toward the surface beneath the ridges and some small portion of the asthenosphere melts, giving rise to magma. *Magma* is defined as molten rock material that forms when temperatures rise and melting occurs in the mantle or crust. The magma that forms beneath the midocean ridges rises upward to the top of the lithosphere where it cools and hardens to form new oceanic crust (Fig. 1.11).

Another disparity in the analogy between a plate and a conveyor is that at the midocean ridge two plates are moving away in opposite directions. The central rift valley that marks the center of a midocean ridge is actually the surface expression of the join between the two plates. Because plates move, or spread outward, away from the midocean ridge,

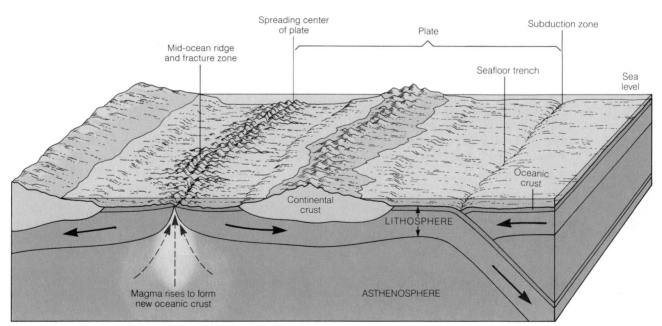

FIGURE 1.11 Section through the Earth's outer layers, showing how magma (dashed arrows) moves from the asthenosphere upward into spreading centers in the ocean floor, and cools there to form new oceanic crust. To accommodate the new materials, the lithosphere (solid arrows) moves away from the fracture zone and eventually sinks slowly down into the asthenosphere again, where it is reheated and eventually mixed again with the mantle.

the new, growing edge of a plate is called a *spreading center.*

Removal of Oceanic Crust at a Subduction Zone

Formation of new oceanic crust is a continuous process. Movement of lithosphere away from the oceanic ridge, like the movement of a conveyor belt, is a continuous process also. Near the spreading edge the lithosphere is thin, and it has a low density because it is heated and expanded by the rising magma. As the lithosphere moves away from the spreading edge it cools, contracts, and becomes denser. Also, the depth of the boundary between the lithosphere and the asthenosphere increases. Finally, at a distance of a thousand or more kilometers from the spreading edge, the lithosphere and its capping of oceanic crust is so cool it is more dense than the hot, weak asthenosphere below and it starts to sink downward. Like a conveyor belt, old lithosphere disappears back into the mantle. The edges along which plates of lithosphere turn down into the mantle, called *subduction zones,* are marked by deep trenches in the seafloor.

As the moving strip of lithosphere sinks slowly through the asthenosphere into the depths of the mantle, it passes from view. Consequently, what happens next is still largely conjecture. On one point, however, we can be quite certain: The lithospheric plate does not turn under, as a conveyor belt does, and reappear at the spreading edge; rather, it is reheated and slowly remixed with the material of the mantle.

Continental Crust

The process described is for plates of lithosphere capped by oceanic crust. All oceanic crust is geologically young because old crust is all returned to the mantle. Unlike oceanic crust, continental crust is not recycled into the mantle; it takes a shorter trip that ends more suddenly. Continental crust is lighter and less dense than even that part of the mantle in the hot asthenosphere. As a result, continental crust is too buoyant to be dragged downward on top of the sinking lithosphere. So, in continent-sized pieces, such crust moves from place to place on the Earth's surface, much as an ice floe floats on a lake or river. Movement stops when one continental mass collides with another, or a plate changes direction when a new spreading center splits the continental crust apart. Because continental crust does not sink down into the mantle, most of the evidence concerning ancient plates and their motions is recorded in the scars carried by ancient continental rocks.

Orogeny. Evidence can be found in the long, more-or-less linear belts of highly deformed rocks in mountain ranges such as the Alps, Appalachians, and Urals. The term *orogeny* refers to the tectonic processes by which large regions of the crust are deformed and uplifted to form such mountain ranges. The causes of orogenies were obscure before plate tectonics established that such mountain ranges form along the margins where continental masses collide.

Continental Splitting. Further evidence can be found in rocks on both sides of the Atlantic Ocean. When a new spreading center splits a large mass of continental crust, a new strip of growing oceanic crust separates the two pieces. Africa and Europe on one side, with the Americas on the other, provide an example. Two hundred and fifty million years ago there was no Atlantic Ocean. Instead, the continents that now border it were joined together into a single huge continent (Fig. 1.12). The place where New York now stands was then as far from the sea as central Mongolia is today.

About 200 million years ago a new spreading center formed. We do not yet fully understand why this occurred, but presumably it involved changes in the mantle below. The spreading center split the ancient continent into the pieces we see today. These fragments then drifted slowly into their present positions. At first the Atlantic Ocean was a narrow body of water that separated North America from Europe and North Africa. As movement continued the ocean widened and lengthened, splitting South America from Africa and then growing to its present form. The Atlantic is still growing wider by about 5 cm each year.

Evidence is abundant to mark where the torn edges formerly fitted together. In one region, pieces of mountain ranges that once formed a long, narrow mountain belt, like the Rocky Mountains of today, have been pulled apart so that now they lie on the two sides of the Atlantic. If these pieces are fitted back together the continental slopes on each side of the ocean, and the now deeply eroded mountains, fit like the matched pieces of a jigsaw puzzle (Fig. 1.12). The line of match follows the present Mid-Atlantic Ridge.

Transform Faults

Besides spreading centers and subduction zones there is a third kind of plate edge, analogous to the sides of a conveyor belt, along which plates simply slip past each other. These edges of slipping are great vertical fractures—or to use a term defined in chapter 15, **_transform faults_**—that cut right down through the lithosphere. One transform fault, much in the public eye because of the threat of earthquakes along it, is the San Andreas Fault in California. This fault separates the American Plate, on which San Francisco sits, from the Pacific Plate, on which Los Angeles sits. As the two plates slide past each other, Los Angeles is slowly moving northwest toward San Francisco.

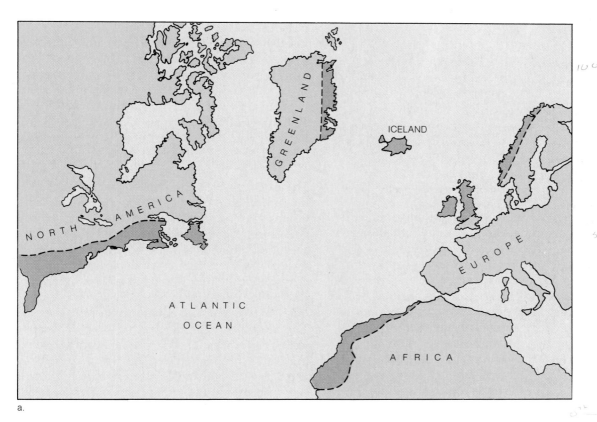

a.

b.

FIGURE 1.12 Opening of the Atlantic Ocean. (a) Rock in eroded fragments of similar mountain belts (brown)—each 350 to 470 million years old—is found on both sides of the Atlantic Ocean. (b) When continents are moved and fitted together as they were 200 million years ago, the fragments are seen to form a continuous belt. The reconstruction provides evidence that the present continents were once part of a larger landmass broken up by the moving lithosphere. Note that Iceland is not present in the reconstruction. It is a young landmass and is a piece of the ocean ridge that marks the line along which the continental separation occurred. (*Source:* Adapted from P. M. Hurley, 1968.)

TABLE 1.2 *Estimated Global Water Inventory*

Reservoir	Volume (thousand km³)	Volume (%)
Rivers	1.25	0.0001
Atmosphere	13	0.001
Soil moisture	67	0.005
Freshwater lakes	125	0.009
Saline lakes and inland seas	104	0.008
Groundwater (to 4 km depth)	8350	0.615
Glacier ice	29,200	2.15
Oceans	1,320,000	97.2
Total	1,357,860	100.

THE EXTERNAL LAYERS

Above the crust sit the ocean, lakes, streams, and other bodies of water, and above them the atmosphere. The interface between the crust and the water + air is a region of intense activity, for it is here that erosion occurs. Erosion continually breaks down rock and moves the broken particles around. As a result, the crust is covered by an irregular blanket of loose rock debris termed the *regolith.* Most of the world's plants and animals live either on the regolith or close to the water–air interface. It is helpful to think of the water, the air, living matter, and the regolith as external layers or envelopes that surround the solid Earth.

The Hydrosphere

The *hydrosphere* is the "water sphere" embracing the world's oceans, lakes, streams, underground water, and all the snow and ice, including glaciers.

Most of the water in the hydrosphere resides in the oceans, but significant amounts are also to be found in lakes, groundwater, and ice sheets (Table 1.2). The hydrosphere is discussed in more detail in chapters 9 and 10.

That portion of the hydrosphere that is ice, snow, and frozen ground is defined as the *cryosphere.* It is discussed in detail in chapter 12.

The Hydrologic Cycle

The day-to-day and long-term cyclic changes that we can observe in the Earth's hydrosphere are collectively known as the *hydrologic cycle* (Fig. 1.13). The cycle is powered by heat from the Sun which evaporates water from the ocean and land surface. The water vapor thus produced enters the atmosphere and moves with the flowing air. Some of it condenses and is precipitated as rain or snow back into the ocean or onto the land. Rain falling on the ground surface may either drain off in streams, percolate into the ground, or be evaporated back into the air where it is further recycled. Snow may remain on the ground for one or more seasons until it melts and the meltwater flows away. Snow that nourishes glaciers remains locked up much longer, through many years or even thousands of years, but eventually it too melts or evaporates and returns to the oceans. Part of the water in the ground is taken up by plants, which return water to the atmosphere through transpiration.

The Atmosphere

The *atmosphere* is the sphere that consists of the mixture of gases that together we call air. It penetrates into the ground, filling the openings, small and large, that are not already filled with water. The atmosphere is always moving, as we are well aware when we feel a wind blowing. The basic reason that the atmosphere is always in motion is that more of the Sun's heat is received per unit of land surface near the equator than near the poles. The heated air near the equator expands, becomes lighter, and rises, like the rise of steam above boiling water in a teakettle. High up, it spreads outward toward both poles. On the way it gradually cools, becomes heavier, and sinks. This air, chilled in higher latitudes, forms a return flow toward the equator. The returning air replaces the warm rising air and, in turn, is heated and rises.

The Coriolis Effect

The Earth's rotation modifies what could otherwise be a simple, convective circulation pattern in the atmosphere. The *Coriolis effect,* named after the nineteenth-century French mathematician who first analyzed it, causes any body that moves freely with respect to the rotating solid Earth to veer toward the right in the Northern Hemisphere and toward the left in the Southern Hemisphere. This is true regardless of the direction in which the body may be moving. Flowing water (such as an ocean current) and flowing air (wind) respond to it. So too do small bodies such as projectiles, if they travel over long paths.

The Coriolis effect breaks up the simple general flow of air between the equator and the poles into belts (Fig. 1.14). At about latitude 30° in the Northern Hemisphere some of the high-level, north-flowing equatorial air descends toward the Earth's

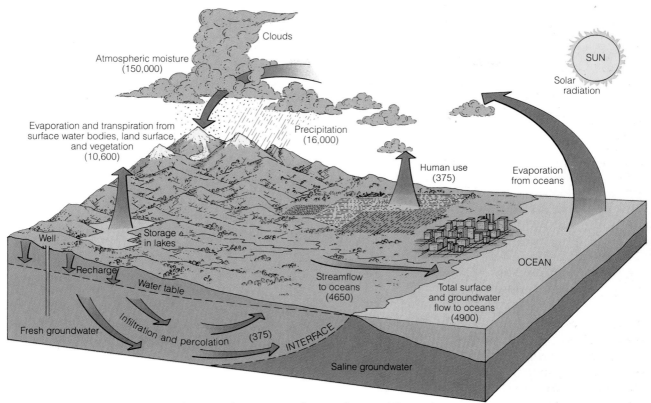

FIGURE 1.13 Hydrologic cycle, showing the amount of water that participates in the hydrologic cycle in the conterminous United States in millions of cubic meters of water a day. (*Source:* After U.S. Geological Survey, 1984.)

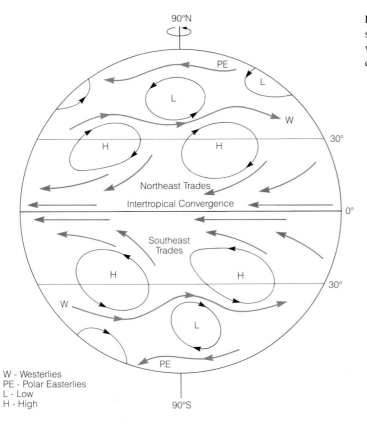

W - Westerlies
PE - Polar Easterlies
L - Low
H - High

FIGURE 1.14 The Earth's planetary wind belts, shown schematically. Red arrows show prevailing path of warm low-latitude winds, while blue arrows show cool high-latitude air flow.

solid surface. As it descends, this cold dry air becomes warmer. This means that near that latitude, all around the world, climates are warm and dry. As a result, much of the arid land in both the Northern and Southern Hemispheres is centered between latitudes 15° and 35° (Fig. 11.1).

Climate

Climate is the average weather of a place or area, and its variability, over a period of years. It is determined by a variety of factors, but those of greatest importance are temperature, precipitation, cloudiness, and windiness.

The distribution of climatic zones in belts that are approximately parallel to lines of latitude is interfered with and distorted by the pattern of oceans, continents, high mountains, and plateaus. Therefore, average temperature and precipitation vary greatly from one place to another. When a change,

however slight, occurs in this vast and complex system, local climates in many areas are likely to be affected.

The understanding of climatic change is made especially difficult because of the complexity of its interacting components (Fig. 1.15). The Sun's heat drives the system and constitutes a primary external factor, but at the same time the atmosphere, biosphere, hydrosphere, cryosphere, and lithosphere can all produce changes that will affect the climatic balance (chapter 12).

The Biosphere

The **biosphere** (the "life sphere"), is the totality of the Earth's living matter and, in addition, dead plants and animals that have not yet been completely decomposed. It embraces innumerable living things, large and small, grouped into millions of different species. We humans are one of the

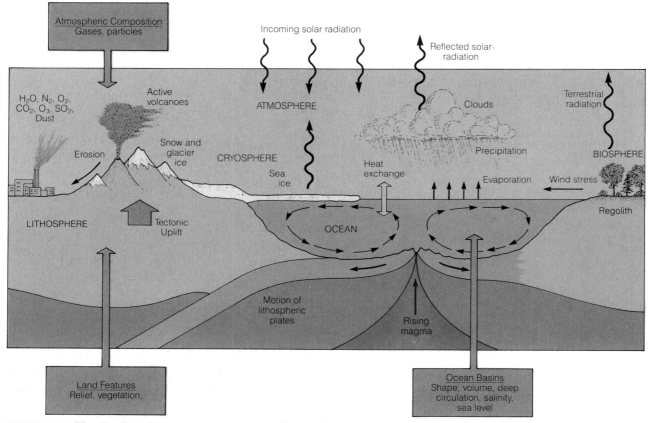

FIGURE 1.15 The Earth's climate system consists of several interacting subsystems: the atmosphere, cryosphere, oceans, lithosphere, and biosphere. Solar energy drives the system, although some of the incoming radiation is reflected back into space from clouds, snow, ice, and atmospheric pollutants. Tectonic movements affect surface relief and the geometry of continents and ocean basins, whereas volcanic and industrial gases affect atmospheric composition. (*Source:* Modified from Gates, 1979.)

species. The composition of the biosphere is distinctive; its chief constituents are compounds of carbon, hydrogen, and oxygen, although it includes other chemical elements as well.

The Regolith

Bedrock is the continuous mass of solid rock that makes up the crust. The atmosphere and hydrosphere react with the bedrock and cause it to break down into the regolith. We can observe the process whenever bedrock is exposed at the Earth's surface. Such exposed rock (or sediment) constitutes an *exposure* or *outcrop*.

Not all regolith remains in the place where it forms. Some of it has been moved and set down in a new site. It is on its way (discontinuously) to the ocean. Regolith that has been transported by any of the external processes is called **sediment** ("settling") (Fig. 1.16).

In some places, where bare bedrock is exposed at the surface, regolith is absent. In other places regolith is 100 m or more thick. In most land areas the upper part of the regolith forms soil (see chapter 7) in which plants grow, including the crops that are the principal food of human populations.

INTERACTIONS BETWEEN THE INTERNAL AND EXTERNAL LAYERS

The Earth's outer layers are places of intense and continual activity. Water and air penetrate the regolith and far into the crust. Chemical reactions and physical disintegration of rock proceed continually as the atmosphere, biosphere, and hydrosphere combine to alter and break down the crust.

When rock weathers to form regolith, some of the more soluble constituents dissolve and eventually concentrate in the ocean. This is the origin of many of the salts in seawater. When raindrops form they dissolve gases from the atmosphere and carry them down to the Earth's surface where they react to form new minerals in the soil. Since material is continually transferred between the Earth's spheres, why should the composition of the atmosphere be constant? Why doesn't the sea become saltier, or fresher? Why does rock 2 billion years old have the same composition as rock only 2 million years old? The answers to these questions are the same. Just as water passes through the hydrologic cycle, so do chemical elements and rock particles also follow cyclic paths. "Wheels within wheels" is an apt way to describe the movement.

Cyclic Movements

The operation of the Earth's natural processes, both internal and external, and the interactions between the various compositional spheres, all involve distinct **cycles.** However, if we carefully measure all of the parts of a cycle, we find that in any sphere (such as the hydrosphere or atmosphere) the materials added are balanced by those removed. Thus, as the cycles roll onward, the spheres maintain their sizes and compositions, or, if changes do occur, they tend to happen exceedingly slowly, often over millions of years.

Some of the most important questions to be answered about human interactions with the environment concern the way we are changing the natural cycles. Activities such as the burning of fossil fuels and the clearing of forests are adding carbon dioxide to the atmosphere more rapidly than it can be removed by natural processes. As a result the composition of the atmosphere is changing quite rapidly. Intensive agriculture with massive additions of fertilizers to the soil, and the mining of ever-larger amounts of mineral resources are slowly but surely changing the composition and size of the regolith. How nature will respond to such

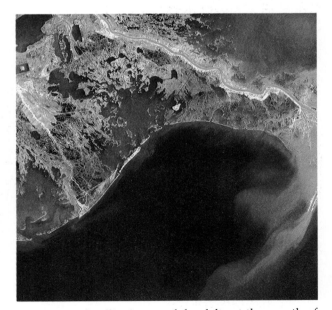

FIGURE 1.16 Satellite image of the delta at the mouth of the Mississippi River. The apparent sharpness of the line that separates the regolith (red) from the hydrosphere (blue) is misleading because the two spheres blend gradually. Off the river's mouth are plumes of turbid water, loaded with suspended particles of rock derived from erosion of the land. When the suspended particles settle, they build up the delta.

changes brought about by human activities is a topic of intense study today.

The Rock Cycle

The Rock Families

Igneous Rock.   The most important of the three major rock families is the **igneous rock** family (named from the Latin word *igneus,* meaning fire). Igneous rocks are formed by the cooling and consolidation of magma.

Sedimentary Rock.   Some products of weathering are soluble and are carried away in solution by streams and rivers, but most weathering products are loose particles that are carried away in suspension. Both the dissolved and suspended materials can be deposited later as sediment and eventually become **sedimentary rock,** which is any rock formed by chemical precipitation or by sedimentation and cementation of mineral grains transported to a site of deposition by water, wind, or ice. Sedimentary rocks constitute the second rock family.

Metamorphic Rock.   The final major rock family is **metamorphic rock** (from the Greek words *meta,* meaning change, and *morphe,* meaning form: hence, change of form). Metamorphic rocks are those rocks whose original form has been changed by reactions in the solid state as a result of high temperature, high pressure, or both. Metamorphism, the process that forms metamorphic rocks, is somewhat analogous to cooking. When meat is put in the oven, it undergoes a series of chemical reactions as a result of the increased temperature. As a result, cooked meat looks and tastes very differently from raw meat. When sedimentary or igneous rocks are subjected to raised temperature and pressure, they undergo chemical reactions too and are said to have been metamorphosed.

Frequency of Rock Types

Most rock in the crust formed initially from magma. It is estimated, for example, that 95 percent of all rock in the crust is either igneous rock or metamorphic rock derived from rock that was originally igneous. However, as seen in Figure 1.17 most of the rock at the Earth's surface is sedimentary. The difference arises because sediments are products of external weathering processes, and as a result they are draped as a thin veneer over the largely igneous crust below. The distribution of rock types is a consequence of the rock cycle.

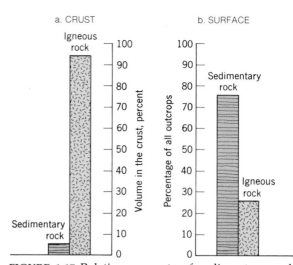

FIGURE 1.17 Relative amounts of sedimentary and igneous rock. Metamorphic rocks are considered to be either sedimentary or igneous, depending on their origin. (a) The great bulk of the crust consists of igneous rock (95 percent), but sedimentary rock (5 percent) forms a thin covering at and near the surface. (b) The extent of sedimentary rock outcropping at the surface is much larger than that of igneous rock, so 75 percent of all rock seen at the surface is sedimentary and only 25 percent is igneous.

The internal processes that form magma, and that in turn lead to the formation of igneous rock, interact with external processes through erosion. When a body of rock is subjected to erosion, the eroded particles form sediment. The sediment may eventually become cemented, usually by substances carried in groundwater, and thereby converted into new sedimentary rock. In places where such rock settles it can reach depths at which pressure and heat cause new compounds to grow, thus forming metamorphic rock. Sometimes sufficiently high temperatures are reached that melting occurs and magma is formed. The new magma can then move upward through the crust, where it can cool and form another body of igneous rock. Eventually, the new body of igneous rock can be uncovered and subjected to erosion. The igneous rock is then attacked by weathering and the broken up weathering products start once more on their way to the sea.

A Recurring Cycle

So far as we can tell, the sequence of events described above has occurred again and again throughout the Earth's long history. The events are just one of many sequences by which materials in

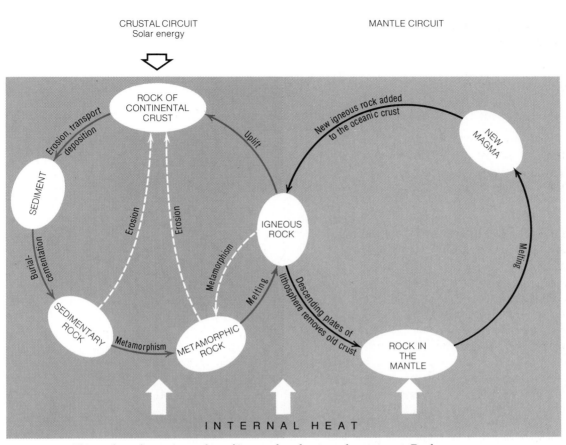

FIGURE 1.18 The rock cycle, an interplay of internal and external processes. Rock material in the continental crust can follow any of the arrows from one phase to another. At one time or another it has followed all of them. Within the mantle circuit, magma rises from a depth and forms new igneous rock in the lithosphere. The old lithosphere descends again to the mantle where it is eventually remixed. Reservoirs are labeled in capital letters; paths representing processes are labeled in lowercase letters.

the Earth's internal layers interact with materials in the external layers. Indeed, it seems possible that the compositions and balances of each of the Earth's layers, whether internal or external, depend to some extent on all the other layers.

Many details of the interaction between internal and external layers are still subjects of research. An important part of the interaction involves tectonics. The internal processes that move the Earth's surface up and down and cause plates of lithosphere to move laterally are tectonic activities. Without tectonism, rock would not continually be uplifted and exposed to erosion.

The Cycle of the Continental Crust. The cycle that most directly connects internal and external layers is the **rock cycle** depicted in Figure 1.18. The rock cycle has two parts, one involving continental crust, the other involving oceanic crust. The example in-

volving igneous rock and sedimentary rock, described above, is simply one circuit among many that occur in the rock cycle of the continental crust.

As shown in Figure 1.18, other circuits involve bodies of sedimentary rock that are never metamorphosed, never melted, or never even deeply buried before they are uplifted and eroded. Whether the circuits are long or short, the continental crust is continually being recycled. Because the mass of the continental crust is large, the average time a rock takes to complete the cycle, from erosion, through rock, to erosion again, is long. Time estimates vary, and they are difficult to make, but the average age of all rock in the continental crust seems to be about 650 million years.

The Cycle of the Oceanic Crust. The rock cycle of the oceanic crust is faster than that of the continental crust. The magma that rises to form new

FIGURE 1.19 Siccar Point, Berwickshire, Scotland. Layers of sedimentary rock, originally horizontal, were bent and tilted into vertical layers during uplift. Erosion developed a new land surface which, on submergence, became the surface on which younger sediments were laid. The interface between the two rock units is called a surface of unconformity. The vertical layers are about 450 million years old. Those lying above the unconformity, named the Old Red Sandstone, are 370 million years old. At this locality, in 1788, James Hutton first realized the significance of unconformity and the fact that it records a cycle of sedimentation, uplift, and erosion that is repeated again and again. The realization led him, in turn, to the Principle of Uniformitarianism.

oceanic crust forms hot igneous rocks that react with seawater. In the reaction process, some of the constituents in the hot rock, such as calcium, are dissolved in the seawater. Other constituents already in the seawater, such as magnesium, are deposited in the igneous rock. This is one way through which the mantle interacts directly with the hydrosphere and indirectly with the continental crust.

When sinking lithosphere carries old oceanic crust back down into the mantle, the oceanic crust is eventually remixed into the mantle. The most ancient oceanic crust of the ocean basins is only about 180 million years old and the average age of all oceanic crust is only 60 million years. This means that the time taken for a circuit through the oceanic rock cycle, from magma, through igneous rock, to

remixing in the mantle, is very much less than the time for a circuit through the continental crust.

Uniformitarianism

The first person to realize the significance of the rock cycle was James Hutton (1726–1797), a Scottish geologist. Hutton examined the evidence in the rocks (Fig. 1.19) and concluded that there was neither evidence of a beginning to the rock cycle, nor any sign of an end. Rather, he pointed out, evidence is abundant that the same processes we observe today have all operated in the past. During the early years of the 19th Century, Hutton's findings were developed into the **Principle of Uniformitarianism.** It states that the same external and internal processes we recognize in action today have been operating throughout most of the Earth's history. This principle provides us with an important capability. We can examine any rock, however old, and compare its characteristics with those of similar rock forming today in a particular environment. We can then infer that the old rock very likely formed in the same sort of environment. Therefore, the Principle of Uniformitarianism provides a first and very significant step in understanding the Earth's history.

Uniformitarianism has played a tremendously important role in the development of geology as a science. However, on occasions it has led to misinterpretations. During the nineteenth century, geologists tried to estimate the duration of the rock cycle by estimating the thickness of all sediments laid down through geological time. It was assumed that rates of deposition remained constant and equal to today's rate of deposition. Thus it was thought to be a simple calculation to estimate the time needed to produce all the sediments. The results we now know were greatly in error. One of the reasons for the error was the assumption of the constancy of geological rates.

The more that is learned of the Earth's history, and the more extensively the timing of past events is determined through radiometric dating (chapter 6), the clearer it becomes that the rates of the cycles have not always been the same. The evidence is strongly against constancy; some rates were once more rapid, others much slower.

One reason that the rate of the rock cycle has changed through time is that the Earth is very slowly cooling down as its internal heat is lost. The Earth's internal temperature is maintained, in part, by natural radioactivity. Early in the Earth's history more radioactive atoms were present than there are today, so more heat must have been produced than is produced at present. Internal processes, which are all driven by the Earth's internal heat, must have been more rapid than they are today. It is possible that 3 billion years ago oceanic crust was created at a faster rate than it is now, and that continental crust was uplifted and eroded at a faster rate. Either or both actions would cause the rock cycle to speed up. It is probable, therefore, that even though the cycles have been continuous, none has maintained a constant rate through time.

Essay

A DRILL HOLE THROUGH THE CRUST?

A challenging idea was broached during the 1960s—why not drill a hole all the way through the crust and get samples of the mantle? A few drill holes for oil and gas had reached depths approaching 9 km, so it seemed to be technically feasible to drill through the thin oceanic crust. The idea was treated with enthusiasm by many in the geological community, and Project Mohole, as it came to be called, was soon launched. But after early tests, and the drilling of some shallow holes into the deep ocean floor, the project was abandoned. The reasons were as much political and financial as they were technical. In the place of Project Mohole there emerged a less ambitious venture for drilling many shallower holes into the seafloor. The Deep Sea Drilling Project (DSDP) was active until 1985, nearly 20 years, and the results it produced were extraordinary. Because of DSDP, and its successor, Ocean Drilling Program (ODP), the geology of the ocean basins has been revealed as never before. Indeed, it is sometimes said we now know the geology of the oceanic crust better than we do that of the continental crust.

The continental crust is thick and it contains within it the record of much that has happened on the Earth during the past 4 billion years. Yet we have only scratched the surface of the continental crust; the deeper portions remain unsampled and hidden from view. New plans are now afoot for deep-drilling programs, but this time they are programs designed to drill down through the hard igneous and metamorphic rocks of the continental crust.

The first country that successfully mounted a deep-drilling program was the U.S.S.R. While American scientists were planning their oceanic drilling program in the 1960s, the Soviets were planning a continental drilling program. Among the sites they selected was one in the Kola Peninsula, in northwestern Russia, not far from Finland (Fig. B1.1). There, ancient crystalline rocks crop out, and by drilling into them it has been possible to get answers to questions such as how rock properties change with depth, how deeply fluids might penetrate, and what differences exist between the geology seen at the surface and the geology predicted at depth. Drilling has been slow, but the Kola superdeep well has now passed 12 km and is still slowly moving ahead. Every extra meter pushes it to a new record and contributes to the development of new technology. The Soviet results are very exciting, and it is hardly surprising that other countries are planning to follow their lead. West Germany, France, and the United States each have deep-drilling projects in advanced stages of planning.

The average thickness of continental crust is 40 km, so a 12-km hole is less than a third of the way through. Will it ever be possible to drill as deep as 40 km? The answer is probably yes, but there are many technical difficulties to be overcome. Temperatures as high as 600°C will be encountered, pressures vastly in excess

SUMMARY

1. The solid Earth is compositionally layered. It consists of a core, mantle, and crust, each differing in composition.

2. The crust consists of two parts: oceanic crust with an average thickness of 8 km, and continental crust with an average thickness of about 40 km.

3. The Earth is also layered with respect to its physical properties, in particular strength. The lithosphere, approximately the outer 100 km of the solid Earth, consists of rock that is strong and relatively rigid. Beneath the lithosphere, down to a depth of 350 km, is the asthenosphere, a region of the mantle where high temperatures make rock weak and easily deformed. Beneath the asthenosphere is the mesosphere, where mantle rocks become gradually stronger.

FIGURE B1.1 Derrick housing for the world's deepest drill hole. The hole is 250 km north of the Arctic circle near Murmansk in the U.S.S.R. The drilling derrick is enclosed to protect it from the Arctic winter. Started in 1970, the drill hole is now approaching 13 km in depth.

of those previously encountered in drilling must be managed, and somehow a way must be found to put tools down the hole without having to stop drilling and pull out all of the equip- ment every time a change or a new drilling bit is needed. Just imagine how long it would take to extract all of the drilling equipment from a hole 30 km deep.

4. The lithosphere consists of fragments, or plates, each of which slides slowly over the asthenosphere at rates up to 12 cm/yr.

5. The lithosphere is presently divided into six large plates and many small ones. At times in the past both more plates and fewer have been present.

6. Plates of lithosphere move, causing continents to move and ocean basins to open and close. Moving plates are the most important factors shaping the face of the Earth.

7. The external layers outside of the solid Earth layers are called, respectively, the hydrosphere, atmosphere, biosphere, and regolith.

8. The Earth's active processes can be divided into internal processes, driven by internal heat, and external processes, driven by the Sun's heat.

9. The internal and external layers continually interact leading to repeated cycles in which materials flow both from one place to another within a layer, and from one layer to another.

10. The hydrological cycle, driven by heat energy from the Sun, is the repeated, cyclic movement of water between ocean, air, and land. The movement occurs through evaporation, wind transport, precipitation of snow and rain, stream flow, and percolation.

11. The rock cycle in the continental crust involves magma, which solidifies and forms igneous rock. The igneous rock is eroded creating sediment, which is deposited in layers that become sedimentary rock. Burial may lead to changes in temperature and pressure, forming metamorphic rock. Eventually temperatures and pressures may become so high that rock melts and forms new magma.

12. The rock cycle in the oceanic crust interacts with that in the continental crust through the agency of plate tectonics.

13. On a time scale of human observation, cycles do not change through natural causes, so the compositions of the Earth's layers remain approximately in balance. Over periods of thousands or millions of years cycles may slowly change.

14. Human activities such as the burning of fossil fuel, intensive farming, and clearing of forests are affecting the cycle of carbon dioxide at a measurable rate.

IMPORTANT WORDS AND TERMS TO REMEMBER

abyssal plain (p. 11)
asthenosphere (p. 8)
atmosphere (p. 16)

bedrock (p. 19)
biosphere (p. 18)

Celsius scale (p. 5)
climate (p. 17)
conduction (p. 5)
continental crust (p. 8)
continental drift (p. 12)
continental rise (p. 9)
continental shelf (p. 9)
continental slope (p. 9)
convection (p. 5)
convection current (p. 5)
core (p. 8)
Coriolis effect (p. 16)
crust (p. 8)
cryosphere (p. 16)
cycle (p. 19)

energy (p. 4)
erosion (p. 4)
exposure (p. 19)

external processes (p. 4)

geothermal gradient (p. 5)

heat (p. 4)
heat flow (p. 5)
hydrologic cycle (p. 16)
hydrosphere (p. 16)

igneous rock (p. 20)
inner core (p. 7)
internal processes (p. 4)

kinetic energy (p. 4)

lithosphere (p. 8)

magma (p. 13)
mantle (p. 8)
materials (p. 4)
mesosphere (p. 8)
metamorphic rock (p. 20)
Mid-Atlantic Ridge (p. 11)
midocean ridge (p. 11)

oceanic crust (p. 8)

orogeny (p. 14)
outcrop (p. 19)
outer core (p. 7)

plate tectonics (p. 11)
potential energy (p. 4)
processes (p. 4)

radiation (p. 5)
regolith (p. 16)
rock (p. 4)
rock cycle (p. 20)

sediment (p. 19)
sedimentary rock (p. 20)
spreading center (p. 14)
subduction zone (p. 14)
submarine canyon (p. 9)

tectonics (p. 11)
temperature (p. 4)
transform fault (p. 15)
trench (p. 11)

*Uniformitarianism
(Principle of)* (p. 22)

QUESTIONS FOR REVIEW

1. How do external processes differ from internal ones?
2. What is the most important form of energy so far as the Earth's internal and external processes are concerned and what are the major sources of energy?
3. Heat is transferred within the Earth in two principal ways; what are they?
4. Describe the Earth's compositional layers.
5. Why does the Earth have a layering of physical properties?
6. What is the asthenosphere and how do its physical properties differ from those of the lithosphere.
7. Why do continents stand high above the floors of the ocean basins?
8. Describe the principal topographic features of the region where the oceanic crust meets the continental crust.
9. What is plate tectonics? Briefly describe why plate tectonics provides an answer for the drifting of continents?
10. How do midocean ridges form and what role do they play in plate tectonics?
11. Name the three kinds of edges that bound plates of lithosphere.
12. Identify the major reservoirs and fluxes in the hydrologic cycle.
13. How is the regolith formed?
14. Briefly describe the rock cycle. How long has the rock cycle been operating?
15. What evidence led James Hutton to discover the Principle of Uniformitarianism?

Minerals

Crystals of potassium feldspar (green) and quartz (gray) from Pike's Peak, Colorado. Feldspar and quartz are the two most abundant minerals in the Earth's crust. The specimen is 20 cm across.

MINERALS AND THEIR CHEMISTRY

Pick up a rock or some sand, soil, or gravel; you will be holding a handful of minerals. Wherever you look you will see minerals or materials which are derived from minerals.

Most minerals are found in abundant quantities and have neither commercial value nor any particular use. A few, like diamonds and rubies, are rare and prized for their beauty. Others, though still few in number, are the raw materials for industry and the basis for national wealth.

Minerals and Human History

Empires have been won and lost for minerals, and powerful countries have collapsed when deposits of valuable minerals were exhausted.

The Romans conquered most of Europe and the Near East in their search for minerals containing copper, gold, tin, iron, silver, and lead. They built a great empire by using the mineral wealth they found. When the mineral deposits were exhausted, or were captured by local tribes, Rome was deprived of its major sources of wealth and its empire slowly declined.

During the eighteenth and nineteenth centuries, Britain prospered because rich deposits of coal, iron, copper, tin, and other resources supported its Industrial Revolution. Now those mineral resources are exhausted.

The United States, also, has prospered as a result of its abundant mineral wealth. Yet, as some of its resources are now facing exhaustion, many wonder what will happen to the United States. Others argue that, as supplies of some minerals run out, it should be possible to develop alternative sources, currently not in use, and that in some cases it may be possible to substitute other materials.

Mineralogy and Petrology

As stated in chapter 1, rocks are aggregates of mineral matter which carry the record of the Earth's history. To decipher that history and to understand how the Earth works, it is necessary to investigate two important subfields of geology:

Mineralogy is the branch of geology that deals with the classification and properties of minerals;

Petrology is the branch of geology closely related to mineralogy that deals with the occurrence, origin, and history of rocks.

We will now examine the two most important characteristics of minerals.

1. *Composition,* which is the kinds of chemical elements present and their proportions.
2. *Structure,* which is the way in which the atoms of the chemical elements are packed together.

Because most minerals contain several kinds of atoms, it is helpful to start by discussing what atoms are and the ways in which they combine.

Atoms

Chemical Elements. Asked to analyze a rock, a chemist would report the kinds and amounts of the chemical elements present, because **chemical elements** are the most fundamental substances into which matter can be separated by chemical means. At present 104 elements are known, and 88 of them occur naturally. Each is separately named and identified by a symbol, such as H for hydrogen, Ag for silver, Si for silicon, and U for uranium. The known elements and their symbols are listed in appendix B.

Each elemental substance consists of a large number of identical particles called atoms. An **atom** is the smallest individual particle that retains all the properties of a given chemical element.

We cannot actually see an atom of hydrogen, or an atom of lead, or an atom of any other element, even with the most powerful microscopes, because individual atoms are so small—about 10^{-10} m in diameter. When we handle a pure chemical element we are seeing instead an aggregation of a vast number of identical atoms. A cube of pure silver 1 cm on an edge contains about 58×10^{21} atoms, a number so large that its significance is almost impossible to grasp. A faint idea of its magnitude may be conceived by imagining that we had somehow spread the cube of silver thinly and evenly over the entire face of the Earth. Each square centimeter of the Earth's surface would then contain approximately 10,000 atoms.

The Atomic Nucleus. Everything on the Earth is composed of atoms, but atoms in turn are built up from still smaller **subatomic particles.** The principal subatomic particles are **protons** (which have positive electrical charges), **neutrons** (which, as their name suggests, are electrically neutral), and **electrons** (which have negative electrical charges that balance exactly the positive charges of protons). Protons and neutrons are dense but very tiny particles and they aggregate together to form the core

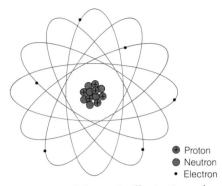

● Proton
● Neutron
• Electron

FIGURE 2.1 Schematic illustration of an atom of carbon-12 (^{12}C). The nucleus contains six protons and six neutrons. Moving in orbits around the nucleus are six electrons.

or *nucleus* of an atom. The sum of the protons plus neutrons in the nucleus is called the *mass number.* Protons give a nucleus a positive charge and the number of protons in the nucleus of an atom is called the *atomic number.* Electrons are even tinier particles; they move, like a distant and diffuse cloud, in orbits around the nucleus (Fig. 2.1).

Elements are catalogued systematically by atomic number, beginning with hydrogen, which has an atomic number of 1 because it has 1 proton. Hydrogen is followed by helium, which has 2 protons—and so the list continues up to the heavier elements such as uranium, which has 92 protons. For an atom to be electrically neutral, the number of orbiting electrons must balance the number of protons in the nucleus. If the number of orbiting electrons is smaller or larger than the number of protons, a net positive or negative charge results.

Isotopes. All atoms having the same atomic number are atoms of the same element. The number of neutrons that accompany the protons in a nucleus can vary, within small limits, without markedly affecting the chemical properties. Any chemical element may have several *isotopes,* which are atoms having the same atomic number but differing numbers of neutrons and thus differing mass numbers. For one element 10 isotopes have been discovered, and all elements have at least 2 isotopes. All of the known, naturally occurring isotopes are listed in appendix B.

Radioactivity. Not all combinations of protons and neutrons are completely stable. Some isotopes transform spontaneously, forming new isotopes and different elements in the process. The transformation process by which an unstable atomic nucleus spontaneously changes to another nucleus is

called *radioactivity.* As we shall see in chapter 6, the rate at which a given radioactive isotope transforms is constant; the rate is a characteristic property of that isotope. Because the transformation rate is constant, it can be used as a kind of geological clock.

The most common radioactive isotopes in the Earth are potassium-40 (written ^{40}K, meaning an isotope with a total of 40 neutrons and protons in the nucleus), uranium-235 (^{235}U), uranium-238 (^{238}U), thorium-232 (^{232}Th), and Carbon-14 (^{14}C). When a radioactive atom transforms to a new atom, a tiny amount of energy is released as heat. This heat energy is referred to as the Earth's radiogenic heat. It is the energy that drives all internal activities.

Radioactive isotopes of potassium, thorium, and uranium (together with the isotopes of a few less common elements such as rubidium) are widely distributed in tiny amounts through the crust and mantle, and their rates of transformation are very slow. Even so, they generate sufficient heat to maintain the high temperature of the Earth's interior.

Compounds and Ions

Compounds

A few minerals, such as metallic gold and platinum, are single chemical elements. Most minerals are combinations of atoms of different chemical elements that are bonded together to form *compounds.* The way elements combine to form compounds is determined by the orbiting electrons; it is not controlled by the nucleus of an atom and thus isotopes of an element all combine the same way.

Energy-Level Shells. Electrons are confined to specific orbits which are arranged at predetermined distances from the nucleus. Because the electrons in each orbit have a specific amount of energy characteristic for that orbit, the orbit distances are commonly called *energy-level shells.* The maximum number of electrons that can occupy a given energy-level shell is fixed. Shell 1, closest to the nucleus, is small and can accommodate only 2 electrons; shell 2, however, can accommodate 8 electrons; shell 3, 18; and shell 4, 32.

Ions

When an energy-level shell is filled with electrons, it is very stable, like an evenly loaded boat. To fill their energy-level shells and so reach a stable configuration, atoms share or transfer electrons among themselves. The movement of electrons naturally upsets the balance of electrical forces, for an atom that loses an electron has lost a negative electrical charge and, therefore, has a net positive charge, while one that gains an electron has a net negative charge. An atom that has excess positive or negative charges caused by electron transfer is called an *ion.* When the charge is positive (meaning that the atom gives up electrons), the ion is called a *cation;* when negative, an *anion.* The convenient way to indicate ionic charges is to record them as superscripts. For example, Li^{+1} is a cation that has given up an electron, while F^{-1} is an anion that has accepted an electron. Compounds contain one or more elements that are cations and one or more that are anions. For a compound to be stable the sum of the positive charges on the cations and the negative charges on the anions must equal zero.

Molecules

A lithium atom has shell 1 filled, but has only one electron in shell 2. The lone outer electron is loosely held and easily transferred to an element such as fluorine, which has seven electrons in shell 2, needing only one more to be completely filled. In this fashion, if the lithium and fluorine are in close proximity both the lithium and fluorine finish with filled shells, and the resulting positive charge on the lithium and the negative charge on the fluorine bind the two atoms together. An example of the way transfer of an electron leads to formation of a compound is illustrated in Figure 2.2. Lithium and fluorine form the compound lithium fluoride, which is written LiF to indicate that for every Li atom there is a counterbalancing F atom. The combined pair of Li and F ions is called a molecule of lithium fluoride. A *molecule* is the smallest unit that retains all the properties of a compound. Properties of molecules are quite different from the properties of their constituent elements. The elements of sodium (Na) and chlorine (Cl) are highly toxic, for example, but the compound sodium chloride (NaCl) is a compound that is essential for human health.

Bonds

Noble Gases

Some elements occur naturally with completely filled energy-level shells. They occur as individual, electrically neutral atoms and show little or no tendency to react with other elements and form compounds. Because they are so unreactive, elements with normally filled outer shells are called *inert* or *noble gases.* The noble-gas elements are helium, neon, argon, krypton, xenon, and radon.

All elements other than the noble gases readily bond with other atoms and form compounds. Atoms form bonds with like or unlike atoms. However, regardless of the pairing, when atoms transfer or share electrons, they finish with a net positive or negative charge, depending on whether they give up or receive the transferred electrons. The manner in which electrons are transferred or shared leads to several distinctive kinds of bonds.

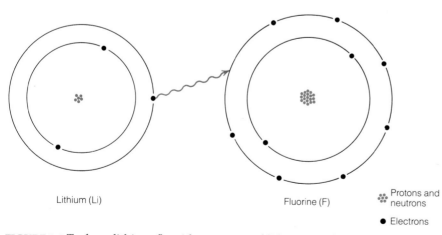

Lithium (Li)   Fluorine (F)

Protons and neutrons

● Electrons

FIGURE 2.2 To form lithium fluoride, an atom of lithium combines with an atom of fluorine. The lithium atom transfers its lone outer-shell electron to fill the fluorine atom's outer shell, creating an Li^{+1} cation and a F^{-1} anion in the process. The electrostatic force that draws the lithium and fluorine ions together is an ionic bond.

Ionic Bonds

When electrons are transferred completely from one atom to another a cation and an anion result.

Even though cations and anions can exist as free entities, an electrostatic force exists between them because they have opposite electrical charges. The electrostatic attraction draws negatively and positively charged ions together and is called an *ionic bond.* The minerals halite (NaCl) and fluorite (CaF_2) are examples of ionically bonded compounds.

Covalent Bonds

The force that arises when two atoms share one or more electrons, and the energy-level shells of both atoms are filled, is called *covalent bonding.* One common substance in which covalent bonds occur is water (H_2O). The outer shell of an oxygen atom has six electrons but requires eight for maximum stability. A hydrogen atom has one electron but requires two for maximum stability. How this is accomplished is shown in Figure 2.3. Many common minerals have either covalent or ionic bonding or a combination of the two.

Metallic Bonds

A small number of minerals display a third kind of bonding called the *metallic bond,* which is really a variation of covalent bonding in which more elec-trons are present than needed to fill the energy level shells. The atoms are held together by covalent bonds. The additional electrons are free to diffuse through the structure, sometimes drifting, sometimes replacing an electron forming a covalent bond. The drifting electrons are very loosely held and are the reason that metals have special properties such as high electrical and thermal conductivity, opacity, and ductility. Gold, silver, platinum, copper, and other naturally occurring metals contain metallic bonds.

Van der Waals Bonds

Lastly, a special kind of bond exists called the *van der Waals bond.* It does not involve the transfer or sharing of electrons but is instead a weak electrostatic attraction that arises because certain ions and atoms are distorted from a spherical shape. The van der Waals bond is a weak bond—much weaker than ionic or covalent bonds—but it plays an important role in the structure of certain minerals, of which graphite is a good example. Graphite contains only atoms of carbon and it has a sheetlike structure; carbon atoms are bonded covalently within the sheets but the sheets are held together by van der Waals bonds. These van der Waals bonds are so weak they are easily broken. Graphite feels slippery when you rub it between your fingers because the rubbing breaks the van der Waals bonds and the sheets slide easily past each other.

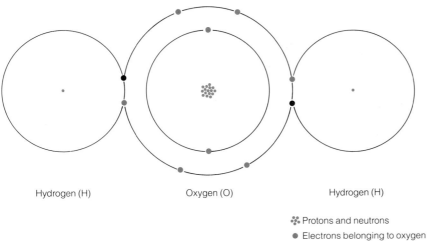

Hydrogen (H) Oxygen (O) Hydrogen (H)

 Protons and neutrons
Electrons belonging to oxygen
Electrons belonging to hydrogen

FIGURE 2.3 Two atoms of hydrogen form covalent bonds with an oxygen atom through sharing of electrons. The oxygen atom thereby has its most stable configuration with eight electrons in the outer shell, and each of the hydrogen atoms fills its outer shell with two electrons, making water (H_2O).

Complex Ions

Sometimes two different atoms form such strong bonds that they seem to act as a single atom. An especially strongly bonded pair is said to form a **complex ion.** Complex ions act in the same way as single ions, forming compounds by bonding with other elements. For example, carbon and oxygen combine to form the very stable carbonate anion $(CO_3)^{-2}$. Other important examples are the sulfate $(SO_4)^{-2}$, nitrate $(NO_3)^{-1}$, and silicate $(SiO_4)^{-4}$ anions.

Organic Compounds. Two broad classes of compounds are recognized. **Organic compounds** are made from carbon and hydrogen, with or without other elements such as nitrogen and oxygen. Organic compounds can form by direct combination of carbon and hydrogen, but most come directly or indirectly from the activities of living organisms. Mixtures of organic compounds are called *organic matter.*

Inorganic Compounds. All other matter is said to be inorganic and its compounds are **inorganic compounds.** Minerals are inorganic compounds. One property of minerals, therefore, is that they are chemical compounds or (in a few cases such as metallic gold and copper, or sulfur) single chemical elements. However, chemical composition is insufficient to define a mineral uniquely; other properties must be considered as well. The most important of the additional properties is crystal structure.

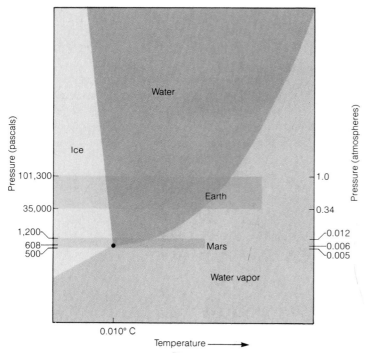

FIGURE 2.4 The control of temperature and pressure on the state of H_2O. In the region marked water, H_2O is in the liquid state, within the ice region it is solid, and in the vapor region it is in the gaseous state. At temperatures and pressures defined by the boundaries between regions, two states can coexist. For example, along the boundary separating ice and water, the solid and liquid states coexist—that is, along the line water coexists with ice. At one specific temperature and pressure the three lines of coexistence meet, and at the triple point ice, water, and water vapor coexist. The triple point for H_2O is at 0.010°C and 608 Pa. The pressures at sea level and at the top of Mount Everest bound the range of pressures at the Earth's surface. By comparison the range of pressures at the surface of Mars is much smaller.

Crystal Structure and States of Matter

States of Matter

Compounds and elements can exist in any of the three *states of matter*—solid, liquid, or gas. H_2O is a familiar example of a compound that can form a *liquid* (water), a *solid* (ice), and a *gas* (water vapor) under suitable conditions at the surface of the Earth. The two factors that control the state of H_2O are temperature and pressure. It is clear from Figure 2.4 that at high temperatures and low pressures water vapor is the stable state for H_2O, while ice forms at low temperatures and high pressures. The temperatures and pressures over which different compounds change from one state to another differ greatly, but the same general statement is true for all substances—low temperatures and high pressures favor the solid state, high temperatures and low pressures favor the gaseous state, while the liquid state occurs in intermediate ranges of temperature and pressure.

Crystal Structure

With minerals we are dealing entirely with the solid state because all minerals are solids. Whereas atoms in gases and liquids are randomly jumbled, the atoms in almost all solids are organized in regular, geometric patterns, like eggs in a carton. The geometric pattern that atoms assume in a solid is called the **crystal structure,** and solids that have a crystal structure are said to be **crystalline.** Solids that lack crystal structures are **amorphous** (Greek for without form). The crystal structure of a mineral is a unique property of that mineral. All specimens of a given mineral have identical structures.

Ionic Radius. The packing of atoms in the mineral galena, PbS, the most common lead mineral, is shown in Figure 2.5. The sulfurs, the anions, are larger than the leads, the cations. The size of ions is commonly stated in terms of the **ionic radius,** which is the distance from the center of the nucleus to the outermost shell of orbiting electrons.

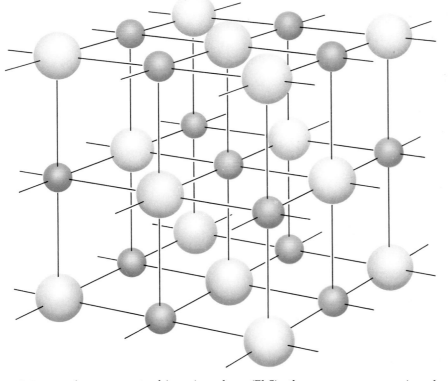

FIGURE 2.5 Arrangement of ions in galena (PbS), the most common mineral containing lead. The packing arrangement is repeated continuously through a crystal, and ions are so small that a cube of galena 1 cm on its edge contains 10^{22} ions each of lead and sulfur. The ions are shown pulled apart along the marked lines to demonstrate how they fit together. Pb (green) is a cation with a charge of $+2$, S (blue) is an anion with a charge of -2. To maintain a charge balance between the atoms, there must be an equal number of Pb and S atoms in the structure.

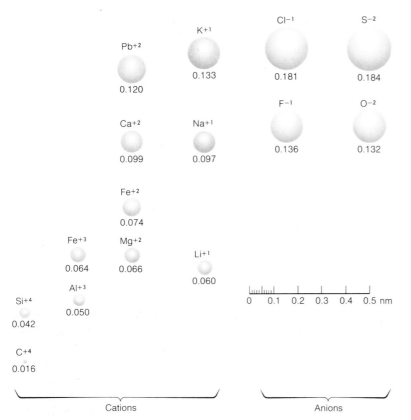

FIGURE 2.6 Ionic radii of some common ions range from C^{+4} at lower left to S^{-2} at upper right. Ions are arranged in vertical groups based on charge, from $^{+4}$ at left to $^{-2}$ at right. Note the anions tend to have larger ionic radii than cations. Ions in each of the pairs Si^{+4} and Al^{+3}, Mg^{+2} and Fe^{+2}, and Na^{+1} and Ca^{+2} are about the same size and commonly substitute for each other in crystal lattices. Radii are expressed in nanometers (nm).

Anions tend to have large radii; cations tend to be small. Most of the volume of a crystal structure is taken up by the large anions, and as a result the crystal structures of minerals are determined largely by the packing arrangements of anions. The radii of some common ions are shown in Figure 2.6 and are given in nanometers (10^{-9} m), a unit of length commonly used for atomic measurement.

Ionic Substitution. It is apparent from Figure 2.6 that certain ions have the same electrical charge and are nearly alike in size. For example, Fe^{+2} and Mg^{+2} have radii of 0.074 nm and 0.066 nm respectively. Because of their similarity in size and charge, ions of Fe^{+2} are often interchanged with ions of Mg^{+2} in magnesium-bearing minerals. The structure of the magnesium mineral is not changed as a result of the substitution, but of course the chemical composition of the mineral is affected.

The substitution of one atom for another in a random fashion throughout a crystal structure is *ionic substitution* or, to use an older term still used by many geologists, *solid solution.*

The way ionic substitutions are indicated in chemical formulas can be demonstrated with the mineral olivine, Mg_2SiO_4. When Fe substitutes for Mg in olivine the formula is written $(Mg,Fe)_2SiO_4$, which indicates that the Fe substitutes for the Mg, but not for any other atoms in the structure. Variations in mineral composition caused by solid solution are often large and, as will become apparent, they are important in the formation of common minerals and rocks.

Definition of a Mineral

A *mineral* is any naturally formed, solid, chemical substance having a definite chemical composition and a characteristic crystal structure. The words *naturally formed* are included in order to exclude the vast number of man-made substances.

TABLE 2.1 *Examples of Polymorphs, with the Most Common Mineral Listed First*

Composition	Mineral Name
C	Graphite
	Diamond
$CaCO_3$	Calcite
	Aragonite
FeS_2	Pyrite
	Marcasite
SiO_2	Quartz
	Cristobalite
	Tridymite

TABLE 2.2 *The Most Abundant Chemical Elements in the Continental Crust*

Element	Weight (%)
Oxygen (O)	45.20
Silicon (Si)	27.20
Aluminum (Al)	8.00
Iron (Fe)	5.80
Calcium (Ca)	5.06
Magnesium (Mg)	2.77
Sodium (Na)	2.32
Potassium (K)	1.68
Titanium (Ti)	0.86
Hydrogen (H)	0.14
Manganese (Mn)	0.10
Phosphorus (P)	0.10
All other elements	0.77
Total	100.00

Some naturally occurring solid compounds do not fulfill the definition of a mineral because they lack either a definite composition, a characteristic crystal structure, or both. Examples are natural glasses and resins, both of which have wide and variable composition ranges and are *amorphous*. Another example is opal, which has a more or less constant composition, but is amorphous. The term *mineraloid* is used to describe such mineral-like substances.

Polymorphs

Each mineral has a unique crystal structure. Some compounds are known to form two or more different minerals because the atoms can be packed to form more than one kind of crystal structure. The compound $CaCO_3$, for example, forms two different minerals. One is *calcite*, the mineral of which limestone and marble are composed; the other is *aragonite*, most commonly found in the shells of clams, oysters, and snails. Calcite and aragonite have identical compositions, but entirely different crystal structures. A compound that occurs in more than one crystal structure is called a ***polymorph*** (many forms). Some common polymorphs are listed in Table 2.1.

Composition

Approximately 3000 minerals are known. Most have been found in the crust, because it is the accessible part of the Earth. A few minerals have been identified only in meteorites, and two new ones were discovered in the Moon rocks brought back by the astronauts. The total number of minerals may seem large, but it is tiny by comparison with the astronomically large number of ways that all the naturally occurring elements might combine to form compounds. The reason for the disparity between observation and theory becomes apparent when the abundances of the chemical elements are considered. Only 12 elements are sufficiently common so that they comprise 0.1 percent by weight, or more, of the Earth's crust. The 12 abundant elements (Table 2.2), collectively, make up 99.23 percent of the mass of the crust. The crust is constructed of a limited number of minerals in which one or more of the 12 abundant elements is an essential ingredient.

Rather than forming distinct minerals, the many scarce elements tend to occur by ionic substitution; they substitute for more abundant elements in common minerals. For example, if a grain of the mineral olivine is analyzed, it would be determined that in addition to Mg, Fe, Si, and O, trace amounts of Cu, Ni, Co, Mn, and many other elements are present as ionic substitutes for the Mg or Fe in the structure.

Minerals containing less common elements than the abundant 12 certainly do occur, but only in small amounts. Most of the very scarce elements only form minerals under special and restricted circumstances. In fact, hafnium, rhenium, and a few other elements are so rare that they are not known to form minerals under any circumstances—they only occur by ionic substitution.

Kinds of Anions. Referring to Table 2.2, it is apparent that two elements, oxygen and silicon, make up more than 70 percent of the crust. Oxygen forms a simple anion, O^{-2}, and silicon forms a simple cation, Si^{+4}, but oxygen and silicon together form an exceedingly stable, complex ion, the ***silicate anion*** $(SiO_4)^{-4}$. Minerals that contain the silicate anion are called *silicate minerals.* In view of the abundances of silicon and oxygen, it is not surprising that silicate minerals are the most common naturally occurring, inorganic compounds. Compounds that contain simple O^{-2} anions, called oxides, are the second most abundant group of minerals. Other natural compounds, although less common than silicates and oxides, are sulfides, chlorides, carbonates, sulfates, and phosphates, for which the anions are, respectively, S^{-2}, Cl^{-1}, $(CO_3)^{-2}$, $(SO_4)^{-2}$, and $(PO_4)^{-3}$.

ROCK-FORMING MINERALS

A few minerals—no more than 20 kinds—are so common that they account for more than 95 percent of all the minerals in the continental and oceanic crusts. These are the *rock-forming minerals,* so called because essentially all common rocks contain one or more of them. We have already seen that silicates and oxides are the most common minerals of the crust, but that silicates are more abundant than oxides. Most rock-forming minerals, therefore, are silicates.

The Silicates

The Silicate Tetrahedron

The four oxygen atoms in the silicate anion are tightly bonded to the single silicon atom. Oxygen is a large anion (Fig. 2.6), while silicon is a small cation. The oxygen ions pack into the smallest space possible for four large spheres. As can be seen in Figure 2.7, the four oxygens sit at the corners of a tetrahedron and the small silicon ion sits in the space between the oxygens at the center of the tetrahedron. Therefore, the shape of the silicate anion is a tetrahedron and the structures and properties of silicate minerals are determined by the manner in which silicate tetrahedra pack together.

Each silicate tetrahedron has four unsatisfied negative charges. This is so because the silicon ion (Si^{+4}) has a charge of $+4$ while oxygen (O^{-2}) has a charge of -2. Each oxygen satisfies one of its negative charges by a bond to the silicon ion at the center of the tetrahedron, leaving each oxygen with one unsatisfied negative charge. To make a stable compound, therefore, each oxygen must satisfy its remaining negative charge. This can happen in two ways.

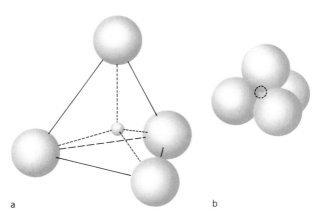

a   b

FIGURE 2.7 Silicate tetrahedron. (a) Expanded view showing large oxygen anions at the four corners, equidistant from a small silicon cation. Dotted lines show bonds between silicon and oxygen ions; solid lines outline the tetrahedron. (b) Tetrahedron with oxygen ions touching each other in natural positions. Silicon (dashed circle) occupies central space.

1. The first way for the charges to be satisfied is for the oxygens of the tetrahedron to form bonds with cations, just as if the tetrahedron was a simple anion. In such a structure, the complex ions are separated from each other because they are completely surrounded by cations. An example of this is found in olivine, in which Mg^{+2} cations balance the charges. Each oxygen is bonded to one silicon and three magnesium atoms (Fig. 2.8).

2. The second, and entirely different, way by which oxygen charges can be satisfied is for two adjacent tetrahedra to share an oxygen. A shared oxygen is bonded thereby to two silicons, and its charges are balanced as a consequence; the two tetrahedra, now joined at a common apex, form an even larger anion unit. As shown in Figure 2.9, when two silicate tetrahedra share a single oxygen, the result is a large, complex anion with the formula $(Si_2O_7)^{-6}$. Only one common mineral, epidote, contains this simplest form of oxygen sharing. However, just as more and more beads can be strung on a necklace, so can *polymerization*, the process of linking silicate tetrahedra into large anion groups, be extended to form huge units.

Most common silicate minerals have anions with complicated polymerizations. If a tetrahedron shares more than one oxygen with adjacent tetrahedra, large circular groups, endless chains, sheets, and three-dimensional networks of tetrahedra can all be formed. However, an important restriction to the polymerization process must always be met—two adjacent tetrahedra can never share more than

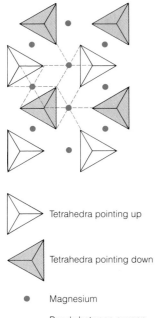

Tetrahedra pointing up

Tetrahedra pointing down

● Magnesium

– – – – Bonds between oxygen and magnesium atoms.

FIGURE 2.8 Structure of olivine. Isolated silicate tetrahedra, here shown in a geometric form in which an oxygen ion sits at each apex and a silicon at the center of each tetrahedra. By viewing the structure from above, we see that one-half of the tetrahedra point up and one-half point down. The magnesium ions lie between the tetrahedra, and each is bonded to six oxygens. Each oxygen is bonded to three magnesiums. Oxygen has a charge of -2, one-half of which is balanced by a silicon ion; magnesium has a charge of $+2$ so that each of the six oxygens to which it is bonded receives a charge of $+1/3$. Because each oxygen is bonded to three magnesiums, the structure is electrostatically balanced and the formula is Mg_2SiO_4.

one oxygen. Stated another way, tetrahedra only join at their apexes, never along the edges or faces. The common polymerizations, together with the rock-forming minerals forming them, are shown in Figure 2.10 and are discussed below in order of increasing complexity of polymerization.

Olivine. Two important minerals contain isolated silicate tetrahedra. The first is **olivine,** a glassy-looking mineral that is usually pale green in color. As previously discussed, olivine has a range of compositions because Fe^{+2} can substitute readily for Mg^{+2}, giving rise to the general formula $(Mg,Fe)_2SiO_4$. Olivine sometimes occurs in such flawless and beautiful crystals that it is used as the gem, *peridot,* but it is also one of the most important minerals in the Earth, being a very common constituent of rocks of the oceanic crust and upper part of the mantle.

a   b

FIGURE 2.9 Two silicate tetrahedra share an oxygen, thereby satisfying some of the unbalanced electrical charges and, in the process, forming a larger and more complex anion group. (a) The arrangement of oxygen and silicon ions in a double tetrahedra, giving the complex anion $(Si_2O_7)^{-6}$. (b) A geometric representation of the double silicate tetrahedra.

Garnet. The second important mineral with isolated silicate tetrahedra is **garnet.** As with olivine, garnets have a range of compositions due to ionic substitution. Garnet has the complex formula $A_3B_2(SiO_4)_3$, where A can be any of the cations Mg^{+2}, Fe^{+2}, Ca^{+2}, and Mn^{+2}, or any mixture of them, while B can be either of the triply charged cations Al^{+3} or Fe^{+3}. Garnet is characteristically found in metamorphic rocks of the continental crust. One of the most striking features of garnet is its tendency to form beautiful crystals. The iron-rich variety of garnet, called *almandine,* is deep red and is well known as a gemstone. Another important property of garnet is its hardness, which makes it useful as an abrasive for grinding and polishing.

Pyroxene and Amphibole. **Pyroxenes** and **amphiboles** are two silicate mineral groups that contain continuous chains of silicate tetrahedra. They differ in that pyroxenes are built from a polymerized chain of single tetrahedra, each of which shares two oxygens, while the amphiboles are built from double chains of tetrahedra equivalent to two pyroxene chains in which half the tetrahedra share two oxygens and the other half share three oxygens. These relations can be seen clearly in Figure 2.10. Both the pyroxene and amphibole chains are bonded together by cations such as Ca, Mg, and Fe. The cations bond to two or more oxygens in adjacent chains that have unsatisfied charges. The general formula for pyroxene is $AB(SiO_3)_2$, where A and B can be any of a number of cations, the most important of which are Mg^{+2}, Fe^{+2}, Ca^{+2}, and Mn^{+2}. The pyroxenes are most abundantly found in rocks of the oceanic crust and mantle, but they also occur in many rocks of the continental crust. The most common pyroxene is a shiny black variety called *augite* which has the complex formula $(Ca,Na)(Mg,Fe,Al,Ti)[(Si,Al)O_3]_2$.

Amphiboles have, perhaps, the most complicated formula of all the rock-forming minerals. The general formula is $A_2B_5(Si_4O_{11})_2(OH)_2$, in which A is most commonly either Ca^{+2}, Mg^{+2}, or Na^{+1}, and B is usually Mg^{+2} or Fe^{+2}. Even this complicated formula does not completely describe the composition of the most abundant variety of amphibole, *hornblende*, a dark green to black mineral that looks very much like augite and, because of ionic

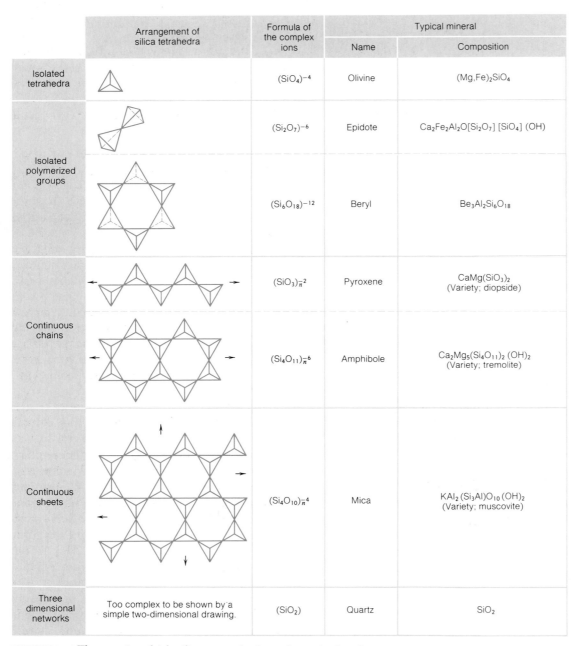

	Arrangement of silica tetrahedra	Formula of the complex ions	Typical mineral	
			Name	Composition
Isolated tetrahedra		$(SiO_4)^{-4}$	Olivine	$(Mg,Fe)_2SiO_4$
Isolated polymerized groups		$(Si_2O_7)^{-6}$	Epidote	$Ca_2Fe_2Al_2O[Si_2O_7][SiO_4](OH)$
		$(Si_6O_{18})^{-12}$	Beryl	$Be_3Al_2Si_6O_{18}$
Continuous chains		$(SiO_3)_n^{-2}$	Pyroxene	$CaMg(SiO_3)_2$ (Variety; diopside)
		$(Si_4O_{11})_n^{-6}$	Amphibole	$Ca_2Mg_5(Si_4O_{11})_2(OH)_2$ (Variety; tremolite)
Continuous sheets		$(Si_4O_{10})_n^{-4}$	Mica	$KAl_2(Si_3Al)O_{10}(OH)_2$ (Variety; muscovite)
Three dimensional networks	Too complex to be shown by a simple two-dimensional drawing.	(SiO_2)	Quartz	SiO_2

FIGURE 2.10 The way in which silicate tetrahedra polymerize by sharing oxygens determines the structures and compositions of the rock-forming silicate minerals. Polymerizations other than those shown are known but do not occur in common minerals. The most important polymerizations are those that produce chains, sheets, and three-dimensional networks.

substitution, has the approximate formula $Ca_2Na(Mg,Fe)_4 [Si_6(Al,Fe,Ti)_2O_{22}] (OH)_2$.

Clays, Micas and Chlorites. **Clays, micas,** and **chlorites** are related minerals that have a polymerized sheet of silicate tetrahedra as their basic building unit. The sheet is formed by each tetrahedron sharing three of its oxygens with adjacent tetrahedra. This leaves a single, unbalanced oxygen in each tetrahedron, leading to the general anion formula $(Si_4O_{10})_n{}^{-4}$. In the simplest case, the electrical charges that remain are balanced by Al^{+3} cations, leading to the formula $Al_4Si_4O_{10}(OH)_8$ for the clay mineral *kaolinite.*

The Al^{+3} ions in kaolinite hold the polymerized sheets together by bonding to the oxygens with unsatisfied charges. In the cases of the micas and chlorites, other ions besides aluminum are present to hold the sheets together. The bonds between silicon and oxygen in the tetrahedra are stronger than the other cation–oxygen bonds that hold the sheets together. As a result, the clays, micas, and

chlorites all display a very pronounced direction of breakage parallel to the sheets (Fig. 2.11).

A new principle must be mentioned to explain the composition of mica. Al^{+3} ions can replace the Si^{+4} ions in silicate tetrahedra by ionic substitution without affecting the property of polymerization. Because Al^{+3} has a smaller charge than Si^{+4}, a substituted tetrahedron has an extra negative charge to be satisfied. The charge cannot be fully satisfied by polymerization, so extra cations must be added to the crystal structure. Approximately one quarter of the tetrahedra in micas contain Al^{+3} instead of Si^{+4} ions. To make up the charge imbalance cations such as K^{+1}, Mg^{+2}, and even some extra Al^{+3} must be added outside of the tetrahedra. The variety of mica called *muscovite,* for example, has the formula $KAl_2(Si_3Al)O_{10}(OH)_2$, while *biotite* has the formula $K(Mg,Fe)_3(Si_3Al)O_{10}(OH)_2$.

Chlorite is a complex sheet structure mineral that is usually green in color. It derives its name from a Greek word meaning green. The unbalanced charges of the polymerized sheet are satisfied by bonding to Mg^{+2}, Fe^{+2}, and Al^{+3} to give the general formula $(Mg,Fe,Al)_6(Si,Al)_4O_{10}(OH)_8$. Chlorite is a common alteration product from other minerals that contain iron and magnesium—such as biotite, hornblende, and augite.

Quartz. The only common mineral composed exclusively of silicon and oxygen is quartz. It is the classic example of a crystal structure that has all its charges satisfied by polymerization of the tetrahedra into a three-dimensional network—meaning that all the oxygens are shared.

Quartz characteristically forms six-sided crystals (Fig. 2.12), and is found in many beautiful colors. It is one of the most widely used gem and ornamental minerals. Common names for some gemstone varieties of quartz are *rock crystal* (colorless), *citrine* (yellow), and *amethyst* (violet). Quartz is a particularly abundant mineral in rocks of the continental crust. Indeed, it is so abundant that certain sedimentary rocks are composed entirely of quartz.

Certain varieties of quartz, formed by precipitation from water solution, are so fine grained they almost appear amorphous and can only be shown to have the internal crystal structure characteristic of minerals through the use of high-powered microscopes, X-ray machines, and other research tools. One common name given to these microcrystalline forms of quartz is *chalcedony.* Other varietal names are *agate,* if it has color banding (Fig. 2.13), or *flint* (gray) and *jasper* (red), if the color is uniform. *Opal* is an amorphous form of silica.

FIGURE 2.11 Perfect cleavage of the mica mineral, muscovite, shown by thin, plane flakes into which this six-sided crystal has been split. The cleavage flakes suggest leaves of a book, a resemblance embodied in the name "books of mica." The cleavage plane is parallel to the polymerized sheet of silica tetrahedra.

FIGURE 2.12 Two quartz crystals with the same crystal forms. Although the size of the individual faces differ markedly between the two crystals, it is clear that each face on one crystal is parallel to an equivalent face on the other crystal. It is a fundamental property of crystals that, as a result of the internal crystal structure, the angles between adjacent faces are identical for all crystals of the same mineral.

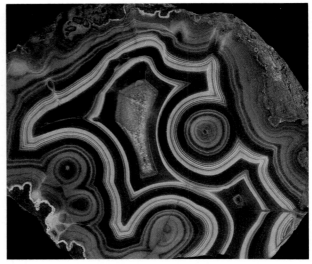

FIGURE 2.13 Agate, a color-banded microcrystalline variety of quartz formed by precipitation of SiO_2 in an open space from groundwater. Color banding is due to minute amounts of impurities. The sample is 10 cm across.

Feldspars. The name feldspar is derived from two Swedish words, *feld* (field) and *spar* (mineral). Early Swedish miners were familiar with feldspar in their mines, and found the same mineral in the abundant rocks they had to pick from the fields around their homes before they could plant crops. They were so struck by the abundance of feldspar that they chose a name to indicate that their fields seemed to be growing an endless crop of minerals. Feldspar is indeed the most common mineral in the Earth's crust. It accounts for about 60 percent of all minerals in the continental crust, and together with quartz it comprises about 75 percent of the volume of the continental crust. Unlike quartz, which is rare in rocks of the oceanic crust, feldspar is also abundant in rocks of the seafloor.

Like quartz, feldspar has a structure formed by polymerization of all the oxygen atoms in the silicate tetrahedra. Unlike quartz, however, some of the tetrahedra contain Al^{+3} substituting for Si^{+4}; so, as in the micas, other cations must be added to the structures to balance the charge.

Feldspar is a complex mineral that has a wide range of compositions. The varietal names and limiting (or ideal) compositions of common feldspar are **potassium feldspar,** $K(Si_3Al)O_8$, **albite** $Na(Si_3Al)O_8$, and **anorthite** $Ca(Si_2Al_2)O_8$. Potassium feldspar has several polymorphs—*orthoclase, microcline,* and *sanidine*—but the structural differences between them are subtle.

The most important ionic substitution in feldspar involves the substitution of Ca^{+2} for Na^{+1}. That is

possible because, as can be seen in Figure 2.6, the two ions are much closer in size than either is to the size of K^{+1}. However, Ca^{+2} and Na^{+1} have different charges, so the actual substitution scheme involves the coupled substitution of two ions: $(Na^{+1} + Si^{+4})$ for $(Ca^{+2} + Al^{+3})$. The substitution is so effective that **plagioclase** is a variety of feldspar with an unbroken range of composition from albite to anorthite.

Other Minerals

Although silicates are the most abundant minerals on the Earth, a number of others—principally **oxides, sulfides, carbonates, phosphates,** and **sulfates**—are common enough to be called rock-forming minerals.

Oxides. Some common oxide minerals are the compounds of iron *magnetite* (Fe_3O_4) and *hematite* (Fe_2O_3), the oxide of titanium, *rutile* (TiO_2), and of course *ice*, the oxide of hydrogen (H_2O). Oxides are important as ore minerals. The principal sources of iron, chromium, manganese, uranium, tin, niobium, and tantalum are oxide minerals.

bond 4 oxygen

Sulfides. The most common sulfide minerals are *pyrite* (FeS_2), *pyrrhotite* (FeS), *galena* (PbS), *sphalerite* (ZnS), and *chalcopyrite* ($CuFeS_2$). The sulfide minerals are exceedingly important as ore minerals, being the principal sources of copper, lead, zinc,

bond w/sulfur

TABLE 2.3 *The Common Rock-Forming Minerals*

Silicates	Oxides	Sulfides	Carbonates	Sulfates	Phosphates
Olivines	Hematite	Pyrite	Calcite	Anhydrite	Apatite
Pyroxenes	Magnetite	Sphalerite	Aragonite	Gypsum	
Augite	Rutile	Galena	Dolomite		
Amphiboles	Ice	Chalcopyrite			
Hornblende					
Garnet					
Quartz	The common ferromagnesian minerals are				
Feldspars					
Potassium feldspar	Augite				
Plagioclase	Biotite				
Micas	Chlorite				
Muscovite	Hornblende				
Biotite	Olivine				
Chlorites					
Clays					
Kaolinite					

nickel, cobalt, mercury, molybdenum, silver, and many other metals.

Carbonates. The complex carbonate anion $(CO_3)^{-2}$ forms three important common minerals: *calcite, aragonite,* and *dolomite.* We have already seen that calcite and aragonite have the same composition, $CaCO_3$, and are polymorphs. Calcite is much more common than aragonite. Dolomite has the formula $CaMg(CO_3)_2$.

Phosphates. *Apatite* is the single important phosphate mineral containing the complex anion $(PO_4)^{-3}$. It has the general formula $Ca_5(PO_4)_3(F,OH)$. This is the substance from which our bones and teeth are made. It is also a common mineral in many varieties of rocks and the main source of phosphorus used for making phosphate fertilizers.

Sulfates. Sulfate minerals contain the complex anion $(SO_4)^{-2}$. Although many sulfates are known, only two are common, and both are calcium sulfate minerals: *anhydrite,* $CaSO_4$; and *gypsum,* $CaSO_4 \cdot 2H_2O$. Both form when seawater evaporates; they are the raw materials used for making plaster of all kinds. Plaster of paris got its name from a quarry near Paris where a very desirable, pure-white form of gypsum was mined centuries ago.

The Ferromagnesian Minerals

All rocks contain minerals and/or mineraloids, but only a few kinds of minerals form the mass of the crust. The common rock-forming minerals are listed in Table 2.3, which can be referred to as we proceed to discuss rocks and, in later chapters, how rocks are put together to form the Earth. One subgrouping of silicate minerals is commonly referred to as the *ferromagnesian minerals,* indicating they contain iron and/or magnesium as essential constituents. The ferromagnesian minerals include olivine, pyroxene, amphibole, chlorite, and biotite. As we shall see in the next chapter, some of the ferromagnesian minerals are important in the characterization of igneous rocks.

THE PROPERTIES OF MINERALS

The properties of minerals are determined by their compositions and their crystal structures. Once we know which properties are characteristic of which minerals, we can use those properties to identify the minerals. Simple property tests are reliable and unambiguous because the number of common minerals is small. It is not necessary to analyze a mineral chemically or to determine its internal crystal structure in order to identify most common ones. The properties most often used to identify minerals are obvious ones such as color, external shape of crystal, and hardness—plus some less obvious properties, such as luster, cleavage, and specific gravity. Each property is briefly discussed below and an extensive table of individual mineral properties is given in appendix C.

Crystal Form and Growth Habit

Crystal Form

When a mineral grows freely (without obstruction from adjacent minerals), it forms a characteristic geometric solid that is bounded by symmetrically arranged plane surfaces. The characteristic geometric solid is called the *crystal form* of a mineral.

The plane faces of a crystal are an external expression of the strict internal geometric arrangement of the constituent atoms. Each plane surface corresponds to a plane of atoms in the crystal structure. The individual atoms are too small to be observed directly, even with the best microscopes, but their ordered array can be sensed and measured in a number of ways.

Unfortunately, crystals are rare in nature because minerals do not usually grow into open, unobstructed space. When well-formed crystals are found, however, an examination of the crystal reveals much about the crystal structure and immediately aids in identification.

The sizes of individual crystal faces differ. Under some circumstances a mineral may grow a long, thin crystal; under others, it may grow a short, fat one. Superficially, the two crystals may look very different; however, the unique characteristic of crystals is not the relative sizes of the individual crystal faces, but the angles between the faces. The angle between any designated pair of crystal faces in a mineral is constant and is the same for all specimens of the mineral, regardless of the overall shape. Two crystals of quartz are shown in Figure 2.12. One is flattened, the other elongate, but it is clear that the same sets of crystal faces occur on both minerals. The sets of faces are parallel; therefore, the angle between any two equivalent faces must be the same on each crystal.

Growth Habit

Every mineral has a characteristic crystal form. Some have such distinctive forms we can use the property as an identification tool without having to measure angles between faces. For example, the mineral pyrite is commonly found as intergrown cubes (Fig. 2.14) with markedly striated faces. Although crystals are rare, certain minerals, of which pyrite and garnet are examples, usually do grow in well-shaped crystals with distinctive crystal forms. Such distinctive growth patterns are called *growth habits.*

Even though most minerals do not develop well-shaped crystals, many have distinctive growth hab-

FIGURE 2.14 Distinctive crystal form of pyrite, FeS_2. One characteristic mineral habit of pyrite is cube-shaped crystals with pronounced striations on the cube face. The largest crystals in the photograph are 3 cm on an edge. The specimen is from Bingham Canyon, Utah.

its that can be used as an aid to identification. For example, Figure 2.15 shows asbestos, a variety of the mineral serpentine that characteristically grows as fine, elongate threads.

Cleavage

The tendency of a mineral to break in preferred directions along bright, reflective plane surfaces is called *cleavage.*

If we break a mineral with a hammer, or drop the specimen on the floor so that it shatters, many

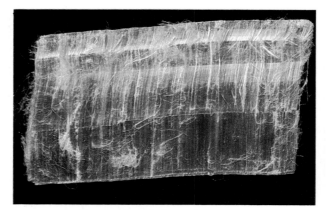

FIGURE 2.15 Some minerals have distinctive growth habits even though they do not develop well-formed crystal faces. Asbestos, a variety of the mineral serpentine, grows as fine, cottonlike threads that can be separated and woven into fireproof fabric.

FIGURE 2.16 Relation between crystal structure and cleavage. (a) Halite, NaCl, has well-defined cleavage planes; it breaks into fragments bounded by perpendicular faces. (b) The crystal structure in the same orientation as the cleavage fragments shows that the directions of breakage are planes in the crystal in which sodium and chlorine atoms occur in equal numbers.

of the broken fragments are seen to be bounded by rough fractures. However, careful examination will reveal that some surfaces are smooth, plane faces. In exceptional cases, such as halite, most of the breakage surfaces will be smooth, plane faces so that the fragments resemble small crystals. A closer look shows that all fragments break along similar planes.

The plane surfaces along which cleavage occurs are governed by the crystal structure (Fig. 2.16). They are planes along which the bonds between atoms are relatively weak. Because the cleavage planes are direct expressions of the crystal structure, they are valuable guides for the identification of minerals.

Many minerals have distinctive cleavage planes. One of the most distinctive is found in *muscovite* (Fig. 2.11). Clay minerals also have distinctive cleavage, and it is this easy cleavage direction that makes them feel smooth and slippery when rubbed between the fingers. Another mineral with a highly distinctive cleavage is calcite, which breaks into perfect rhombs. Besides micas, clays, and calcite, a number of other common minerals such as feldspar, amphibole, pyroxene, and galena have distinctive cleavages.

Luster

The quality and intensity of light reflected from a mineral produce an effect known as **luster.** Two minerals with almost identical color can have quite different lusters. The most important lusters are described as *metallic*, like that on a polished metal surface; *vitreous*, like that on glass; *resinous*, like that of resin; *pearly*, like that of pearl; and *greasy*, as if the surface were covered by a film of oil.

Color and Streak

Color

The color of a mineral is one of its striking properties, but unfortunately is not a very reliable means of identification. Color is determined by several factors, but the main cause is variation in composition. Ionic substitution causes the compositions of minerals to vary within small ranges. Some elements can create strong color effects even when they are present in very small amounts. For example, the mineral corundum (Al_2O_3) is commonly white or grayish in color, but when small amounts of Cr have replaced Al by solid solution, this mineral is blood red, forming *ruby,* a prized gem variety of corundum. Similarly, when small amounts of Fe and Ti are present, the corundum is deep blue and another prized gemstone, *sapphire,* is the result (Fig. 2.17).

Streak

Color in opaque minerals with metallic lusters can be very confusing because the color is partly a property of grain size. One way to reduce errors of judgment where color is concerned is to prepare a **streak,** which is a thin layer of powdered mineral

FIGURE 2.17 The many colors of corundum (Al_2O_3), here cut as gemstones, are caused by tiny amounts of trace elements entering the crystal structure by ionic substitution. The red color of ruby is produced by Cr^{+3} substitution for Al^{+3}, and the blue color of sapphire is produced by Fe^{+3} and Ti^{+4}.

TABLE 2.4 *Mohs' Scale of Hardness*

	Relative Number in the Scale	Mineral	Hardness of Common Objects
Decreasing	10.	Diamond	
	9.	Corundum	
	8.	Topaz	
	7.	Quartz	
	6.	Potassium feldspar	Pocketknife; glass
	5.	Apatite	
	4.	Fluorite	Copper penny
	3.	Calcite	Fingernail
	2.	Gypsum	
	1.	Talc	

made by rubbing a specimen on a nonglazed porcelain plate. The powder diffuses light and gives a reliable color effect that is independent of the form and luster of the mineral specimen. Red streak characterizes hematite whether the specimen itself is red and earthy, like the streak, or black and metallic, like magnetite. Limonite gives a brown streak, and magnetite a black streak. For many minerals, particularly those that have greasy or vitreous lusters, the streak is an undiagnostic white.

Hardness

The term **hardness** refers to the relative resistance of a mineral to scratching. It is a distinctive property of minerals. Hardness, like crystal form and cleavage, is governed by crystal structure and by the strength of the bonds between atoms. The stronger the bonds, the harder the mineral. Degree of hardness can be decided in a relative fashion by determining the ease or difficulty with which one mineral will scratch another. Talc, the basic ingredient of most body ("talcum") powders, is the softest mineral known, and diamond is the hardest. Mohs' relative hardness scale between talc (number 1) and diamond (number 10) is divided into 10 steps, each marked by a common mineral (Table 2.4). These steps do not represent equal intervals of hardness, but the important feature of the hardness scale is that any mineral on the scale will scratch all minerals below it. Minerals on the same step of the scale are just capable of scratching each other. For convenience, we often test relative hardness

by using a common object such as a penny, or a penknife, as the scratching instrument, or glass as the object to be scratched.

Density and Specific Gravity

Density

Another obvious physical property of a mineral is its density, which in practical terms means how heavy it feels. We know that equal-sized baskets of feathers and of rocks have different weights: feathers are light, rocks are heavy. The property that causes this difference is **density,** or the average mass per unit volume. The units of density are numbers of grams per cubic centimeter (g/cm^3). Minerals with a high density, such as gold, have their atoms closely packed. Minerals with low density, such as ice, have loosely packed atoms.

Minerals are divided into a heaviness or density scale. Gold has a density of 19.3 g/cm^3 and feels very heavy, but many others such as galena (PbS) and magnetite (Fe_3O_4) which have densities of 7.5 g/cm^3 and 5.2 g/cm^3, respectively, also feel heavy by comparison with most silicate minerals which have densities of 2.5–3.0 g/cm^3.

Specific Gravity

Density is not commonly measured directly. The specific gravity is usually measured instead. It is the ratio of the weight of a substance to the weight of an equal volume of pure water. **Specific gravity** is a ratio of two weights, so it does not have any units. Because the density of pure water is 1 g/cm^3, the specific gravity of a mineral is numerically equal to its density. Specific gravity can be approximated by comparing different minerals held in the hand. Metallic minerals such as galena feel "heavy," whereas nearly all others feel "light."

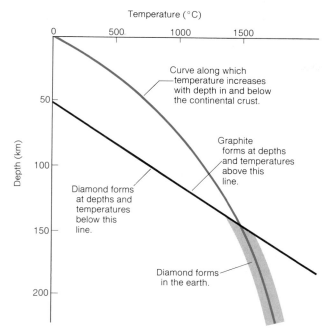

FIGURE 2.18 Line separating conditions of temperature and pressure of overlying rock (here plotted as depth) at which the two polymorphs of carbon, diamond and graphite, grow. At a pressure equal to that at a depth of 145 km, the diamond–graphite line intersects the curve depicting the way the Earth's temperature changes with depth. Diamond can form only at 145 km or more below the surface—at depths well into the mantle.

MINERALS AS INDICATORS OF THE ENVIRONMENT OF THEIR FORMATION

Minerals should not be regarded merely as objects of beauty or sources of economic materials. Contained within their makeup are the keys to the conditions under which they—and the rocks they are in—have formed. The study of minerals, therefore, can provide insight into the chemical and physical conditions in regions of the Earth that are inaccessible to direct observation and measurement.

An understanding of the growth environments of minerals has come largely through studying minerals in the laboratory. By suitable experiments, for example, scientists have been able to define the temperatures and pressures at which a diamond forms rather than its polymorph, graphite (Fig. 2.18). Because it is possible to infer how temperature and pressure increase with depth in the Earth, we can state with certainty that rocks in which diamonds are found are samples of the mantle from at least 145 km below the Earth's surface.

Minerals are also used to obtain information about former environments of the Earth's surface. Past climates, for example, can be deciphered from the kinds of minerals formed during erosion. The composition of seawater in past ages can be determined from the minerals formed when seawater evaporated and deposited its salts. Rather than elaborating many examples at this point, we turn next to an examination of rocks, for rocks are assemblages of minerals, and the ways assemblages occur are even more informative than the ways in which individual minerals occur.

ROCK

Texture

At first glance rocks seem confusingly varied. Some appear platy or distinctly layered and display pronounced, flat cleavage surfaces of mica. Others are coarse, evenly grained, and lack layering; yet, they may still contain the same kinds of minerals present in the platy, micaceous rock. Studying a large number of rock specimens soon makes it clear that no matter what kind of rock is being examined—sedimentary, metamorphic, or igneous—the differences between samples can be described in terms of two kinds of small-scale features. The first feature is *texture*, by which is meant the overall appearance that a rock has because of the size, shape, and arrangement of its constituent particles. For example, the mineral grains may be flat and parallel to each other, giving the rock a pronounced platy, or flaky, texture—like a pack of playing cards. In addition, the various minerals may be unevenly distributed and concentrated into specific layers. The rock texture is then both layered and platy. Specific textural terms are used for each of the rock families and will be introduced at the appropriate place in the next three chapters.

Mineral Assemblage

The second small-scale feature in a rock is the assemblage of minerals present. A few kinds of rock contain only one mineral, but most rocks contain two or more of the rock-forming minerals. The specific minerals, and the amounts present, reflect the composition of the rock.

The varieties and abundances of minerals present in rocks, commonly called *mineral assemblages,* are important pieces of information for interpreting a rock record. The composition of the crust and the mantle are distinctly different and their temperatures and pressures also differ. As a result, mineral assemblages of rocks formed in the mantle are quite different from those formed in the crust and may be used to decipher the origin of the rocks.

Lithology

The systematic description of rocks in terms of mineral assemblage and texture is termed **lithology.** When lithologies are discussed, two useful terms, **megascopic** and **microscopic,** are used to describe the textures and mineral assemblages of rocks. Megascopic refers to those features or rocks that can be perceived by the unaided eye, or by the eye assisted by a simple lens that magnifies up to ten times. Microscopic refers to those features of rocks that require high magnification in order to be viewed. Commonly, examination of a microscopic texture requires the preparation of a special *thin section* that

must be viewed through a microscope. A thin section is prepared by first grinding a smooth, flat surface on a piece of rock. The flat surface is then glued to a glass slide and the rock fragment is ground to a slice so thin that light passes through it easily. The appearance of the same rock on a polished surface and in a thin section is shown in Figure 2.19.

WHAT HOLDS ROCK TOGETHER?

We need not have had much experience in order to realize that some kinds of rock hold together with great tenacity, whereas other kinds are easily broken apart. The most tenacious rocks are igneous

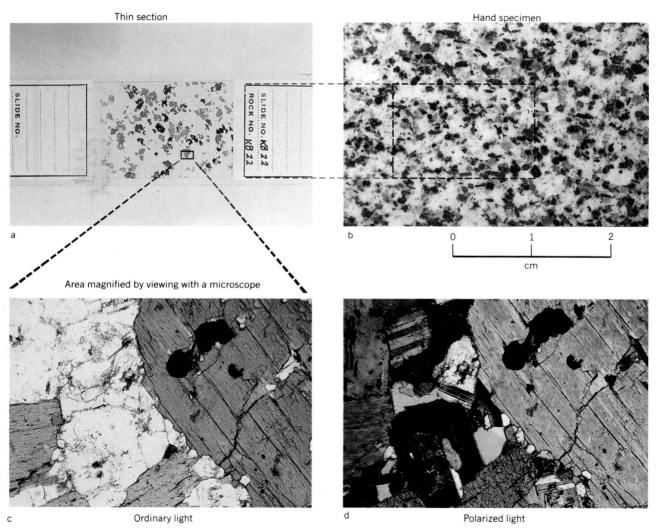

FIGURE 2.19 In the study of rocks, polished surfaces and thin slices reveal textures and mineral assemblages to great advantage. The specimen here is an igneous rock containing quartz, plagioclase, hornblende, and biotite.

and metamorphic, because these possess intricately interlocked mineral grains. The growing minerals crowd against each other, filling all spaces and forming an intricate, three-dimensional jigsaw puzzle. A similar interlocking of grains holds together steel, ceramics, and bricks.

The forces that hold together the grains of sedimentary rocks are less obvious. Sediment starts out as a loose aggregate of particles and is transformed into sedimentary rock in four ways.

1. *Pressure.* Overlying sediment or vibrations of the ground arising from earthquakes can force the irregularly shaped grains in a sediment into a tight, coherent mass. The interlocking of grains that results from this kind of packing is not strong, as in igneous rock, but under some circumstances it can hold sediment together.

2. *Deposition of a cement.* Water that circulates slowly through the open spaces between grains deposits new materials such as calcite, quartz, and iron oxide, which cement the grains together.

3. *Compaction.* The weight of overlying deposits can squeeze water out of deeply buried sediment, compacting the sediment, and reducing the pore space. Compaction forces small grains close together and makes more effective the capillary forces exerted by remaining films of intergranular water.

4. *Recrystallization.* As sediment becomes deeply buried, its mineral grains begin to be recrystallized and the newly growing grains interlock and form strong aggregates. The process is like that which occurs when ice crystals in a snow pile recrystallize to form a compact mass of ice.

Essay

MINERALS AND SOCIETY

Underlying modern society is a complex industrial network that fabricates and distributes all of the materials, food, and clothes needed to keep the world's 5 billion people alive, and, for the most part, reasonably healthy. The industrial network must be fed by raw materials drawn from the Earth, and underpinning all of the resources are minerals and rocks. To supply their needs the peoples of the world now mine and process several hundred different rock and mineral products. They range from rare minerals such as diamond, used in jewelry and for special grinding and cutting tools, to chalcopyrite, which is the main ore mineral of copper, through gypsum, from which plaster is made, to the clays used in brick making, and the crushed limestone that farmers put on their fields to counteract soil acidity.

When we add up all of the many kinds of mineral products used by modern society the amounts are truly enormous (Fig. B2.1). In 1987 it is estimated that the 5 billion people in the world consumed an enormous 18 billion metric tons of newly mined mineral substances. If this material were somehow piled up into a cube, each edge of the cube would be 2000 m (about 1.25 miles) in length. Each year the world population grows larger and each year a few more people manage to raise their living standards a little, and thereby consume resources at a slightly greater rate. Rising living standards and growing resource consumption seem to go hand in hand.

Can the minerals in the Earth's crust sustain both a growing population and a high standard of living for everyone? This difficult question has many experts worried. The group of minerals they worry most about are those used as sources of metals—minerals such as galena, sphalerite, and chalcopyrite, the main ore minerals of lead, zinc, and copper respectively. Metals, the experts point out, begin the chain of resource exploitation. Without metals

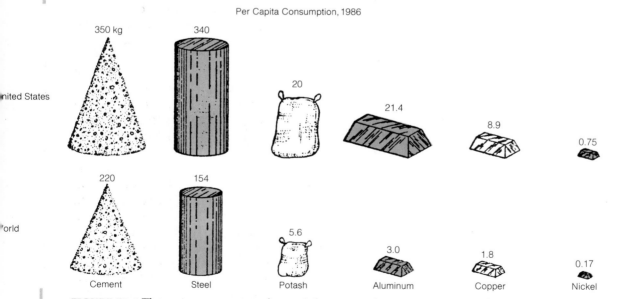

FIGURE B2.1 The average amount of material consumed per person each year, called the per capita consumption, is greater in an industrially advanced country such as the United States than it is for all the rest of the world (*Source:* After data from the U.S. Bureau of Mines.)

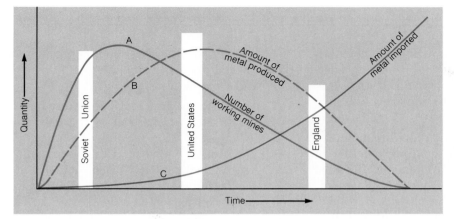

FIGURE B2.2 Changing metal production illustrated by the histories of three countries. The number of mines rises rapidly (curve A) but declines when the rate of mine exhaustion exceeds the discovery rate. The amount of metal produced (curve B) also rises, then falls when mines become exhausted. Curve C represents the growing imports of metal when internal producton fails to meet needs. As time passes, the position of a country moves from left to right. In the late nineteenth century Britain was about where the United States is today, at which time the United States was at about the present position of the Soviet Union.

one cannot make machines. Without machines the chemical energy of coal or oil cannot be converted into useful mechanical work. Without mechanical energy the tractors that pull plows must grind to a halt; trains and trucks must stop running; and indeed our whole industrial complex must become still and silent.

Experts lack the hard data needed to draw reliable conclusions about the adequacy of mineral resources. Optimistic experts point to the great success our technological society has enjoyed over the past two centuries as ever more remarkable discoveries have been made. If mineral resources are limited, they suggest, we will find ways to get around the limits by substitution or through the discovery of new technologies. The opposite and more pessimistic opinion is held by many geologists. Technologically advanced societies have faced local mineral resource limits in the past, they point out, but the solution has usually been to import new supplies from elsewhere rather than trying to develop substitutes or develop effective recycling measures.

Some of the countries in Europe, such as Britain, were once great suppliers of metals

(Fig. B2.2). Today most of the mines are closed, the minerals are mined out, and the industry of Britain runs on raw materials from abroad. The United States, too, was once self-sufficient in most minerals and was a supplier of many minerals to other, less fortunate countries. Slowly the situation has changed and the United States has become a net importer of minerals, relying on supplies from countries with newer mining industries—such as Australia, Chile, South Africa, and Canada. The only large industrial country that can still supply most of the mineral resources it needs is the Soviet Union. But eventually the Soviet mines and those of Australia and other countries will be depleted of their minerals too. Where then does society turn?

This question must be answered in the forseeable future. It is highly likely that within the lifetimes of the people who read this book, mineral resource limitations will occur. Which minerals, and therefore which metals, will first be in short supply is still an open question. That it will happen, however, is no longer an open question. How society will cope and respond is one of the great social and scientific issues still to be solved.

SUMMARY

1. Minerals are naturally formed, solid, chemical elements or chemical compounds having a definite chemical composition and a characteristic crystal structure. Crystal structure is the geometric array of atoms.

2. Chemical compounds, and therefore minerals, are formed through the bonding together of atoms of different chemical elements. The most important kinds of bonding in minerals are ionic and covalent bonds, which involve, respectively, the transfer and sharing of orbiting electrons.

3. Approximately 3000 minerals are known, but of these about 20 make up more than 95 percent of the Earth's crust and are called the rock-forming minerals.

4. The composition of some minerals varies because of ionic substitution by which an ion in a crystal structure can be replaced by a foreign ion with a like ionic charge and ionic radius.

5. Some compounds are found with the same composition but different crystal structures. Each structure is a separate mineral. Minerals having the same compositions but differing structures are called polymorphs.

6. Silicates are the most common minerals, followed by oxides, carbonates, sulfides, sulfates, and phosphates.

7. The basic building block of silicate minerals is the silicate tetrahedron, a complex anion in which an Si^{+4} ion is bonded to four O^{-2} ions. The four O^{-2} ions sit at the apexes of a tetrahedron, with the Si^{+4} at its center. Adjacent silicate tetrahedra can bond together to form larger complex anions by sharing an oxygen. The process is called polymerization.

8. The feldspars are the most abundant family of minerals in the Earth's crust, comprising approximately 60 percent of the mass of the crust. Quartz is the second most common mineral in the crust but is more abundant in the continental crust than in the oceanic crust.

9. The principal properties used to characterize and identify minerals are crystal form, mineral habit, cleavage and fracture, luster, color and streak, hardness, and specific gravity.

10. The differences between rocks can be described in terms of the mineral assemblage, and the texture imparted by the constituent particles present.

IMPORTANT WORDS AND TERMS TO REMEMBER

amorphous (p. 35)
anion (p. 32)
atom (p. 30)
atomic number (p. 31)

carbonate (p. 42)
cation (p. 32)
chemical element (p. 30)
cleavage (p. 44)
complex ion (p. 34)
compound (p. 31)
covalent bond (p. 33)
crystal form (p. 44)
crystalline (p. 35)
crystal structure (p. 35)

density (p. 46)

electron (p. 30)

energy-level shell (p. 31)

ferromagnesian mineral (p. 43)

growth habit (p. 44)

hardness (p. 46)

inert gas (p. 32)
inorganic compound (p. 34)
ion (p. 32)
ionic bond (p. 33)
ionic radius (p. 35)
ionic substitution (p. 36)
isotope (p. 31)

lithology (p. 48)
luster (p. 45)

mass number (p. 31)
megascopic (p. 48)
metallic bond (p. 33)
microscopic (p. 48)
mineral (p. 36)
mineral assemblage (p. 47)
mineralogy (p. 30)
mineraloid (p. 37)
molecule (p. 32)

neutron (p. 30)
noble gas (p. 32)
nucleus (p. 31)

organic compound (p. 34)
oxide (p. 42)

petrology *(p. 30)*
phosphate *(p. 42)*
polymerization *(p. 38)*
polymorph *(p. 37)*
proton *(p. 30)*

radioactivity *(p. 31)*
rock-forming mineral *(p. 38)*

silicate *(p. 38)*
silicate anion *(p. 37)*
silicate minerals *(p. 38)*
silicate tetrahedron *(p. 38)*
solid solution *(p. 36)*
specific gravity *(p. 46)*
streak *(p. 45)*
subatomic particles *(p. 30)*

sulfate *(p. 42)*
sulfide *(p. 42)*

texture *(p. 47)*
thin section *(p. 48)*

van der Waals bond *(p. 33)*

taste like salt?   *measurable/ comparable/*

MINERAL NAMES TO REMEMBER

agate *(p. 41)*
albite *(p. 42)*
amphibole *(p. 39)*
anhydrite *(p. 43)*
anorthite *(p. 42)*
apatite *(p. 43)*
aragonite *(p. 43)*
augite *(p. 40)*

biotite *(p. 41)*

calcite *(p. 43)*
chalcedony *(p. 41)*
chalcopyrite *(p. 42)*

chlorite *(p. 41)*
clay *(p. 41)*
corundum *(p. 46)*

diamond *(p. 47)*
dolomite *(p. 43)*

feldspar *(p. 42)*
flint *(p. 41)*

galena *(p. 42)*
garnet *(p. 39)*
graphite *(p. 47)*
gypsum *(p. 43)*

halite *(p. 45)*
hematite *(p. 42)*
hornblende *(p. 40)*

ice *(p. 35)*

jasper *(p. 41)*

kaolinite *(p. 41)*

magnetite *(p. 42)*
mica *(p. 41)*
muscovite *(p. 41)*

olivine *(p. 39)*
orthoclase *(p. 42)*

plagioclase *(p. 42)*
potassium feldspar *(p. 42)*
pyrite *(p. 42)*
pyroxene *(p. 39)*
pyrrhotite *(p. 42)*

quartz *(p. 41)*

rutile *(p. 42)*

sphalerite *(p. 42)*

QUESTIONS FOR REVIEW

1. What is a mineral?
2. How do chemical elements join together to form compounds?
3. Describe some properties that can be used to characterize minerals besides the composition and crystal structure.
4. What are polymorphs? Identify the polymorphs of $CaCO_3$.
5. What are isotopes? Do they affect a mineral's properties?
6. What is ionic substitution? Does it influence any of the properties of minerals?
7. What are rock-forming minerals? Approximately how many are there? What is the most common rock-forming mineral?
8. Describe the structure, composition, and ionic charge of the complex silicate anion.

9. Describe how silicate anions join together to form silicate minerals.
10. Describe the polymerization of silicate anions in the following minerals: the pyroxenes; the micas; the feldspars.
11. Name five common rock-forming minerals that are not silicates, and say what kinds of anion each contains.
12. What are the two properties used to describe differences between rocks?
13. Describe the four ways by which loose aggregates can be transformed into rock.
14. What holds the mineral grains in metamorphic and igneous rocks together?

CHAPTER 3

Magmas, Igneous Rocks, Volcanoes, and Plutons

The town of Furnas sits in a volcanic caldera on the island of São Miguel, Azores. The town takes its name from local hot springs and fumaroles.

MAGMA

The melting of large masses of rock to form magma, and the subsequent solidification of magma to form igneous rock are two of the most important processes that happen to the Earth. Similar melting and solidification processes have occurred on all the rocky planets and rocky moons in the solar system. The formation of magma is the most important process that all of the rocky planets in the solar system have in common.

Magma that reaches the Earth's surface and flows as hot streams or sheets is called *lava.* By observing the eruption of lava, we can draw three important conclusions concerning magma.

1. Magma is characterized by a *range of compositions* in which silica (SiO_2) is always predominant.
2. Magma is characterized by *high temperatures.*
3. Magma has the properties of a liquid, including the *ability to flow.* This is true even though some magma is almost as stiff as window glass. Most magma is a mixture of crystals and liquid (often referred to as *melt*).

Composition

The composition of magma is controlled by the most abundant elements in the Earth—Si, Al, Fe, Ca, Mg, Na, K, H and O. Because O is the most abundant anion, it is usual to express compositional variations of magmas in terms of oxides, such as SiO_2, Al_2O_3, CaO, and H_2O. The most abundant oxide component is SiO_2 (silica).

Chemical analyses of magma indicate that three narrow ranges of composition predominate and hence that three distinct types of magma are more common than all others. The first type of magma contains about 50 percent SiO_2, the second about 60 percent, while the third contains about 70 percent SiO_2. The names of common igneous rocks derived from the three magmas are, respectively, basalt, andesite, and granite (Fig. 3.1). For convenience the three magmas are referred to as *basaltic, andesitic,* and *granitic* respectively. The three are not formed in equal abundance. Approximately 80 percent of all magma erupted by volcanoes is basaltic, while andesitic and granitic magmas are each about 10 percent of the total.

Gases Dissolved in Magma

Small amounts of gas are dissolved in all magma. Gases are usually not major compositional constituents, but even small amounts of gas can influence the properties of magma. Typical magmas have gas contents ranging from 0.2 to about 3 percent. The principal gas is water vapor, which, together with carbon dioxide, accounts for more than 98 percent of all gases emitted from volcanoes. Other gases include nitrogen, chlorine, sulfur, and argon which are rarely present in amounts exceeding 1 percent.

Temperature

The temperature of magma is difficult to measure but it can be done during volcanic eruptions. Volcanoes are dangerous places, however, and scientists who study them are not anxious to be roasted alive. Measurements must be made from a distance using optical devices. Magma temperatures determined in this manner range from 1040° to 1200°C. Experiments on synthetic magmas in the laboratory suggest that under some conditions, magma might even form at temperatures as low as 625°C.

Viscosity

Dramatic pictures of lava flowing rapidly down the side of a volcano prove that some magmas are very fluid. Basaltic lava moving down a steep slope on Mauna Loa in Hawaii has been clocked at a speed of 16 km/h. Such fluidity is very rare. Flow rates are more commonly measured in meters per hour or even meters per day. Magma containing 70 percent or more SiO_2 flows so slowly that movement can hardly be detected. The properties of such magma are more akin to those of solids than to liquids.

The internal property of a substance that offers resistance to flow is called *viscosity.* The more viscous a magma, the less fluid it is. Viscosity of a magma depends on temperature and composition (especially the silica and dissolved-gas contents).

Effect of Temperature on Viscosity

The effect of temperature on viscosity is simple to understand. As with asphalt, thick oil, or ice cream, rising temperature leads to greater fluidity. The higher the temperature, therefore, the lower the viscosity and the more readily a magma flows. A very hot magma erupted from a volcano may flow readily, but it soon begins to cool, becomes more viscous, and eventually slows to a complete halt (Fig. 3.2).

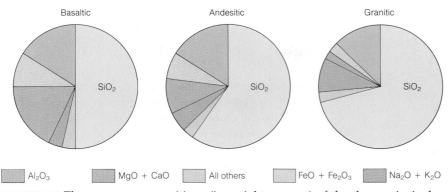

Basaltic · Andesitic · Granitic

Al$_2$O$_3$ · MgO + CaO · All others · FeO + Fe$_2$O$_3$ · Na$_2$O + K$_2$O

FIGURE 3.1 The average compositions (in weight percent) of the three principal kinds of magma.

Effect of Silica Content on Viscosity

The effect of silica content on viscosity is not so obvious as the effect of temperature. The $(SiO_4)^{-4}$ tetrahedra that occur in silicate minerals (chapter 2) occur also in magmas. Just as they do in minerals, the tetrahedra polymerize by sharing oxygens. However, unlike the tetrahedra in minerals, those in magma form irregularly shaped groupings of chains, sheets, and networks. As the average number of tetrahedra in the polymerized groups becomes larger, the magma becomes more resistant to flow and behaves increasingly like a solid. The number of tetrahedra in the groups depends on the silica content of the magma. The higher the silica content, the larger the polymerized groups.

The Origin of Magma

The making of a magma requires that temperatures be very high—so high, in fact, that we usually think of rock as being fireproof. So high too, that one of the important turning points in the history of geology was the demonstration by a Scot, James Hall, almost 200 years ago, that common rocks could be melted and that such melts had the properties of magma erupted from volcanoes.

Geothermal Gradient

No longer is there a question as to whether or not a rock will melt; rather, the questions are, what kind of rock melted, did all of the rock melt, or

FIGURE 3.2 The way a lava flows is controlled by viscosity. Two different flows are visible. They have the same basaltic composition. The lower flow is a smooth ropy surfaced lava, locally called *pahoehoe*. It formed from a hot, gas-charged, highly fluid lava. The upper lava, which was cooler and gas-poor, was very viscous and slow moving. It broke into hot, rubbly fragments called *aa*. The aa was erupted from Kilauea Volcano, Hawaii, in 1972.

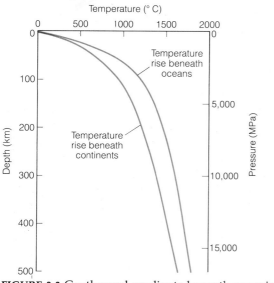

FIGURE 3.3 Geothermal gradients beneath oceanic crust and continental crust. The graph is drawn so that depth and pressure on the vertical axis increase downward, as they do in the Earth.

only part of it, how does temperature increase with depth, and at what depth does melting occur? The answers are not straightforward and in order to approach them, it is necessary to consider the geothermal gradient (which is the rate of increase of temperature downward in the Earth [Fig. 3.3]), and the way pressure increases with depth.

Geothermal gradients in the continental crust, and beneath the crust in the mantle, differ from those in and beneath the oceanic crust. This is so because the rocks in the two crusts differ signifi-

cantly in their capacities to serve as thermal blankets to the mass of hot mantle rocks below.

The Effect of Pressure on Melting

As can be seen in Figure 3.3, temperatures beneath both the oceanic and continental crusts rise to about 1000°C at rather shallow depths. Measurements made on lavas prove that some magmas are fluid at 1000°C, so an immediate question is, "Why isn't the Earth's mantle entirely molten?" The answer is that the pressures are too great, and pressure influences melting temperatures. As the pressure rises, the temperature at which a compound melts also rises. For example, albite ($NaAlSi_3O_8$) melts at 1104°C at the Earth's surface, where the pressure is 0.1 MPa, but at a depth of 100 km, where the pressure is 35,000 times greater, the melting temperature is 1440°C. Therefore, whether a particular rock melts and forms a magma at a specified place and depth in the Earth depends both on the geothermal gradient and on the effect that pressure has on the melting properties of the rock.

The effect of pressure on melting is straightforward provided a mineral is dry (Fig. 3.4a). When water or water vapor is present, however, a complication enters. Wet minerals melt at lower temperatures than dry minerals of the same composition because water dissolves in the melt. Furthermore, as the pressure rises, the effect of water increases. This is so, because the higher the pressure, the greater the amount of water that will dissolve in the melt. Therefore, increasing pressure decreases still further the temperature at which a wet mineral starts to melt. This is exactly opposite to the effect of pressure on the melting of a dry

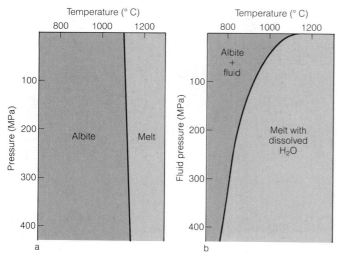

FIGURE 3.4 Influence of pressure on the melting temperature of albite ($NaAlSi_3O_8$). (a) Dry-melting curve. Increasing pressure raises the melting temperature. (b) Wet-melting curve. H_2O dissolves in the melt and decreases the melting temperature.

mineral. The effect of water on the melting of albite can be judged by comparing Figures 3.4a and b. The effects of pressure and water on the melting properties of a mixture of minerals and rocks is the same as the effects of pressure and water on the melting on a single mineral, albite.

Partial Melting

If a rock melts completely, the resulting magma must have the same chemical composition as its parent. However, rock is a mixture of minerals, and a mixture does not melt at one specific temperature as a single mineral does. Rather, a rock melts over a temperature interval that may be as much as 500°C (Fig. 3.5). Once a rock reaches the temperature at which melting starts, a small quantity of magma will form. The magma has a different composition from the unmelted residue of rock. Suppose now that the magma from the partially melted rock is squeezed out and separated from the residue. Both the magma and the residual rock will have compositions that differ from each other and from the starting composition of the parent rock. The process of forming magmas with differing compositions through the incomplete melting of rocks is known as ***magmatic differentiation by partial melting*** (Fig. 3.6). It is not difficult to see that the composition of a magma that develops by partial melting will depend both on the composi-

tion of the parent rock and the percentage of the rock that melts.

Basaltic Magma

Igneous rocks of the oceanic crust are formed from basaltic magma. The oceanic crust is thin and too cool for melting to occur. Therefore, we must conclude that the mantle is the source of basaltic magma.

The dominant minerals found in igneous rocks in the oceanic crust are olivine, pyroxene, and feldspar. Each of them is water-free. This fact suggests that basaltic magma is probably a dry, or water-poor magma. Indeed, all evidence from observations of basaltic lava during eruption suggests that the water content of basaltic magma rarely exceeds 0.2 percent. It must be concluded, therefore, that basaltic magma originates by some sort of dry, partial-melting process in the mantle.

Much debate has centered on the question of the exact chemical composition of the mantle. Rarely can we actually observe mantle rocks, but in a few places it is possible to find fragments carried up and ejected from volcanoes. From such evidence it appears that the upper portion of the mantle contains rocks rich in olivine and garnet. Such rocks are called ***garnet peridotites.***

Laboratory experiments on the dry, partial-melting properties of garnet peridotite show that at pressures and temperatures equivalent to those

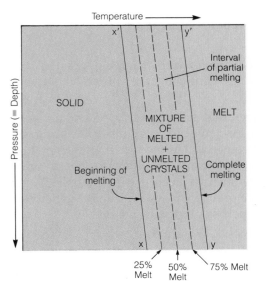

FIGURE 3.5 Dry melting of rock containing several kinds of minerals. The pressure effects are similar to those shown in Figure 3.4a. Line x–x' marks the onset of melting, curve y–y' the completion of melting. Between the two lines is a region in which melt and a mixture of unmelted crystals coexist.

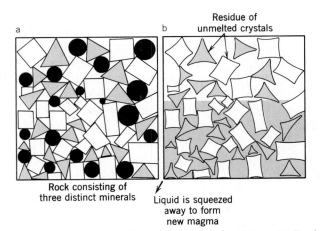

FIGURE 3.6 Magma formed by partial melting. (a) Rock consisting of three distinct minerals. (b) The first mineral that starts to melt will dissolve a small portion of the other minerals. The composition of the newly formed liquid, therefore, differs from the bulk composition of the remaining minerals. When the liquid is squeezed out, it forms magma of one composition and leaves a residue of unmelted crystals having a different composition.

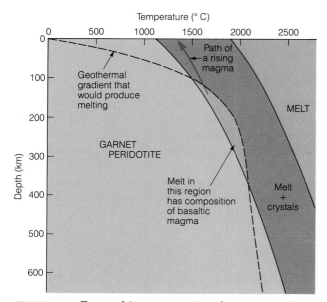

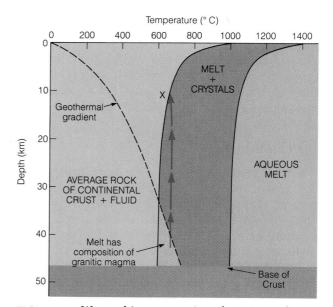

FIGURE 3.7 Dry-melting properties of garnet peridotite and the generation of basaltic magma. Partial melting of the peridotite occurs at depths between 100 and 350 km. The melt that forms has the composition of basaltic magma. A basaltic magma will rise toward the surface, following a depth–temperature curve like that shown by the red line.

FIGURE 3.8 Wet-melting properties of continental crust and the generation of granitic magma. Melting commences at a pressure equivalent to a depth of about 35 km. The melt that forms has the composition of granitic magma. The highly viscous magma slowly rises and follows a depth–temperature path like that shown by the red curve. At point *X* the magma reaches the freezing curve and is completely solid.

reached at a depth of 100 km below the surface—that is, at pressures and temperatures reached in the asthenosphere—a 10 to 15 percent partial melt will yield a magma of basaltic composition (Fig. 3.7). While this leaves unanswered, for the moment, the question of a heat source, and why basaltic magma should develop in some places but not others, we can nevertheless confidently accept the conclusion that basaltic magma forms by dry partial melting of rocks in the upper mantle.

Once a body of basaltic magma has formed, it will start to rise because it is less dense than the rock around it. The rate of rising can be quite rapid. One line of evidence comes from earthquakes. As basaltic magma rises, it widens fractures and sometimes causes new ones in the rock it is passing through. The opening of fractures causes small earthquakes that can be detected by sensitive devices on the surface. Evidence from Hawaii indicates that basaltic magma can rise at rates as fast as a kilometer a day. Under such circumstances, the rate of rising is much faster than the rate of cooling. The pressure–temperature path followed by a rising basaltic magma will, therefore, be something like that shown in Figure 3.7. The higher a basaltic magma rises, the further away it will be from its solidification temperature. Therefore, not surprisingly, a lot of basaltic magma manages to

rise through the crust without solidifying. When it reaches the surface, it is erupted as lava.

Granitic Magma

Two observations suggest an origin for granitic magma:

1. Volcanoes that extrude magma of granitic composition are confined to regions of continental crust.

2. Volcanoes that erupt magma of granitic composition also give off a great deal of water vapor, and igneous rocks formed from granitic magma contain significant quantities of water-bearing minerals such as mica and amphibole. Water therefore probably plays a role in the generation and properties of granitic magma.

These two points of evidence suggest (1) that the source of granitic magma lies within the continental crust, and (2) that the origin involves some sort of wet partial melting. Laboratory experiments bear these suggestions out. When, in the laboratory, water-bearing rocks of the continental crust start to melt, the composition of the liquid that forms is granitic. As seen in Figure 3.8, the wet-melting curve for a rock of average crustal composition in-

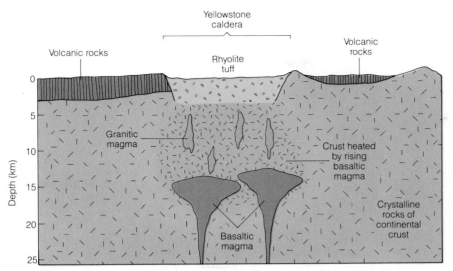

FIGURE 3.9 Simplified section through the Yellowstone caldera. Basaltic magma rising from the mantle fills magma chambers (red) and locally melts crustal rocks to form magma of granitic composition. Eruption of the magma to form volcanic rock leaves the magma chamber partly empty and the roof collapses to form a caldera.

tersects the geothermal gradient at a depth of 35 to 45 km, a depth near the base of the continental crust. A 30 percent partial melt of continental crust has the composition of a granitic magma. The melting can be produced in two ways. First, ordinary, water-bearing, crustal rocks will start to melt if pushed down to these depths during continental collisions. Second, if a heat source such as a body of basaltic magma rising up from the mantle locally raises the temperature of the continental crust, partial melting will occur. This is what has happened beneath Yellowstone Park (Fig. 3.9).

Once a granitic magma has formed, it starts to rise. However, it rises slowly because it has a high SiO_2 content and is very viscous. As it slowly rises the pressure on it will decrease. As discussed earlier, the effectiveness of water in reducing the melting temperature is diminished by reduced pressure. A rising magma formed by wet partial melting will therefore tend to solidify and form an igneous rock underground unless it increases in temperature. But a rising magma traverses bodies of cool rock, and there is no source of heat to cause an increase in temperature. As a result, the depth–temperature path of a rising body of granitic magma brings it closer and closer to its solidification temperature. Therefore, most granitic magmas solidify underground rather than reaching the surface and forming lavas.

Andesitic Magma

The chemical composition of andesitic magma is close to the average composition of the continental crust. Igneous rocks formed from andesitic magma are commonly found in the continental crust. From these two facts it might be supposed that andesitic magma forms by the complete melting of a portion of continental crust. Some andesitic magma may indeed be generated in this way, but andesitic magma is also extruded from volcanoes that are far from continental crust. In those cases, the magma must be developed either from the mantle or from the oceanic crust. Laboratory experiments provide a possible answer.

In the laboratory, partial melting of wet oceanic crust under suitably high pressure yields a magma of andesitic composition. An interesting hypothesis suggests how this might happen in nature. When a moving plate of lithosphere plunges back into the asthenosphere, it carries with it a capping of oceanic crust saturated by seawater. The plate heats up and eventually the wet crust starts to melt. Wet partial melting that starts at a pressure equivalent to a depth of 80 km produces a melt having the composition of andesitic magma (Fig. 3.10).

There are a number of details concerning the melting process of wet oceanic crust that remain to be deciphered, but two pieces of evidence are very supportive of the idea that much of the andesitic magma forms in this manner. The first concerns active volcanoes. When the locations of active volcanoes in and around the Pacific Ocean are plotted, it is apparent that a well-defined line separates regions where andesitic magma is observed from regions where it is not found. The line is called the ***Andesite Line*** (Fig. 3.11). On the Pacific side of the line, and inside the main ocean basin, andesitic

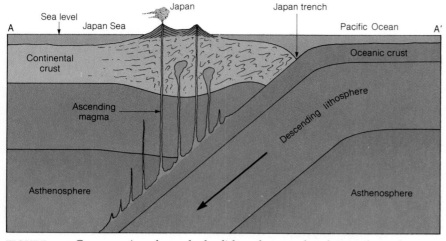

FIGURE 3.10 Cross section through the lithosphere and asthenosphere demonstrating the probable way by which andesitic magma is formed by wet partial melting of oceanic crust. Melting starts at a depth of 80 km. Andesitic magma rises, creating volcanoes on the Japanese islands.

magma is unknown. All volcanoes inside the line erupt basaltic magma. Outside the line, both andesitic and basaltic magma are common. The Andesite Line coincides closely with subduction zones, which, as stated in chapter 1, mark the very places where plates of lithosphere sink back into the asthenosphere. Therefore, if andesitic magma does form by the partial melting of wet oceanic crust,

the process occurs at the very places on the Earth where such crust is found at suitable depths and temperatures for wet partial melting to occur.

The second line of evidence comes from the distribution of andesitic volcanoes with respect to the subduction zone. On the upper surface of a plate of lithosphere, a subduction zone is marked by the presence of a deep-sea trench (Fig. 3.10). Beyond

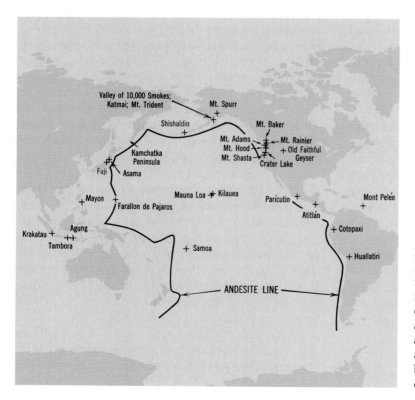

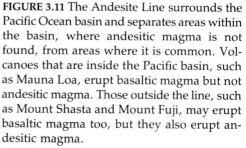

FIGURE 3.11 The Andesite Line surrounds the Pacific Ocean basin and separates areas within the basin, where andesitic magma is not found, from areas where it is common. Volcanoes that are inside the Pacific basin, such as Mauna Loa, erupt basaltic magma but not andesitic magma. Those outside the line, such as Mount Shasta and Mount Fuji, may erupt basaltic magma too, but they also erupt andesitic magma.

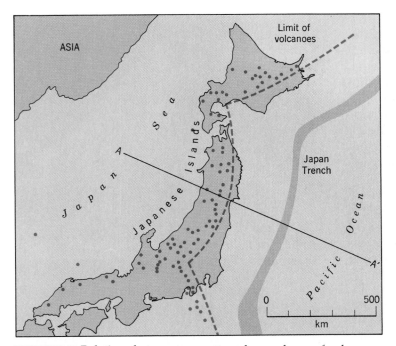

FIGURE 3.12 Relations between ocean trenches and arcs of volcanoes erupting andesitic magma. Arc-shaped Japanese islands are parallel to the Japan Trench. Andesitic volcanoes active during the last million years are also confined behind arcuate boundaries. The cross section depicted in Figure 3.10 is approximately along the line A–A'.

the subduction zone, the lithosphere sinks into the asthenosphere, carrying with it its capping of wet oceanic crust. A distance equivalent to a depth of about 80 km marks the edge of an arcuate cluster of volcanoes (Fig. 3.12).

Solidification of Magma

Literally hundreds of different kinds of igneous rock have been found. Most are rare, but the fact that they exist emphasizes an important point: A magma of a given composition can crystallize into many different kinds of igneous rock. This is true because magma is a complicated liquid. It does not solidify into a single compound, as water does when it freezes to form ice. Solidifying magma forms several different minerals, and those minerals start to crystallize from the cooling magma at different temperatures. The process is just the opposite of partial melting. Therefore, a cooling magma consists of a mixture of already crystallized minerals and still uncrystallized melt. Because different minerals begin to crystallize at different temperatures, the composition of the remaining melt changes continually as the temperature changes. If at any stage

during crystallization the melt becomes separated from the crystals, a magma with a brand-new composition results, while the crystals left behind form an igneous rock with a quite different composition.

Magmatic Differentiation by Fractional Crystallization

There are a number of ways by which crystal-melt separations can occur. For example, compression can squeeze melt out of a crystal-melt mixture. Another mechanism involves the sinking of the dense, first crystallized minerals to the bottom of a magma chamber thereby forming a solid mineral layer covered by melt. However a separation occurs, the compositional changes that occur in magmas by the separation of early formed minerals from residual liquids are called *magmatic differentiation by fractional crystallization.*

Bowen's Reaction Series

It was a Canadian-born scientist, N. L. Bowen, who first recognized the importance and complexities of magmatic differentiation by fractional crystallization. He knew that plagioclases crystallized from basaltic magma are usually calcium-rich (anorthitic), while those formed from granitic magma are commonly sodium-rich (albitic). Andesitic magma, he observed, tends to crystallize plagioclases of intermediate composition. Bowen also knew that plagioclases in many igneous rocks have concentric zones of differing compositions such that the innermost, and therefore earliest formed core, is anorthitic in composition, and that successive layers are increasingly albite-rich (Fig. 3.13a).

Bowen's experiments provided a common explanation for these observations. He discovered that even though the composition of the first plagioclase that crystallized from a magma of basaltic composition was calcium-rich, the composition changed toward a more sodic composition as crystallization proceeded, and the ratio of crystals to melt increased. This meant that all the plagioclase crystals, even the earliest ones formed, must have continually changed their compositions as the magma cooled.

Continuous Reaction Series. As explained in chapter 2, the plagioclase solid-solution series involves a coupled substitution in which $Ca^{+2} + Al^{+3}$ are replaced in the structure by $Na^{+1} + Si^{+4}$. Bowen referred to such a continuous change of mineral composition in a cooling magma as a ***continuous reaction series,*** by which he meant that even though the composition changed continually, the crystal structure remained unchanged. The process by which this occurs is controlled by the rates at which the four ions, Ca^{+2}, Al^{+3}, Si^{+4}, and Na^{+1}, can diffuse through the plagioclase structure. Such diffusion is an exceedingly slow process. A complete chemical balance (commonly referred to as chemical equilibrium) is rarely attained because the cooling rates of magmas are faster than diffusion rates. As a result, zoned plagioclase crystals are formed. The anorthite-rich inner zones are out of chemical equilibrium with the albite-rich outer zones, and with the residual magma. Bowen pointed out that the existence of zoned crystals has important implications. When anorthite-rich cores are present, the remaining melt is necessarily richer in albite than it would have been if equilibrium had been maintained. According to Bowen, the anorthite-rich cores are another example of magmatic differentiation by fractional crystallization. If, in a partially crystallized magma containing zoned crystals, the melt were somehow squeezed out of the crystal mush, the result would be an albite-rich magma, and the residue would be an anorthite-rich rock.

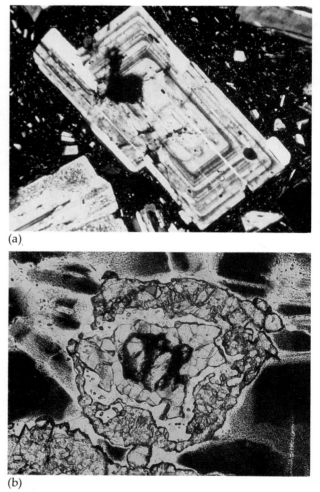

(a)

(b)

FIGURE 3.13 Textures illustrating Bowen's reaction series. (a). Zoned plagioclase crystal in andesite, proof of continuous reaction, photographed in polarized light to enhance the zoning. Bands near center are anorthite-rich, progressing to albite-rich at the rim. The crystal is 2 mm long. (b). A grain of olivine in a gabbro surrounded by reaction rims of pyroxene and amphibole demonstrates discontinuous reaction. The diameter of the outer rim is 3mm.

Discontinuous Reaction Series. Bowen identified several sequences of reactions besides the continuous reaction series of the feldspars. One of the earliest minerals to form in a cooling basaltic magma is olivine. Continued cooling changes the olivine composition slightly by solid solution, but eventually a point is reached where the olivine reacts with silica in the melt to form a more silica-rich mineral, pyroxene (Fig. 3.13b). An idealized example of such a reaction is:

$$\underset{\text{Olivine}}{Mg_2SiO_4} + \underset{\text{Silica in magma}}{SiO_2} \rightarrow \underset{\text{Pyroxene}}{2MgSiO_3}$$

At a still lower temperature, pyroxene reacts to form amphibole, which contains more silica than pyroxene, and then the amphibole reacts in turn

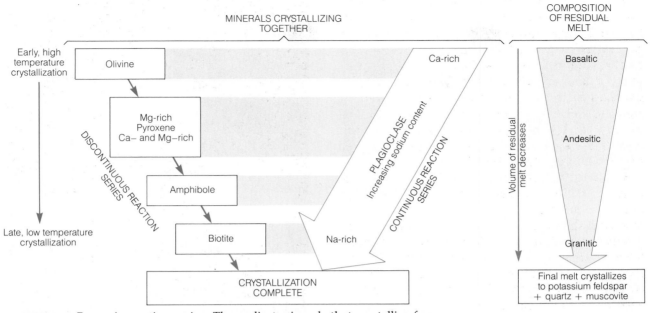

FIGURE 3.14 Bowen's reaction series. The earliest minerals that crystallize from a magma of basaltic composition are olivine and calcium-rich plagioclase (anorthite). As cooling and crystallization proceed, olivine (upper left) reacts with the remaining liquid to form pyroxene. Pyroxene in turn reacts to form amphibole, and amphibole forms biotite. The early plagioclase that co-crystallizes with olivine is calcium-rich, but as cooling proceeds, the early plagioclase reacts with the residual melt and continually changes its composition, becoming more sodium-rich. The composition of the residual melt in contact with the crystallized minerals becomes increasingly silica-rich, and eventually the final small fraction of melt has the composition of a granitic magma.

to form an even more siliceous mineral, biotite. Such a series of reactions, where early formed minerals form entirely new compounds through reactions with the remaining liquid, is called a *discontinuous reaction series*. If a core of olivine is shielded from further reactions by a rim of pyroxene, the remaining liquid will be more silica-rich than it would be if equilibrium were maintained and all the olivine were converted to pyroxene. If partial reactions occurred in both continuous and discontinuous reaction series, Bowen reasoned that differentiation by fractional crystallization in a basaltic magma could, under some circumstances, even produce a residual magma with a granitic composition (Fig. 3.14).

It is now known that Bowen's conclusions are correct so far as the reaction series is concerned, and that the reactions are vitally important in producing rock types of many compositions from basaltic magma. However, large volumes of granitic magma do not seem to form by fractional crystallization. The main evidence against Bowen's idea is simply that, even under ideal conditions, no more than 10 percent of the volume of basaltic magma can be differentiated to granitic magma. Yet bodies

of igneous rock formed by crystallization of granitic magma are often of immense size—hundreds of thousands of cubic kilometers of rock. Basaltic magma chambers where the differentiation could occur are simply not large enough to have produced such huge masses of granite. Another line of evidence comes from the distribution of granitic magma. It always occurs in the continental crust. But if granitic magma formed by direct differentiation of basaltic magma we would surely expect to find some in the oceanic crust, for it is there that basaltic magma is most common. The principal manner by which granitic magma forms, therefore, must be through partial melting of continental crust.

KINDS OF IGNEOUS ROCK

Magma, like most liquids, is less dense than the solid from which it forms. Therefore, once formed, the low-density magma will exert an upward pressure on the enclosing rocks, and will slowly push its way upward.

Most magma solidifies below the surface. We refer to igneous rock formed in such a manner as

FIGURE 3.15 Textures seen in thin sections of (a) basalt, (b) diabase, (c) gabbro, and (d) basalt porphyry. All these rocks have the same composition. Basalt, an aphanite, is fine grained because it cooled very rapidly. Gabbro, a phanerite, is coarse grained and cooled slowly. *Diabase* is a term used for a fine- to medium-grained gabbro that cooled at intermediate rates. Basalt porphyry contains phenocrysts of plagioclase set in a matrix that is so fine the grains can barely be resolved with a microscope. The coarse phenocrysts formed during slow cooling at depth; the matrix formed when the partly crystallized magma was suddenly extruded as a lava. In each case the field of view is the same: 7 mm across.

intrusive igneous rock. Some magma does, of course, reach the surface where it flows out as lava, solidifies, and forms *extrusive igneous rock.*

Texture

Phanerites and Aphanites

Phanerites. In general, intrusive igneous rocks tend to be coarse grained. Igneous rocks in which the component mineral grains are distinguishable megascopically are called **phanerites.** Such rocks are coarse grained because magma that solidifies below the surface tends to cool slowly and has sufficient time to form large, clearly visible mineral grains a millimeter or larger in diameter.

Aphanites. By contrast, magma that cools rapidly forms very fine-grained and even glassy rocks.

Natural glass results from lava that cools and solidifies too rapidly for its atoms to organize themselves into minerals. Most extrusive igneous rocks are crystalline but very fine grained, or they are a mixture of glass and fine-grained minerals. Such rocks are called **aphanites,** defined as igneous rocks in which the component grains cannot be readily distinguished with the naked eye or even with the aid of a simple hand lens.

Porphyries

One special class of igneous rock has a distinctive mix of large and small grains and thus is partly phaneritic, partly aphanitic. A rock containing such a texture is called a **porphyry,** meaning any igneous rock consisting of coarse mineral grains scattered through a mixture of fine mineral grains (Fig. 3.15). The isolated large crystals in a porphyry are called

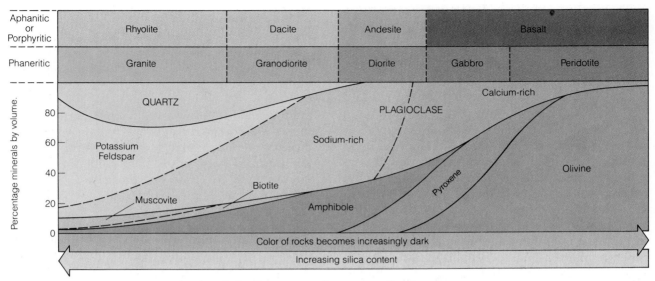

FIGURE 3.16 The proportions of minerals in the common igneous rocks. Boundaries between rock types are not abrupt but gradational, as suggested by the broken lines. To determine the composition range for any rock type, project the broken lines vertically downward, then estimate the percentages of the minerals by means of the numbers at the edge of the diagram.

phenocrysts, and they form in the same way that coarse crystals in phanerites do—by slow cooling of magma deep in the crust. The fine-grained, commonly aphanitic *groundmass* (or *matrix*) that encloses phenocrysts is evidence that the partly cooled magma underwent a later, rapid stage of cooling, perhaps by being moved rapidly upward during crystallization. In the new setting, the magma cooled more rapidly, and as a result the later mineral grains, which form the groundmass, are all tiny. Most phenocrysts have well-developed crystal forms because they grow within a fluid and do not encounter interference from other crystals growing near them, as seen in Figure 3.15c. The shapes of mineral grains in phanerites, in aphanites, and in the groundmass of porphyries, are irregular and intensely interlocked, as can be seen in Figure 3.15a, b, and c. This is so because during the final stages of mineral growth and crystallization all the mineral particles are crammed against each other, thus preventing the formation of smooth crystal faces. Instead, the interlocking network of grain boundaries acts like a giant, three-dimensional jigsaw and makes an igneous rock coherent and solid.

Mineral Assemblages

Once it has been determined whether an igneous rock has a phaneritic, aphanitic, or porphyritic texture, a name can be given to any specimen on the basis of its mineral assemblage. To see how this is done it is convenient to employ the diagram shown in Figure 3.16.

All common igneous rocks are composed of one or more of these six minerals: quartz, feldspar (both potassium feldspar and plagioclase), mica, amphibole, pyroxene, and olivine. When the percentage of each mineral in a rock specimen has been estimated, the correct place in Figure 3.16 is located and the corresponding rock name determined. The aphanitic or phaneritic name is selected, whichever is appropriate. If a rock has a porphyritic texture, we use the term determined by the mineral assemblage as an adjective, using the grain size of the groundmass for the noun. For example, if the groundmass is aphanitic, we would refer to a rhyolite porphyry; but if it is phaneritic, we refer to a granite porphyry. It should always be kept in mind when using Figure 3.16 to determine the name of an igneous rock that nature does not draw sharp boundaries between igneous rocks; every conceivable gradation in texture and composition can be found.

The boundary line between granodiorite and diorite in Figure 3.16 is drawn on the presence or absence of quartz. Quartz is present in granodiorite but is absent (or present only at a trace level) in diorite. The boundary line between diorite and gabbro is drawn where the dark-colored ferromagnesian minerals reach 50 percent of the total and exceed the light-colored feldspars. The boundary is carried through between the aphanitic equivalents of diorite and gabbro—andesite and basalt respectively—on the basis of color.

Varieties

Granite and Granodiorite

Feldspar and quartz are the chief minerals in *granite* and *granodiorite.* A mica, either biotite or muscovite, is usually present and many granites contain scattered grains of hornblende. Commonly, the biotite and hornblende are in nearly perfect crystals; this suggests that they crystallized first from the magma. Feldspar started crystallizing next and the grains crowded against and hampered each other in growth. Quartz crystallized last and so is molded around the angular grains of the earlier minerals. This interlocking arrangement of visible mineral grains characteristic of granite is called *granular texture.*

The term *granite* is applied only to quartz-bearing rocks in which potassium feldspar is predominant. The name *granodiorite* applies to similar rocks in which plagioclase is the chief feldspar. Without special equipment the differences in feldspars are not always easily recognized, and in a general study the term *granitic* is extended to this whole group of rocks.

Huge intrusive masses of granite and granodiorite are common in continental crust and both originate through crystallization of granitic magma formed by wet partial melting of continental crust. Neither granite nor granodiorite is formed in the oceanic crust.

Pegmatite. An igneous rock that has abnormally large mineral grains is called a *pegmatite.* The term is used for rocks, usually of granitic composition, in which average grain diameters are 2 cm or larger. Individual mineral grains up to several meters long have been discovered in some pegmatites. Pegmatites are mined for muscovite which can sometimes be found in large sheets up to a meter across and for minerals that yield valuable elements such as lithium and beryllium.

Diorite

The chief mineral in *diorite* is plagioclase feldspar. Quartz is usually absent and the ferromagnesian minerals are generally more abundant than in granodiorite. Diorite forms many large intrusive masses, but it is not so abundant as granitic rocks.

Gabbro

Dark diorite grades into *gabbro* in which the ferromagnesian minerals may exceed 50 percent of the rock. The chief dark minerals in gabbro are pyroxene and olivine.

Peridotite and Anorthosite. Some phaneritic igneous rocks related to gabbro contain 90 percent or more olivine in which case we refer to them as *peridotites;* others contain 90 percent or more plagioclase, and are called *anorthosites.*

Rhyolite and Dacite

A fine-grained rock with phenocrysts of quartz is either a *rhyolite* or a *dacite.* The quartz indicates an excess of silica and, therefore, a close chemical kinship to granite and grandiorite. Rhyolites are the aphanitic equivalents of granites, dacites are the aphanitic equivalents of granodiorites. Rhyolites and dacites usually have phenocrysts of both quartz and feldspar. Phenocrysts of biotite are also common. Colors of the groundmass range from nearly white to shades of gray, yellow, red, or purple.

Andesite

Andesite is an aphanitic igneous rock, often similar in appearance to dacite, but lacking quartz phenocrysts. Usually it has phenocrysts of plagioclase and dark-colored ferromagnesian minerals. Common colors are shades of gray, purple, and green, but some andesites are very dark, even black.

Andesite is extremely abundant as a volcanic rock, especially around the margins of the Pacific Ocean. The name comes from the Andes of South America.

Basalt

Basalt is a fine-grained rock, sometimes porphyritic, that appears dark even on freshly broken thin edges. Common phenocrysts are plagioclase, pyroxene, and olivine. Common colors are black, dark brown or green, and very dark gray. The name *basalt* is an ancient one, derived from the Latin *basaltes,* but it is a misnomer because *basaltes* was the name used by the Romans for a dark-gray limestone from Ethiopia.

LAVA, PYROCLASTS, AND VOLCANOES

Magma that reaches the Earth's surface is lava. It tends to flow out as hot sheets which cool rapidly in contact with the atmosphere. As a result extrusive igneous rocks tend to be fine grained and even glassy. Many extrusive igneous rocks also display distinctive features caused by dissolved gas escaping rapidly from solution in the magma.

Obsidian

Obsidian is a lustrous, glassy igneous rock which displays a distinctive pattern on a broken surface. The fracture pattern consists of a series of smooth, curved surfaces, and is known as **conchoidal fracture** (Fig. 3.17).

Most obsidians appear dark, even black. Because many of them correspond in chemical composition to rhyolite and granite, they seem to contradict the rule that rocks with a high silica content are light colored. But rhyolite obsidian chipped to a thin edge appears white, even transparent. The dark coloring results from a small amount of dark mineral matter distributed evenly in the glass. By contrast, basalt obsidian is opaque or very dark colored, even on a thin edge.

Pumice

When the gas dissolved in a magma comes out of solution it forms tiny bubbles. In rapidly cooled viscous magmas a glassy froth called a **pumice** will sometimes result. Because the cavities of most of the glassy bubbles in pumice remain sealed, pumice will often float on water.

Vesicles and Amygdules

The gas escaping from a low-viscosity lava tends to bubble away into the atmosphere rather than forming a pumice. However, in a rapidly cooled lava the upper and lower parts of the flow are commonly *vesicular*, meaning they are filled with small openings or **vesicles** made by the final escaping gases as the temperature of the magma dropped and the viscosity increased rapidly (Fig. 3.18a). Vesicular basalts are common. In many flows the vesicles have later been filled with calcite, quartz, or some other mineral deposited by heated groundwater. Vesicles filled by secondary minerals are called **amygdules;** a basalt containing them is called an **amygdaloidal basalt** (Fig. 3.18b).

a

FIGURE 3.17 Obsidian from the Jemez Mountains, New Mexico. The specimen is glass of a rhyolitic composition. The curved ridges are typical of the fracture pattern in glass broken by a sharp blow. The specimen is 10 cm across (W. Sacco).

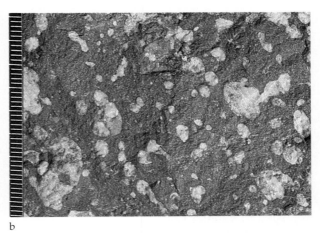

b

FIGURE 3.18 Textures in basalts. (a) Vesicular basalt. Note the phenocrysts of yellowish-green olivine. (b) Amygdaloidal basalt formed when secondary minerals such as calcite and zeolites fill vesicles. Each specimen is 4.5 cm across.

a

b

FIGURE 3.19 Tephra. (a) Large spindle-shaped bombs, up to 50 cm in length, cover the surface of a cinder cone on Haleakala Volcano, Maui. (b) Intermediate-sized tephra called lapilli. Two layers of tephra are separated by a soil layer. The upper layer was erupted from Mount St. Helens in 1800 A.D., the lower one in 1480 A.D.

Pyroclasts and Tephra

Gases sometimes escape from a magma in such a violent fashion that pieces of solid rock are ripped off the walls of the volcanic vent. The sticky magma may also splatter and shatter into small, hot fragments. Fragments of rock and magma extruded violently from a volcano are called *pyroclasts* (named from the Greek words *pyro,* meaning heat or fire, and *klastos,* meaning broken; hence, hot, broken fragments). Loose assemblages of pyroclasts are called *tephra* (Fig. 3.19). The terms used to describe tephra of different size—bombs, lapilli, and ash—are listed in Table 3.1.

The word *ash* is somewhat misleading because ash means, strictly, the solid that is left after something flammable, such as wood, has burned. But the fine particles thrown out by volcanoes look so like true ash that it has become a convenient cus-

TABLE 3.1 *Names for Tephra and Pyroclastic Rock*

Average Particle Diameter (mm)	Tephra (unconsolidated material)	Pyroclastic Rock (consolidated material)
>64 mm	Bombs	Agglomerate
2–64 mm	Lapilli	Lapilli tuff
<2 mm	Ash	Ash tuff

tom to use the word for this material too. Individual ash particles are often called *volcanic shards* (Fig. 3.20).

Pyroclastic Rock

Pyroclastic rocks are transitional between igneous and sedimentary rocks. They are rocks formed through the cementation or welding of tephra, and are named either *agglomerate* or *tuff.* As seen in Table 3.1, the terms *agglomerate, lapilli tuff,* and *ash tuff* describe the textures of pyroclastic rocks. Mineral assemblages are used to determine the composition of such pyroclastic rocks, just as they are with other igneous products. For example, we refer to a rock of appropriate composition as an andesitic lapilli tuff if it is aphanitic, and dioritic lapilli tuff if it is phaneritic.

Conversion of Tephra to Pyroclastic Rock

Conversion of tephra to pyroclastic rock can come about in two ways. The first, and most common way, is through the addition of a cementing agent introduced by groundwater (Fig. 3.21a). The most common cementing agents are calcite and quartz.

FIGURE 3.20 Ash-sized volcanic shards of rhyolitic composition. The glass fragments, here photographed through a microscope, are the main constituent of the volcanic ash called Obispo Tuff, found in California. The specimen is 6 mm across.

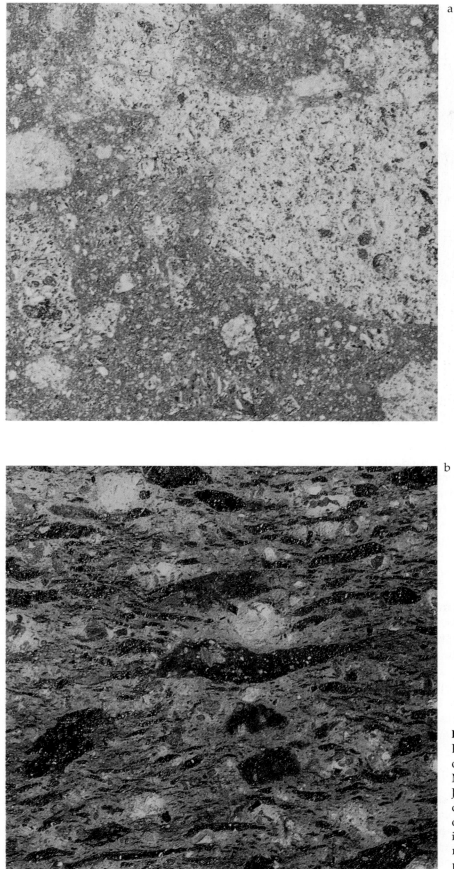

a

b

FIGURE 3.21 Two kinds of tuff. (a) Lapilli tuff, formed by cementation of lapilli and ash, from Clark County, Nevada. (b) Welded tuff from the Jemez Mountains, New Mexico. The dark patches are glassy fragments of pumice flattened during welding. Note the fragments of other rocks in the specimen. Both samples are 5 cm across.

FIGURE 3.22 Mauna Kea, a shield volcano on Hawaii, as seen from Mauna Loa. The view is almost directly north. A pahoehoe flow is in the foreground on the northeast flank of Mauna Loa.

The second way tephra is transformed to pyroclastic rock is through the welding of hot, glassy ash particles. *Welded tuff* is the name applied to pyroclastic rocks, the glassy fragments of which were plastic and so hot when deposited that they fused to form a glassy rock. Very commonly the glassy fragments are pieces of pumice which become flattened and compacted during the fusion process, giving welded tuffs a rather distinctive texture that resembles a flow pattern (Fig. 3.21b).

Volcanoes

When magma reaches the Earth's surface, it does so through a *volcano,* a vent from which igneous matter, solid rock debris, and gases are erupted. The term *volcano* comes from the name of the Roman god of fire, Vulcan, and it immediately conjures up visions of sheets of lava pouring out over the landscape. In fact, volcanoes, and the eruption process, are much more varied than is commonly realized.

Shield Volcanoes

The volcano easiest to visualize is one built up of successive flows of very fluid lava. Such lavas are capable of flowing great distances down gentle slopes, and of forming thin sheets of nearly uniform thickness. Eventually the pile of lava builds up a *shield volcano,* which is a broad, dome-shaped edifice (convex upward) with a surface slope of only a few degrees (Fig. 3.22).

The slope of a shield volcano is slight near the summit because the magma is hot and very fluid. It will readily run down a very slight slope. The further the lava flows down the flank, the cooler and less fluid it becomes, and the steeper a slope must be in order for it to flow. Slopes typically range from less than 5° near the summit, to 10° on the flanks.

Shield volcanoes are characteristically formed by the eruption of basaltic lava—the proportions of ash and other fragmental debris are small. Hawaii, Tahiti, Samoa, the Galápagos, and many other oceanic islands are the upper portions of large shield volcanoes.

Pyroclastic Cones

The lavas of some volcanoes, particularly those of rhyolitic, dacitic, and andesitic compositions, are so highly viscous that gas can only escape from them with great violence, ejecting quantities of pumice, cinders, volcanic ash, and other pyroclasts. As the debris showers down, a *pyroclastic cone* consisting entirely of pyroclastic debris is built around the vent. The slope of the cone is determined by the angle of repose of the debris (Fig. 3.23). Fine ash will stand at a slope angle of 30° to

FIGURE 3.23 Puu Hau Kea, Hawaii, a pyroclastic cone. Mauna Loa, a shield volcano, is visible in the background.

35°, while cinders generally stand at an angle of about 25°. The gradual decrease in the volume of fallout material away from the vent leads to more gentle slopes near the base of the cone.

Stratovolcanoes

Large, long-lived volcanoes of andesitic, dacitic, and rhyolitic composition emit a combination of lava flows and pyroclastic debris. The volume of pyroclastic material generally exceeds the volume of the lava, and so the slopes of the large volcanic cones, which may be thousands of meters high, are steep like those of pyroclastic cones. As the volcano develops, the lava flows act as a cap to slow down erosion of the soft volcanic ash, and thus the volcano may become much larger than a typical pyroclastic cone.

Stratovolcanoes (also called *composite volcanoes*) are defined as volcanoes that emit both fragmental material and viscous lava, and that build up steep conical mounds. Near the summit of a stratovolcano, the slope is about 30°, like that near the summit of a pyroclastic cone. Toward the base, the slopes of stratovolcanoes flatten to about 6° to 10°.

The beautiful, steep-sided cones of stratovolcanoes are among Earth's most picturesque sights (Fig. 3.24). The snow-capped peak of Mount Fuji in Japan has inspired poets and writers for centuries. Mount Rainier and Mount Baker in Washington and Mount Hood in Oregon are majestic examples in North America. Andesites and rhyolites are most commonly found on continents; therefore, composite volcanoes are more common on continents than in the ocean basins.

FIGURE 3.24 Two steep-sided, towering stratovolcanoes in the Aleutians. The larger cone (rear) is Pavlov; the near one is Pavlov's Sister.

Craters, Calderas, and Other Volcanic Features

Certain features associated with pyroclastic cones, shield, and stratovolcanoes give volcanic terrains a unique character. Fractures may split a cone so that lava, ash, or both emerge along its flanks. Small, satellitic ash and spatter cones then develop, peppering the slope of the main volcano like so many small pimples. Gases emerge from small vents, altering and discoloring nearby rocks, and hot springs may form, emitting evil-smelling sulfurous gases.

Craters. Near the summits of most volcanoes is a *crater,* a funnel-shaped depression from which gases, fragments of rock, and lava are ejected.

Calderas. Many volcanoes are marked near their summits by a striking and much larger depression than a crater. This is a *caldera,* a roughly circular, steep-walled basin several kilometers or more in diameter. Calderas originate through collapse following eruption and the partial emptying of a magma chamber. The rapid ejection of magma during a large pyroclastic eruption leaves the chamber from which the volcanic ash was emitted empty or partly empty. The now-unsupported roof of the chamber slowly sinks under its own weight, like a snow-laden roof on a shaky barn, dropping downward on a ring of steep vertical fractures. Subsequent volcanic eruptions commonly occur along these

FIGURE 3.25 Crater Lake, Oregon, occupies a caldera 8 km in diameter that crowns the summit of a once lofty stratovolcano, posthumously called Mount Mazama. Wizard Island, a small pyroclastic cone, formed after the collapse that created the caldera.

fractures, thus creating roughly circular rings of small cones. Crater Lake, Oregon, occupies a circular caldera 8 km in diameter (Fig. 3.25), formed after a great eruption about 6600 years ago. The volcano that erupted has been posthumously called Mount Mazama. What remained of the roof after a pyroclastic outpouring of about 75 km³ of magma then collapsed into the partly empty magma chamber. Yellowstone is also located in a giant caldera that formed following a gigantic pyroclastic eruption 2.2 million years ago (Fig. 3.9). This was one of three pyroclastic eruptions, the last of which took place 600,000 years ago.

Resurgent Cauldrons. A volcano does not cease activity following the formation of a caldera. Magma starts reentering the magma chamber, and in the process causes the uplifting of the collapsed floor of a caldera to form a structural dome. Such a feature is called a ***resurgent cauldron.*** Subsequently, small pyroclastic cones and lava flows build up in the interior of the caldera. Wizard Island, in Crater Lake, is a cone having such an origin.

Lava Domes. When lava is extruded following a major volcanic eruption it tends to be extremely viscous because it has very little dissolved gas left in it. The sticky lava then squeezes out to form a ***lava dome*** (Fig. 3.26), which is a volcanic dome characterized by an upheaved, consolidated conduit filling of lava.

Thermal Springs and Geysers

When volcanism finally ceases, rock in the old magma chamber remains very hot for possibly a million years or more. Descending groundwater that comes into contact with the hot rock becomes heated and tends to rise again to the surface along rock fractures where it forms a ***thermal spring.***

Water temperatures in thermal springs may be as hot as 100°C. Because dissolution is more rapid in hot water than in cold, hot springs are more likely to be unusually rich in mineral matter dissolved from rocks with which they have been in contact. In some springs the mineral content is said to have medicinal properties.

Geysers. A hot spring equipped with a system of plumbing and heating that causes intermittent eruptions of water and steam is a ***geyser.*** The name comes from the Icelandic word *geysir,* meaning to

FIGURE 3.26 A lava dome in the crater of Mount St. Helens, Washington, in May 1982. The plume rising above the dome is steam.

gush, for Iceland is the home of many geysers (Fig. 3.27). Most of the world's geysers outside Iceland are in New Zealand and in Yellowstone National Park.

Old Faithful in Yellowstone, the most famous American geyser, erupts for a few minutes about once an hour, throwing a jet of steam and water high in the air. During intervals between eruptions the emptied geyser is refilled with water, which is then heated to the critical point at which the next eruption is triggered.

FIGURE 3.27 The great Geysir, Iceland, from which all geysers take their name.

Fissure Eruptions on Land

Some lava reaches the surface via elongate fissures through the crust. Extrusion of lava along an extended fracture is called a *fissure eruption.* Such eruptions are characteristically associated with basaltic magma, and the lavas that emerge tend to spread widely and to create flat lava plains called *plateau basalts.* The fissure eruption at Laki in Iceland in 1783 occurred along a fracture 32 km long. Lava flowed 64 km outward from one side of the fracture and nearly 48 km outward from the other side. Altogether it covered an area of 588 km². The volume of the lava extruded was 12 km³, making this the largest lava flow in historic times. It was also one of the most deadly lava flows in history because an estimated 10,000 people were killed. There is good evidence to prove that larger eruptions have occurred in prehistoric times. The Roza flow, a great sheet of basaltic lava in eastern Washington State, can be traced over 22,000 km² and shown to have a volume of 650 km³.

Fissure Eruptions Beneath the Sea

The most extensive volcanic system on the Earth lies beneath the sea. There the features that split the center of the midocean ridges are channelways for the rising basaltic magma that forms new oceanic crust at the edges of plates of lithosphere. Seawater cools the basaltic magma so rapidly that plateau basalts do not form. Rather, two very distinctive, but much smaller lava forms are observed.

Sheeted Flows. Close to a vent, where the lava temperature is highest, thin sheets of lava with rapidly quenched, glassy surfaces develop. These are called *sheeted flows* (Fig. 3.28a). They build up piles of lava in which each sheet may only be 20 cm or so thick.

Pillow Lava. Farther away from a vent, where the temperature of the flowing lava has decreased, pillow structure develops. The term *pillow lava* describes a structure characterized by discontinuous, pillow-shaped masses of lava, ranging in size from a few centimeters to a meter or more in greatest dimension (Fig. 3.28b).

Pillow structure forms when the surface of the viscous lava is quickly chilled. The brittle, chilled surface cracks, making an opening for the still-molten magma inside to ooze out like a strip of toothpaste. In turn, the newly oozed strip chills, its surface cracks, and the process continues. The

a

b

FIGURE 3.28 Lava flows beneath the sea. (a) Sheeted flows of basalt in the central part of the Galápagos Rift. (b) Tubular-shaped pillows of basalt photographed in the central rift of the East Pacific Rise.

end result is a pile of lava pillows that resemble a jumbled pile of sandbags.

Ophiolite Complexes. Because oceanic crust is 10 km or more thick, we cannot see whether it is all made of sheeted flows and pillow lavas. Nor has it yet been possible to sample the entire crust by drilling and dredging. But nature has provided a few samples on land that are believed to be old oceanic crust. They exist because masses of continental crust, each riding on its plate of lithosphere have, at various times in the past, collided with each other. During such collisions fragments of oceanic crust are apparently broken off and caught up in the crumpled edges of the colliding continents. One of the best examples of ancient crust is in a suite of rocks called an *ophiolite complex* exposed on the island of Cyprus. A generalized diagram of an ideal ophiolite complex is shown in Figure 3.29. At the top is a thin veneer of sediment

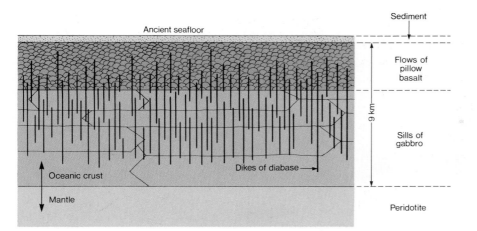

FIGURE 3.29 Idealized section through an ophiolite complex. Flows of pillow basalt cap a thick pike of gabbro sills. Both are intruded by nearly vertical gabbro dikes (black). At the boundary between oceanic crust and mantle, gabbro is in contact with peridotite.

deposited after the igneous activity ceased. Beneath the sediment are layers of basaltic sheeted flows and pillow lavas, and beneath the lavas are many sills consisting of gabbro. Cutting through the sills and lavas are thousands of vertical dikes of gabbro. The dikes form extensive, parallel sheets. Basalt and gabbro have the same composition and are presumably formed from identical magma. Beneath the gabbro sills are rocks of different composition (rocks such as peridotite) that are characteristic of the upper mantle. Therefore, the whole array of rocks in an ophiolite complex is believed to include not only a sample of oceanic crust but also a small sample of the upper mantle as well.

PLUTONS

Beneath every volcano there lies a complex of chambers and channelways through which magma reaches the surface. The magmatic channels of an active volcano cannot be seen, but ancient channelways can be examined when they have been unroofed and laid bare by erosion. The ancient channelways are filled with intrusive igneous rocks because they are the underground sites where magma solidified.

All bodies of intrusive igneous rock, regardless of shape or size, are called ***plutons*** after Pluto, the

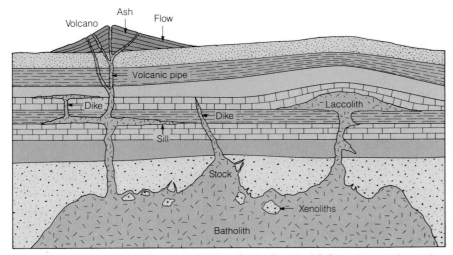

FIGURE 3.30 Diagrammatic cross section through part of the crust to show the various forms assumed by plutons. Many plutons were once connected with volcanoes, and a close relationship exists between intrusive and extrusive igneous rocks.

FIGURE 3.31 Shiprock, New Mexico, a volcanic neck 400 m high. The conical-shaped volcano that once surrounded the volcanic neck has been removed by erosion. The prominent ridges radiating outward are dikes. The view is from the air, looking southwest.

Greek god of the underworld. The magma that forms a pluton did not originate where we now find an intrusive igneous rock. Rather, the magma was intruded upward from the place where it was generated.

Minor Plutons

Plutons are given special names depending on their shapes and sizes (Fig. 3.30). Dikes and sills are tabular bodies.

Dikes. Tabular sheets of intrusive igneous rock cutting across the layering of the intruded rock are called **dikes.** Dikes may occur at any depth; they are commonly steeply inclined or vertical, and they mark ancient channelways of upward-rising magma.

Sills. Tabular sheets of intrusive igneous rock that are parallel to the layering of the intruded rock are called **sills.** Intrusion of a sill requires that all the overlying rocks be lifted upward by an amount equal to the thickness of the sill. Thus, sills tend to be intruded at shallow depths, where the weight of overlying rock is low. Commonly, dikes and sills occur together as part of a network of plutons, as shown in Figure 3.30.

Both dikes and sills can be very large. For example, the Great Dike in Zimbabwe is a tabular body of gabbro nearly 500 km in length and about 8 km wide, with essentially vertical walls. An example of a large and well-known sill can be seen in the cliffs of the Palisades that line the Hudson River opposite New York City. The Palisades sill reaches a thickness of about 300 m, and like the

Great Dike, is a gabbro. Like the Great Dike also, the magma that formed the Palisades sill differentiated by fractional crystallization. In both masses it is possible to observe distinct compositional layers formed by the settling of early formed crystals.

Laccoliths. A variation of a sill is a .**laccolith,** a lenticular intrusive body parallel to the layering of the intruded rocks, above which the layers of the intruded rocks have been bent upward to form a dome.

Volcanic Pipes and Volcanic Necks. A volcanic pipe is the approximately cylindrical conduit of igneous rock forming the feeder pipe immediately below a volcanic vent. When erosion exposes a pipe, a roughly cylindrical mass of igneous rock called a **volcanic neck** is the result. A famous example of a volcanic neck, together with associated dikes, can be seen at Shiprock, New Mexico (Fig. 3.31).

Major Plutons

Batholiths. Some plutons are enormous bodies compared to dikes and sills. A **batholith** is the largest kind of pluton. It is an intrusive igneous body of irregular shape that cuts across the layering of the rock it intrudes. Most batholiths are composite masses that comprise a number of separate intrusive bodies of slightly differing composition. The differences possibly reflect variations in the crustal rocks from which the magma formed. Some batholiths exceed 1000 km in length and 250 km in width—the largest in North America is the Coast Range Batholith of British Columbia and northern

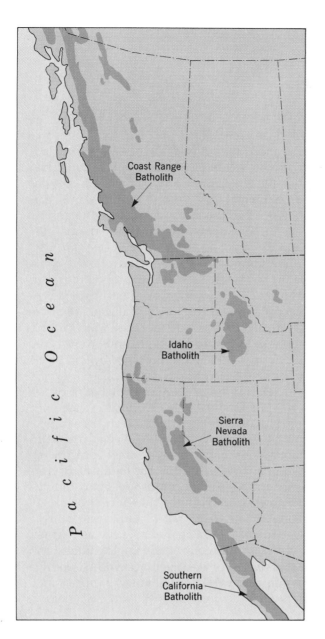

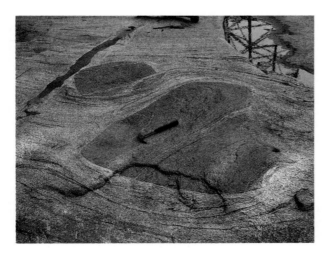

FIGURE 3.32 The Idaho, Sierra Nevada, and Southern California batholiths, largest in the United States, are dwarfed by the Coast Range Batholith in Southern Alaska, British Columbia, and Washington. Each of these giant batholiths formed from magma generated by the partial melting of continental crust, and each intrudes metamorphosed rocks.

Washington, which has a length of about 1500 km (Fig. 3.32).

Geologists commonly reserve use of the term *batholith* for bodies of intrusive igneous rock that have exposures in excess of 100 km².

Stocks. When the exposure is less than 100 km², a cross-cutting pluton is called a **stock.**

It is apparent from Figure 3.32 that a stock may merely be a companion body to a batholith or even the top of a partly eroded batholith. What is not apparent in Figure 3.32, however, is what the bottom of a batholith looks like. Where it is possible to see them, the walls of batholiths tend to be nearly vertical. This had led to a commonly held perception that batholiths extend downward to great depths—possibly even to the base of the crust. Geophysical measurement and studies of very deeply eroded bodies of igneous rock suggest that this perception is incorrect. Batholiths seem to be only 20 to 30 km thick, which is rather small compared to their great widths and lateral extents.

Stoping. Despite their huge sizes, batholiths do move upward. Even though intruded rocks can be pushed upward by the slowly rising magma, some other process must also operate. The rising magma apparently dislodges fragments of the overlying country rock by a process known as **stoping.** Dislodged blocks are more dense than the rising magma and will therefore sink. As sinking proceeds, the fragments may react with and be partly dissolved by the magma.

Xenoliths. Not all fragments dissolve, but instead they may sink all the way and reach the floor of the magma chamber. Any fragment of country rock still enclosed in a magmatic body when it solidifies is known as a **xenolith** (Fig. 3.33) (from the Greek words *xenos*, stranger, and *lithos*, stone).

FIGURE 3.33 Xenoliths of metamorphic rock dislodged from overlying rock by intrusion of the Petersburg Granite. The outcrop is in the bed of the James River, Richmond, Virginia. The size of the xenolith can be judged from the hammer.

Essay

VOLCANOES THAT DISAPPEARED

During the summer of 1883 an Indonesian volcano called Mount Krakatau started to emit steam and ash. Krakatau had erupted about 200 years earlier, but all memory of that event had vanished, and the volcano was considered dormant. During the afternoon of Sunday, August 26, 1883, volcanic activity increased, and on August 27 Krakatau blew up and disappeared during an exceptionally violent pyroclastic eruption. As a telegram of the time tersely reported, "Where once Mount Krakatau stood, the sea now plays." Noise from the paroxysmal explosion was heard on an island in the Indian Ocean, 4,600 km away. Giant waves, 40 m high, spread out from the site of the explosion and crashed into the shores around southeast Asia. Thirty-six thousand people lost their lives.

The eruption of Krakatau was not just a local event. The effect of the eruption was felt around the world. About 20 km³ of tephra was ejected during the eruption, and some of the smallest, dust-sized particles, were blasted as high as 50 km into the stratosphere. Within 13 days the stratospheric ash had encircled the globe and everywhere observers were reporting strangely colored sunsets. Sometimes the sunsets were green, or blue, at other times scarlet, or flaming orange. In November 1883, a vivid red sunset flickered over New York and looked so like the glow from a massive fire that fire engines were called out. It is estimated that the suspended dust and volcanic gases made the atmosphere so opaque to the Sun's rays that the temperature around the Earth dropped an estimated 0.5°C during 1884. It was 5 years before all of the ash had fallen to the ground and the climate returned to nearly normal conditions.

We know a lot about the eruption of Krakatau, but we are still learning about another, and much larger eruption, that happened 68 years earlier, in 1815. Like Krakatau, Tamboro is an Indonesian volcano that was wrongly thought to be dead. When Tamboro rumbled to life in 1815 it produced an eruption of extraordinary violence.

The eruption of Tamboro reached a climax on April 11 and 12, 1815, when an estimated 150 km³ of tephra was ejected, and 50,000 people were killed. Tamboro injected an enormous amount of volcanic dust into the atmosphere. Records of climate for the early years of the nineteenth century are poor, but the summer of 1816 is known to have been exceptionally cold. Crops failed, and frosts occurred in parts of New England as early as August. Starvation stalked the New England states and the countries of western Europe; 1816 is known as the year without a summer.

SUMMARY

1. Igneous rock forms by the solidification and crystallization of magma.

2. Magma forms by the complete or partial melting of rock.

3. Three kinds of magma predominate—they are basaltic, andesitic, and granitic.

4. Basaltic magma forms by dry partial melting of rock in the mantle. Andesitic magma forms by wet partial melting of oceanic crust. Granitic magma forms by wet partial melting of rock in the continental crust.

5. The principal controls on the physical properties of magma are temperature and the SiO_2 and dissolved-gas content. High temperature and low dissolved-gas and low SiO_2 content result in fluid magma, such as basaltic magma. Lower temperature and higher dissolved-gas and higher SiO_2 content result in viscous magma, such as andesitic and granitic magma.

6. The sizes and shapes of volcanic edifices depend on the kind of material extruded, viscos-

It will never be known how many people died indirectly as a result of the eruption of Tamboro but the number is very large. Famine in Ireland as a result of the crop failure of 1816 led to an epidemic of typhus and countless thousands died. The first worldwide cholera epidemic started in Asia late in 1816 and continued to spread until 1833. When it had finally run its course, millions were dead. Many authorities now suspect that the epidemic was a direct result of the eruption of Tamboro.

The geologic record is filled with evidence of eruptions similar to those of Krakatau and Tamboro. During the last few million years, for example, violent pyroclastic eruptions have occurred repeatedly in California, Oregon, Washington, Nevada, New Mexico, Alaska, and Wyoming. The evidence is preserved in layers of ash and tuff. The eruption of Mount St. Helens in 1980 (see cover) released between 1 and 2 km^3 of pyroclastic debris, and that was tiny by comparison with Tamboro. What would be the magnitude of the human disaster if Yellowstone were to erupt again, as it did 2.2 million years ago when 2500 km^3 of fragmental material was ejected (much of it now preserved in the Huckleberry Ridge Tuff), or even if it repeated the eruption of the Lava Creek Tuff, 600,000 years ago, when 100 km^3 of pyroclastic debris was ejected?

During the last thousand years about 520 volcanoes have erupted (Table B3.1). They are widely distributed around the Earth's surface;

TABLE B3.1 *The Most Disastrous Volcanic Eruptions in Recorded History*

Place	Date of Eruption	Estimated Deaths
Tamboro, Indonesia	1815	50,000
Krakatau, Indonesia	1883	36,000
Mont Peleé, Martinique	1902	30,000
Nevado de Ruiz, Colombia	1985	25,000
Kelut, Indonesia	1586	10,000
Laki, Iceland	1783	10,000
Unzen, Japan	1792	10,000

a volcano is always erupting somewhere on the Earth. Many others that have not erupted for the past thousand years will almost certainly erupt during the next thousand. The timing and frequency of eruptions are irregular, making predictions uncertain. Yet, promising developments may help. Scientists have noted that prior to many eruptions, swarms of small earthquakes occur, and slight warping of the ground occurs in the immediate vicinity of the volcano. Through arrays of seismometers, the earthquakes can be monitored from a safe distance. With films sensitive to radiation in the infrared region, cameras mounted on orbiting satellites can post a continuous watch for developing hot spots, warning of areas of likely activity so that closer appraisals can be made and endangered populations alerted.

ity of the lava, and explosiveness of the eruptions.

7. Volcanoes that extrude viscous lava, generally rich in SiO_2, tend to be explosive. Those that extrude fluid lava, generally low in SiO_2, erupt less violently.

8. Viscous magmas are commonly erupted as pyroclasts that build steep-sided pyroclastic cones or stratovolcanoes.

9. Low-viscosity magmas tend to be erupted as lava flows. They build gently sloping shield volcanoes.

10. Widespread sheets of basaltic rock have resulted from fissure eruptions of low-viscosity lava.

11. Processes that separate remaining melt from already formed crystals in a cooling magma lead to the formation of a wide diversity of igneous rocks.

12. Igneous rock may be intrusive (meaning it formed within the crust) or extrusive (meaning it formed on the surface). The texture and grain size of igneous rock indicate how and where the rock cooled.

13. Igneous rocks rich in quartz and feldspar, such as granite, granodiorite, and rhyolite, are characteristically found in continental crust. Basalt, rich in pyroxene and olivine, is derived from magma formed in the mantle, and is common in the oceanic crust.

14. Following a large pyroclastic eruption the summit of a volcano may collapse into the partly empty magma chamber, forming a caldera.

15. All bodies of intrusive igneous rock are called plutons. Special names are given to plutons based on the shapes and sizes of the bodies.

IMPORTANT WORDS AND TERMS TO REMEMBER

agglomerate (p. 70)

amygdaloidal basalt (p. 69)

amygdule (p. 69)

andesite (p. 68)

Andesite Line (p. 61)

anorthosite (p. 68)

aphanite (p. 66)

ash (p. 70)

ash tuff (p. 70)

basalt (p. 68)

batholith (p. 78)

bomb (p. 70)

Bowen's Reaction Series (p. 63)

caldera (p. 73)

composite volcano (p. 73)

conchoidal fracture (p. 69)

continuous reaction series (p. 64)

crater (p. 73)

dacite (p. 68)

diabase (p. 66)

dike (p. 78)

diorite (p. 68)

discontinuous reaction series (p. 65)

fissure eruption (p. 76)

gabbro (p. 68)

garnet peridotite (p. 59)

geothermal gradient (p. 58)

geyser (p. 74)

granite (p. 68)

granitic (p. 68)

granodiorite (p. 68)

granular texture (p. 68)

groundmass (p. 67)

laccolith (p. 78)

lapilli (p. 70)

lapilli tuff (p. 70)

lava (p. 56)

lava dome (p. 74)

magma (p. 56)

magmatic differentiation by fractional crystallization (p. 63)

magmatic differentiation by partial melting (p. 59)

matrix (p. 67)

mineral assemblage (p. 67)

obsidian (p. 69)

ophiolite complex (p. 76)

partial melting (p. 59)

pegmatite (p. 68)

peridotite (p. 68)

phanerite (p. 66)

phenocryst (p. 67)

pillow lava (p. 76)

plateau basalt (p. 76)

pluton (p. 77)

porphyry (p. 66)

pumice (p. 69)

pyroclast (p. 70)

pyroclastic cone (p. 72)

resurgent cauldron (p. 74)

rhyolite (p. 68)

shard (p. 70)

sheeted flow (p. 76)

shield volcano (p. 72)

sill (p. 78)

stock (p. 79)

stoping (p. 79)

stratovolcano (p. 73)

tephra (p. 70)

texture (of a rock) (p. 66)

thermal spring (p. 74)

tuff (p. 70)

vesicle (p. 69)

viscosity (p. 56)

volcanic neck (p. 78)

volcanic shard (p. 70)

volcano (p. 72)

welded tuff (p. 72)

xenolith (p. 79)

zoned crystal (p. 64)

QUESTIONS FOR REVIEW

1. What is the difference between phaneritic and aphanitic igneous rock and how does the difference arise?

2. What is the distinguishing feature of a pyroclastic rock?

3. Do porphyries always contain phenocrysts? What minerals would you expect to find as phenocrysts in a rhyolite porphyry?

4. What is the major difference between the mineral assemblage of a diorite and a granodiorite? Between granite and a granodiorite?

5. What is the origin of pumice? Why will it float on water?

6. How does a lapilli tuff differ from a welded tuff? Would you expect both rocks to form as a result of eruptions from the same volcano?

7. Is the major oxide component of magma SiO_2, MgO, or Al_2O_3? Briefly describe the effect of composition on the fluidity of magma.

8. How hot is magma that erupts from volcanoes? What is the effect of temperature on the viscosity of magma?

9. What does the term *partial melting* mean?

10. How does the magma-forming property of dry partial melting differ from wet partial melting?

11. What is the origin of andesitic magma? With what kind of volcanoes are andesitic eruptions associated?

12. What is magmatic differentiation by fractional crystallization?

13. How does a continuous reaction series differ from a discontinuous series?

14. Why does a shield volcano have a gentle-sided slope, while a stratovolcano is steep sided?

15. How do lava domes form?

16. Describe two kinds of distinctive basaltic lava flows developed by submarine fissure eruptions.

17. How does a dike differ from a sill?

18. How big can batholiths be?

19. By what mechanism are granitic batholiths believed to move upward in the crust?

20. What are xenoliths and how do they form?

CHAPTER 4

Sediments and Sedimentary Rocks

Cobbles on a Pacific beach have been abraded and rounded by constant agitation
in the surf zone.

FROM SEDIMENT
TO SEDIMENTARY ROCK

Like a perpetually restless housekeeper, nature is ceaselessly sweeping regolith off the nonweathered rock below, carrying the sweepings away, and depositing them as sediment in river valleys, lakes, seas, and innumerable other places. We can see sediment being transported by trickles of water after a rainfall and by every wind that carries dust. The mud on a lake bottom, the sand on the beach, even the dust on a windowsill is sediment. Because erosion and deposition of rock particles take place almost continuously, we find sediment nearly everywhere.

When a thick pile of sediment accumulates, the particles near the base are compacted due to the weight of the overlying deposit. Over time, they become cemented together to form a solid aggregate of rock. Most commonly the cement is new mineral substance deposited from water percolating slowly through the spaces between particles of sediment. By compaction and cementation, therefore, sediment is gradually transformed into sedimentary rock.

Sedimentary rock is the rock most commonly found at the Earth's surface, where it forms a thin but extensive blanket over igneous and metamorphic rocks beneath. Its most obvious feature is sedimentary layering, often strikingly exposed on mountainsides, in the walls of canyons, or in artificial cuts along highways or railroads (Fig. 4.1).

SEDIMENTS AND SEDIMENTATION

Clastic Sediment

A close look at sediment beside a stream shows that pebbles or sand grains are simply bits of rocks and minerals. A magnifying glass discloses that the finer sedimentary particles, too, are derived from broken-up rock, but that generally the particles have undergone chemical change. Feldspars, for instance, have been partly altered to clay. All sediment of this kind is known as *clastic sediment*, from the Greek word *klastos* (broken). It consists of

FIGURE 4.1 Multicolored layered sedimentary rocks in Capitol Reef National Park, Utah.

the accumulated particles of broken rock and may include the skeletal remains of dead organisms (Fig. 4.2). Such sediment is also referred to as *detritus,* a general term for loose rock and mineral fragments derived from older rock. A continuous gradation of particle size is found, from the largest boulder down to submicroscopic clay particles. This range of particle size is the primary basis for classifying clastic sedimentary rocks (Table 4.1).

If a sedimentary rock is made up of mineral particles derived from the erosion of igneous rock, how can we tell it is sedimentary and not igneous? In addition to obvious clues such as sedimentary layering, rock texture also provides important clues.

In igneous rock the grains are irregular and interlocked, but in sedimentary rock the particles are commonly rounded and show signs of the abrasion they received during transport (Fig. 4.2). Clastic sedimentary rock also reveals cement that holds the particles together, whereas igneous rock con-

sists of interlocking crystals. Fossils are another important feature. Life cannot tolerate the high temperatures under which igneous rocks form, so the presence of ancient shells or similar evidence of past life is an excellent clue to sedimentary origin.

Chemical Sediment

Certain kinds of rock contain fossils and other sedimentary characteristics, yet seem to be free of clastic sediment. The rock is indeed sedimentary, and the material composing it has been transported. However, the sediment is not clastic because its components were dissolved, transported in solution, and precipitated chemically instead of by mechanical means. Sediment formed by precipitation of minerals from solution in water is *chemical sediment.* It forms in two principle ways.

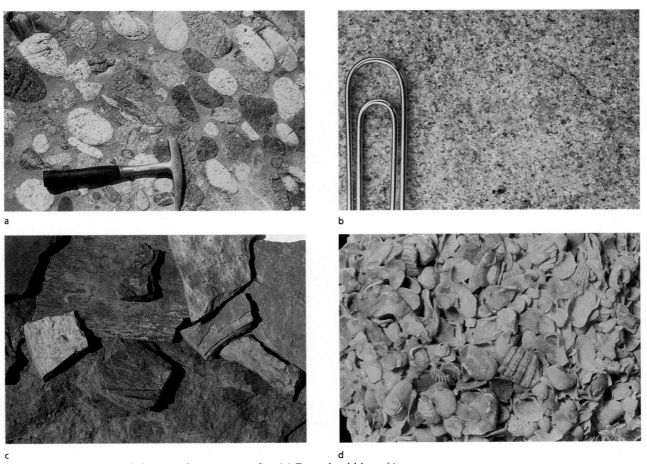

a

b

c

d

FIGURE 4.2 Examples of clastic sedimentary rocks. (a) Round cobbles of igneous and metamorphic rocks cemented in a matrix of sand form a coarse cobble conglomerate. (b) Grains of quartz sand cemented by silica produce a coarse, hard sandstone. (c) Platy fragments of shale are composed of lithified muds. (d) Broken fragments of sea shells cemented together with calcite makes a clastic sedimentary rock called *coquina.*

TABLE 4.1 *Definition of Clastic Particles, Together with the Sediments and Sedimentary Rocks Formed from Them*

Name of Particle	Range Limits of Diameter (mm)[a]	Name of Loose Sediment	Name of Consolidated Rock
Boulder	More than 256	Gravel	Conglomerate and
Cobble	64 to 256	Gravel	sedimentary
Pebble	2 to 64	Gravel	breccia
Sand	1/16 to 2	Sand	Sandstone
Silt	1/256 to 1/16	Silt	Siltstone
Clay[b]	Less than 1/256	Clay	Claystone, mudstone, and shale

Source: After Wentworth, 1922.

[a]Note that size limits of sediment classes are powers of 2, just as are memory limits in microcomputers (for example, 2K, 64K, 128K, 256K, 512K).

[b]Clay, used in the context of this table, refers to particle size. The term should not be confused with clay minerals, which are definite mineral species.

One way consists of biochemical reactions resulting from the activities of plants and animals within the water. For example, tiny plants living in seawater can decrease the acidity of the surrounding water and so cause calcium carbonate to precipitate.

Chemical sediment also forms as a result of inorganic reactions within the water. When the water of a hot spring cools, it may precipitate opal or calcite. Another common example is simple evaporation of seawater or lake water. As the water evaporates, thereby concentrating dissolved matter that is in solution, salts begin to precipitate out and remain as a residue of chemical sediment (Fig. 4.3). The table salt we eat comes mainly from sedimentary rock formed in this way.

Most chemical sedimentary rocks contain only one important mineral and this is used as a basis for classification. Among the most common chemical rocks are rock salt (halite = NaCl) and gypsum. *Limestone,* formed chiefly of the mineral calcite, and *dolostone,* formed mainly when calcite is replaced by the mineral dolomite, can be either clastic or chemical sedimentary rocks.

Transport and Deposition of Clastic Sediments

Transport

Sediment is transported in many ways. It may slide down a hillside or be carried by the wind, by a glacier, or by flowing water. In each case, when transport ceases, the sediment is deposited in a fashion characteristic of the transporting agent. When sediment is transported by sliding or rolling downhill, the result is generally a mixture of particles of all sizes. Much of the sediment carried by a glacier is deposited either beneath the ice or at the glacier's edge. Such sediment also is a mixture of sedimentary particles of all sizes.

FIGURE 4.3 Precipitated salts encrust the surface of Searles Lake playa in southeastern California. The chemical plant on the lakeshore extracts and refines saline minerals.

Deposition

In the transport of sedimentary particles by wind or water, deposition occurs when the flowing water or moving air slows to a speed at which particles can no longer be carried. In a general way, therefore, the size of the grains in sediment moved by wind or water tells us something about the speed of the transporting medium. Coarse-grained sediment indicates deposition from fast-flowing wind or water; fine-grained sediment indicates either that the wind or water was slow moving, or that only fine sediment was available for transport.

Environmental Clues

The size of particles in sedimentary rock and the way they are packed together, as well as other distinctive features, provide evidence about the environment in which the original sediment was deposited. The existence of ancient oceans, coasts, lakes, streams, swamps, and all the other places where sediment accumulates, can be demonstrated from clues in sedimentary rock. Fossils within a sedimentary layer may also provide information about former climates. Some animals and plants are restricted to warm, moist climates whereas others are associated only with cold, dry climates. By using the climatic ranges of modern plants and animals as guides and invoking the Principle of Uniformitarianism, one can infer the general nature of the climate in which similar ancestral forms lived.

Diagenesis

Diagenesis is a term for the changes that affect sediment after its initial deposition, and during and after its slow transformation into sedimentary rock. The first and simplest change is *compaction,* which occurs as the weight of the accumulating sediment forces the rock and mineral grains together. This reduces the pore space and eliminates some of the contained water. Precipitation of dissolved substances by water circulating through pore spaces then bonds the grains together through *cementation.* Calcium carbonate is one of the most common cements (Fig. 4.4), but silica may also bond grains together forming a particularly hard cement. After burial, less stable minerals may change to more stable forms through *recrystallization.* This process is especially common, for example, in porous reef limestone.

Important chemical alterations also affect sediments. In the presence of oxygen (an *oxidizing en-*vironment), organic remains are quickly converted into carbon dioxide and water. If oxygen is lacking (a *reducing environment*), the organic matter does not decay but instead may be slowly transformed into solid carbon in the form of peat or coal. Similarly, organic oils and fats may be converted into carbon-rich residues (*hydrocarbons*).

Features of Sediments and Sedimentary Rocks

Stratification and Bedding

Sedimentary stratification results from a layered arrangement of the particles in sediment or sedimentary rock. Each *stratum* (plural = *strata*) is a distinct layer of rock that accumulated at the Earth's surface. While layering is an obvious feature of most sedimentary rocks, it is seen also in some volcanic rocks (lava flows, pyroclastic deposits) and in many metamorphic rocks. Looking closely at

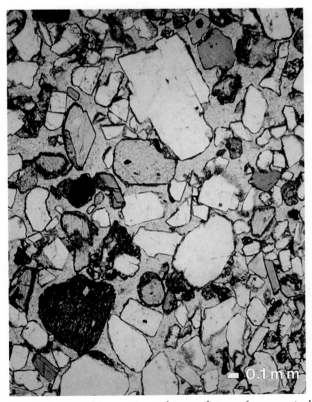

FIGURE 4.4 A thin section of a sandstone from central Washington shows sand grains that are bonded together by calcite cement. Light-colored grains are plagioclase, brownish grains are pyroxene, and the large dark grain is a volcanic rock fragment.

sedimentary rocks that are distinctly stratified, we can see that the strata differ from one another because of differences in some characteristic of the particles or in the way in which they are arranged. For example, one stratum may consist of particles of different diameter from those in another stratum. In a clastic rock, changes of diameter result from fluctuations of energy in a stream, in surf, in wind, in a lake current, or in whatever agent is responsible for the deposit. Such energy changes, usually small, are not the exception but the rule.

The layered arrangement of strata in a body of rock is referred to as **bedding.** Each *bed* within a succession of layered strata can be distinguished from adjacent beds by differences in thickness or character.

Sorting

Sorting is a conspicuous feature of sediments deposited by flowing water or air. Sorting according to specific gravity (ratio of the weight of a given volume of material to the weight of an equal volume of water) is evident in a mineral *placer* (chapter 18). Particles of unusually heavy minerals such as gold, platinum, and magnetite are deposited quickly and concentrated on streambeds or on beaches, whereas lighter particles are carried onward. Most of the particles transported by water or wind, however, are common rock-forming minerals such as quartz and feldspar that have similar specific gravities. Therefore, such particles typically are not sorted according to specific gravity but rather according to size (Fig. 4.5). Long-continued handling of particles by turbulent water and air results in gradual destruction of the weaker particles, leaving behind the particles that can better survive in the turbulent environment. Very commonly the survivor is quartz, because it is hard and lacks cleavage. In this case sorting is based on durability.

Rounding

Mechanically weathered particles broken from bedrock tend to be angular, because breakage commonly occurs along grain boundaries, fractures, and surfaces of stratification. The same particles tend to become smooth and rounded as they undergo transport by water or air, and are abraded by impact with other rock fragments (Fig. 4.5). Figure 4.6 illustrates how sand grains become shaped during transport. In general, the greater the distance of travel, the greater is the degree of rounding.

FIGURE 4.5 Well-rounded grains of quartz sand from the St. Peter Sandstone of Wisconsin have been sorted by size and polished by constant shifting and abrasion in surf along an ancient shoreline.

Parallel Strata

Layers of sediment fall into two classes according to the geometric relations between successive strata. **Parallel strata** are those whose individual layers are parallel. Parallelism indicates that deposition probably occurred in water, and that the activity of waves and currents was minimal (except for the special case of graded layers, described below). Indeed, sediment deposited on lake floors and in the deep sea commonly occurs as parallel layers.

Some sediments display a distinctive alternation of parallel layers having different properties. Such alternation suggests the influence of some naturally occurring rhythm that has influenced sedimentation. A pair of such sedimentary layers deposited over the cycle of a single year is termed a *varve* (Swedish for "cycle"). Varves are most commonly seen forming in deposits of high-latitude or high-altitude lakes where there is a strong contrast in seasonal conditions. In spring and early summer, when a cover of winter ice melts away and the inflow of sediment-laden water increases, coarse sediment is deposited on the lake floor. With the onset of colder conditions in the autumn, stream flow decreases and ice forms over the lake surface. During winter, very fine sediment that has remained suspended in the water column slowly set-

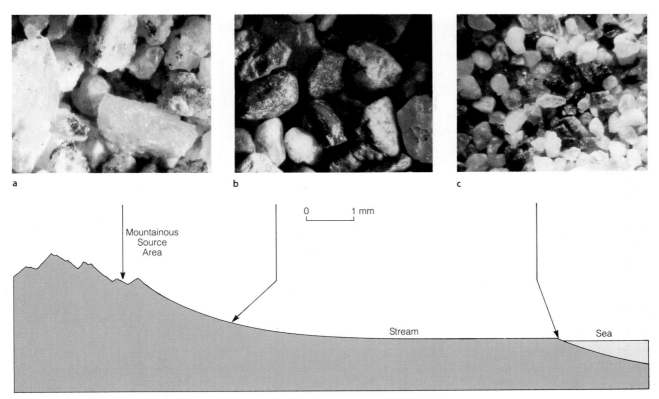

FIGURE 4.6 Rounding and sorting of mineral grains during transport. (a) Mineral grains loosened and separated from igneous and metamorphic rocks by mechanical and chemical weathering have the same angular shapes they assumed when the minerals crystallized in the parent rock. (b) Carried downslope by streams, the sand experiences abrasion and grains become rounder. (c) By the time the sediment reaches the mouth of the stream, it has traveled a great distance and the well-sorted grains have become well rounded.

tles to form a thinner, darker layer above the coarse lighter-colored summer layer. Varves are common in deposits formed near the margins of ice age glaciers (Fig. 4.7) and are also seen in some ancient sedimentary rocks.

Cross Strata

Cross strata are inclined with respect to a thicker stratum within which they occur (Fig. 4.8). All such strata consist of particles coarser than silt and are the work of turbulent flow in streams, in wind, or in waves along a shore. As they move along, the particles tend to collect in ridges, mounds, or heaps in the form of ripples and waves that migrate slowly forward in the direction of the current. As particles continually accumulate on the downcurrent slope of the pile, they produce strata having inclinations as great as 30° to 35°. The direction in which cross strata are inclined, then, is the direction in which the related current of water or air was flowing at the time of deposition.

Arrangement of Particles Within a Stratum

Several kinds of particle arrangements are possible within a single layer of sediment. Each arrange-

FIGURE 4.7 Rhythmically laminated glacial lake sediments exposed in a stream cut in the James Bay Lowland of eastern Canada. Each pair of layers constitutes a varve; in each varve, light colored silty sand deposited during summer months grades upward into darker clay deposited during winter. (S. C. Porter.)

FIGURE 4.8 Cross-stratified sandstone near Kanab, Utah, consists of ancient sand dunes that have been converted to sedimentary rock. Cross strata are inclined to the right, in the direction toward which the ancient winds were blowing.

ment provides information about the conditions under which the sediment was deposited.

Uniform Layers. A layer that consists of particles of about the same diameter is called a ***uniform layer.*** A uniform layer of clastic sediment implies deposition of particles of a single size, with little change in the velocity of the transporting agent. A uniform layer of nonclastic rock implies uniform precipitation from solution, which produces crystalline particles of a single size. A layer that is subdivided into still thinner layers marked off from each other by differences in grain size suggests a transporting agent having a fluctuating velocity.

Graded Bedding. If a mixture of small solid particles having different diameters and about the same specific gravity is placed in a glass of water, shaken vigorously, and then allowed to stand, the particles will settle out and form a deposit on the bottom of the glass. The largest particles settle first, followed by successively smaller ones. The finest may stay in suspension for hours or days before finally settling out. Thus, particle size in the deposit decreases from the bottom upward. This arrangement characterizes ***graded bedding*** in which the particles of each bed grade upward from coarser to finer (Fig. 4.9).

A graded bed can also form from a sediment-laden current. As the current slows down, the heaviest and coarsest particles settle out first, followed by lighter and finer ones.

Nonsorted Layers. The particles in some sedimentary rocks are a mixture of different sizes arranged chaotically, without any obvious order. Such sediments are created, for example, by rockfalls, slow movement of debris down hillslopes, slump-

FIGURE 4.9 Graded bedding in marine sediments deposited in 80 m of water about 10,500 years ago in southwestern Norway. The graded layer passes upward from coarse sand at the base to fine sand above. (I. Aarseth.)

FIGURE 4.10 Stony tillite of an ancient glaciation overlies a glacially abraded surface of older rock at Nooitgedacht, near Kimberley, South Africa.

a

b

FIGURE 4.11 Modern ripples and ancient ripple marks. (a) Ripples forming in shallow water near the shore of Ocracoke Island, North Carolina. (b) Ripple marks on the bedding surface of an ancient quartzite bed in the Baraboo Syncline of south-central Wisconsin.

ing of loose deposits on the seafloor, mudflows, and deposition of debris by glaciers and by floating ice. Some nonsorted sedimentary rocks are given specific names (for example, a *tillite* is a nonsorted rock of glacial origin; Fig. 4.10).

Surface Features on Sedimentary Layers

Various features preserved on the surfaces of strata provide clues about the origin of sedimentary rocks and the environments in which they formed. Bodies of sand that are being moved by wind, by streams, or by coastal waves are often rippled, and such ripples are frequently preserved in sandstones and siltstones as *ripple marks* (Fig. 4.11). Some claystones and siltstones contain layers that are cut by polygonal markings. By comparing them with similar features in modern sediments, we infer that these are *mud cracks,* cracks caused by shrinkage of wet mud as its surface dries (Fig. 4.12). Mud cracks therefore suggest tidal flats, exposed streambeds, playa lakes, and similar environments. Footprints and trails of animals are often associated with ripple marks and mud cracks (Fig. 4.13). Even the impressions of raindrops made during brief, intense showers can be preserved in strata.

Fossils

The remains of animals and plants that are buried with sediments, protected against oxidation and erosion, and preserved through the slow process of conversion to rock, become *fossils* (Fig. 4.14). Fossils provide significant clues about former environments. For example, plant fossils can provide estimates of past precipitation and temperature for sites on land, and fossils of tiny floating organisms can tell as about former sea-surface temperature and salinity conditions. Fossils are also the chief basis for correlation of strata and the construction of the geologic column (chapter 6).

Concretions

Masses of hard rock having distinct boundaries are found enclosed in some sedimentary strata. Such bodies, called *concretions,* consist of material pre-

a

FIGURE 4.13 Tracks of a three-toed dinosaur exposed on the surface of a sandstone bed in the Painted Desert near Cameron, Arizona. All the tracks in the picture belong to a single species.

b

FIGURE 4.12 Modern and ancient mud cracks. (a) Mud cracks formed at the surface of a dry lake floor in Death Valley, California. (b) Ancient mud cracks preserved on the surface of a mudstone bed exposed at Ausable Chasm, New York.

cipitated from solution, commonly around a nucleus. They range in diameter from less than a centimeter to 2 m or more (Fig. 4.15). Their shape may be spheroidal or may range through a variety of bizarre forms, many with remarkable symmetry, to elongate bodies that parallel the stratification of the rock. Perfectly preserved fossils are found at the centers of some concretions.

Color

The color of fresh sedimentary rock is determined by the colors of the minerals and rock fragments

FIGURE 4.14 A fossil trilobite (*Olenellus getzi*) is exposed on the bedding plane of an ancient marine sandstone.

FIGURE 4.15 Round concretions the size of cannon balls litter the surface near Lake Powell, Arizona, where they have weathered out of the enclosing Navajo Sandstone.

that compose it. Iron sulfides and organic matter, buried with sediment, are responsible for most of the dark colors in sedimentary rocks. Reddish and brownish colors result mainly from the presence of iron oxides, occurring either as powdery coatings on mineral grains or as very fine particles mixed with clay.

The color of weathered outcrops of sedimentary rock may differ from the same rock when fresh. For example, a sandstone that is pale gray on freshly broken surfaces may have a surface coating of iron oxide in weathered outcrop that gives it a yellowish-brown color. In this case, the surface color is derived from chemical weathering of iron-bearing mineral grains in the rock.

SEDIMENTARY FACIES AND DEPOSITIONAL ENVIRONMENTS

Sedimentary Facies

If we examine a sequence of exposed sedimentary rocks, most likely changes will be seen in their character as we move up from one layer to the next above. These differences reflect changes in depositional conditions at a particular place through time. However, if any single deposit is traced away from the initial outcrop, it may change laterally. Most sedimentary strata change character from one area to another as a result of changes in the conditions under which the sediments accumulated. A diversity of environments is seen when traveling across the edge of a continent and into the adjacent ocean basin (Fig. 4.16). Within each natural zone distinctive sediments and associated organisms are found which serve to identify that depositional environment.

A distinctive group of characteristics within a body of sediment that differs from those elsewhere in the same body is a *sedimentary facies.* Each facies may be represented, for example, by distinctive grain size, grain shape, stratification, color, depositional structures, or fossils. Adjacent facies merge into one another either gradually or abruptly, depending on the relationships between the two former depositional environments (Fig. 4.17).

Nonmarine Environments

Sediment derived from the mechanical and chemical breakdown of rocks is moved inexorably toward the sea. En route it is transported by ice, water, or wind. The sediment may be reworked repeatedly, by one or several of these agencies, before reaching its final resting place where it is slowly converted into sedimentary rock.

Stream Sediments

Streams constitute the principal agency for transporting sediment across the land. Their deposits can be seen nearly everywhere and constitute *alluvium,* the general name given to sediment deposited by streams in nonmarine *fluvial environments* (Fig. 4.16). The sediment differs from place to place depending on the type of stream, the energy available for work, and the nature of the sedimentary load. A typical large, smoothly flowing stream may deposit well-sorted layers of coarse and fine particles as it swings back and forth across its valley. Fine silt and clay will be deposited on the adjacent floodplain during spring floods, and organic sediments will accumulate in abandoned sections of the channel. By contrast, a stream issuing from the front of a mountain glacier may divide into an intricate system of interconnected channels that change in size and direction as the volume of water fluctuates and the stream copes with an

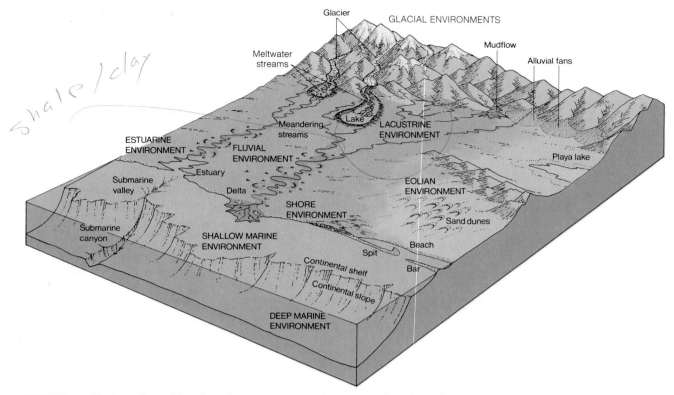

shale/clay

FIGURE 4.16 Various depositional environments occurring across the edge of a continent and the adjacent margin of an ocean basin.

abundance of debris. The resulting sediments will consist largely of cross-cutting channel deposits.

Such differences in the texture and structure of sediments, and of rocks formed from them, can tell us a great deal about past environments of deposition. So too can associated plant and animal fossils sometimes tell us whether the enclosing sediments represent, for example, the floodplain of a subtropical river or a desert alluvial fan.

Lake Sediments

Lake sediments accumulate chiefly in two places in **lacustrine environments** (Fig. 4.16): the lakeshore and lake floor. Lakeshore deposits consist of generally well-sorted sand and gravel that form beaches, as well as bars and spits across the mouths of bays. Where a stream enters a lake, its sediment load will be dropped as its speed and transporting ability

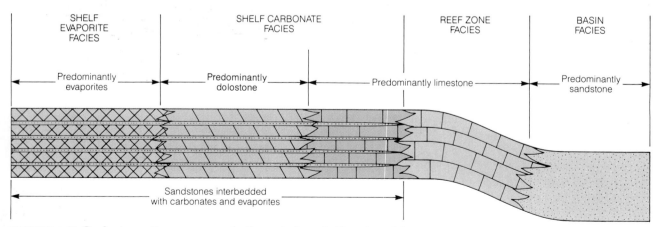

FIGURE 4.17 Geologic section across a shallow marine shelf and reef into an adjacent deep basin showing the gradational relationship among sedimentary facies. (*Source:* After Motts, 1968.)

are suddenly decreased. The resulting deposit is a *delta* (chapter 10). It consists of inclined and generally well-sorted layers of the delta front that pass downward and outward into thinner, finer, and even-laminated layers on the lake floor (Fig. 4.7).

Glacial Sediments

Sedimentary debris eroded and transported by a glacier is either deposited at the base of the glacier or is released at its edge as melting occurs. The sediment is then subjected to further reworking by meltwater. Debris deposited directly from ice commonly forms a heterogeneous mixture of particles ranging in size from clay up to large boulders and consisting of all the rock types over which the ice has passed (Fig. 4.10). Such sediment characteristically is nonsorted and nonstratified, in contrast to most other nonmarine sediments. The rock fragments typically are poorly rounded and often angular (Fig. 4.18a), except in cases where they were rounded before being picked up and reworked by the glacier.

Sediments deposited next to a glacier by meltwater tend to be sorted, just as are waterlaid sediments from nonglacial environments. Both stream sediments and lake sediments are commonly associated with ice-laid deposits in glaciated terrains. Such sediments often are discontinuous and may change character abruptly as melting ice causes lakes to drain and meltwater streams to experience sudden changes in volume and direction of flow.

Wind-Transported Sediments

Both wind activity and the geologic results of it are commonly referred to as **eolian,** after *Aeolus,* the Greek god of wind. Sediment carried by the wind tends to be finer grained than that moved by other erosional agents because air is much less dense than water or ice. Sand grains are easily moved where strong winds occur, as along seacoasts and in deserts, and where vegetation is too discontinuous to stabilize the land surface. In such **eolian environments** (Fig. 4.16), the sand piles up to form dunes consisting of well-sorted sand grains and having bedding similar to that in a delta. Individual grains typically have a frosted appearance (Fig. 4.18b) due to the repeated impacts they receive as they bounce along. Using these characteristics, dune sands can easily be identified in the rock record (Fig. 4.8).

Silt picked up and moved by the wind can be deposited as a blanket of fine sediment across the landscape (chapter 11). Such sediment is thickest

and coarsest near its source and becomes progressively thinner and finer with increasing distance

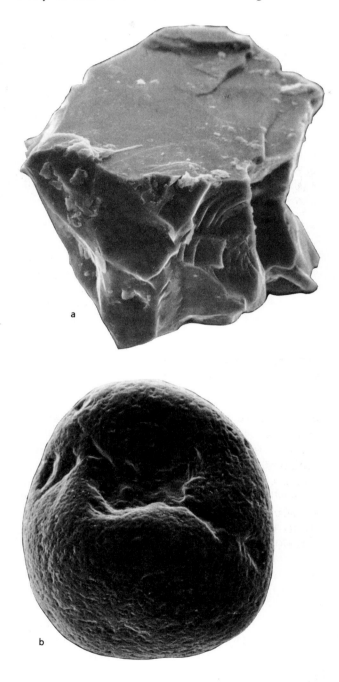

FIGURE 4.18 Surface features of sand grains, seen on enlarged pictures taken with a scanning electron microscope, aid in differentiating among transporting agencies. (a) Surface of a quartz grain (0.1-mm diameter) that has been crushed and abraded during transport at the bed of a Swiss glacier displays distinctive concoidal fractures. (b) Surface of a wind-transported quartz grain (0.5-mm diameter) from south-central Libya has a distinctive pitted appearance caused by mechanical chipping as grains impact one another during strong sandstorms.

downwind. Although common as a sediment in many parts of the world, wind-blown silt is virtually unknown as a sedimentary rock, probably because it is easily eroded from the landscape and therefore is unlikely to be widely preserved.

Sediments of the Continental Shelves

The world's rivers continuously transport detritus to the edges of the land where much of it then accumulates on the continental shelves, sometimes to great thicknesses. In part spurred by the search for large undersea reservoirs of oil and gas, geologists have learned a great deal about the sediments accumulating on the shelves.

Near-Shore Sediments

Stream sediments that reach the edge of the sea accumulate near the mouths of rivers or are carried either seaward or laterally along the coast by currents. In the process, they are reduced in size by abrasion and are sorted.

Estuarine Sediments. Much of the load transported by a large river may be trapped in an *estuary,* a semienclosed body of coastal water within which seawater is diluted with freshwater (Fig. 4.16). Coarse sediment settles close to land and may quickly fill an estuary. In cases where an estuary is slowly subsiding, a thick body of estuarine sediment may accumulate.

Deltaic Sediments. Marine deltas are built where streams deposit their load at the edge of the sea. Large deltas are complex sedimentary deposits consisting of coarse stream-channel sediments and finer-grained interchannel sediments that grade progressively offshore into finer sediments deposited on the seafloor (chapter 10).

Beach Sediments. Ocean beaches consist of the coarser fraction of whatever range of rock particles is contributed by erosion of sea cliffs or by rivers. Quartz, the most durable of common minerals in continental rocks, is a typical component of beach sands. In general, beach sediments are better sorted than stream sediments of comparable coarseness and typically display cross stratification. Dragged back and forth by the surf and turned over and over, particles of beach sediment become rounded by abrasion. Although beach gravel and gravelly alluvium may be similar in appearance, on many beaches pebbles and cobbles assume a distinctive flattened or discoidal shape (see opening photo).

Offshore Sediments

Fine-Grained Sediment. The outward flow of fresh water through an estuary is often substantial and may extend out across the submerged continental shelf as a distinct layer overlying denser, salty marine water. Some fine-grained sediment thereby reaches the outer shelf in suspension where it either settles slowly to the bottom or is ingested by floating organisms and falls to the bottom as fecal pellets. The nature of such deposits is now well known, largely as a result of the search for offshore oil and gas. In some places sediment as much as 14 km thick has accumulated over the last 70 to 100 million years. To build the whole pile, an average of less than a millimeter of sediment need have been deposited each year.

Coarse Sediment. Most coarse marine sediment is deposited within 5 to 6 km of the land and is dispersed by longshore currents. Coarse sediment is also found at greater distances offshore as far as the seaward limits of the shelves. Its observed patchy distribution may be partly the work of localized currents, but to a greater extent it is the result of changing sea level. At times when the sea fell below its present level, the shoreline migrated seaward across the shelves, exposing new land. Bodies of coarse sediment deposited near shore or on the land at such times subsequently were submerged as sea level rose across the shelves. Such sediment is referred to as *relict sediment,* for it is not in equilibrium with present environments, but rather is a relict of past conditions. It has been estimated that as much as 70 percent of the sediment cover on the continental shelves can be classified as relict sediment.

Shelf Strata. When world sea level stands high against the continents, the submerged shelves trap most of the detritus that reaches them from the adjacent land and prevent it from reaching the deep ocean. Of course some detritus never reaches the shelf. It is trapped by inland basins, where it is buried and preserved. However, analysis of ancient strata shows that the amount of sediment retained on land is much less than the amount carried to the shelf. Only an estimated 10 percent of the sediment reaching the shelf remains in suspension long enough to arrive in the deep sea, so it is clear that the great bulk of the Earth's sedimentary strata is shelf strata whose sediment originated within the continents themselves. The shelves, in effect, conserve continental crust which is continually recycled within the continental realm.

FIGURE 4.19 Oblique aerial view of the Bahama Banks, a broad carbonate shelf off the southeastern United States. In places, low islands of carbonate sediment rise above sea level.

Carbonate Shelves and Reefs

Carbonate sediments of biogenic origin accumulate where the influx of land-derived sediment is minimal and where the climate and sea-surface temperatures are warm enough to promote the abundant growth of carbonate-secreting organisms. Most carbonate sediment on carbonate shelves consists of sand-sized skeletal debris, together with inorganic precipitates. Coarser debris is found mainly near reefs or in areas of turbulence and strong currents. Carbonate sediments near the landward margins of such shelves often are mixed with clastic debris from the land.

One type of carbonate shelf consists of a wide protected lagoon in which fine carbonate muds accumulate. The lagoon is bordered on its seaward margin by a protective reef inhabited by reef-building corals and coralline algae. The shelf off eastern Australia, with its Great Barrier Reef, is a good example.

A second type, of which the Bahama Banks are an example (Fig. 4.19), is an open carbonate platform that lacks a bordering reef. On such a platform, surface currents tend to winnow fine particles and move them to deeper water, leaving a coarse lag of carbonate sediment on the shelf. The surface deposits of carbonate shelves, like those of most other continental shelves, include both modern sediment and relict sediment that formed when sea level was lower than today.

Marine Evaporite Basins

Where ocean water in a restricted basin lies within an area of very warm climate, evaporation of the water can cause deposition of *evaporite.* Such a sedimentary deposit is composed chiefly of minerals precipitated from a saline solution through evaporation. Halite and gypsum are common examples.

The Mediterranean Sea is such a basin. Were it not for continuous inflow of Atlantic water at its western end, the Mediterranean would gradually decrease in volume due to evaporation, which is especially high in its eastern half. It has been estimated that if deprived of new water, evaporation would cause the landlocked sea to dry up completely in about 1000 years. In the process, a layer of salt about 70 m thick would be precipitated.

Extensive evaporites underlie the Mediterranean Basin. These are regarded as evidence of former periods when high evaporation, together with a continuous inflow of Atlantic water to supply the necessary salt, allowed evaporites to accumulate to a thickness of 2 to 3 km.

FIGURE 4.20 Deep-sea turbidite beds that have been tilted on end and exposed in a wave-eroded bench along the coast of the Olympic Peninsula, Washington.

Sediments of the Continental Slope and Rise

Along most of their lengths, the shallowly submerged continental shelves pass abruptly into continental slopes which descend to depths of several kilometers. Sediments that reach the shelf edge are poised for further transport down the slope and onto the adjacent continental rise.

Turbidity Currents and Turbidites

Thick bodies of coarse sediment of continental origin lie at the foot of the continental slope at depths as great as 4 to 5 km. Such occurrences were an unsolved mystery until it appeared that they could be explained by *turbidity currents,* gravity-driven currents consisting of dilute mixtures of sediment and water having a density greater than the surrounding water.

Such currents have been produced in the laboratory in water-filled tanks into which a dense mixture of water, silt, and clay has been introduced. They also have been documented moving across the floors of lakes and reservoirs. In the oceans, turbidity currents have been set off by earthquakes, landslides, and major coastal storms. Off the mouths of large rivers, they may be set in motion by major floods.

The accumulated evidence leads us to believe that turbidity currents are very effective geologic agents on continental slopes where they can develop velocities greater than those of the swiftest streams on land. Some reach a velocity of more than 90 km/h and transport up to 3 kg/m^3 of sediment, spreading it as far as 1000 km from the source.

A turbidity current typically deposits a graded layer of sediment called a *turbidite* (Fig. 4.20). Such a graded layer is the result of rapid, continuous loss of energy in the transporting agent. This would occur in a turbidity current as it slows down, allowing successively finer sediment to be deposited.

At any site on the continental rise or adjacent deep-sea plain, a turbidite is deposited very infrequently, perhaps only once every few thousand years. In these places, far distant from the source region, the deposits are mainly thin-bedded layers a few millimeters to 30 cm thick. Although deposition is infrequent, over millions of years turbidites can slowly accumulate to form vast deposits beyond the continental realm.

Deep-Sea Fans

Some large canyons that are cut into continental slopes are aligned with the mouths of big rivers such as the Amazon, Congo, Ganges, and Indus. The mouths of most such canyons lead into huge fan-shaped features at the base of the continental slope that spread downward and outward to the deep seafloor (Fig. 4.21). These features are *deep-sea fans.* The surfaces of some are marked by channels as much as 200 m deep.

The sediments of deep-sea fans, sampled by coring, prove to have been derived mostly from the land. They include fragments of land plants, as well as fossils of shallow- and deep-water marine organisms. Also present are many graded layers, consisting of mixtures ranging from clay particles up to small pebbles, which are interpreted as turbidites.

Deep-sea fans are a major exception to the generalization that final deposition of land-derived sediment in the ocean is largely confined to the continental shelves. When shelves are exposed at times of lowered sea level and rivers extend across them nearly to the continental slope, the stage is set for the rapid building of deep-sea fans.

Sediments of the Deep Seafloor

Analyses of samples brought up by coring devices have made it possible to sort out the various sources

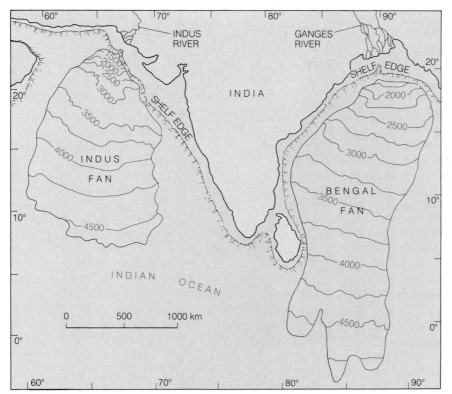

FIGURE 4.21 The Indus and Bengal deep-sea fans have been built by the Indus and Ganges-Bramaphutra rivers, respectively, on the seafloor adjoining the Indian subcontinent. Water depths are in meters. (*Source:* After Kolla and Coumes, 1985.)

from which seafloor sediment is derived. The study of great numbers of samples indicates clearly that all the sediments are mixtures; no one body of sediment comes entirely from a single source.

Land-Derived Sediment

Terrigenous sediment (Latin for "earthborn") is sediment derived from sources on land. Such sediment is carried to the sea by rivers, eroded from coasts by wave action, transported by wind (fine dust and volcanic ash), and released from floating ice. The bulk of terrigenous sediment eventually accumulates as turbidite deposits along continental rises and adjacent deep-sea plains.

Reddish or brownish clay, largely of terrestrial origin, consists mostly of clay minerals, quartz, and micas. The clay is confined to high latitudes and to depths of more than 4 km. Its color is due to gradual oxidation as the particles slowly settle to the seafloor.

Iceberg-rafted sediments are restricted to high latitudes where icebergs, mainly from Greenland and Antarctica, melt and drop their load of sediment to the seafloor.

Deep-Sea Oozes

Pelagic sediment (from *pelagos*, Greek for "the sea") is sediment consisting of material of marine organic origin. It is largely composed of microscopic shells and skeletons of marine animals and plants and is commonly referred to as deep-sea ooze (Fig. 4.22).

Calcareous ooze occurs over broad areas at low to middle latitudes where warm sea-surface temperatures favor the growth of carbonate-secreting organisms. The minute shells of these organisms settle and slowly accumulate over wide areas of the seafloor. However, calcareous oozes are not found in these same latitudes where the water is 4 to 5 km deep. This is because cold deep-ocean waters are under high pressure and contain more dissolved carbon dioxide that shallower waters. Consequently, they can readily dissolve any carbonate that reaches their level.

Over much of the North Pacific and part of the central South Pacific, where the ocean floor lies at great depths, carbonate ooze is lacking. Nearly all the calcareous material settling in these areas is dissolved before reaching the bottom. Instead, the surface is covered by reddish clay composed of

FIGURE 4.22 Skeletons of foraminifera (smooth globular forms), radiolaria (coarse-meshed objects), and rod-shaped sponge spicules from a deep-sea ooze, photographed by a scanning electron microscope. Fossils are from a core collected in the western Indian Ocean during a Deep Sea Drilling Project cruise.

extremely fine material derived mainly from continental sources.

Other parts of the oceans are floored with *siliceous ooze*, most notably in the equatorial Pacific, in part of the Indian Ocean, and in a belt encircling the Antarctic region. These are areas of high biologic productivity, in part related to upwelling of waters having a high nutrient content. In such regions, siliceous organisms predominate and become a primary component of the deep-sea sediments.

Volcanic Sediment

Deep-sea sediments locally include a component of *volcanic sediment* shed from submarine volcanoes and volcanic islands, as well as ash produced by explosive volcanic eruptions and spread across the sea.

Essay

TAKING THE TEMPERATURE OF THE ANCIENT OCEAN

We have become used to seeing maps of current national and world temperature patterns in our daily papers and projected on television news. How interesting it would be to see similar maps of remote geologic times so we could learn more about past conditions at the Earth's surface. In fact, such maps are now being prepared, based in part on unique evidence preserved in sediments of the deep sea.

The upper parts of cores taken from seafloors throughout the world's oceans consist of soft sediments that typically contain multitudes of tiny marine fossils. Most of these organisms were pelagic forms that lived in the surface waters. When they died, their shells rained down on the seafloor in vast numbers. The sedimentation rate, however, is extremely slow, so that it may take a thousand years or

more for a single centimeter of sediment to accumulate.

The percentages of different species change downward through a sediment core, typically shifting back and forth from predominantly warm-water types to cold-water types. Studies of living pelagic species show that as surface-water temperature changes from equator to pole, the species composition also changes; the assemblage living in warm tropical waters differs in composition from that living in cooler middle latitudes and also from that of cold polar waters. By identifying the species present at any level in a sediment core and comparing the fossil assemblage with modern ones, an estimate of former surface ocean temperature can be obtained. In practice, geologists sample a level in a core that represents a spe-

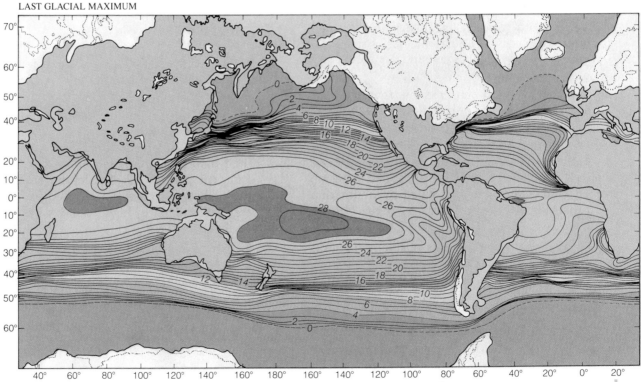

LAST GLACIAL MAXIMUM

FIGURE B4.1 Map showing reconstructed August sea-surface temperatures (°C) during the last glacial age, about 18,000 years ago, based on fossil assemblages in deep-sea cores. Cold polar water extended far south of its present limit in the North Atlantic and northwest Pacific, whereas waters remained warm at low latitudes. White areas show extent of contemporary major ice sheets and mountain glaciers. (*Source:* After CLIMAP Project Members, 1981.)

cific time interval (for example, the time of maximum expansion of glaciers during the last glacial age) and determine the sea-surface temperature at the core site from the preserved fossils. Data from hundreds of cores scattered widely over the oceans have been used to construct a global map of sea-surface temperatures for the last glacial maximum (Fig. B4.1). This map can be compared with a similar map of present ocean temperatures to determine the differences between the glacial-age ocean and the modern one. In this way, it has been shown that some regions, like the North Atlantic, were as much as 14°C colder during glacial times than they are today, while other regions, mainly in the subtropics, experienced little or no change.

SUMMARY

1. Sediment is transported by streams, glaciers, wind, slope processes, and ocean currents. After deposition, it experiences compaction and cementation as it is transformed into sedimentary rock.

2. Clastic sediment consists of fragmental rock debris resulting from weathering, together with the broken remains of organisms. Chemical sediment forms where substances carried in solution are precipitated.

3. Various arrangements of the particles in strata are seen in parallel strata and cross strata, uniform layers, graded layers, and nonsorted layers.

4. Particles of sediment become rounded and sorted during transport by water and air, but not during transport by glaciers.

5. Most sedimentary strata are built of continental detritus that is transported to the submerged continental margins. Some is trapped in basins on land where it is deposited by nonmarine processes, and a small percentage reaches the deep sea.

6. Depositional environments of nonmarine and shallow-marine sediments can be inferred from such properties as texture, degree of sorting and rounding, character of stratification, and types of contained fossils.

7. Most land-derived sediment reaching the continental shelves is deposited close to shore where it is reworked by longshore currents. Extensive areas are covered by relict sediments deposited at times of lower sea level.

8. Carbonate shelves are found in low latitudes where warm waters promote growth of carbonate-secreting organisms and little or no sediment is contributed from the continents.

9. Thick evaporite deposits can accumulate in restricted marine basins where evaporation is high and continuous inflow provides a supply of saline water.

10. By depositing turbidites, turbidity currents have built large deep-sea fans at the base of the continental slope.

11. Chief kinds of sediments on the deep seafloor beyond the continental rise are brownish or reddish clay, calcareous ooze, and siliceous ooze. Their distribution is largely related to surface water temperature, water depth, and surface productivity. Iceberg-rafted sediments are largely restricted to high latitudes near major ice sheets.

12. An extensive unit of strata may possess several facies, each determined by a different depositional environment. The boundaries between facies may be abrupt or gradational.

13. Sediment is constantly recycled, nearly always within the continental realm.

IMPORTANT WORDS AND TERMS TO REMEMBER

alluvium (p. 95)

bed (p. 90)

bedding (p. 90)
breccia (p. 88)

calcareous ooze (p. 101)
carbonate shelf (p. 99)

cementation *(p. 89)*
chemical sediment *(p. 87)*
clastic sediment *(p. 86)*
claystone *(p. 88)*
compaction *(p. 89)*
concretions *(p. 93)*
conglomerate *(p. 88)*
coquina *(p. 87)*
cross strata *(p. 91)*

deep-sea fans *(p. 100)*
delta *(p. 97)*
detritus *(p. 87)*
diagenesis *(p. 89)*
dolostone *(p. 88)*

eolian *(p. 97)*
eolian environments *(p. 97)*
estuary *(p. 98)*
evaporite *(p. 99)*

fecal pellets *(p. 98)*
fluvial environments *(p. 95)*

fossils *(p. 93)*

graded bedding *(p. 92)*

hydrocarbons *(p. 89)*

lacustrine environments *(p. 96)*
lagoon *(p. 99)*
limestone *(p. 88)*

mud cracks *(p. 93)*
mudstone *(p. 88)*

oxidizing environment *(p. 89)*

parallel strata *(p. 90)*
pelagic sediment *(p. 101)*
placer *(p. 90)*

recrystallization *(p. 89)*
reducing environment *(p. 89)*
reef *(p. 99)*
relict sediment *(p. 98)*
ripple marks *(p. 93)*

sandstone *(p. 88)*
sedimentary facies *(p. 95)*
sedimentary stratification *(p. 89)*
shale *(p. 88)*
siliceous ooze *(p. 102)*
siltstone *(p. 88)*
sorting *(p. 90)*
strata *(p. 89)*
stratification *(p. 89)*
stratum *(p. 89)*

terrigenous sediment *(p. 101)*
tillite *(p. 93)*
turbidite *(p. 100)*
turbidity currents *(p. 100)*

uniform layer *(p. 92)*

varve *(p. 90)*
volcanic sediment *(p. 102)*

QUESTIONS FOR REVIEW

1. On what basis are sediments and sedimentary rocks classified?

2. What reactions can lead to precipitation of chemical sediments?

3. What features in a sediment or sedimentary rock are responsible for stratification?

4. Describe the processes involved in the conversion of sediment to sedimentary rock.

5. Why are gold and platinum, eroded from bedrock, often concentrated in stream placers?

6. Describe cross stratification and graded layers, and explain what they tell about conditions of deposition.

7. What clues do surface features on bedding surfaces of sedimentary rocks provide regarding depositional environments?

8. How would you tell sediments of an alluvial facies from those of a lacustrine or an eolian facies?

9. If estuaries are generally shallow bodies of water, how can you explain the occurrence of thick estuarine accumulations in the sedimentary record?

10. How can the occurrence of widespread relict sediments on the continental shelves be explained?

11. If the Mediterranean Sea, which is an enclosed marine basin, were to evaporate completely, less than 100 m of evaporite sediments would be deposited. How, then, can you explain the presence of continuous evaporite deposits under the Mediterranean that are more than 2 km thick?

12. What distinctive features of the sediments in deep-sea fans provide clues about the way in which the sediments were transported?

13. What factors explain the distribution of calcareous and siliceous oozes on the floors of the ocean basins?

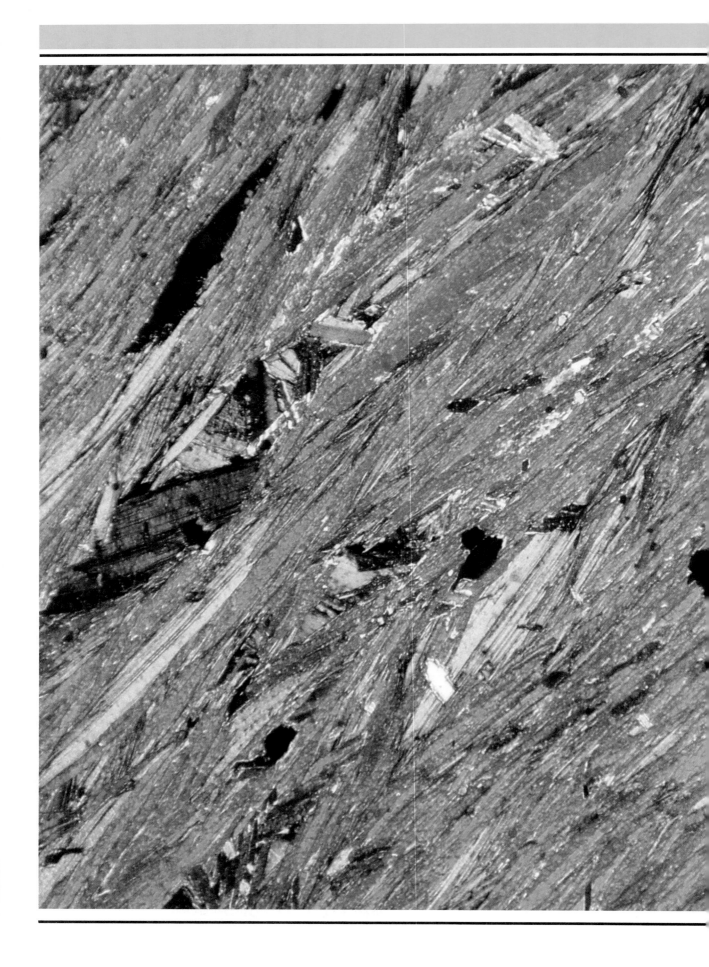

Metamorphism and Metamorphic Rocks

Muscovite schist as seen in a thin section and viewed in polarized light. Grains are oriented so that the cleavage is perpendicular to the direction of maximum stress.

THE MEANING OF METAMORPHISM

Meta is an old Greek prefix which means change. Another old Greek word is *morphe,* and as explained in chapter 1, when used as a combining word, it means form or shape. Thus the term *metamorphism* means change of form or shape. The term is used to describe all changes in mineral assemblage and rock texture, or both, that take place in rocks in the solid state within the Earth's crust as a result of changes in temperature and pressure.

Igneous, pyroclastic, or sedimentary rock, once formed, may be subjected to a new set of conditions through such events as compression caused by continental collision, heating by a large igneous intrusion, or deep burial. As a result, new textures and new mineral assemblages may develop. Changes start happening to mineral grains almost as soon as particles in a sediment are deposited. As explained in chapter 4, the early changes are said to be the effects of *diagenesis.* Indeed, it is a common practice to group all of the changes that occur below about 200°C under the term *diagenesis.* When effects are caused by temperatures in excess of about 200°C, and by pressures above the moderate pressures equivalent to those produced by a few thousand meters of overlying rock, the changes are said to be the effects of metamorphism.

CONTROLLING FACTORS IN METAMORPHISM

The mineral assemblage of a rock undergoing metamorphism plays a controlling role in the new mineral assemblage. So do changes in temperature and pressure because these are the two principal causes of metamorphism. The effects of temperature and pressure are not entirely straightforward because their effectiveness is strongly influenced by such things as the presence or absence of fluids in rocks, the duration of heating, how long a rock is subjected to high pressure, and whether it is simply compressed or whether it is twisted and broken during metamorphism.

Chemical Reactivity Induced by Fluids

The innumerable open spaces between the grains in a sedimentary rock, or the tiny fractures that can be seen in many igneous rocks, are called *pores.* All pores are filled by a watery fluid. The fluid is never pure water, and at high temperature it is more likely to be a vapor than a liquid. Nevertheless, the *intergranular fluid,* for that is its best designation, plays a vital role in metamorphism. The fluid always has dissolved within it small amounts of gases, such as carbon dioxide, and salts, such as sodium chloride, plus traces of all the mineral constituents, such as quartz, that are present in the enclosing rock.

When the temperature and pressure change, so does the composition of the intergranular fluid. Some of the dissolved constituents move from the fluid to the new minerals forming in the metamorphic rock. Other constituents move in the other direction, from the minerals to the fluid. In this way the intergranular fluid serves as a transporting medium, or "juice," that speeds up chemical reactions in much the same way that water in a stewpot speeds up the cooking of a tough piece of meat.

When intergranular fluids are absent, or present only in tiny traces, metamorphic reactions are very slow. When a "dry" rock is heated, few changes occur because the growth of new minerals means that atoms must move by diffusing through the solid minerals. Diffusion through solids is an exceedingly slow process. If, somehow, an intergranular fluid is introduced, perhaps because the rock is crushed, or otherwise deformed, diffusion of the atoms from one place to another can take place through the intergranular fluid. This is a vastly faster process, and as a result, new minerals grow rapidly and the metamorphic effects are pronounced.

Grade of Metamorphism

The term used for rock changed at low temperature and low pressure is *low-grade metamorphism.* Such rock contains many hydrous minerals. As pressure increases and metamorphism proceeds, the amount of pore space decreases and the intergranular fluid is slowly driven out of the rock. As the temperature of the rock increases, hydrous minerals (meaning they contain H_2O in their structure), such as clays and chlorite, recrystallize to water-free (anhydrous) minerals, such as feldspars and pyroxenes. In the process, water is released. The released water joins the intergranular fluid and is also slowly driven out of the metamorphic rock. For this reason, rock subjected to high temperature and high pressure, for which the term *high-grade metamorphism* is used, contains fewer hydrous minerals.

Prograde Metamorphism. The metamorphic changes that occur while temperatures and pressures are rising, and while abundant intergranular fluid is present, are termed *prograde metamorphic effects.*

Retrograde Metamorphism. Changes that occur as temperature and pressure are declining, and after much of the intergranular fluid has been expelled, are called *retrograde metamorphic effects.* Not surprisingly, because of the lack of fluids, retrograde metamorphic effects happen less rapidly, and are less pronounced, than prograde effects. Indeed, it is only because retrograde reactions happen so slowly that we see high-grade metamorphic rocks at all. If retrograde reactions were rapid, all metamorphic rocks would react back to clays and other low-grade minerals.

Temperature

When a mixture of flour, salt, sugar, yeast, and water is baked in an oven, the high temperature causes a series of chemical reactions—new compounds grow and the final result is a loaf of bread. When rocks are heated, new minerals grow and the final result is a metamorphic rock. In the case of the rocks, the source of heat is the Earth's internal heat. Rock can be heated simply by burial, or by a nearby igneous intrusion. But burial and the process of intrusion can also cause a change in pressure. Therefore, whatever the cause of the heating, metamorphism can rarely be considered to be entirely due to the rise in temperature. The combined effects of changing temperature and pressure must be considered together.

Pressure

The combined influence of temperature and pressure on the melting properties of rocks and minerals was discussed in chapter 3 (see Fig. 3.10, for example). Metamorphic transformations are also controlled by the dual effects of temperature and pressure. Because the compositions of the rocks being metamorphosed and the temperature and pressure of metamorphism range widely, the mineral assemblages in metamorphic rocks also range widely. By comparing mineral assemblages seen in nature with those produced synthetically in the laboratory, we can delineate the ranges of pressure and temperature conditions under which metamorphism occurs in the crust (Fig. 5.1).

So far pressure has been discussed as if it were equal in all directions, as it is in a liquid. But rock has strength, and can withstand pressures that are

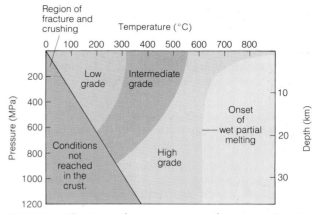

FIGURE 5.1 Regions of temperature and pressure (equivalent to depth) under which metamorphism occurs in the crust.

greater in one direction than another. When the deformation of a solid is discussed, the term *stress* is used instead of pressure. Stress has the connotation of direction, as will be discussed more completely in chapter 15. It is possible to stress a solid more strongly in one direction than another. The textures in many metamorphic rocks record inhomogeneous stress during metamorphism. The sheet-structure minerals that crystallize during metamorphism, micas and chlorites, grow so that the polymerized $(Si_4O_{10})^{-4}$ sheets are perpendicular to the direction of maximum stress (Fig. 5.2). Indeed, the presence in a rock of a texture resulting from parallel flakes of mica is strong evidence that the rock has been metamorphosed.

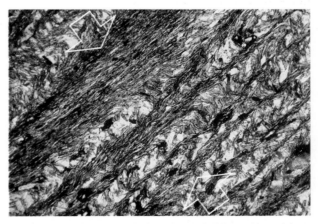

FIGURE 5.2 Thin section of a metamorphic rock revealing the parallel arrangement of mica grains. Direction of maximum stress is indicated by arrows. The sample is 1 cm wide.

Time

All chemical reactions require a certain amount of time to proceed to completion. Some reactions, such as the burning of methane gas (CH_4) to yield carbon dioxide and water, happen so rapidly that they create explosions. At the other end of the scale are reactions that require millions of years to proceed to completion. Many of the chemical reactions that occur in rocks undergoing metamorphism are of the latter kind. No reliable ways have yet been developed to determine exactly how long a given metamorphic rock has remained at a given temperature and pressure. However, it can be readily demonstrated in the laboratory that high temperature, high pressure, and long reaction times produce large mineral grains. Thus, it is possible to draw the interesting general conclusion that coarse-grained rocks are the products of long-sustained metamorphic conditions (possibly over millions of years) at high temperatures and pressures, while fine-grained rocks are products of lower temperatures, lower pressures, and shorter reaction times.

METAMORPHIC RESPONSES TO CHANGES IN TEMPERATURE AND PRESSURE

Foliation

In a few metamorphic rocks the only textural change that occurs as a result of increased temperature and pressure is an increase in grain size. Such changes are rare and most metamorphic rocks develop additional, and conspicuous, directional textures. As metamorphism proceeds, and the sheet-structure minerals such as muscovite and chlorite start to grow, the minerals are oriented so that the sheets are perpendicular to the direction of maximum stress. The new, parallel flakes of mica produce a texture called *foliation,* named from the Latin word, *folium,* meaning leaf (Fig. 5.3). Foliation may be pronounced or it may be subtle, but when present it provides strong evidence of metamorphism.

Slaty Cleavage

During the earliest stages of mineral growth, pressure is caused by the weight of the overlying rock. The new sheet-structure minerals, and therefore the foliation, tend to be parallel to the bedding planes of a sedimentary rock. But with deeper burial, or when lateral compression deforms the flat

FIGURE 5.3 A planar fabric (foliation) produced in granite by metamorphism. The foliation is caused by a parallel orientation of mica grains. The outcrop is in Namaqualand, South Africa. The pocket knife is 6 cm long.

sedimentary layers into folds, the sheet-structure minerals and the foliation are no longer parallel to the bedding planes (Fig. 5.4). Regardless of the orientation of the original bedding, metamorphic rocks break readily in the direction of the foliation. When the rocks are so fine grained that the new mineral grains can only be seen with the microscope, the breakage property is called *slaty cleavage,* which is defined as the property by which a rock breaks into platelike fragments along flat planes (Fig. 5.5).

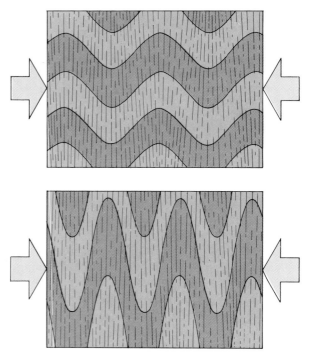

FIGURE 5.4 Two examples of slaty cleavage in folded strata. Arrows indicate direction of maximum stress.

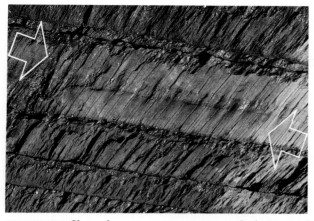

FIGURE 5.5 Slaty cleavage cuts across nearly horizontal bedding. The cleavage is developed in the Martinsburg Formation, which was subjected to low-grade metamorphism. Arrows indicate the direction of maximum stress. The sample is 70 cm across and is near Palmerton, Pennsylvania.

Schistosity

Slaty cleavage develops at low grades of metamorphism. Under intermediate and high grades of metamorphism, grain sizes increase and individual mineral grains can be seen with the naked eye. Foliation remains but it is no longer a flat plane.

Intermediate and high-grade metamorphic rocks tend to break along wavy, or slightly distorted surfaces, reflecting the presence, and orientation, of grains of quartz, feldspar, and other minerals. Such breakage directions arise from the **schistosity**, a term derived from the Latin, *schistos*, meaning cleaves easily, and referring to the parallel arrangement of coarse grains of the sheet-structure minerals, like mica and chlorite, formed during metamorphism under conditions of differential stress.

Mineral Assemblages

Metamorphism produces new mineral assemblages as well as new textures. As temperature and pressure rise, one new mineral assemblage follows another. For any given rock composition, each assemblage is characteristic of a given range of temperature and pressure. A few of these minerals are rarely found (or not at all) in igneous and sedimentary rocks. Their presence in a rock is usually evidence enough that the rock has been metamorphosed. Examples of these metamorphic minerals are chlorite, serpentine, epidote, and talc. An illustration of the way mineral assemblages change with grade of metamorphism is given in Figure 5.6.

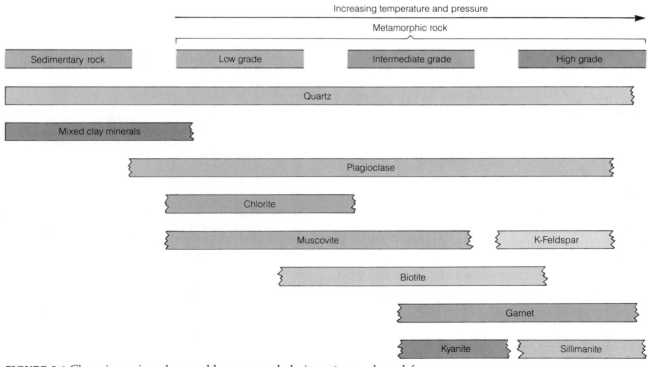

FIGURE 5.6 Changing mineral assemblages as a shale is metamorphosed from low to high grade. Kyanite and sillimanite are polymorphs (Al_2SiO_5) that are found only in metamorphic rocks.

KINDS OF METAMORPHIC ROCK

The naming of metamorphic rocks is based partly on texture, partly on mineral assemblage. The most widely used names are those applied to metamorphic derivatives of shales and basalts. This is so because shales and basalts are, respectively, the most abundant sedimentary and igneous rock types.

Metamorphism of Shale and Mudstone

Slate

The low-grade metamorphic product of shale or mudstone is **slate.** The minerals usually present in both shale and mudstone include quartz, clays of various kinds, calcite, and possibly feldspar. Under conditions of low-grade metamorphism, muscovite and/or chlorite crystallize. Although the rock may still look like a shale, the tiny new mineral grains produce slaty cleavage.

Slate is a word derived from Old French, *slat,* referring to the useful properties such rock has as a roofing material.

Phyllite

Continued metamorphism to intermediate grade produces both larger grains of mica and a changing mineral assemblage; the rock develops a pronounced foliation and is called **phyllite** (from the Greek, *phyllon,* a leaf). In a slate it is not possible to see the new grains of mica with the unaided eye but in a phyllite they are just large enough to be visible.

Schist and Gneiss

Still further metamorphism beyond that which produces a phyllite, leads to a coarse-grained rock with pronounced schistosity, called **schist.** A comparison of the differing grain sizes in slate, phyllite, and schist can be seen in Figure 5.7. At higher grades of metamorphism, minerals start to segregate into separate bands. A high-grade rock, with coarse grains and pronounced foliation, but with layers of micaceous minerals segregated from layers of minerals such as quartz and feldspar, is called a **gneiss** (pronounced nice, from a word in Old High German, *gneisto,* meaning to sparkle) (Fig. 5.8).

The names slate and phyllite describe textures and are commonly used without adding mineral names as adjectives. The names of the coarse-grained

a

b

c

FIGURE 5.7 The grain sizes in these specimens of (a) slate, (b) phyllite, and (c) schist show how mineral growth continues during metamorphism. The three rocks are photographed at the same magnification and have the same chemical composition.

rocks, schists and gneisses, are also derived from textures but in these cases mineral names are usually added as adjectives; for example, we refer to

FIGURE 5.8 A coarse-grained gneiss. Minerals present are feldspar and quartz (both light colored), and biotite (dark). The parallel orientation of biotite grains proves that the rock is metamorphic in origin. The specimen is 8 cm across.

FIGURE 5.9 Amphibolite resulting from metamorphism of a pillow basalt. Compare Figure 3.28b. Pillow structure was deformed during metamorphism but can be discerned by the borders of pale-yellow epidote formed from the original glassy rims of the pillows. The outcrop is in Namibia.

a quartz–plagioclase–biotite–garnet gneiss. The difference arises because minerals in coarse-grained rocks are large enough to be seen and readily identified.

Metamorphism of Basalt

Greenschist

The main minerals in basalt are olivine, pyroxene, and plagioclase, each of which is anhydrous. When a basalt is subjected to metamorphism under conditions where H_2O can enter the rock and form hydrous minerals, distinctive mineral assemblages develop. At low grades of metamorphism, assemblages such as chlorite + albite + epidote + calcite form. The resulting rock is equivalent in metamorphic grade to a slate, but has a very different appearance. It has pronounced foliation, but it also has a very distinctive green color because of its chlorite content; it is termed *greenschist.*

Amphibolite and Granulite

When a greenschist is subjected to temperature and pressure equivalent to intermediate grade metamorphism, chlorite is replaced by amphibole; the resulting rock is generally coarse grained, and is called an *amphibolite* (Fig. 5.9). Because amphibole has a chain structure, rather than a sheet structure, the effect of differential stress is to cause the amphibole to grow as elongate grains. The grains tend to line up so that their long axes are parallel and

point in the direction of minimum stress. A rock that has a parallel arrangement of elongate mineral grains is said to possess a *lineation.* At highest grade metamorphism, amphibole is replaced by pyroxene and the rock developed is called a *granulite.*

Metamorphism of Limestone and Sandstone

Marble and *quartzite* are, respectively, the metamorphic derivatives of limestone and sandstone. Neither limestone nor quartz sandstone (when pure) contain the necessity ingredients to form sheet- or chain-structure minerals. As a result marble and quartzite commonly lack foliation.

Marble

Marble consists of a coarsely crystalline, interlocking network of calcite grains. During recrystallization of a limestone, bedding planes, fossils, and other features of sedimentary rocks are largely obliterated. The end result, as shown in Figure 5.10a, is an even-grained rock with a distinctive, somewhat sugary texture. Pure marble is snow white in color and consists entirely of pure grains of calcite. Such marbles are rare, even though one may not think so looking at the large number of marble gravestones and statues in cemeteries. Most marble contains impurities such as organic matter, pyrite, limonite, and small quantities of silicate minerals, that impart various colors.

a

b

FIGURE 5.10 Textures of nonfoliated metamorphic rocks seen in thin sections and viewed in polarized light. Notice the interlocking grain structure produced by recrystallization during metamorphism. Each specimen is 2 cm across. (a). Marble, composed entirely of calcite. All vestiges of sedimentary structure have disappeared. (b). Quartzite. Faint traces of the original rounded quartz grains can be faintly seen in some of the grains.

Quartzite

Quartzite is derived from sandstone by the filling in of the space between the original grains with silica, and by recrystallization of the entire mass (Fig. 5.10b). Commonly, the ghostlike outlines of the original sedimentary grains can still be seen, even though recrystallization may have rearranged the original grain structure completely.

KINDS OF METAMORPHISM

The processes that result from changing temperature and pressure, and that cause the metamorphic changes observed in rocks, can be grouped under

FIGURE 5.11 Mechanical deformation has caused the quartzite pebbles in this conglomerate in Namibia, originally round, to become flattened and elongate. The pocketknife is 6 cm long.

the terms *mechanical deformation* and *chemical recrystallization.* Mechanical deformation includes grinding, crushing, and the development of new textures (Fig. 5.11). Chemical recrystallization includes all the changes in mineral composition, in growth of new minerals, and the loss of H_2O and CO_2 that occur as rock is heated. Different kinds of metamorphism reflect the different levels of importance of the two groups of processes.

Cataclastic Metamorphism

Purely mechanical effects do sometimes occur without any changes in mineral chemistry, but they are rare and usually localized. For example, in intensely stressed masses of coarse-grained rocks such as granite, individual mineral grains may be shattered and pulverized. This sort of deformation occurs in brittle rocks and is called *cataclastic metamorphism* (Fig. 5.12).

Contact or Thermal Metamorphism

Contact metamorphism, which is also known as *thermal metamorphism,* occurs adjacent to bodies of hot igneous rock that are intruded into cooler rock of the crust. Such metamorphism happens in response to a pronounced increase in temperature but without extensive mechanical deformation. The temperature of the igneous rock may be as high as 1000°C, that of the intruded rocks only 200° to 300°C. Rock adjacent to the intrusive becomes heated and metamorphosed, developing a well-defined shell, or *metamorphic aureole,* of altered rock (Fig. 5.13).

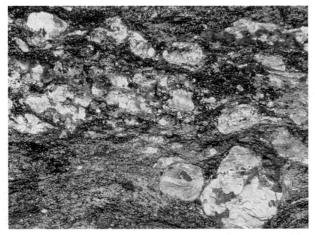

FIGURE 5.12 Cataclastic metamorphism of a granitic dike in biotite-rich gneiss. Brittle granite, rich in feldspar (pink) and quartz (light gray), was fragmented during deformation while the gneiss was ductile and tended to flow plastically. The field of view is about 50 cm across.

Metamorphic Aureole

The width of an aureole of contact metamorphosed rock depends on the size of the intrusive body and the amount of fluid present. With a small intrusive, such as a dike or sill a few meters thick, the width of the metamorphic aureole may be only a few centimeters. A large intrusion contains much more heat energy than a small one. When the intrusion is large, perhaps a kilometer or more in diameter, the aureole may reach a hundred meters or more in width.

Within a metamorphic aureole it is usually found that several different and roughly concentric zones of mineral assemblages can be identified. Each zone is characteristic of a certain temperature range. Immediately adjacent to the intrusion where temperatures are high, we find anhydrous minerals such as garnet and pyroxene. Beyond them are found hydrous minerals such as epidote and amphibole, and beyond them, in turn, micas and chlorites. The exact assemblage of minerals in each zone depends, of course, on the chemical composition of the intruded rock as well as the temperature and pressure reached during metamorphism.

Hornfels

Hornfels is a distinctive rock type produced by contact metamorphism. It is a hard, fine-grained rock, composed of an interlocked mass of uniformly sized mineral grains. Most commonly, hornfels is derived by metamorphism of shale. Most hornfels is fine grained because it forms rapidly under circumstances where insufficient heating time is available for large mineral grains to grow. For this reason, it is generally presumed that development of hornfels happens adjacent to shallow intrusive bodies.

Burial Metamorphism

Sediments, together with interlayered pyroclastic and volcanic rocks, may attain temperatures of 300°C or more when buried deeply in a sedimentary basin. Abundant pore water is present in buried sediment and it helps new minerals to grow. But the fabric of the metamorphic rock that results from simple **burial metamorphism** may look like that of an essentially unaltered sedimentary rock because there is little mechanical deformation involved. The fam-

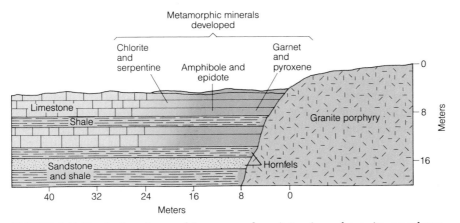

FIGURE 5.13 Contact metamorphism around an intrusion of granite porphyry near Breckenridge, Colorado. The new minerals form a series of zones, or aureoles, around the intrusive.

ily of minerals that particularly characterize the conditions of burial metamorphism are the zeolites. The **zeolites** are a group of silicate minerals with fully polymerized crystal structures containing the same chemical elements as feldspars, but also containing water. Burial metamorphism, which is the first stage of metamorphism following diagenesis, is usually observed in deep sedimentary basins, such as trenches on the margins of tectonic plates. As temperatures and pressures increase, burial metamorphism grades into regional metamorphism.

Regional Metamorphism

The most common metamorphic rocks of the continental crust occur through areas of tens of thousands of square kilometers and the process that forms them is called **regional metamorphism.** Unlike contact or burial metamorphism, regional metamorphism involves a considerable amount of mechanical deformation in addition to chemical recrystallization. As a result, regionally metamorphosed rocks tend to be distinctly foliated. Slate, phyllite, schist, and gneiss are the most common varieties of regionally metamorphosed rocks, and they are usually found in mountain ranges, or eroded mountain ranges. Because such mountain ranges form as a result of collision between fragments of continental crust, regional metamorphism is a consequence of plate tectonics. Greenschists and amphibolites are also products of regional metamorphism. They tend to be found where segments of ancient oceanic crust of basaltic composition have been incorporated into the continental crust and later metamorphosed.

Consider what happens when a pile of strata is subjected to horizontal compressive stress. The strata become folded and buckled. The folding and buckling causes the crust to become locally thickened. As a result the rocks near the bottom of the thickened pile are subjected to elevated pressure and temperature. New minerals start to grow. However, rocks are poor conductors of heat; so the heating-up process can be very slow. The temperatures reached depend on both depth and duration of burial of rock in the folded pile. If the folding and thickening is very slow, heating of the pile keeps pace with the temperature of adjacent parts of the crust (that is, a normal continental geothermal gradient is maintained). However, if burial is very fast, as it is with sediment dragged down in a subduction zone, the pile has insufficient time to heat up and conditions of high pressure, but rather low temperatures, prevail.

Metamorphic Zones

The first geologists to make a systematic study of a regionally metamorphosed terrain did so in the Scottish Highlands. They observed that rocks having the same overall chemical composition (that of a shale), could be subdivided into a sequence of zones, each zone having a distinctive mineral assemblage. Each assemblage, in turn, was characterized by the appearance of new minerals. They selected characteristic *index minerals,* which proceeding from low-grade rocks to high-grade rocks, marked the appearance of each new mineral assemblage. Their index minerals were, in order of appearance, chlorite, biotite, garnet, staurolite, kyanite, and sillimanite. By plotting on maps the places where each of the index minerals first appeared in rocks having the chemical composition of shale, the workers in the Scottish Highlands defined a series of isograds.

Isograds. An **isograd** is a line on a map connecting points of first occurrence of a given mineral in metamorphic rocks. The concept of isograds is now widely used by those who study metamorphic rocks of all kinds; it is just as applicable to burial and contact metamorphism as it is to regional metamorphism.

The regions on a map between isograds are known as **metamorphic zones.** We speak of the chlorite zone, the biotite zone, and so forth, and it is the zones that are commonly depicted on maps showing the relationships between metamorphic rocks (Fig. 5.14).

Metamorphic Facies

Careful study of metamorphic rocks around the world has demonstrated that the chemical compositions of rocks are little changed by metamorphism. The main changes that do occur are the addition or loss of volatiles such as H_2O and CO_2. However, the principal constituents of rocks remain fixed. The changes brought about during metamorphism, then, are changes in the mineral assemblage, not changes in the overall chemical composition of the rocks. The conclusion to be drawn from this fact is that the mineral assemblages observed in the metamorphic derivatives of common sedimentary and igneous rocks are controlled by the temperature and pressure of metamorphism to which the rocks are subjected. Based on this conclusion, the famous Finnish geologist Pennti Eskola proposed, in 1915, the concept of **metamorphic facies.** The concept is that contrasting assemblages

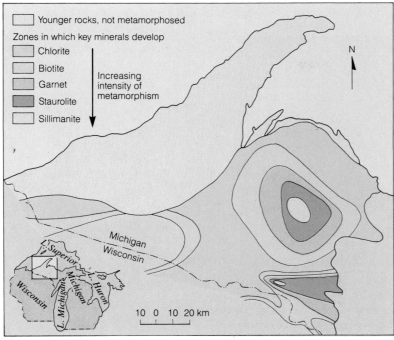

a

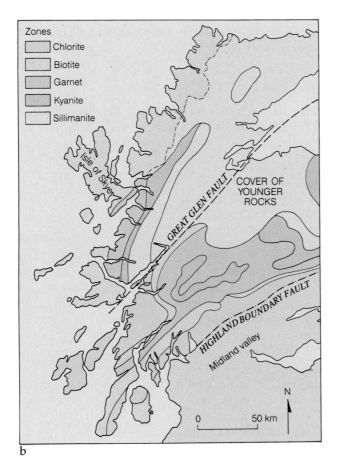

b

FIGURE 5.14 Metamorphic zones resulting from regional metamorphism. (a) Michigan. (*Source:* After James, 1955.) (b) Scottish Highlands.

of minerals that reach equilibrium during metamorphism within a specific range of physical conditions belong to the same metamorphic facies. Eskola drew his conclusions from studies of metamorphosed basalts that were interlayered with rocks of entirely different composition.

An analogy with cooking is appropriate; think of a large roast of beef. When it is carved, one sees that the center is rare, the outside well done, and in between is a region of medium-rare meat. The differences occur because the temperature was not uniform throughout. The center, or "rare-meat" facies, is a low-temperature zone; the outside, or "well-done" facies, is a high-temperature zone.

Metamorphic facies were originally described in terms of recurring mineral assemblages, to each of which there was assumed to be a specific set of temperature and pressure conditions. The realization that temperature, pressure, and rock composition each play a role in determining the mineral assemblage provided the link needed to allow conditions of metamorphism to be determined through laboratory experiments, and eventually to prove Eskola's suggestion. The concept is now applied to a very wide range of temperatures and pressures. The principal metamorphic facies, together with geothermal gradients to be expected under three differing geological conditions, are shown in Figure 5.15.

Because Eskola was studying metamorphosed basalts when he proposed the metamorphic facies

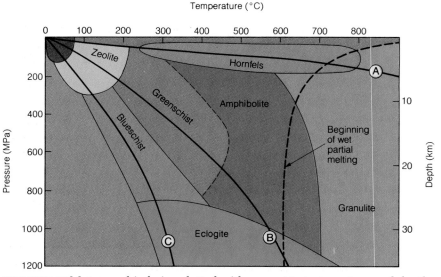

FIGURE 5.15 Metamorphic facies plotted with respect to temperature and depth. Curve A is a typical thermal gradient around an intrusive igneous rock that is causing contact metamorphism. Curve B is a normal continental geothermal gradient. Curve C is the geothermal gradient developed in a subduction zone.

concept, most of the names he gave to metamorphic facies reflect the mineral assemblages developed in rocks of basaltic composition. It is important to remember, however, as shown in Table 5.1, that mineral assemblages are just as much a result of rock composition as they are of the temperature and pressure of metamorphism. When comparing mineral assemblages of rocks subjected to differing grades of metamorphism, therefore, one must be certain that they have the same overall chemical composition.

The Place of Metamorphism

One of the triumphs of the revolution in geology brought about by plate tectonics is that it provides,

TABLE 5.1   *Characteristic Minerals of Differing Metamorphic Facies for Selected Rocks[a]*

Facies Name	Precursor Rock Type	
	Basalt	Shale
Granulite	Pyroxene, Plagioclase, Garnet	Biotite, K-Feldspar, Quartz, Andalusite
Amphibolite	Amphibole, Plagioclase, Garnet, Quartz	Garnet, Biotite, Muscovite, Sillimanite, Quartz
Epidote-Amphibolite	Amphibole, Epidote, Plagioclase, Garnet, Quartz	Garnet, Chlorite, Muscovite, Biotite, Quartz
Greenschist	Chlorite, Amphibole, Plagioclase, Epidote	Chlorite, Muscovite, Plagioclase, Quartz
Blueschist	Blue-Amphibole, Chlorite, Ca-rich Silicates	Blue-Amphibole, Chlorite, Quartz, Muscovite, Lawsonite
Eclogite	Pyroxene (variety Jadeite), Garnet, Kyanite	Not Observed
Hornfels	Pyroxene, Plagioclase	Andalusite[b], Biotite, K-Feldspar, Quartz
Zeolite	Calcite, Chlorite, Zeolite (variety Laumontite)	Zeolites, Pyrophyllite, Na-Mica

[a]For temperature and pressure conditions of each facies, refer to Figure 5.15.
[b]Andalusite is a *polymorph* of kyanite and sillimanite.

for the first time, an explanation for the distribution of metamorphic zones in regionally metamorphosed rocks. Regional metamorphism occurs at the subduction boundary of a plate, as shown in Figure 5.16.

Burial metamorphism is believed to occur in the lower portions of the thick piles of sediment that accumulate on the continental shelf and continental slope. Such metamorphism is known to be happening today in the great pile of sediments accumulated in the Gulf of Mexico, off the mouth of the Mississippi River.

The temperatures and pressures characteristic of *blueschist* and *eclogite facies* metamorphism are reached when crustal rocks are dragged down by a rapidly subducting plate. Under such conditions pressure increases more rapidly than temperature and as a result rock is subjected to high pressure and relatively low temperature. Rocks subjected to blueschist and eclogite facies metamorphism are

widespread in the Coast Ranges of California. Blueschist metamorphism is probably happening today along the subducting margin of the Pacific Plate, where it plunges under the coast of Alaska and the Aleutian Islands.

The metamorphic conditions characteristic of *greenschist* and *amphibolite facies* metamorphism occur where crust is thickened by continental collision, or heated by rising magma. It is the most common setting for metamorphism and such rocks can be observed throughout the Appalachians and the Alps. Such metamorphism is no doubt occurring today beneath the Himalaya where the continental crust is thickened by collision, and beneath the Andes where it is heated by rising magma. If the crust is sufficiently thick, rocks subjected to amphibolite facies metamorphism can reach temperatures at which wet partial melting commences, and metamorphism passes into magmatism.

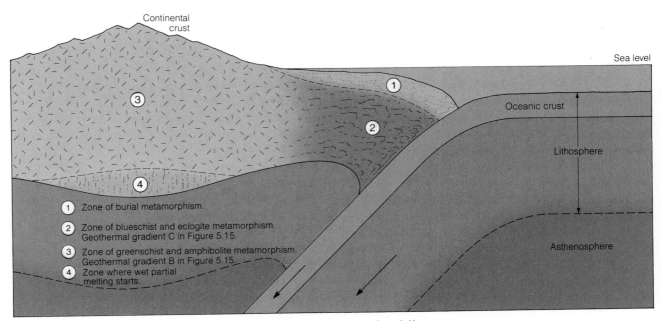

FIGURE 5.16 Diagram of a convergent plate boundary showing the different regions of metamorphism.

Essay

PRESSURE–TEMPERATURE–TIME PATHS FOR
METAMORPHIC ROCKS

Rock subjected to a high grade of metamorphism develops a distinctive mineral assemblage. But in order to reach the temperature and pressure at which high-grade metamorphism occurs, the rock must first have been subjected to lower temperatures and pressures and thus previously must have contained other mineral assemblages. Prograde metamorphism is a dynamic process and mineral assemblages replace one another in succession.

One of the triumphs of modern scientific instrumentation has been the development of techniques by which tiny mineral fragments can be examined and analyzed. In many metamorphic rocks microscopic relicts of those earlier mineral assemblages remain (Fig. B5.1); by analyzing them scientists can decipher the way pressure (P) and temperature (T) changed with time (t) during metamorphism. In the language of the scientists who carry out the research, it is possible to determine the $P–T–t$ path of a metamorphic rock. Surprising as it may seem, it turns out that metamorphic rocks are rarely subjected simultaneously to the highest pressures and highest temperatures.

FIGURE B5.1 A grain of garnet in a garnet-biotite gneiss. The small inclusions are relicts from a earlier mineral assemblage formed under P-T conditions different from those that formed the garnet-biotite gneiss. The garnet grain is 2 mm across.

SUMMARY

1. Metamorphism involves changes in mineral assemblage and rock texture occurring in the solid state as a result of changes in temperature and pressure.

2. Mineral assemblages and rock textures change continually as temperature and pressure change.

3. Mechanical deformation, recrystallization, and chemical reactions are the processes that affect rock during metamorphism.

4. The presence of intergranular fluid greatly speeds up metamorphic reaction.

5. Foliation, slaty cleavage, and schistosity arise from parallel growth of minerals formed during metamorphism.

6. Heat given off by bodies of intrusive igneous rock causes contact metamorphism and creates contact metamorphic aureoles.

7. Regional metamorphism is the result of plate tectonics. Regionally metamorphosed rocks are produced along subduction and collision edges of plates.

8. Rocks of the same chemical composition (and subjected to identical metamorphic environments) react to form the same mineral assemblages.

This can only mean that the *P–T–t* paths of rocks undergoing metamorphism are only approximated by the geotherms shown in Figure 5.15. The actual *P–T–t* paths must be more complicated as shown in Figure B5.2.

Consider the case of rock subjected to metamorphism along a plate subduction boundary. The rate of burial, owing to compression, and therefore the rate of pressure increase can be so rapid that thermal equilibrium cannot be maintained. The maximum pressure (*P*) is reached at point B (Figure B5.2), before the maximum temperature (*T*). When the pressure is released a little, and uplift starts, radiogenic heating will still be in progress, and thus the highest temperature and the highest-grade mineral assemblages will be produced at point C. Uplift, like burial, tends to be rapid, so the *P–T–t* path for a body of rock subjected to retrograde metamorphism will follow the curve C → A.

By a further advance of modern scientific instrumentation, it is sometimes possible to obtain radiometric ages for the metamorphic mineral assemblages (chapter 6). Knowing the *P–T–t* path of metamorphism, and two or more time points on the path, we can calculate *P–T–t*

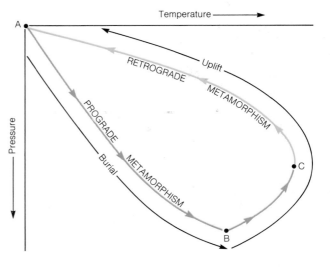

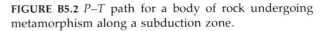

FIGURE B5.2 *P–T* path for a body of rock undergoing metamorphism along a subduction zone.

rates and thereby compare the subduction rate of former tectonic plates with present subduction rates. They turn out to be very similar. Bodies of metamorphic rock are turning out to be sensitive monitors of large-scale tectonic processes.

IMPORTANT WORDS AND TERMS TO REMEMBER

amphibolite (p. 113)
amphibolite facies (p. 114)

blueschist facies (p. 119)
burial metamorphism (p. 115)

cataclastic metamorphism (p. 114)
chemical crystallization (p. 114)
contact metamorphism (p. 114)

eclogite facies (p. 119)

foliation (p. 110)

gneiss (p. 112)

granulite (p. 113)
greenschist (p. 113)
greenschist facies (p. 119)

high-grade metamorphism (p. 108)
hornfels (p. 115)

intergranular fluid (p. 108)
intermediate grade metamorphism (p. 111)
isograd (p. 116)

lineation (p. 113)
low-grade metamorphism (p. 108)

marble (p. 113)
mechanical deformation (p. 114)
metamorphic aureole (p. 114)
metamorphic facies (p. 116)
metamorphic zones (p. 116)
metamorphism (p. 108)
mineral assemblage (p. 111)

phyllite (p. 112)
pore (p. 108)
prograde metamorphic effects (p. 109)

quartzite (p. 113)

regional metamorphism (p. 116)
retrograde metamorphic effects (p. 109)

schist (p. 112)
schistosity (p. 111)
slate (p. 112)
slaty cleavage (p. 110)
stress (p. 109)

thermal metamorphism (p. 114)

zeolites (p. 116)

QUESTIONS FOR REVIEW

1. What factors control metamorphism?
2. How and why does slaty cleavage form?
3. What is schistosity? How does it differ from lineations?
4. What is the difference between a schist and a gneiss?
5. How does a quartzite differ from a sandstone?
6. What is a metamorphic aureole?
7. What is regional metamorphism?
8. What is the geological setting of regional metamorphism? Can you name two places in the world where regional metamorphism is probably happening today?
9. What is burial metamorphism? Can you suggest some place on the Earth where it is probably happening today?

10. What controls the development of metamorphic zones?
11. What is the metamorphic facies concept and why does it help in the study of metamorphic rocks?
12. Under what conditions of pressure and temperature does blueschist facies metamorphism occur? What is the geologic environment where such temperatures and pressures are found?
13. Name three minerals that are found only in metamorphic rocks.
14. What are the two main volatiles that are added to, or lost from, rocks undergoing metamorphism?
15. What is cataclastic metamorphism? How would you distinguish it from contact metamorphism of an impure limestone?

C H A P T E R 6

Stratigraphy and the Geologic Column

The name of the Cretaceous Period is derived from widespread white chalk that crops out in sea cliffs along the English Channel, like these from along the English coast at Beachy Head.

STRATIGRAPHY

Scholars who study the history of ancient civilizations find the story of human events increasingly difficult to reconstruct the farther back in time they go. Geologists are historians too. They try to piece together the history of the Earth from geologic records of past events. Like the human record, the geologic record becomes less complete and more difficult to interpret with increasing age.

The historical information that geologists have to work with is largely in the form of stratified rocks that crop out at the Earth's surface or that can be penetrated by drilling. If we examine the rocks that are exposed in the upper walls of the Grand Canyon (Fig. 6.1) where the Colorado River has cut nearly 2 km into the Earth's crust, we can see many nearly horizontal layers. These strata formed one atop the other as sediment accumulated on the floor of a shallow sea. Such rocks contain important clues about past environments at and near the Earth's surface. If their sequence and age can be deter-

mined, they provide a basis for reconstructing much of Earth history.

The study of strata is called *stratigraphy.* Knowledge of stratigraphic principles and relationships, and of the relative ages of rock sequences, make it possible to work out many of the fundamental principles of physical geology.

Original Horizontality

Most sedimentary rocks are laid down beneath the sea, generally in relatively shallow waters. Under such conditions, each new sedimentary layer is laid down almost horizontally over older ones. This observation is consistent with the *law of original horizontality* which states that water-laid sediments are deposited in strata that are horizontal or nearly horizontal, and parallel or nearly parallel to the Earth's surface. From this generalization we can infer that rock layers now inclined, or even folded, must have been disturbed since the time when they were deposited in a horizontal position.

FIGURE 6.1 The Grand Canyon, where the Colorado River is joined by Shinumo Creek. Flat-lying strata are separated from underlying gently dipping Precambrian strata by an angular unconformity.

Stratigraphic Superposition

Toward the end of a cold winter it is often possible to see a layer of old snow that is compact and perhaps also dirty, overlain by fresh, looser clean snow deposited during the latest snowstorm. Here are two layers, or strata, that were deposited in sequence, one above the other. The very simple principle involved here also applies to layers of sediment and sedimentary rock. Known as *the principle of stratigraphic superposition,* it states that in any sequence of sedimentary strata, not later overturned, the order in which they were deposited is from bottom to top.

Relative Ages of Strata

The principle of stratigraphic superposition implies a scale of relative time, by which the relative ages of two strata can be fixed, according to whether one of the layers lies above or below the other. It does not allow us to determine the absolute age of any stratum in years, for the stratigraphic relationships only tell us the age of one relative to the other.

The importance of relative ages determined through stratigraphic superposition was first forcefully presented and widely introduced to science by William Smith, an English land surveyor, shortly before the beginning of the nineteenth century. His profession gave him an ideal opportunity to observe not only the landscape but the rocks that underlie it. While surveying for the construction of new canals in western England, he observed many sedimentary strata and soon realized that they lay, as he put it, "like slices of bread and butter" in a definite, unvarying sequence (Fig. 6.2). He became familiar with the physical characteristics of each layer and with the sequence of the layers. By looking at a specimen of sedimentary rock collected from anywhere in southern England, he could name the layer from which it had come and, of course, the position of the layer in the sequence.

In areas subject to tectonic deformation, overturned bedding may be present. In such cases, criteria must be sought to determine whether a succession of beds forms a normal sequence of deposits or is instead in reverse order. Three ways by which this can be done, using graded bedding, cross-bedding, or ripple marks, are shown in Figure 6.3.

Breaks in the Stratigraphic Record

The Earth's crust is always changing, always moving, and those changes dictate that processes of sedimentation and erosion shift both in their intensity and location through time. Rates of change can also differ greatly from place to place. For example, in the deep sea slow and continuous sedimentation may prevail for millions or tens of millions of years with little apparent change. On the continents, by contrast, sedimentation is disrupted periodically by environmental changes that lead to

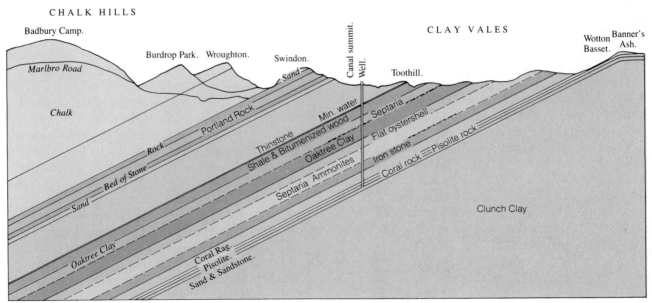

FIGURE 6.2 Stratigraphic section constructed by William Smith showing the succession of strata in north Wilshire, England.

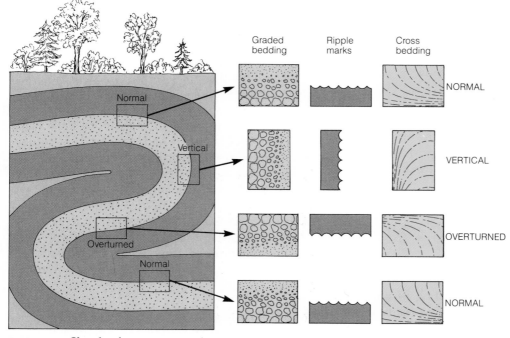

FIGURE 6.3 Sketch of an exposure showing in cross section a number of folded strata and some sedimentary features that are useful in determining whether the strata are normal (right side up), vertical, or overturned.

intervals of erosion or nondeposition. Such changes affect not only the emergent lands but submerged continental margins as well where most of the sediment from the continents accumulates. Because erosion destroys some of the stratigraphic record, the part that is preserved is often incomplete and marked by discontinuities where intervals of geologic time, some brief and others very long, are not represented by deposits.

Unconformity and Hiatus

An **unconformity** is a substantial break or gap in a stratigraphic sequence that marks the absence of part of the rock record. It records a change in either environmental conditions that caused deposition to cease for a considerable time, or erosion that resulted in loss of part of an earlier-formed depositional record, or a combination of both. An unconformity is a physical feature that can be identified in a rock sequence. It is usual to refer to the lapse in time recorded by an unconformity as a *hiatus* (Fig. 6.4).

Kinds of Unconformities

Unconformities can form as a result of local or regional uplift of land masses, fluctuations of sea level, and changes in climate that affect the behavior of streams, glaciers, and other depositional systems. The possible variations in crustal movement, erosion, and sedimentation are numerous, so there are several kinds of unconformities found in crustal rocks.

Three types of unconformity can be seen in the Grand Canyon (Fig. 6.5). The most obvious is the **angular unconformity**, which is marked by angular discordance between older and younger rocks (Figs. 6.1 and 6.5). It normally implies that older strata were deformed and then truncated by erosion before the younger layers were deposited across them. By mapping, geologists have recognized three different levels in the walls of the canyon where the sedimentary rocks are separated by a **disconformity**, which is an irregular surface of erosion between parallel strata. At the base of the sedimentary section is a **nonconformity**, where strata unconformably overlie igneous or metamorphic rocks (Fig. 6.5).

A study of unconformities brings out the close relationship between crustal movements, erosion, and sedimentation. All of the Earth's land surface is a potential surface of unconformity. Some of today's surface will be destroyed by erosion, but some will be covered by sediment and preserved as a record of the present landscape. Vigorous erosion

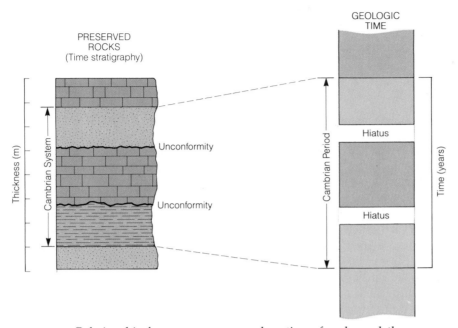

FIGURE 6.4 Relationship between a preserved section of rocks and the corresponding geologic time interval. Unconformities mark hiatuses, times for which no depositional record remains.

is now taking place where there has been recent uplift of the land. Erosion by streams is laying bare the records of Earth history in old rocks, and in doing so it is destroying some of those records. Meanwhile, the eroded material is being carried away and deposited elsewhere. Thus, in a sense, accumulation in one place compensates for destruction in another. The many surfaces of unconformity exposed in rocks of the Earth's crust are records of former land surfaces formed when a portion of the crust was uplifted by the Earth's internal forces and exposed to erosion. Preservation of such

a surface of erosion occurs following downwarping and deposition of sediment as a result of external processes. Unconformities testify that interactions between the internal and external processes have been going on throughout the Earth's long history.

STRATIGRAPHIC CLASSIFICATION

Every rock stratum can tell us something about the physical and biological character of a part of the Earth at some time in the geologic past. If someone

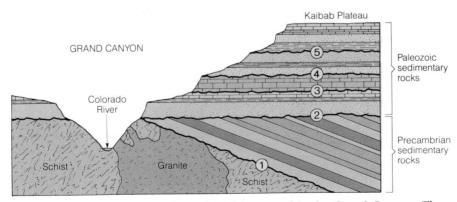

FIGURE 6.5 Geologic section through rocks exposed in the Grand Canyon. The lowest unconformity (1), separating tilted sedimentary strata from older crystalline rocks, is a nonconformity. An angular unconformity (2) separates the tilted strata from horizontally layered strata above, while three disconformities (3, 4, and 5) are seen still higher in the section.

were to start counting strata they would quickly realize that the rock record is like a vast library of knowledge consisting of thousands upon thousands of volumes.

Rock Stratigraphy

It is easy to identify the sedimentary rock directly above the major unconformity in the western part of the Grand Canyon as a sandstone (Fig. 6.1), but a thorough study must distinguish it from other sandstones.

Rock-Stratigraphic Units. One respect in which any stratum differs from all others is its position in the vertical sequence of strata. Hence, it is given a designation by which its position is fixed and by which it can be catalogued and referred to. Such a unit is called a ***rock-stratigraphic unit,*** defined as a body of rock having a high degree of overall lithologic homogeneity.

The basic unit used in designation of rock stratigraphy is the ***formation,*** which is a stratum, or collection of strata, distinctive enough on the basis of physical properties to constitute a distinctive, recognizable unit for geologic mapping. A formation must be thick and extensive enough to be shown on a geologic map, and it must be readily distinguishable from the strata immediately above and below. Each formation is given a name. In North America it typically is the name of a geographic locality near which the unit is best exposed (for example, Lexington Sandstone, Fox Hills Sandstone, Green River Formation). Igneous and metamorphic rocks are also designated by geographic names (for example, San Juan Tuff, Conway Schist).

Formations can be subdivided into successively smaller rock-stratigraphic units (Table 6.1). ***Members,*** which are subdivisions of formations, can be further subdivided into ***beds.*** Several formations can be assembled into larger units called ***groups.*** Such a classification scheme is similar to that used by a librarian who classifies books into a succession of categories that are progressively more specific.

Time Stratigraphy

It is difficult to comprehend the immensity of geologic time, measured in millions and billions of years, and yet it must be dealt with as geologists seek to unravel the story of the Earth from its preserved strata. An interval of geologic time can have meaning only in the context of rocks that were deposited during that interval. Because geologic time is an

TABLE 6.1 *Hierarchy of Rock-Stratigraphic Units*

Group
Formation
Member
Bed

abstract concept (whereas rocks are material objects that can be handled and studied), it is necessary to classify parcels of time separately from parcels of rock representing those times.

Time-Stratigraphic Units. A unit representing all the rocks that formed during a specific interval of geologic time is called a ***time-stratigraphic unit.*** Each of its boundaries, upper and lower, is everywhere the same age. A rock-stratigraphic unit, such as a formation, is defined only on the basis of its material characteristics, and its boundaries lie where a recognizable change in physical properties occurs. A time-stratigraphic unit, in contrast, may include more than one rock type and its boundaries may not necessarily coincide with a formational boundary.

Time-stratigraphic units are traditionally based on the fossil assemblages they contain and are ranked so as to represent progressively shorter time intervals. Rocks comprising a ***system,*** the primary unit, represent a time interval sufficiently great that such units can be used all over the world. Most systems encompass time intervals of tens of millions of years (Table 6.2). Names for systems arose from the early studies of rock strata, mainly in Europe, and often derive from geographic localities (see below). Larger groupings of two or more systems are called ***erathems.*** Rocks representing successively smaller intervals of time within a system are referred to as ***series*** and ***stages*** (Table 6.2); while such units have the potential for worldwide application, more commonly they are used within and between geographic provinces and continents.

TABLE 6.2 *Primary Time-Stratigraphic Units, Equivalent Geologic Time Units, and Their Average Length During the Last Half Billion Years of Earth History*

Time-Stratigraphic Units	Geologic Time Units	Average Length (million yr)
Erathem	Era	190
System	Period	52
Series	Epoch	18
Stage	Age	7

Geologic Time Units

Intervals of geologic time are based on the rock record as expressed by time-stratigraphic units. Units of geologic time are nonmaterial. Each merely corresponds to the time represented by a particular time-stratigraphic unit. Like such units, they are defined to include various intervals of time. For example, a geologic *period* embraces the time during which a geologic system accumulated, while an *epoch* equates with a series, and an *age* is equivalent to a stage (Table 6.2).

Because the geologic record is incomplete and punctuated by numerous unconformities, time-stratigraphic units in a certain area may not contain a complete depositional record for the corresponding geologic time interval (Fig. 6.4). However, many of the gaps can be bridged by piecing together sequences of strata from different geographic areas, thereby providing us with a more complete picture of Earth history.

Diachronous Boundaries

Depositional environments of sediments tend to change character laterally, as well as through time. A particular sedimentary facies, on the basis of which a formation is defined, may accumulate over different time intervals in different places. This means that the age of its upper and lower boundaries may differ from one place to another. In other words, the unit is said to possess *diachronous boundaries,* defined as boundaries that vary in age in different areas.

Time Lines. Figure 6.6 shows an example in which a delta is building progressively seaward. In the various depositional environments, marsh sediments are accumulating above a coarse, sandy facies that passes down the front slope of the delta into silty, clayey beds. These in turn grade laterally at the base of the delta front into thin, clayey facies offshore. The successive *time lines* (lines of constant age), which match the present surface, slope downward through the delta. Facies boundaries, on the other hand, cross time lines, reflecting the progressive displacements of the various depositional environments in the seaward direction. If the delta sediments are preserved in the geologic record and converted to sedimentary rock, the boundaries of the formations that might then be recognized on the basis of the several sedimentary facies would not be the same age everywhere. Instead, they would be progressively younger in the direction that the delta was built.

CORRELATION OF ROCK UNITS

William Smith's discovery that strata containing similar fossils are similar in age, no matter where they occur, was not considered by him to reflect any particular scientific principle; it was purely practical. Nevertheless, it opened the door to the correlation of sedimentary strata over increasingly wide areas. *Correlation* means the determination of equivalence, in geologic age and position, of the succession of strata found in two or more different areas. Smith correlated strata, on the dual basis of

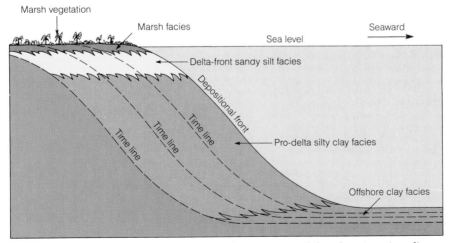

FIGURE 6.6 Diagrammatic section through a growing delta showing time lines and marking successive positions of the delta surface passing through different sedimentary facies. If eventually converted to rock, the resulting formations will be younger in the direction of delta growth.

physical similarity and fossil content, initially over distances of several kilometers, and later over tens of kilometers. By means of fossils alone it ultimately became possible to correlate through hundreds and then thousands of kilometers.

Correlation involves two main tasks. One is to determine the ages, relative to one another, of units exposed in local sections within an area being studied. Then the ages of the units, relative to a standard scale of geologic time, must be found. To accomplish these goals, a geologist employs various physical and biological criteria; one is not necessarily more dependable or precise than the others.

Physical Criteria

Where sedimentary rocks are exposed continuously within an area, individual strata may be traceable for considerable distances. Such correlation based on continuity of strata is generally straightforward and reliable. But eventually strata thin and die out, or merge with adjacent strata, thereby complicating and severely limiting this method of correlation. If, instead of tracing strata, formations are used, more reliable correlations can generally be made because by definition formations are larger units and mappable over broad regions. In doing so it is necessary to make a basic assumption, namely, that a formation is essentially the same age throughout. However, as stated above, formation boundaries may be diachronous. Correlation based on physical tracing of rock units must therefore be done cautiously, keeping in mind the potential pitfalls involved.

Continuous exposures are not common so one is faced with correlating between widely spaced outcrops. A stratum may be deformed and eroded so that only parts of it remain. The physical matching of remnants of a formation over a broad region generally involves the use of rock characteristics that permit the unit to be distinguished from others. Obvious criteria include gross lithology, mineral content, grain size, grain shape and orientation, sedimentary structures, color, and response of the rock or sediment to weathering. Correlation based on any of these is likely to be reliable over short distances, but it generally becomes less reliable through longer distances because the physical characteristics often change laterally (a change of facies).

Key Beds

A thin and generally widespread bed with characteristics so distinctive that it can be easily rec-ognized is called a *key bed.* Such beds can be useful in correlating major rock sections. A correlation may be greatly strengthened if several key beds can be identified and traced from one outcrop to the next. In areas of volcanic activity, ash layers can serve as distinctive key beds for purposes of regional correlation (Fig. 6.7).

Well Logs

Most sedimentary rocks of the continents are not exposed to view, so information about them can be obtained only by drilling. Energy companies attempting to assess potential petroleum resources beneath the land surface employ various techniques to determine the nature and distribution of subsurface rock units. Numerous test wells drilled in sedimentary basins give access to buried strata. Instruments lowered into the open wells provide continuous measurements of electrical properties of the subsurface units. The resulting signals can be compared from well to well and a correlation of strata made based on the similarity of the records (Fig. 6.8).

Biological Criteria

We call a fossil that can be used to identify and date the strata in which it is found, and is useful for local correlation of rock units, an *index fossil.* To be most useful, an index fossil should have common occurrence, a wide geographic distribution, and a very restricted range in age. The best examples are swimming or floating organisms which underwent rapid evolution and which quickly became widely distributed. If a distinctive index fossil is recognizable at an outcrop it can provide a rapid and reliable means of correlation (Fig. 6.9). While some genera and individual species permit long-range correlation of rocks in different sedimentary basins or even on different continents, more often close dating and correlation involves using assemblages of fossils of as many different types as possible.

GEOLOGIC COLUMN

Detailed stratigraphic studies throughout the world during the past 150 years have made it possible to assemble an increasingly long and complex *geologic column.* This is a composite columnar section combining in chronological order the succession of known strata, fitted together on the basis of their

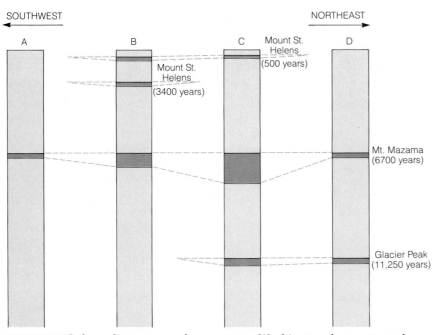

FIGURE 6.7 Lake sediment cores from western Washington demonstrate the use of distinctive ash layers from Mount St. Helens, Mount Mazama, and Glacier Peak volcanoes for regional correlation. Because the age of organic matter associated with the layers has been obtained by radiocarbon dating, the layers can be used to assess the ages of sediments associated with them.

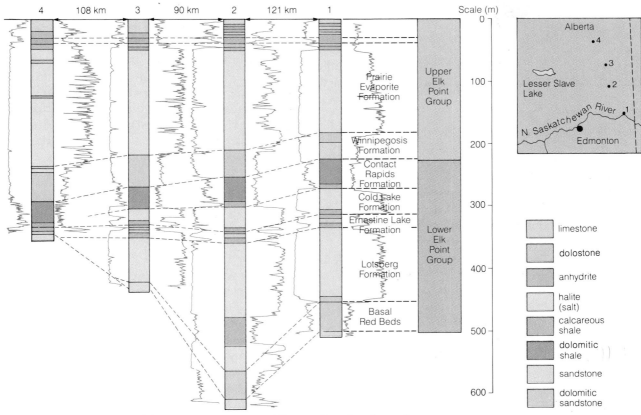

FIGURE 6.8 Correlation of formations in eastern Alberta, Canada, through use of electrical properties of sedimentary strata in boreholes. Curves to the left of holes record self-potential, whereas those to the right record electrical resistivity.

Locality 1   Locality 2   Locality 3

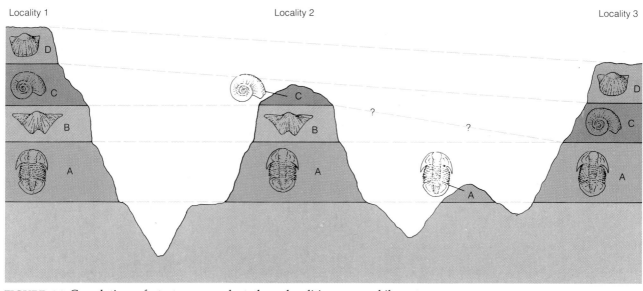

FIGURE 6.9 Correlation of strata exposed at three localities, many kilometers apart, on the basis of similarity of the fossils they contain. The fossils show that at Locality 3 stratum B is missing because C directly overlies A. Either B was never deposited there, or it was deposited and later removed by erosion before the deposition of layer C.

fossils or other evidence of relative or actual age. This worldwide standard is being constantly added to and refined as more rock units are described, mapped, and dated.

GEOLOGIC TIME SCALE

Standard names have evolved for the subdivision of geologic time (and also the rocks on which these time units are based). Those which can be used worldwide are eons, eras, periods, and epochs (Table 6.3).

The *geologic time scale* is a sequential arrangement of geologic time units, as currently understood, and it is divided into four major eons, principally on the basis of the observable presence or absence and type of life-forms that characterized each.

Eons. An *eon* is the largest time interval into which geologic time is divided. The term *Hadean* (Greek for beneath the earth) is given to the oldest eon. This is the earliest part of the Earth's history, an interval for which no rock record is known. However, rocks of this age are present on other planetary bodies of the solar system whose earliest crustal rocks have been little modified since they accumulated. *Archean* (Greek for ancient) rocks, the oldest we know of on the Earth, contain mi-

croscopic life-forms of bacterial character. *Proterozoic* (Greek for "earlier life") rocks include evidence of multicelled organisms that lacked preservable hard parts. Understandably, the record of the ancient Archean and Proterozoic is not as well known as the record of younger rocks because many of the ancient rocks have been intensely deformed, metamorphosed, and eroded. *Phanerozoic* (Greek for visible life) rocks often contain plentiful evidence of past life in the form of well-preserved hard parts. Most examples of fossils that we see displayed in museums or illustrated in books are of Phanerozoic age.

Eras. Geologic **eras** encompass major spans of time that also are defined on the basis of the life forms found in the corresponding rocks. No formal eras are yet widely recognized for Archean and Proterozoic rocks, but the Phanerozoic Eon is divided into the *Paleozoic* (Greek for old life), *Mesozoic* (Greek for middle life), and *Cenozoic* (Greek for recent life) eras, each name reflecting the relative stage of development of the life of these intervals (Table 6.3). Paleozoic forms of life progress from marine invertebrates to fishes, amphibians, and reptiles. Early land plants also appeared, expanded, and evolved. The Mesozoic saw the rise of the dinosaurs, which became the dominant vertebrates on land. Toward the end of that era, primitive mammals appeared and later dominated the Cenozoic Era. The Mesozoic also witnessed the evolution of flowering plants, while during the Cenozoic Era grasses appeared and became an important food for grazing mammals.

Periods. The geologic **periods** have a more haphazard nomenclature. They were defined over an interval of nearly 100 years on the basis of strata

TABLE 6.3 *Geologic Column, Time Scale, Major Worldwide Subdivisions, Ages of Boundaries, and Origin of Names*

Subdivisions Based on Strata/Time				Radiometric Dates[a] (millions of years ago)	Origin of Names of Periods of Paleozoic and Mesozoic and Epochs of Cenozoic
Eonothem/*Eon*	Erathem/*Era*	System/*Period*	Series/*Epoch*		
Phanerozoic	Cenozoic	Quaternary[b]	Holocene	0	Greek for wholly recent
				0.01	
			Pleistocene	1.6	Greek for most recent
		Tertiary[b]	Pliocene	5.3	Greek for more recent
			Miocene	23.7	Greek for less recent
			Oligocene	36.6	Greek for slightly recent
			Eocene	57.8	Greek for dawn of the recent
			Paleocene	66.4	Greek for early dawn of the recent
	Mesozoic	Cretaceous			Chalk (Latin = *creta*) in southern England and northern France
		Jurassic		144	Jura Mountains, Switzerland and France
		Triassic		208	Threefold division of rocks in Germany
				245	
	Paleozoic	Permian	Numerous units	286	Province of Perm, Russia
		Pennsylvanian[c]		320	State of Pennsylvania
		Mississippian[c]		360	Mississippi River
		Devonian		408	Devonshire, county of southwest England
		Silurian		438	Silures, ancient Celtic tribe of Wales
		Ordovician		505	Ordovices, ancient Celtic tribe of Wales
		Cambrian		570	Cambria, Roman name for Wales
Proterozoic[d]		No subdivisions in wide use			
Archean[d]				2500	
Hadean[e]				3800(?)	
				~ 4650	

Handwritten annotations: "Ice ages" and "6.1" near Pleistocene; "Q" near Quaternary; "T" near Tertiary; "Extinction" and "X → Dinos" near Paleocene/Cretaceous boundary; "K" Cretaceous, "J" Jurassic, "R" Triassic; "Fossil record disappear" and "carboniferous" brace near Pennsylvanian/Mississippian; "Pm", "Cp", "Cm", "D", "S", "O", "C" letter symbols; "pre-cambrian" near Proterozoic.

Source: Based largely on data from Palmer, 1983.
[a]Time divisions are not drawn to uniform scale.
[b]Derived from eighteenth- and nineteenth-century geologic time scale that separated crustal rocks into a fourfold division of Primary, Secondary, Tertiary, and Quaternary, based largely on relative degree of induration and deformation.
[c]Mississippian and Pennsylvanian are equivalent to Lower and Upper Carboniferous Period of Europe (named for abundance of coal in these rocks).
[d]Proterozoic plus Archean are equivalent to Precambrian.
[e]No rocks of this eon are known on Earth, but they exist on other planetary bodies in the solar system.

that crop out in Britain, Germany, Switzerland, Russia, and the United States. The names of the periods and corresponding systems within the Paleozoic and Mesozoic eras are partly geographic in origin, but in some cases they are based on characteristics of the strata in the place of original study (Table 6.3). Well-exposed rocks of these ages are found on nearly all the continents.

Epochs. The **epochs** of the Tertiary Period were defined in a piecemeal fashion. Studies of marine strata in sedimentary basins of France and Italy led English geologist Charles Lyell to subdivide the rocks into groupings based on the percentage of their fossils that are represented by still-living species (Table 6.3). Each of the various periods of the Paleozoic and Mesozoic eras are also subdivided into epochs, the names of which are primarily geographic in origin. They are used mainly by specialists concerned with detailed studies of these strata and their contained fossils.

The names of the geologic time scale constitute the standard time language of geologists the world over. Through their use one can begin to comprehend numerous details of Earth history that have led to the discovery of many of the important principles of physical geology which are discussed in this book.

MEASURING GEOLOGIC TIME

The scientists who worked out the geologic column were challenged by the question of time. They knew the order in which the different systems had formed but they also wished to know whether the sediments in each system had accumulated during the same length of time. They sought answers to questions such as these: "How much time elapsed between the end of the Cambrian and the beginning of the Permian?" "How long was the Tertiary?" The question of time is as important as the geologic column itself. An answer is important to questions such as the age of the Earth, the time during which the rock cycle has been operating, the age of the ocean, how fast mountain ranges rise, and how long humans have inhabited the Earth.

During the nineteenth century many attempts were made to subdivide the geologic column in a scale of years by indirect methods. One widely used method consisted of estimates of the time during which the rock cycle has been at work. If it is assumed that sediment has always been deposited at rates equal to today's rate, it is possible, at least

in theory, to say how much time was needed to accumulate all the sediment now preserved in sedimentary strata. Several difficulties exist. First, rates of deposition are not everywhere constant. Second, in every pile of sediment gaps occur representing intervals where deposition ceased (hiatuses) and there is no way to estimate the duration of these gaps. Third, in rock older than Cambrian very few fossils exist which can help put strata in their proper sequence. Because of such difficulties, early estimates of the duration of the rock cycle were unreliable and tended to be much too short. They were also highly variable. Estimates of the age of the Earth, based on the total thickness of sedimentary layers, ranged from 3 million years to 1.5 billion years!

What was needed in order to resolve the dilemma was a way to measure geologic time by some process that runs continuously, that is not reversible, that is not influenced by other processes and other cycles, and that leaves a continuous record without gaps in it. At the end of the nineteenth century (1896) the discovery of radioactivity provided the needed method. By 1904, the American scientist Benjamin Boltwood had demonstrated that the radioactivity of uranium could be used to determine the time of formation of uranium minerals. That discovery opened the door to radiometric dating, a new and reliable means of measuring geologic time.

Radiometric Dating

Natural Radioactivity

Most of the isotopes of the chemical elements found in the Earth are stable and not subject to change. However, a few are radioactive because of an instability in the nucleus. The nucleus of a radioactive isotope is said to transform spontaneously to a nucleus of either a more stable isotope of the same chemical element or to an isotope of a different chemical element. The rate of transformation is different for each isotope. Even though the process is one of transformation—from an unstable nucleus to a more stable one—it has become common practice to call the process radioactive decay. An atomic nucleus undergoing radioactive decay is said to be a **parent**; the product arising from radioactive decay is called a **daughter product.**

Many of the radioactive isotopes that were once in the Earth have decayed away and are no longer present. This is so because their rates of spontaneous decay were fast. A few radioactive isotopes that transform very slowly are still present, however, and it is the slow steady decay of these remaining isotopes that can be used for radiometric dating. It is the same slow radioactive decay process that creates the heat which keeps the Earth's interior hot.

Rates of Decay

Careful study of radioactive elements in the laboratory has shown that decay rates are unaffected by changes in the chemical and physical environment. Radioactive decay is a property of the atomic nucleus, whereas the chemical and physical environments affect the orbiting electrons. This is a particularly important point because it leads to the conclusion that rates of radioactive decay are not influenced by geologic processes.

Each radioactive element decays according to a distinct and measurable timetable, but all decay timetables follow the same basic law. Consider a radioactive isotope that transforms to a daughter product. The number of decaying parent atoms continually decreases while the number of daughter atoms continually increases. The proportion, or percentage, of atoms that decay during one unit of time is constant. However, although the proportion is constant, the actual number keeps decreasing because the parent atoms are being used up.

In a mineral sample that contains atoms of a radioactive isotope, the overall radioactivity (the sum of the radioactivity of all the parent atoms remaining in the sample) decreases continually (Fig. 6.10). Because the rate of decrease is measured as a percentage of the number of atoms left undecayed, it is usual to designate a decay rate in terms of the **half-life** of the parent, meaning the time required to reduce the number of parent atoms by one half. The time units marked in Figure 6.10 are half-lives. Of course they are of equal length, just as years are. But at the end of each one, the number of atoms that decay (and, therefore, the combined radioactivity of the sample) has decreased by exactly half.

While the proportion of parent atoms declines, the proportion of daughter atoms increases. Figure 6.10 shows that the growth of daughter atoms just matches the decline of parent atoms. When the number of remaining parent atoms is added to the number of daughter atoms, the sum is the number of parent atoms that a mineral sample started with. That fact is the key to the use of radioactivity as a means of measuring time and determining ages.

Potassium–Argon ($^{40}K/^{40}Ar$) Dating

We have selected one of the naturally radioactive isotopes, potassium-40 (^{40}K), to illustrate how minerals can be dated. Potassium has three natural isotopes: ^{39}K, ^{40}K, and ^{41}K. Only one, ^{40}K, is radioactive and its half-life is 1.3 billion years. Twelve

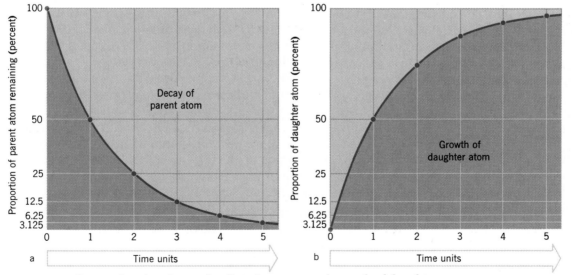

FIGURE 6.10 Curves showing decay of radioactive atoms and growth of daughter products. (a) At time 0, a sample consists of 100 percent radioactive parent atoms. During each time unit, half the atoms remaining decay to daughter atoms. (b) At time 0, no daughter atoms are present. After one time unit corresponding to a half-life of the parent atoms, 50 percent of the sample has been converted to daughter atoms. After two time units, 75 percent of the sample is daughter atoms, 25 percent parent atoms. After three time units, the percentages are 87.5 and 12.5, respectively.

percent of the ^{40}K atoms decay to ^{40}Ar, an isotope of the gas argon. The remaining 88 percent of the ^{40}K atoms decay to produce ^{40}Ca. Careful measurements have shown that the ratio of ^{40}Ar to ^{40}Ca daughter atoms is always the same. When a potassium-bearing mineral crystallizes from a magma, or grows within a metamorphic rock, it traps a sample of ^{40}K in its crystal structure. ^{40}Ca and ^{40}Ar can also be trapped by atomic substitution for other atoms, or in submicroscopic fractures.

Atoms of calcium are tightly bound in the crystal because they form chemical bonds with other atoms present. However, argon is an element that has unusual atomic properties. Because its orbiting electron shells are filled, atoms of argon do not readily form chemical bonds. Thus, the daughter atoms are not chemically bound in minerals. This in turn means that at high temperature, argon, unlike calcium, rapidly diffuses out of a mineral and does not stay trapped. When the argon content of a mineral is measured, therefore, what is determined is the ^{40}Ar accumulated during the time since a mineral started trapping and retaining argon. Although ^{40}Ar is present in a magma when a mineral crystallizes, a mineral rarely retains any initial argon because magmatic temperatures are above trapping temperatures. All the ^{40}Ar atoms in a potassium-bearing mineral, therefore, must come from decay of ^{40}K and must have accumulated since the temperatures fell below the trapping temperature. All that now has to be done is to measure the amount of parent ^{40}K that remains and the amount of ^{40}Ar that has been trapped. The half-life of ^{40}K being known, it is a straightforward matter to calculate the radiometric age—the length of time a mineral has contained its built-in radioactivity clock.

Dating by ^{40}K is not limited to minerals such as muscovite that contain potassium as a major element. Even minerals that contain small amounts of potassium substituting for other elements by atomic substitution will serve the purpose. Thus, hornblende, a calcium–iron–magnesium silicate, can be used because it generally contains a small quantity of potassium. Some rocks contain several different minerals that can be used for dating, and then it is possible to use the "whole rock" for dating.

K/Ar dating is most successfully applied to volcanic and pyroclastic rocks because their crystallization and cooling is rapid. As a result, they have formation ages that are essentially coincident with their trapping ages. Because argon analyses can be performed with great accuracy, and because contamination by initial argon at the time of crystal-lization is generally not a problem, the method can be used, under ideal circumstances, for volcanic rocks as young as 50,000 years. For this reason, K/Ar dating has proved very useful in studies of archaeology as well as geology.

Other Radiometric Dating Methods

Many naturally radioactive isotopes can be used for radiometric dating, but six isotopes predominate in geologic studies. These are the radioactive isotopes of uranium, thorium, potassium, rubidium, and carbon. They occur widely in different minerals and rock types, and they have a very wide range of half-lives so that many geologic materials can be dated radiometrically (Table 6.4).

Radiocarbon (^{14}C) Dating. Among the radiometric dating methods listed in Table 6.4, the one based on ^{14}C (also known as radiocarbon) is unique for two reasons. The first is that the half-life of ^{14}C is short by comparison with the half-lives of ^{40}K, ^{87}Rb, and the isotopes of uranium. The second reason is that the amount of daughter product cannot be measured.

Radiocarbon is continuously created in the atmosphere through bombardment of nitrogen-14 (^{14}N) by neutrons created by cosmic radiation. ^{14}C, with a half-life of 5730 years, decays back to ^{14}N. The ^{14}C mixes with ordinary carbon ^{12}C and ^{13}C and diffuses rapidly through the atmosphere, hydrosphere, and biosphere. Because the rates of mixing and exchange are rapid compared with the half-life, the proportion of ^{14}C is nearly constant throughout the atmosphere. As long as the production rate remains constant, the radioactivity of natural carbon remains constant because rate of production balances the rate of decay.

While an organism is alive and is taking in carbon from the atmosphere, it contains this balanced proportion of ^{14}C. However, at death the balance is upset, because replenishment by life processes such as feeding, breathing, and photosynthesis ceases. The ^{14}C in dead tissues continually decreases by radioactive decay. The analysis for the radiocarbon date of a sample involves only a determination of the radioactivity level of the ^{14}C it contains. This is done by measuring the radiation emitted as a by-product of radioactive decay. The daughter product, ^{14}N, cannot be measured because it leaks away and because of atmospheric contamination. In order to measure an age with ^{14}C, it is necessary to make two assumptions: (1) that the rate of production of ^{14}C has been constant throughout the

TABLE 6.4 *Some of the Principal Isotopes Used in Radiometric Dating*

Isotopes		Half-Life of Parent (years)	Effective Dating Range (years)	Minerals and Other Materials That Can Be Dated
Parent	Daughter			
Uranium-238	Lead-206	4.5 billion	10 million–4.6 billion	Zircon
				Uraninite and pitchblende
Uranium-235	Lead-207	710 million		
Thorium-232	Lead-208	14 billion		
Potassium-40	Argon-40	1.3 billion	50,000–4.6 billion	Muscovite
	Calcium-40			Biotite
				Hornblende
				Whole volcanic rock
Rubidium-87	Strontium-87	47 billion	10 million–4.6 billion	Muscovite
				Biotite
				Potassium-feldspar
				Whole metamorphic or igneous rock
Carbon-14	Nitrogen-14	5,730 ± 30	100–70,000	Wood, charcoal, peat, grain, and other plant material
				Bone, tissue, and other animal material
				Cloth
				Shell
				Stalactites
				Groundwater
				Ocean water
				Glacier ice

last 70,000 years or so (the range of time to which the short half-life of ^{14}C limits the usefulness of the method); and (2) that all samples start with the same ^{14}C/^{12}C ratio.

The ^{14}C method is based partly on the assumption that ^{14}C in the atmosphere has been constant. This is not strictly true, as has been shown by careful radiocarbon dating of tree rings of known age spanning the last 8000 years. The result has been the development of calibration curves that permit radiocarbon dates to be converted to calendar ages.

Because of its application to organisms (by dating fossil wood, charcoal, peat, bone, and shell material) and its short half-life, radiocarbon has proved to be enormously valuable in establishing dates for prehistoric human remains and for recently extinct animals. In this way it is of extreme importance in archaeology. It is also of great value in dating the most recent part of geologic history, particularly the latest of the glacial ages. For example, the dates of many samples of wood taken from trees overrun by the advance of the latest of the great ice sheets and buried in the rock debris thus deposited, show that the ice reached its greatest extent in the Ohio–Indiana–Illinois region about 18,000 to 21,000 years ago. It is even possible to date ice such as that in the Greenland ice sheet directly, for as the ice forms, bubbles of air are trapped in it. The car-

bon dioxide in the air bubbles can be liberated in the laboratory and dated, providing an age for the time of ice formation.

Similarly, radiocarbon dates afford the means for determining rates of geologic processes, such as: the rate of advance of the last ice sheet across Ohio; the rates of rise of the sea against the land while glaciers melted throughout the world; average rates of circulation of water in the deep ocean; the rates of local uplift of the crust that raised beaches above sea level; and even the frequency of volcanism.

Time and the Geologic Column

Through the various methods of radiometric dating, the dates of solidification of many bodies of igneous rock have been determined. Many such bodies have identifiable positions in the geologic column, and because of this it becomes possible to date, approximately, a number of the sedimentary layers in the column.

The standard units of the geologic column consist of sedimentary strata containing characteristic fossils, but the typical rocks from which radiometric dates (other than radiocarbon dates) are determined are igneous rocks. It is necessary, therefore, to be sure of the relative time relations between an igneous body that is datable and a sedimentary

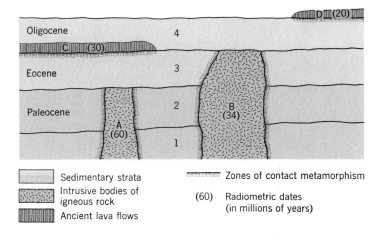

Sedimentary strata
Intrusive bodies of igneous rock
Ancient lava flows

Zones of contact metamorphism
(60)   Radiometric dates (in millions of years)

FIGURE 6.11 Idealized section illustrating the application of radiometric dating to the geologic column. For method, see the text discussion.

layer whose fossils closely indicate its position in the column.

Figure 6.11 shows in an idealized manner how apparent ages of sedimentary strata are approximated from the apparent ages of igneous bodies. The age of a stratum is bracketed between bodies of igneous rock, the apparent ages of which are known. In Figure 6.11 four sedimentary strata, whose geologic ages are known from their fossils, are separated by surfaces of erosion. Related to the strata are two intrusive bodies of igneous rock (A, B) and two sheets of extrusive igneous rock (C, D). From the apparent dates of the igneous bodies and the geologic relations shown, we can draw the following inferences about the ages of the sedimentary strata.

Stratum	Age (millions of years)	Interpretation
4	<34<30>20	Age lies between 20 and 30 million years
3	<60>34>30	Age lies between 34 and 60 million years
2	>60>34	Age of both is more
1	>60>34	than 60 million years

To separate 1 from 2, dates from other localities are needed. Dates from igneous rocks elsewhere could also narrow the possible ages of 3 and 4. Through this combination of geologic relations and radiometric dating methods (Fig. 6.12) it has been possible to fit a scale of time to the geologic column. The scale is being continually refined.

It is a great tribute to the work of geologists during the first half of the nineteenth century that the geologic column they established by the ordering of strata into relative ages has been fully confirmed by radiometric dating. Comparisons between the numbers column and the names column in Table 6.3 show this.

Table 6.3 demonstrates the match between the grouping of strata, divided into successively smaller subdivisions called systems, series, and stages, and the corresponding time units called periods, epochs, and ages. It is possible to speak of the time units, of course, whether or not the actual dates are known. It would also be correct to speak of events that occurred in the Devonian Period (or simply in Devonian time) based on the fossil record, even if it were not known that the dates of that period fall between 415 and 360 million years ago.

AGE OF PLANET EARTH

Table 6.3 indicates that the earliest record comes from the great assemblage of metamorphic and igneous rocks formed during the Archean and Proterozoic Eons, collectively known as Precambrian rocks. Of the many radiometric dates obtained from them, the youngest are around 600 million years, the oldest about 3.8 billion years. The Precambrian unit of the geologic column, then, existed during a *minimum* time equal to 3.8 billion minus 600 million years, or 3.2 billion years—a span about five times as long as the time elapsed since the Precambrian unit ended.

Given that some Precambrian rocks are nearly 4 billion years old, the beginning of the Earth's history must be still farther back in time. The oldest radiometric dates have been obtained on individual mineral grains from clastic sedimentary rocks in Australia. Dates that are almost as old—3.8 billion

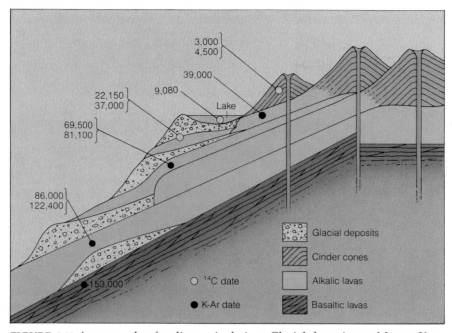

FIGURE 6.12 An example of radiometric dating. Glacial deposits on Mauna Kea Volcano, Hawaii, are interlayered with lava flows and cinder cones that have been dated by the K–Ar method. The youngest glacial deposit is bracketed by radiocarbon dates; it is younger than organic matter associated with cinders that are 22,150 years old, but older than sediments at the bottom of a lake that are 9080 years old.

years—have been obtained for a boulder of granite from a layer of conglomerate in Greenland. The existence of granite proves both that continental crust was present and that the rock cycle was operating 3.8 billion years ago. Further confirmation comes from another of the very ancient Precambrian rocks, a body of granite in South Africa. Although itself an igneous rock, this ancient granite contains xenoliths of quartzite. At an earlier time, before it became enveloped by the granite magma, the quartzite must have been part of a layer of sandstone. Before that, it must have been part of a layer of loose sand. And even earlier still, an igneous rock must have been subjected to weathering and erosion to produce the grains of sand. Clearly, therefore, the rock cycle must have been operating in its present manner well before the granite magma solidified. Hence, as far back as we can see through the geologic column, we find evidence of the rock cycle and, because we see ancient sediment that must have been transported by water, we know that when that sediment was deposited there must have been a hydrosphere.

We have been speaking of the oldest rock so far discovered. How much older might our planet be?

Strong evidence suggests that the Earth formed at the same time as the Moon, the other planets, and meteorites (small independent bodies that have "fallen" onto the Earth). Through various methods of radiometric dating and, in particular, the Rb/Sr and U/Pb systems, it has been possible to determine the ages of meteorites and of "Moon dust" (brought back by astronauts) as 4.6 billion years. By inference, the time of formation of the Earth, and indeed of all the other planets and meteorites in the solar system, is believed to be 4.6 billion years ago. Lead isotopes have been used to check this conclusion in the following manner.

Iron meteorites that are devoid of stony matter contain trace amounts of lead but are free of uranium. This is so because uranium does not replace metallic iron by atomic substitution, but lead does. When iron meteorites were formed, therefore, they incorporated a little lead but no uranium. Accordingly, the amount of ^{206}Pb and ^{207}Pb in iron meteorites must have remained unchanged from the time the meteorites formed. The only way a change could occur would be for ^{238}U and ^{235}U (the parents of ^{206}Pb and ^{207}Pb, respectively) to be present in the iron meteorite. The $^{206}Pb/^{207}Pb$ ratio in an iron meteorite is 0.903. Therefore, this must have been the $^{206}Pb/^{207}Pb$ ratio of the cosmic dust cloud from which meteorites and the planets formed.

Stony meteorites and stony planets do contain uranium. The earth is a stony planet, and its $^{206}Pb/^{207}Pb$ ratio is no longer 0.903. By measuring the ratio of $^{206}Pb/^{207}Pb$ in large samples of deep-sea clays, in seawater, and other well-mixed terrestrial materials, it can be estimated that today the $^{206}Pb/^{207}Pb$ ratio for the whole Earth is 1.186. The ratio is slowly changing because the half-lives of ^{238}U and ^{235}U are, respectively, 4.5 billion and 710 million years. ^{235}U is disappearing faster than ^{238}U.

This means that the further back we go in the Earth's history, the smaller must have been the ratio of $^{238}U/^{235}U$. It follows then, that the further we go back the smaller must have been the ratio of the daughter products, $^{206}Pb/^{207}Pb$. It is possible to calculate that 2 billion years ago the ratio was 1, and that at 4.6 billion years ago it was 0.903, identical with the ratio in iron meteorites. While this is not absolute proof that the Earth formed 4.6 billion years ago, it is very strong evidence in favor of the idea.

THE MAGNETIC POLARITY TIME SCALE

The Earth is like a giant magnet. It has an invisible magnetic field that permeates everything. The exact cause of the Earth's magnetism is not understood, but it is known that whatever the cause is, it is located in the core. The most likely explanation is that the magnetism is controlled by fluid motions in the molten iron of the outer core, and produced by electric currents flowing there. When an electric current flows, it does so by movement of electrons or other charged particles such as protons and ions. If an electric current flows in a wire, an invisible magnetic field, surrounds the wire. Similarly, if a wire moves through a magnetic field, an electric current flows in the wire. Electricity and magnetism are two aspects of the same phenomenon.

Thermoremanent Magnetism

Magnetite and certain other iron-bearing minerals can become permanently magnetized. This property arises because orbital electrons spinning around a nucleus are equivalent to an electric current and create a tiny atomic magnetic field (Fig. 6.13). In minerals that can become permanent magnets, the atomic magnets line up in parallel arrays and rein-

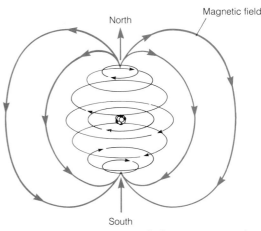

FIGURE 6.13 Movement of electrons around an atomic nucleus has the same effect as movement of electrons in a wire—they create a magnetic field.

force each other. In nonmagnetic minerals, the atomic magnets are oriented in random directions.

Curie Point. Above a temperature called the **Curie point,** all permanent magnetism is destroyed because the thermal agitation of atoms is such that permanent magnetism is impossible. The Curie point for magnetite is about 500°C. Above that temperature, the magnetic fields of all the atoms are randomly oriented and cancel each other out. Below the Curie point, the magnetic fields of adjacent atoms influence each other. Within small regions, or domains, of the solid, the magnetic fields reinforce each other (Fig. 6.14). When an external magnetic field is present, all magnetic domains that are parallel to the magnetic field become larger and expand at the expense of adjacent, nonparallel domains. Quickly, the parallel domains become predominant and a permanent magnet is the result.

Consider what happens when lava cools. All the minerals crystallize at temperatures above about 700°C—well above the Curie points of any magnetic minerals present. As the crystallized lava continues to cool, the temperature will drop below 500°C, the Curie point for magnetite. When it does so, all the magnetite grains in the rock become tiny permanent magnets owing to the Earth's magnetic field. The magnetic poles of all the magnetite grains will have the same polarity as the Earth's field. So long as that lava lasts (until it is destroyed by weathering or metamorphism), it will carry a record of the Earth's magnetic field at the moment it passed through the Curie point. Permanent magnetism that is a result of thermal cooling is called

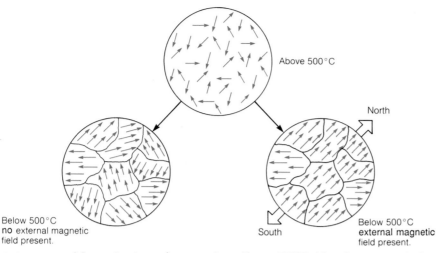

FIGURE 6.14 Magnetization of magnetite. Above 500°C (the Curie point) the thermal motion of atoms is so great that the magnetic poles of individual atoms, shown as arrows, point in random directions. Below 500°C atoms in small domains influence one another and form tiny magnets. In the absence of an external field, the domains are randomly oriented. In the presence of a magnetic field (lower right), most pole directions tend to be parallel to that of the external field and magnetite becomes permanently magnetized. The example in the lower right is that for a magnetite grain in a crystallizing lava.

thermoremanent magnetism. It is the most important kind of *paleomagnetism.*

The Polarity-Reversal Time Scale

From a study of thermoremanent magnetism in lavas it was discovered early in the twentieth century that some rocks contain a record of reversed polarity. That is, when their paleomagnetism was measured, some lavas indicated a south magnetic pole where the north magnetic pole is today, and vice versa (Fig. 6.15). The ages of lavas can be accurately determined using radiometric dating techniques, especially the $^{40}K/^{40}Ar$ method. Through combined radiometric dating and magnetic polarity measurements in thick piles of lava extruded over several million years, it has been possible to determine when magnetic polarity reversals occurred (Fig. 6.16).

Careful worldwide studies of polarity reversals through measurements of thermoremanent magnetism in lavas have established two important facts. The first fact concerns the speed with which a reversal occurs. During a 1000- to 5000-year interval, the magnetic field slowly dies down to a very low intensity, the poles move erratically and fluctuate widely around the globe, then the magnetic field rapidly builds up again with the poles reversed. The second fact proves that polarity reversals are global effects, not local phenomena. A record of a reversal occurs on all continents and in all ocean basins at the same time. The most important fact concerns the actual record of polarity reversals. A detailed record of all changes back to the Jurassic

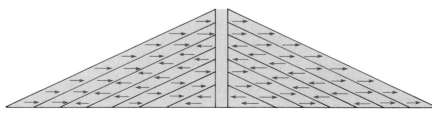

FIGURE 6.15 Lavas retain a record of the polarity of the Earth's magnetic field at the instant they cool through the Curie point. A pile of lava flows, like those in the volcanoes of the Hawaiian Islands, may record several field reversals, each of which can be dated using potassium–argon dating principles.

Period has now been assembled, and still-earlier reversals are the topic of ongoing research.

Magnetic Chrons and Subchrons.   The polarity record for the past 20 million years is shown in Figure 6.16. Periods of predominantly normal polarity (as at present), or predominantly reversed polarity, are called *magnetic chrons.* The four most recent chrons have been named for scientists who made great contributions to studies of magnetism: Brunhes, Matuyama, Gauss, and Gilbert, respectively. It is apparent from Figure 6.16 that many of the magnetic reversals are short-lived and that the reversal pattern does not have any obvious regularity. It is also apparent that short-term reversals sometimes occur during a chron. Such short-term magnetic reversals are called *magnetic subchrons.*

Use of magnetic reversals for geological dating differs from other dating methods. Magnetic reversal records all look the same in the rock record.

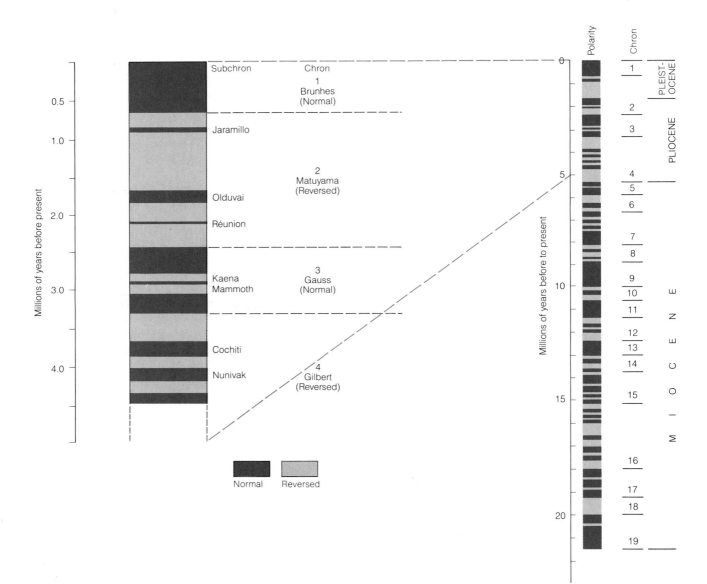

FIGURE 6.16 Polarity reversals back to 20 million years. Periods of normal polarity, as today, and periods of primarily reversed polarity are called chrons. Nineteen chrons have been identified to the beginning of the Miocene. Within each chron, one or more subchrons may occur. During a normal chron, subchrons are times of reversed polarity. The youngest subchrons have been given names based on the location of samples in which they were discovered. Polarity reversals are dated back to the mid-Jurassic, approximately 162 million years ago.

When evidence of a magnetic reversal is found in a sequence of rocks, the problem is to know which of the many reversals it actually is. Additional information is needed. When a continuous record of reversals can be found, starting with the present, it is simply a matter of counting backward. This is the technique used in the dating of oceanic crust (chapter 16).

Depositional Remanent Magnetism

Sedimentary rocks acquire weak but permanent magnetism through the orientation of magnetic grains during or after sedimentation. As clastic sedimentary grains settle through ocean or lake water, or even as dust particles settle through the air, any magnetite particles present will tend to orient themselves parallel to the magnetic lines of force. Remanent magnetism acquired through processes of sedimentation is called *depositional remanent magnetism.*

Depositional remanent magnetism has proved to be a very sensitive and important dating technique. When fossils are present, an approximate age can be given to sedimentary rocks. Knowing the approximate age of a sediment, geologists can determine the exact age from the magnetic reversals. Sediment cores recovered from the seafloor can be dated very accurately using a combination of fossils and magnetic reversals.

With so many dating methods now available, geologists are starting to obtain quantitative answers to many questions which only a few years ago could only be approached in a descriptive way (Fig. 6.17). Magnetic reversals can even provide a way to measure rates of sedimentation in the world ocean.

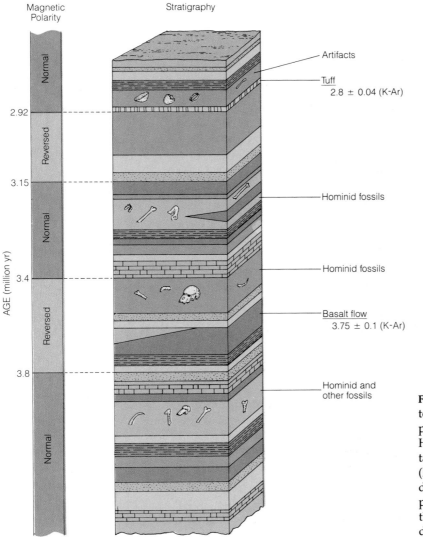

FIGURE 6.17 Example of several dating techniques brought to bear on a geological problem. Dates for sediments from the Haddar area of northern Ethiopia containing fossils of ancestral human beings (hominids) have been obtained by K–Ar dating of basalt and tuff, as well as by paleomagnetism. The dates indicate that the early hominids lived in this region during the Pliocene Epoch (Table 6.3).

Essay

UNCONFORMITIES, TECTONICS, AND SEA-LEVEL
FLUCTUATIONS

In the early 1970s, scientists involved in the
search for new supplies of oil and gas began
accumulating consistent but puzzling evidence that suggested some unconformities were
more widespread than they had suspected.
Some seemed to be worldwide. The evidence
came from seismic stratigraphy which involves the use of high-resolution seismic exploration techniques. The data obtained are
like those in Figure B6.1. In such a profile, the
lines depict the arrangement of subsurface rock
units that act as reflectors of seismic waves
generated either by explosions, or by the
pounding of air hammers. Seismic reflection
profiles, with their patterns of smooth and
broken lines, make it possible to interpret the
regional interrelations of strata, and their
structure, thickness, and probable depositional environment, as well as the presence,
character, and distribution of unconformities.
In this manner, large sequences of strata have
been identified that are separated by major
unconformities. Some unconformities can be
traced widely across continents and into strata
on submerged continental margins. Such
widespread unconformities either record major crustal uplifts or are related to changes of
sea level. They seem to reflect events in Earth
history that have worldwide significance. The
evidence is still being evaluated, but it favors
widespread changes in sea level rather than
simultaneous crustal uplifts on several continents.

From studies of unconformities and marine
strata a curve of sea-level fluctuations during
the Phanerozoic Eon has been developed (Fig.
B6.2). The evidence suggests that over much
of the last 600 million years major portions of
the continents were submerged by shallow seas
in which marine sediments accumulated.

Plate tectonics provides a possible explanation for changes in the shapes and volumes
of the ocean basins through time and for the
unconformities that record the changes. Plate

FIGURE B6.1 Colors in this vertical seismic reflection profile across southern Texas represent
different velocities of sound traveling through
the section. An ancient carbonate reef (yellow), at
a depth of about 6 km, dips gently toward the
Gulf of Mexico.

velocities have varied considerably in the past. Times of high plate velocities are times when new oceanic crust is being created at a rapid rate and hence times when volcanism along the midocean ridge was very active. At times of very active volcanism, the midocean ridge system was large because it was thermally expanded. The size of the ocean basins would have been correspondingly reduced. That would force ocean water to spill out of the ocean basins and rise against the adjoining lands. The resulting sea-level fluctuation could have been as much as 500 m.

The volume of the ocean basin could also change through crustal thickening along zones where continents collide. Such thickening would effectively reduce the area of continental crust and cause a corresponding increase in the area of oceanic crust. The result would be a world fall of sea level.

Worldwide unconformities have still not been explained. At present all we know is that they occur and that more than one explanation is possible. The times of greatest marine invasion took place during the early Paleozoic Era and during the Cretaceous Period. These intervals are represented by extensive marine sedimentary rocks on nearly all the continents. The stratigraphic record indicates that erosion of the lands and the continental margins occurred during times when the sea contracted and retreated from the land. During periods of marine regression, as such times are known, the resulting sediment is carried to the deep sea. Erosion surfaces produced during regression were subsequently buried by sediment during marine transgressions, at which time the sea expanded and once again covered the land. Some of the regressions coincide with boundaries between periods and between epochs, suggesting that global sea-level changes may account for some of the primary stratigraphic boundaries.

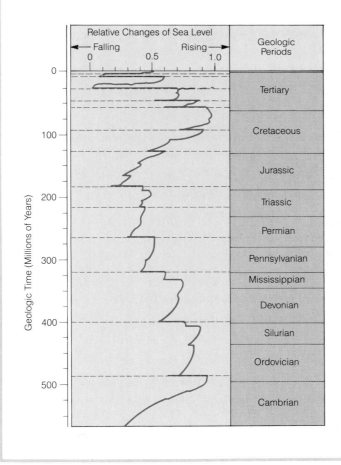

FIGURE B6.2 Reconstructed record of sea-level fluctuations during the last 600 million years.

SUMMARY

1. Strata provide a basis for reconstructing Earth history and past surface environments. Most strata were horizontal when deposited and accumulated in sequence from bottom to top.

2. An extensive unit of strata may possess several facies, each determined by a different depositional environment. The boundaries between facies may be abrupt or gradational.

3. Unconformities are physical breaks in a stratigraphic sequence marking a hiatus. Angular discordance implies disturbance of rocks prior to deposition of overlying strata.

4. A formation is a fundamental rock unit for field mapping distinguished on the basis of its distinctive physical characteristics and named for a geographic locality. Formations may be assembled into groups and subdivided into members and beds.

5. Systems are rock sequences that accumulated during a specified time interval. They may be grouped into erathems and subdivided into series and stages.

6. Geologic time units are based on time-stratigraphic units and represent the time intervals during which the corresponding rock units accumulated.

7. Although formational boundaries may coincide with boundaries of time-stratigraphic units, often their boundaries are diachronous.

8. Correlation of strata is based on physical and biological criteria that permit demonstration of time equivalence. Reliability of correlation is greatest if several criteria are used.

9. The geologic column, pieced together over more than a century and a half, is a composite section of all known strata, arranged on the basis of their contained fossils or other age criteria.

10. The geologic time scale is a hierarchy of time units established on the basis of corresponding time-stratigraphic units. Systems and periods are based on type sections or type areas in Europe and North America. The geologic time scale constitutes the global standard to which geologists correlate local sequences of strata.

11. Decay of radioactive isotopes of various chemical elements is the basis of radiometric dating.

12. Potassium–argon (^{40}K/^{40}Ar) dating can be used both to determine the formation age of a mineral or in some cases of whole rocks.

13. The age of the Earth, determined by uranium–lead dating, is 4.6 billion years.

14. Radiocarbon dating is only effective in relatively young materials (less than 70,000 years).

15. Remanent magnetism and the polarity-reversal time scale are particularly useful for dating oceanic crust, lavas, and young sedimentary rocks.

16. A sedimentary rock layer can be dated radiometrically by being bracketed between two bodies of igneous rock to which the radiometric method can be applied.

IMPORTANT WORDS AND TERMS TO REMEMBER

age (p. 131)
angular unconformity (p. 128)
Archean (p. 134)

bed (p. 130)

Cenozoic (p. 134)
chron (p. 144)
correlation (of strata) (p. 131)
Curie point (p. 142)

daughter product (p. 136)
depositional remanent magnetism (p. 145)

diachronous boundaries (p. 131)
disconformity (p. 128)

eon (p. 134)
epoch (p. 131)
era (p. 134)
erathem (p. 130)

formation (p. 130)

geologic column (p. 132)
geologic time scale (p. 134)
geologic time units (p. 131)

group (of formations) (p. 130)

Hadean (p. 134)
half-life (p. 137)
hiatus (p. 128)

index fossil (p. 132)

key bed (p. 132)

magnetic chron (p. 144)
magnetic subchron (p. 144)
member (p. 130)
Mesozoic (p. 134)

nonconformity (p. 128)

original horizontality (law of) (p. 126)

paleomagnetism (p. 142)
Paleozoic (p. 134)
parent (radioactive) (p. 136)
period (geologic) (p. 135)
Phanerozoic (p. 134)

polarity reversal (p. 143)
Proterozoic (p. 134)

radiometric dating (p. 136)
rock-stratigraphic unit (p. 130)

series (p. 130)
stage (p. 130)
strata (p. 126)
stratigraphy (p. 126)
stratigraphic superposition (principle of) (p. 127)
subchron (p. 144)
system (p. 130)

thermoremanent magnetism (p. 142)
time line (p. 131)
time-stratigraphic unit (p. 130)

unconformity (p. 128)

QUESTIONS FOR REVIEW

1. How do the law of original horizontality and the principle of stratigraphic superposition help geologists unravel the history of deformed belts of sedimentary rock?

2. What geologic events are implied by an angular unconformity? By a disconformity?

3. How does a rock-stratigraphic unit differ from a time-stratigraphic unit?

4. How are strata correlated from place to place?

5. What is the geologic column?

6. The geologic time scale is divided into four eons. Name them in the correct order, starting with the most ancient.

7. What are the three eras of the Phanerozoic Eon?

8. What are features that make radioactivity an ideal way to measure geologic time?

9. Define a radiometric age and name three dating schemes.

10. How has ^{14}C been used to date the advance of the last great ice sheet in central North America?

11. What is the Curie point and why is it important for paleomagnetism?

12. Polarity reversals of the Earth's magnetic field are recorded in rocks in two different ways. What are they?

13. How can magnetic polarity reversals be used to determine the timing of past geologic events?

14. What method might be used to determine the radiometric age of a rhyolite tuff suspected of being about 3 million years old?

QUESTIONS FOR REVIEW OF PART 1

1. Describe the rock cycle and discuss three lines of evidence that prove the cycle is operating on the Earth.

2. What determines the lower boundaries of the lithosphere, the asthenosphere, and the mesosphere? Draw a diagram to show the inferred layering in the Earth based on (a) chemical composition and (b) physical properties.

3. Granite batholiths tend to intrude rocks that have been subjected to regional metamorphism. Can you suggest why this is so?

4. Would you expect to find the same kinds of minerals in the beach sands on Hawaii and Cape Cod? Give a reason for your answer.

5. Describe the sedimentary characteristics that would permit you to distinguish between fine sand deposited (a) on the floor of a lake; (b) on an ocean beach; and (c) in a desert sand dune.

6. What criteria would you use to distinguish between the following pairs of rocks: (a) quartz–biotite–plagioclase gneiss and granite; (b) mudstone and andesitic tuff; (c) marble and limestone?

7. What criteria would you use to distinguish between the following materials: (a) chalcedony and obsidian; (b) garnet and rhyolite; (c) shale and schist?

8. Why does magma rise up toward the surface? Is there a difference between magma and lava?

9. How is the eruptive style of volcanoes related to plate boundaries? Which is the most dangerous plate boundary as far as the potential loss of human life from volcanic eruption is concerned?

10. The pebbles in conglomerates are commonly gneisses and quartzites, but rarely are they marbles or coal. Why?

11. What properties favor the concentration of minerals such as garnet, tourmaline, and diamond in clastic sediments? If you were a prospector looking for diamonds, would you test shales, sandstones, or limestones?

12. How and why do fossils provide a means of determining the relative age of different strata? Why are fossils rarely found in igneous and metamorphic rocks?

13. How do sedimentary facies differ from metamorphic facies?

14. What kinds of events in Earth history led to distinctive features in the stratigraphic record that permit boundaries of geologic eras and periods to be defined?

15. Describe the process of radioactive decay and give an example of how it provides a basis for estimating the age of a rock or sediment.

16. How do changes in the magnetic properties of rocks and sediments permit ages to be assigned to horizons within a stratigraphic sequence?

A plume of steam rises above the
glacier-clad summit of Mount Erebus,
part of an active volcanic archipelago
in West Antarctica.

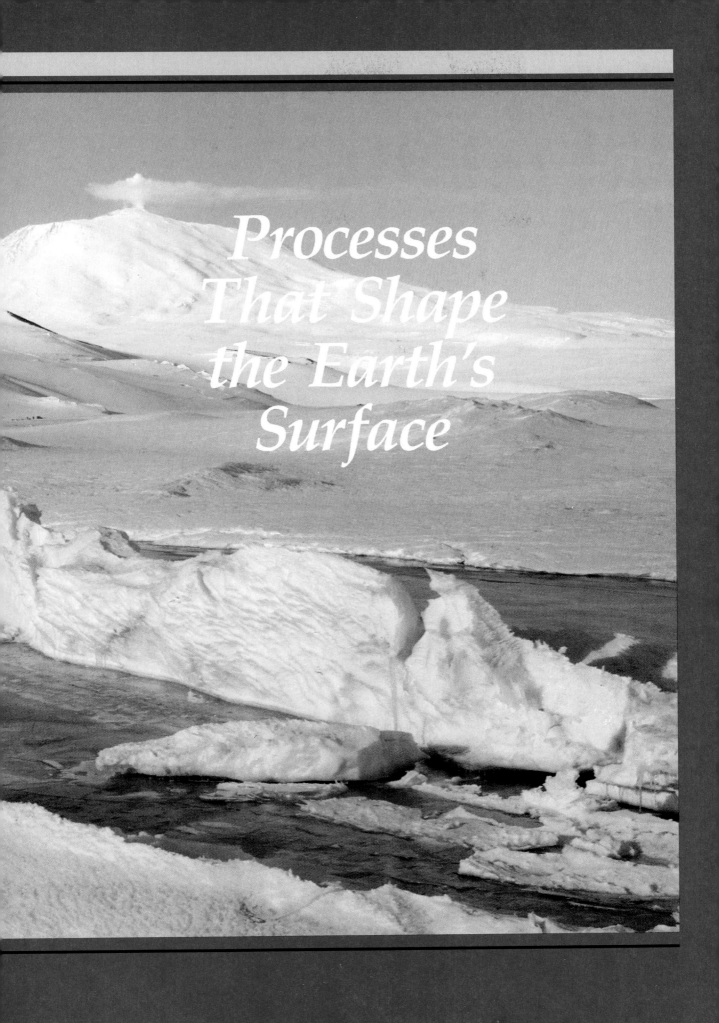

*Processes
That Shape
the Earth's
Surface*

Weathering and Soils

This granitic bedrock on the flank of Gondola Ridge in Antarctica has been cavernously weathered, producing a cliff face that resembles Swiss cheese.

SURFACE PROCESSES

The previous chapters have presented a broad look at the Earth's general character—its solid surface, its gaseous, liquid, icy, and biologic envelopes. We have examined its crust (composed of minerals and rocks), the energy that drives its internal and external activities, the matter that moves gradually through the rock cycle, and the steady march of time, which is so long that even very slow processes have accomplished enormous results.

We will next examine the external processes in more detail, for they are the most easily visible. Together they form a chain in which rock is broken up, transported as sediment, and deposited to form strata. The logical processes to start with are those by which rock is decomposed and disintegrated, leaving the resulting pieces ready to begin their journey downslope.

a

b

c

FIGURE 7.1 Three dated gravestones in a New England cemetery show the strong influence of rock composition on rate of chemical weathering in a humid climate. (a) Marble, consisting of very soluble calcite and exposed for 171 years, is greatly corroded. The entire surface is rough and the inscription is hard to read. (b) Medium-grained sandstone exposed for 196 years contains feldspar and micas. This rock is much less soluble than marble, but is roughened overall. (c) Very insoluble fine-grained slate, exposed for 275 years, is almost unaltered. Incised lettering is sharp and clear.

WEATHERING

People have long sought stone that would be durable for buildings, tombstones, and other structures, but their success has been mixed. The durability of rock varies with climate, composition, texture, and degree of exposure to weather. If gravestones made of firm rock begin to crumble within only a few centuries (Fig. 7.1), what would happen to rock exposed to the atmosphere through thousands or millions of years?

Fast or slow, mechanical and chemical alteration occurs wherever the lithosphere and atmosphere meet. Their contact, however, is not easily drawn. It is a zone rather than a surface. It extends downward into the ground to whatever depth air and water penetrate. In this critical zone both hydrosphere and biosphere are also involved. Within it the rock constitutes a porous framework, full of fractures, cracks, and other openings, some of which are very small but all of which make the rock vulnerable. This open framework is continually being attacked, both chemically and physically, by water solutions. Given sufficient time, the result is conspicuous alteration of the rock.

When exposed to the atmosphere, no rock (whether bedrock or man-made structures of stone) escapes the effects of *weathering,* the chemical alteration and mechanical breakdown of rock materials during exposure to air, moisture, and organic matter. The results of such alteration are often seen in highway cuts and other large excavations that expose the bedrock. In Figure 7.2 loose, unorga-

nized, earthy regolith in which the texture of fresh rock is no longer apparent (zone 1) grades imperceptibly downward, into rock that has been altered but still retains its organized appearance (zone 2), and into fresh unaltered bedrock (zone 3). It is evident from such exposures that alteration of the fresh rock progresses from the surface downward.

Processes of Weathering

A close look at the bedrock in Figure 7.2 would show that near the bottom of the exposure (zone 3) the cleavage surfaces of feldspar grains in the gneiss flash brightly between grains of quartz. Higher up (zone 2) such surfaces are lusterless and stained. Near the top (zone 1) the grains of quartz, although still distinguishable, are separated by soft, earthy material that no longer resembles the former feldspar which has largely decomposed. The changes that have occurred result mainly from *chemical weathering,* which is the decomposition of rocks.

In some places, however, regolith consists of fragments identical to the adjacent bedrock. The mineral grains are fresh or only slightly altered. This relationship is commonly seen in the piles of loose rock fragments at the base of bedrock cliffs from which the debris obviously has been derived. When compared with the bedrock, the coarse rock fragments show little or no chemical change, implying that bedrock can be broken down not only chemically but also mechanically. Although we consider *mechanical weathering,* the disintegration of rocks, as being distinct from chemical weath-

① Loose, earthy regolith; texture and structures disappear as rock particles are slowly churned by roots, worms, and other agents.

② Bedrock weakened by chemical alteration.

③ Fresh, unaltered granite gneiss with crystalline texture, wavy foliation, and joints.

FIGURE 7.2 Weathering profile showing gradation upward from fresh bedrock (gneiss) to earthy regolith.

ering, the two processes generally work hand in hand and their effects are inseparably blended.

Mechanical Weathering

Mechanical weathering of rock is common in nature and is brought about by the development of fractures and growth of ice or salt within them, the heat of fires, and the activities of plants and animals.

Development of Joints

Rock masses buried deeply beneath the ground surface are subjected to enormous pressures due to the weight of overlying rock. As erosion wears down the surface, the weight and pressure are reduced. The rock may adjust to this unloading by upward expansion. As it does so, closely spaced fractures can develop parallel to the land surface so that the rock resembles a stacked deck of cards (Fig. 7.3a). These are *joints,* defined as fractures along which no observable movement has occurred. The slabs of rock between joints may be no more than 10 cm thick near the ground surface, but they become thicker with depth as the distance between joints increases. Generally they disappear below a depth of about 50 m. Rarely do joints occur singly. Most commonly they form a widespread group of parallel fractures called a *joint set* (Fig. 7.3b).

One class of joints is restricted to tabular bodies of igneous rock—such as dikes, sills, lava flows, and welded tuffs—that cooled rapidly close to the surface. When such a body of igneous rock cools, it contracts and may fracture into pieces, in much the same way that a very hot glass bottle, plunged into cold water, contracts and shatters. Unlike shattered glass, cooling fractures in igneous rock form regular patterns. For joints that split igneous rocks into long prisms or columns the term *columnar joints* is used.

Joints not only physically break up an otherwise solid rock mass but act as passageways by which rainwater can enter the rock and promote further weathering.

Crystal Growth

Groundwater percolating through fractured rocks contains ions that may precipitate out of solution to form salts. The force exerted by salt crystals growing within rock cavities or along grain boundaries can be enormous and results in disaggregaor rupturing of rocks. Such effects can be seen in desert regions where precipitation of salts results from the evaporation of rising groundwater. Acid rain falling on industrial cities also can leach ions from masonry, bricks, and mortar. As salts precipitate, the growing crystals can disfigure and weaken buildings and important cultural monuments (Fig. 7.4).

FIGURE 7.3a Well-developed sheetlike jointing in massive granite forming stair-step surface on a mountainside in the Sierra Nevada, California.

FIGURE 7.3b This exposure of granite in Howe Sound, British Columbia, displays three directions of breakage. When the granite cooled it was a solid mass without any fractures. Subsequently three sets of joints developed, the most prominent being the vertical surfaces that break up this roadside outcrop. The field of view is 4 m wide.

to −15°C. At higher temperatures ice pressures are too low to be very effective, and at lower temperatures the rate of ice growth drops because the water necessary for crack growth is less mobile.

Frost wedging is responsible for the production of most rock debris seen on the slopes of high mountains. At lower altitudes it is likely to be most important in places where the number of yearly freeze–thaw cycles reaches a maximum.

Effects of Heat

Some geologists have speculated that daily heating of rock in bright sunlight followed by a comparable cooling each night should have a disruptive effect because the common rock-forming minerals expand by different amounts when heated. Surface temperatures as high as 80°C have been measured on desert rocks, and daily temperature variations of more than 40°C have been recorded on rock surfaces. Dark-colored rocks, like basalt, and rocks that do not easily conduct heat inward achieve the highest surface temperatures. Nevertheless, despite a number of careful laboratory experiments, no one has yet demonstrated that such heating and cooling have noticeable physical effects on rocks. These experiments, however, have been carried out only over relatively brief time intervals. Possibly thermal fracturing takes place only after repeated extreme temperature fluctuations over long periods of time.

Fire, on the other hand, can be very effective, as anyone knows who has witnessed the explosive shattering of a rock beside a campfire when it becomes overheated. The heat of forest and brush fires can lead to the spalling off of large rock flakes from exposed bedrock or boulders. Because rock is a relatively poor conductor of heat, an intense fire heats only the thin outer shell, which expands and breaks away. Studies of fire history in forested regions show that large natural fires, most started by lightning, may recur every several hundred years. Over long intervals of geologic time, fires may therefore have contributed significantly to the mechanical breakdown of surface rocks.

Plants and Animals

Seeds germinate in cracks in rocks to produce plants that extend their roots farther into the cracks. As trees grow, their roots wedge apart adjoining blocks of bedrock. In much the same way they also disrupt sidewalks, garden walls, and even buildings (Fig. 7.6). Large trees swaying in the wind can cause cracks to widen, and if blown over can pry rock

FIGURE 7.4 Carved stone head on building in Florence, Italy shows effects of industrial pollution. Acid rainwater has caused disfiguring of the face by discoloration and disintegration of the rock.

Frost Wedging

In climatic regions where temperatures fluctuate about the freezing point for part of the year, water in the ground is subjected to periodic freezing and thawing. When water freezes to form ice, its volume increases by about 9 percent. In addition, as freezing occurs in the pore spaces of a rock, water is strongly attracted to the growing ice, thereby increasing the stresses against the rock. This leads to a very effective physical weathering process known as *frost wedging*, the formation of ice in a confined opening within rock, thereby causing the rock to be forced apart. The high stresses resulting from the increase in volume as ice crystallizes lead to disruption of rocks. These effects are strong enough to force apart not only tiny particles but huge blocks, some weighing many tons (Fig. 7.5). The disruption and breakdown is likely the result of the slow enlargement and extension of cracks as ice crystals grow in them. Frost wedging of rocks is probably most effective at temperatures of −5°

FIGURE 7.5 Blocks of granite on the side of Taylor Valley, Antarctica, have been disrupted by frost wedging.

apart. Although it would be difficult to measure, the total amount of rock breakage done by plants must be very large. Much of it is obscured by chemical decay, which takes advantage of the new openings as soon as they are created.

Large and small burrowing animals (for example, rodents and ants) bring partly decayed rock particles to the surface where they are exposed more fully to chemical action. More than 100 years ago, Charles Darwin made careful observations in his English garden and calculated that every year earthworms bring particles to the surface at the rate of more than 2.5 kg/m² (10 tons per acre). After a study in the basin of the Amazon River, geologist J. C. Branner wrote that the soil there "looks as if it has been literally turned inside out by the burrowing of ants and termites." Although burrowing animals do not break down rock directly, the amount of disaggregated rock moved by them during many millions of years must be enormous. It illustrates again the cumulative effect of small forces acting over vast intervals of geologic time.

Chemical Weathering

Minerals in igneous rocks that formed at high temperatures, and in metamorphic rocks under high pressures and high temperatures, are chemically unstable under conditions at the Earth's surface where both temperature and pressure are much lower. Such minerals break down and their components form new, stable minerals.

The principal active agent of rock decomposition is chemically active water solutions (weak acids).

The effects of chemical weathering are therefore most pronounced in regions where both precipitation and temperature are high, thereby promoting chemical reactions.

FIGURE 7.6 Tree roots force apart the walls of ruined buildings of Angkor Wat in the jungle of Cambodia. In the same manner, roots penetrating cracks in bedrock help force the rock apart.

Effects on Rock-Forming Minerals

As rainwater falls through the atmosphere it dissolves small quantities of carbon dioxide, producing weak carbonic acid. Moving downward and laterally through the soil, this acid solution is strengthened by the addition of more carbon dioxide released from decaying vegetation. The carbonic acid ionizes to form hydrogen ions and bicarbonate ions (Table 7.1). The hydrogen ions are extremely effective in decomposing minerals, for they are so small they can enter a crystal and replace other ions, thereby changing the composition.

Hydrolysis. The effectiveness of the H^{+1} ion is illustrated by the way in which potassium feldspar, a common rock-forming mineral, is decomposed by hydrogen ions in water (Table 7.1). The H^{+1} ions enter the potassium feldspar and replace the potassium ions, which then leave the crystal and pass into solution. Water combines with the remaining aluminum–silicate molecule to create the clay mineral kaolinite. The chemical reaction, in which the H^{+1} or OH^{-1} ions of water replace ions of a mineral, is called **hydrolysis.** It is one of the chief processes involved in the chemical breakdown of common rocks. The resulting kaolinite is a **secondary mineral,** because it was not present in the original rock. Kaolinite is the most conspicuous of the three products of the reaction. It is a common member of the group of very insoluble minerals that constitute clay, and as clay, it accumulates and forms a substantial part of the regolith.

Leaching. Silica released by chemical weathering remains in the clay-rich regolith or is slowly taken into solution and moves away. Many of the potassium ions also escape in solution, and some, together with dissolved silica, find their way through streams to the sea. This matter carried away in solution is said to have been *leached* from the parent rock. **Leaching** is the continued removal, by water solutions, of soluble matter from bedrock or regolith.

The silicate minerals in igneous rocks that crystallize at the highest temperatures (temperatures most different from those at the Earth's surface) tend to be the ones that weather most readily. They include olivine as well as calcium-rich feldspar, pyroxene, and amphibole. Biotite and sodium-rich feldspar are less easily weathered because they crystallize at lower temperatures. Quartz, which crystallizes at a still-lower temperature, is among the most stable rock-forming minerals and expe-

TABLE 7.1 *Common Chemical Weathering Reactions*

1. Production of carbonic acid by solution of carbon dioxide:

$$\underset{\text{Water}}{H_2O} + \underset{\substack{\text{Carbon}\\\text{dioxide}}}{CO_2} \rightleftharpoons \underset{\substack{\text{Carbonic}\\\text{acid}}}{H_2CO_3} \rightleftharpoons \underset{\substack{\text{Hydrogen}\\\text{ion}}}{H^{+1}} + \underset{\substack{\text{Bicarbonate}\\\text{ion}}}{HCO_3^{-1}}$$

2. Chemical weathering of potassium feldspar by hydrolysis:

$$\underset{\substack{\text{Potassium}\\\text{feldspar}}}{4KAlSi_3O_8} + \underset{\substack{\text{Hydrogen}\\\text{ions}}}{4H^{+1}} + \underset{\text{Water}}{2H_2O} \rightarrow \underset{\substack{\text{Potassium}\\\text{ions}}}{4K^{+1}}$$
$$+ \underset{\text{Kaolinite}}{Al_4Si_4O_{10}(OH)_8} + \underset{\text{Silica}}{8SiO_2}$$

3. Oxidation of ferrous iron to form goethite:

$$\underset{\substack{\text{Ferrous}\\\text{oxide}}}{4FeO} + \underset{\text{Water}}{2H_2O} + \underset{\text{Oxygen}}{O_2} \longrightarrow \underset{\text{Goethite}}{4FeO{\cdot}OH}$$

4. Alteration of goethite to form hematite by dehydration:

$$\underset{\text{Goethite}}{2FeO{\cdot}OH} \longrightarrow \underset{\text{Hematite}}{Fe_2O_3} + \underset{\text{Water}}{H_2O}$$

5. Weathering of carbonate rock by carbonic acid:

$$\underset{\substack{\text{Calcium}\\\text{carbonate}}}{CaCO_3} + \underset{\substack{\text{Carbonic}\\\text{acid}}}{H_2CO_3} \longrightarrow \underset{\substack{\text{Calcium}\\\text{ion}}}{Ca^{+2}} + \underset{\substack{\text{Bicarbonate}\\\text{ions}}}{2(HCO_3)^{-1}}$$

riences little obvious chemical decay. Nevertheless, over time, even quartz can be very slowly taken into solution.

Oxidation. Iron is a normal constituent of many common rock-forming minerals, including biotite, augite, and hornblende. When it is released during weathering it is rapidly oxidized from the ferrous form (Fe^{+2}) to the ferric form (Fe^{+3}) if oxygen is present. The result is a new yellowish mineral, **goethite** (Table 7.1). Because water is usually present, goethite is a hydrous mineral that forms through a combination of oxidation and **hydration,** the absorption of water into a crystal structure. Goethite may later be dehydrated, by loss of water, to form **hematite,** a mineral having a brick-red color (Table 7.1). The intensity of these colors in weathered rocks and soils can provide clues to the time that has elapsed since weathering began and to the degree or intensity of weathering.

Effects on Common Rocks

What happens in the weathering of potassium feldspar is a key to understanding the weathering of silicate rocks, such as granite, that contain this mineral. Table 7.2 contrasts the chemical weathering of granite and basalt, showing the resulting min-

TABLE 7.2 *Chemical Weathering of Granite and Basalt*

Rock	Primary Minerals	Weathering Products	
		Residual Minerals	Soluble Ions Retained in Solution
Granite	Feldspars	Clay minerals	Na^{+1} and K^{+1}
	Micas	Clay minerals	K^{+1}
	Quartz	Quartz	
	Fe–Mg minerals	Clay minerals, hematite, and goethite	Mg^{+2}
Basalt	Feldspars	Clay minerals	Na^{+1} and Ca^{+2}
	Fe–Mg minerals	Clay minerals	Mg^{+2}
	Magnetite	Hematite and goethite	

erals that form and ions that are carried away in solution.

Carbonate rocks, such as limestone, are weathered in a different way. Limestone consists mainly of calcium carbonate which is only slightly soluble in pure water. In the presence of weak carbonic acid, the calcium and bicarbonate ions are removed in solution, leaving behind only the nearly insoluble impurities (chiefly clay and quartz) that are always present in small amounts in limestone. As limestone weathers, the regolith that develops from it consists mainly of clay and quartz.

Concentration of Stable Minerals

Not only quartz but other minerals as well are relatively stable at the Earth's surface and so resist destruction by chemical weathering. Minerals such as gold, platinum, and diamond persist in weathered regolith, are eroded, and become sediment. Because some of these minerals have higher densities than common minerals such as quartz, they sink and settle to the beds of streams where they may collect to form a *placer,* a deposit of heavy minerals concentrated mechanically. Some may be sufficiently concentrated to form mineral deposits of economic value (chapter 18).

Weathering Rinds

If a cobble of weathered basalt is cracked open, one commonly will see a discolored rim, or *rind,* surrounding a darker core of fresh unaltered rock (Fig. 7.7). The rind consists of residues resulting from chemical weathering. Similar discolored rinds form on all but the most chemically stable rock types. The rind increases in thickness as weathering slowly attacks the solid core. As a result, rind thickness is a useful measure of the relative age of sediments that contain rocks of the same type and occur in comparable climatic settings.

Exfoliation and Spheroidal Weathering

As some rocks weather they experience **exfoliation,** the spalling off of successive shells, like the skins of an onion, around a solid rock core (Fig. 7.8). In some cases only a single exfoliation shell is present, but there may be 10 or more. When exfoliation begins, the outermost shells are commonly bounded by nearly parallel joint planes and are relatively flat, whereas the innermost are progressively more spherical as the rock is reduced in size and the corners become more and more rounded.

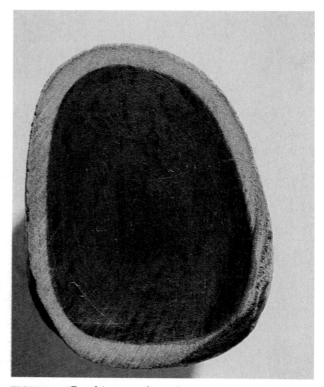

FIGURE 7.7 Basaltic stone from the eastern Cascade Range displays a well-developed weathering rind about 2 mm thick.

FIGURE 7.8 Exfoliating granite boulders near Iferouane in the Aïr Mountains of Niger. Thin sheetlike spalls are seen on the surfaces of the boulders.

Exfoliation is caused by physical or chemical forces that produce differential stresses within a rock. For example, when feldspars are weathered to clay, the volume of weathered rock is greater than the volume of original rock. Stresses thus produced cause shells to separate from the main mass of the rock. A body of rock may also experience a progressive decrease in vertical pressure as a region is uplifted and overlying rocks are removed by erosion.

Beneath the ground surface, chemical weathering can produce concentric or spherical shells of decayed rock that are successively loosened and separated from a solid rock core as water moves along joints and attacks the solid rock from all sides. The results are often seen in newly made roadcuts (Fig. 7.9). Such *spheroidal weathering* produces rounded boulders through progressive decomposition, rather than exfoliation. Spheroidal forms

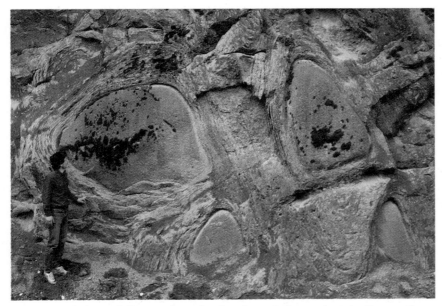

FIGURE 7.9 Spheroidally weathered volcanic rock in roadcut in the northern Cascade Range.

created by this means often lie in distinct rows running in several directions. This pattern results from intersecting joint sets that control the slow movement of water through the rock.

At this point two important relationships should be noted. First, the effectiveness of chemical reactions increases with increased surface area available for reaction. Second, surface area increases simply from subdivision of large blocks into smaller blocks. Subdividing a cube, while adding nothing to its volume, greatly increases the surface area (Fig. 7.10). Repeated subdivision leads to a remarkable result. One cubic centimeter of rock subdivided into particles the size of the smallest clay minerals results in an aggregate surface area of nearly

4000 m². Weathering causes subdivision and so promotes further weathering.

Factors That Influence Weathering

Rock Type and Structure

Because different minerals react differently to weathering processes, rock type clearly must influence decomposition. Quartz is so resistant to chemical breakdown that rocks rich in quartz are also resistant. In many places, hills and mountains that consist of granite or quartzite stand distinctly higher than surrounding terrain underlain by less resistant rocks that contain less quartz.

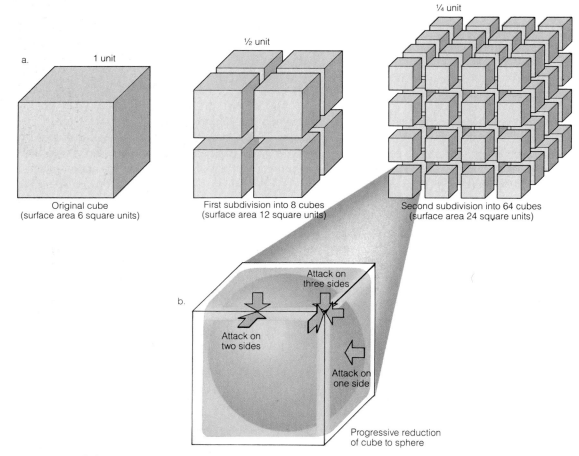

FIGURE 7.10 Subdivision and weathering of rock cubes. (a) Subdivision of a cube into smaller cubes. Each time a cube is subdivided by slicing it through the center of each of its edges, the aggregate surface area doubles. This greatly increases the speed of chemical reaction. (b) Geometry of spheroidal weathering. Solutions that occupy joints separating nearly cubic blocks of rock attack corners, edges, and sides at rates that decline in that order, because the numbers of corresponding surfaces under attack are 3, 2, and 1. Corners become rounded; eventually the blocks are reduced to spheres. Once a spherical form is achieved, energy of attack becomes uniformly distributed over the whole surface, so that no further changes in form occur.

The rate of weathering of a rock is influenced also by its texture and structure. Even if a rock consists entirely of quartz (quartz sandstone, quartzite) but contains closely spaced joints or other partings, it may break down rapidly, especially if attacked by frost processes.

Contrasts in local topography often result from *differential weathering,* weathering that occurs at different rates as a result of variations in the composition and structure of rocks or in intensity of weathering (Fig. 7.11). In a sequence of alternating shale and quartz sandstone, the shale is likely to weather more easily, leaving the sandstone beds standing out in relief. If the beds are horizontal, the result is likely to be a stepped topography, with the sandstone forming abrupt cliffs between more gentle slopes of shale. If the bedding is inclined, the sandstone will stand as ridges separated by linear depressions underlain by shale.

Slope

When a mineral grain is loosened by weathering on a steep slope, it may be washed downhill by the next rain. With the solid products of weathering moving quickly away, fresh bedrock is continually exposed to renewed attack, so that weathered rock extends only to a slight depth beneath the surface. On gentle slopes, however, weathering products are not easily washed away, and in places accumulate to depths of 50 m or more.

Climate

Moisture and heat promote chemical reactions. Not surprisingly, therefore, weathering is more intense and generally extends to greater depths in a warm moist climate than in a dry cold one (Fig. 7.12). Rocks such as limestone and marble, which consist almost entirely of soluble calcite, are very susceptible to chemical weathering in a moist climate and commonly underlie subdued landscapes. In a dry climate, however, the same rocks may form bold cliffs, because with scant rainfall and patchy vegetation, little carbonic acid is present to dissolve the carbonate minerals. In cold climates, chemical weathering proceeds very slowly. In such regions the effects of mechanical weathering are generally more obvious, for over wide areas bedrock is littered with frost-wedged rubble.

Time

Studies of the decomposition of stone in ancient buildings and monuments show that hundreds or even thousands of years are required for hard rock

FIGURE 7.11 Differential weathering of jointed sandstone in Arches National Park, Utah, results in a remarkable topography of rock pinnacles and walls separated by deep clefts.

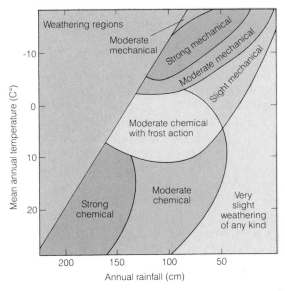

FIGURE 7.12 Climatic control of weathering process. Mechanical weathering is dominant where rainfall and temperature are both low. High temperature and precipitation favor chemical weathering. (*Source:* After L. C. Peltier, 1950.)

to decompose to depths of only a few millimeters. Granite and other hard bedrock surfaces in New England, Scandinavia, the Alps, and elsewhere still display polish and fine grooves made by ice-age glaciers before they disappeared about 10,000 years ago. In such cool temperate climates it must take many tens of thousands of years to create weathered regolith like that shown in Figure 7.2. However, in regions that have been continuously exposed to weathering processes for many millions of years, the zone of weathering often extends much deeper. In some tropical areas, mining operations have exposed bedrock that has been thoroughly decomposed to depths of 100 m or more.

The rates at which rocks weather have been estimated in several ways. First, experimental studies have been designed in which the length of the experiment provides time control, while the processes were speeded up by increasing temperature and available water and by decreasing particle size. Second, studies have been made of the degree of weathering of man-made structures, the ages of which are known. Third, studies of radiometrically dated rock or sediments that have been exposed to weathering for thousands or millions of years can provide estimates of average rates over much longer intervals. The results of such investigations suggest that the rates at which most weathering processes operate tend to decrease steadily with time.

SOILS

Origin

The physical and chemical breakdown of solid rock by weathering processes is the initial step in the formation of soil. However, soil also contains at least a little, and commonly much, organic matter mixed in with the mineral components. This organic fraction is an essential part of the usual definition of *soil:* that part of the regolith which can support rooted plants.

The organic matter in soil is derived from the decay of plant material, partly through the activity of bacteria. Plants are nourished by decayed plant material in the soil as well as by decomposed mineral matter which is drawn upward, in water solution, through their roots. Therefore, plants are involved in the manufacture of their own fertilizer. These activities represent continual cycling of nutrients between the regolith and biosphere. With its partly mineral, partly organic composition, soil forms an important bridge between the Earth's lith-

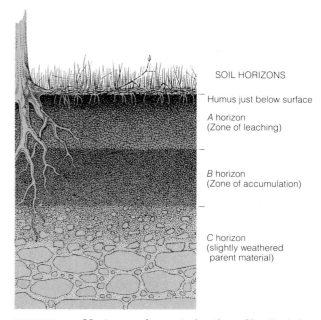

FIGURE 7.13 Horizons of a typical soil profile. Each horizon grades downward into the one below it.

osphere and its teeming biosphere. To people, soil means food; thus it is a fundamental natural resource of every nation.

Soil Profile

As weathering of bedrock and regolith proceeds, soil gradually evolves. Normally it develops characteristic *horizons* that together constitute a ***soil profile,*** the succession of distinctive horizons in a soil from the surface downward to the unaltered parent material beneath it (Fig. 7.13).

The uppermost horizon, called the A horizon, is typically grayish or blackish (at least near its top) because of the addition of ***humus,*** the decomposed residue of plant and animal tissues. The A horizon has lost some of its original substance through the downward transport of clay particles and, more importantly, through the chemical leaching of soluble minerals.

The B horizon is commonly brownish or reddish. It is enriched in clay and iron oxides produced by weathering of minerals within the horizon and also transported downward from the overlying A horizon. The B horizon is often characterized by structure: it breaks into blocks or prisms, each of which may be coated with clay. Although the B horizon generally is penetrated by plant roots, it contains less organic matter than the humus-rich A horizon.

The underlying C horizon does not constitute a part of the soil proper. Instead, it consists of slightly weathered parent material, either bedrock or regolith, in which oxidation has produced a detectable change in color.

Young, or immature, soils lack a B horizon and display only an A horizon overlying a thin oxidized C horizon. As the soil develops, a B horizon appears, initially distinguishable by its color. As clay accumulates, the B horizon develops structure and the soil assumes a mature character. With the further passage of time, the B horizon slowly increases in thickness. Contrasts in the degree of soil development can often be seen as one travels across successively older sedimentary deposits exposed at the surface. For example, immature soils typically are found on sediments laid down during the last 10,000 years, whereas older sediments display mature soils that are progressively more developed with increasing age of the deposits.

Soil-Forming Factors

Differences among soils, commonly reflected by differences in soil profile characteristics, result from the influence of several important soil-forming factors: climate, vegetation cover and soil organisms, parent material, topography, and time. We have already seen that these factors influence weathering. They are also important in the evolution of soils.

Parent materials and topography differ widely and strongly influence the character of soils, especially during the early part of soil development. Climate, which in turn affects soil organisms and vegetation cover, may be an even stronger influence than bedrock in ultimately determining soil character. Under similar climatic conditions the profiles of mature soils developed on different kinds of rock become remarkably alike.

Soils of Extreme Climates

Polar Soils

In cold, high-latitude deserts (chapter 11), soils typically are well drained, lack an obvious A horizon, and display a weak yellowish-brown B horizon under a layer of coarse frost-churned stones (Fig. 7.14a). In wetter environments, matlike tundra vegetation overlies perennially frozen ground that prevents downward percolation of water. The resulting poorly drained conditions lead to soils that are saturated with water and rich in organic matter. Only on well-drained sites in such environments do soils develop having characteristic A and B horizons. Because the cold climate retards chemical processes, high-latitude soils generally do not display a thick, clay-enriched horizon typical of old temperate-latitude soils.

Alkaline Soils

In dry climates, where lack of moisture inhibits the leaching and removal of carbonate minerals, carbonates may accumulate in the upper part of the C horizon. This makes the soil strongly alkaline, in contrast to the acidic soils of humid regions. An important part of the carbonate accumulation results from evaporation of water that rises in the ground, bringing dissolved salts from below. In extensive arid areas of the southwestern United States, carbonates have in this way built up in the soil profile a solid, almost impervious layer of whitish calcium carbonate generally known as *caliche* (Fig. 7.14b).

Lateritic Soils

Soils forming under conditions of extensive rainfall and very warm temperatures are characterized by extreme chemical alteration of the parent material. Some soils of this type have the interesting property of hardening after wetting and drying. The result is a material called **laterite,** a hardened soil horizon characterized by extreme weathering that has led to concentration of secondary oxides of iron and aluminum. Laterites constitute a primary source of **bauxite,** an ore of aluminum, from which the silica has been leached away. Because they are so hard, some lateritic soils can be cut into durable bricks and used for construction (Fig. 7.14c).

Rate of Soil Formation

Although the development of soil is part of the complex process of weathering, soil formation and weathering are not the same thing. Weathering chiefly concerns the decomposition of bedrock, which is very slow and takes place over long intervals of geologic time. The time required to form a soil profile in regolith can be much shorter.

A soil profile can form quickly in some environments. A study in the Glacier Bay area of southern Alaska showed that within a few years after retreat of glaciers an A horizon developed on the newly revegetated landscape (Fig. 7.15). In this area, moderate temperatures and high rainfall promote rapid leaching of parent material. As the plant cover becomes denser, the soil becomes more acid, and leaching is more effective. After about 50 years a

(a)

(b)

FIGURE 7.14 (a) Profile through a weakly developed polar desert soil on Svalbard near 79° N latitude. (F. C. Ugolini). (b) A soil profile in semiarid central New Mexico has a layer of whitish caliche at the top of the C horizon. (c) Blocks of laterite have been used in the construction of this temple at Angkor Wat in the Cambodian jungle.

(c)

B horizon appears and the combined thickness of the A and B horizons reaches about 10 cm. Over the next 165 years, a mature forest develops on the landscape, the A and B horizons increase in thickness to 15 cm, and iron oxides accumulate in the developing B horizon.

In less humid climates, rates of soil formation are slower and it may take thousands of years for a detectable B horizon to appear. B horizons of soils in the midcontinental United States that have developed during the last 10,000 years contain little clay, whereas those dating back about 100,000 years generally display considerable clay enrichment and associated structure. By contrast, glacier-free polar deserts of Antarctica are so dry and cold that sediments more than a million years old have only very weakly developed soils. Deep reddish soils of temperate and subtropical regions probably date to the Tertiary Period and took many millions of years to form.

The lengthy time involved in developing a mature, productive soil emphasizes the potentially disastrous results of soil erosion in agricultural regions. Agricultural soils are a prime natural resource,

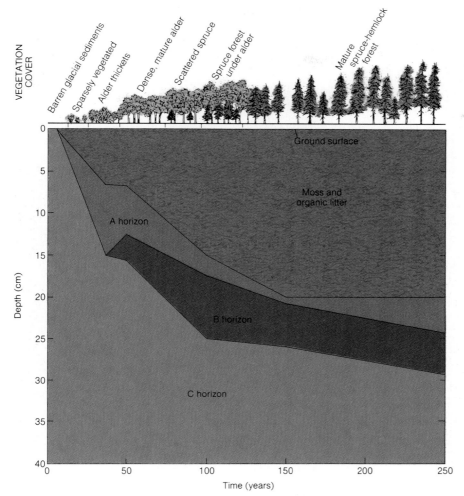

FIGURE 7.15 Progressive soil development in Glacier Bay, Alaska, over the past 250 years. With time, organic litter increases in thickness and A and B horizons develop. For the first 40 years, the A horizon increases in thickness to more than 5 cm but directly overlies the C horizon. The B horizon then begins to develop and after another 60 years has reached a thickness of 10 cm. As vegetation changes, organic litter at the forest floor increases in thickness to about 20 cm after 150 years. The age of each study site was determined by counting the annual growth rings of the oldest surrounding trees. (*Source:* Adapted from data of F. C. Ugolini.)

but once destroyed, they can only be replaced over long intervals of time by weathering activity. Although they might be viewed as renewable resources over the very long term, over the lifetime of individuals, or even nations, they must be considered nonrenewable resources that should be carefully utilized and preserved.

Paleosols

If a landscape is covered by sediments or lava, the surface soil is buried and becomes part of the geo-logic record. The soil is now a ***paleosol,*** defined as a soil that formed at the ground surface and subsequently is buried and preserved (Fig. 7.16). Its top is therefore an unconformity. Paleosols have been identified in rocks and sediments of many different ages, but they are especially common in unconsolidated deposits of the Quaternary Period. Distinctive and widepsread paleosols have been used to subdivide, correlate, and date sedimentary sequences. They also can provide important clues about former landscapes, vegetation cover, and climate.

FIGURE 7.16 A thick reddish-brown paleosol (beside figure) developed on a pumice layer near Guatemala City, Guatemala is overlain by a layer of whitish airfall pumice and pyroclastic flows that underlie other pumice layers capped by thinner brownish paleosols. A fault has displaced the layers about 2 m vertically.

Essay

CAVERNOUS WEATHERING IN ANTARCTICA

Most of the Antarctic continent is buried beneath a gigantic ice sheet as much as 4 km thick. However, in scattered places around its margin there exist small "oases" of bare land. Because of the high latitude and an extremely dry climate, these small nonglaciated areas are miniature cold deserts (chapter 11). The largest areas of bare land are found in the Dry Valleys of the McMurdo Sound Region, directly south of New Zealand. A walk through these valleys dramatizes how weathering in this cold polar environment differs from that in more familiar temperate climatic zones (Fig. B7.1).

Among the most unusual elements of the landscape are free-standing boulders of granite that look like huge pieces of Swiss cheese. Large hollows have been carved in their flanks, so that many are only fragmentary relics of their original shape. Some have overhanging roofs, or rise on thin pedestals a meter or more above the ground surface. Adjacent bedrock valley walls display similar caverns, often of impressive size (chapter frontispiece).

These weird natural shapes are the product of *cavernous weathering.* While not restricted to cold deserts, this phenomenon is very common in the glacier-free zones of Antarctica. The cavernous forms are generally attributed to two related processes. The first is crystallization of salts from solutions confined within pores and fine cracks between mineral grains in a rock; growth of salt crystals exerts enough pressure to separate mineral grains or small fragments at the rock surface. The second involves an agent of erosion and transport. In the Dry Valleys, the agent is the strong and persistent winds blowing off the adjacent ice sheet. The powerful winds, assisted by blowing sand, pluck off the flaking rock fragments and carry them away. Salt weathering generally begins on the underside of a boulder and continues until the resulting hollows meet, thereby causing collapse of the rock. The relative degree of cavernous weathering has proved a useful means of estimating the ages of landscape surfaces in Antarctica, for the older a surface is, the greater is the degree of boulder modification.

FIGURE B7.1 A granite boulder in the Dry Valleys region of Antarctica has been cavernously weathered under an extreme polar climate.

SUMMARY

1. The zone of weathering extends to whatever depth air and water penetrate. Water solutions, which enter the bedrock along joints and other openings, attack the rock chemically and physically causing breakdown and decay.

3. Mechanical and chemical weathering, although involving very different processes, generally work together.

3. Subdivision of large blocks into smaller particles increases surface area and thereby accelerates chemical weathering.

4. Growth of crystals, especially ice and salt, along fractures and other openings in bedrock is a major process of mechanical weathering.

5. Daily heating of rocks by the sun followed by nocturnal cooling may cause little or no breakdown, but intense fires can lead to spalling of rock surfaces.

6. The wedging action of plant roots and the churning of rock debris by burrowing animals can have large cumulative effects over time.

7. Carbonic acid is the prime agent of chemical weathering; heat and moisture speed chemical reactions.

8. Chemical weathering converts feldspars into clay minerals. Quartz is resistant to chemical decay and commonly remains as sand grains.

9. The effectiveness of weathering depends on such factors as rock type and structure, surface slope, local climate, biologic activity, and the time over which weathering processes operate.

10. Soils consist of weathered regolith capable of supporting plants. Soil profiles display distinctive horizons, the character of which depends on such factors as climate, vegetation cover and soil organisms, composition of parent material, topography, and time.

11. The A horizon is rich in organic matter and has lost soluble minerals through leaching. Clay accumulates in the B horizon together with substances leached from the A horizon. Both overlie the C horizon, which is slightly weathered parent material.

12. Soils of the polar deserts generally lack an A horizon. In tundra environments soils tend to be saturated and rich in organic matter.

13. Caliche is a common component of many arid-region soils and forms in the upper part of the C horizion.

14. Laterites are typical of tropical climates, display extreme weathering, and have concentrations of iron and aluminum oxides.

15. Paleosols are buried soils that can provide clues about former topography, plant cover, and climate.

IMPORTANT WORDS AND TERMS TO REMEMBER

A horizon (p. 166)

B horizon (p. 166)
bauxite (p. 167)

C horizon (p. 166)
caliche (p. 167)
carbonic acid (p. 161)
chemical weathering (p. 157)
columnar joints (p. 158)

decomposition (p. 157)
differential weathering (p. 165)
disaggregation (p. 157)
disintegration (p. 160)

exfoliation (p. 162)

frost wedging (p. 159)

goethite (p. 161)

hematite (p. 161)
humus (p. 166)
hydration (p. 161)
hydration rind (p. 00)
hydrolysis (p. 161)

joint (p. 158)
joint set (p. 158)

laterite (p. 167)
leaching (p. 161)

mechanical weathering (p. 157)

paleosol (p. 169)
placer (p. 162)

secondary mineral (p. 161)
soil (p. 166)
soil profile (p. 166)
spheroidal weathering (p. 163)

weathering (p. 157)
weathering rind (p. 162)

QUESTIONS FOR REVIEW

1. What is the difference between decomposition and disintegration of rocks?

2. How do joints affect rock weathering?

3. How does acid rain in industrialized regions contribute to weathering?

4. Why is frost wedging most effective at temperatures between $-5°$ and $-15°C$?

5. Why does a forest fire often cause flakes to spall off exposed boulders?

6. How do burrowing animals influence chemical weathering?

7. What are placers, and how do they form?

8. Why does the physical breakup of a rock increase the effectiveness of chemical weathering?

9. What causes exfoliation of massive rocks?

10. Explain the roles of rock type, structure, slope, climate, and time in weathering.

11. How does regolith developed on limestone differ from that developed on an igneous rock?

12. Describe the horizons in the profile of a well-developed soil.

13. What are the five principal factors that influence soil development?

14. What can you infer about the climate of an area where (*a*) lateritic soils are developing, and where (*b*) caliche is accumulating in a soil profile?

15. Why do paleosols mark unconformities in a sequence of strata?

Mass-Wasting

The massive boulders underlying the resort community of Breuil-Cervinia in the Italian Alps were deposited by a huge prehistoric rockfall from the upper slopes of the Matterhorn, visible here in the distance.

ROLE OF MASS-WASTING

Mass-wasting is the movement of regolith downslope by gravity without the aid of a transporting medium, such as water, ice, or wind. Nevertheless, water plays an important role in mass-wasting. Saturation of regolith with water reduces friction between rock particles thereby making movement easier. This is the main reason why some mass-wasting activities are especially common and effective after long or intense rains. It is not always easy to separate weathering from mass-wasting and other agents of erosion, for they constitute a continuum of processes that interact and overlap. Their end result is the gradual breakdown of solid bedrock and the redistribution of its weathered components.

Gravity

A smooth vegetated slope may appear outwardly stable and show little obvious evidence of geologic activity. Yet if we examine the regolith beneath the surface, we quite likely will find rock particles derived from bedrock that is exposed only in areas that lie farther upslope. We can deduce, therefore, that the particles have moved downslope.

The force that makes the rock particles move is *gravity,* as it pulls persistently on rock debris at the Earth's surface. On a horizontal surface, gravity holds objects in place by pulling on them in a direction perpendicular to the surface. On any slope, gravity can be resolved into two component forces. The *perpendicular component of gravity* (g_p in Fig. 8.1) acts at right angles to the slope and holds objects in place. The *tangential component of gravity* (g_t in Fig. 8.1) acts along and down the slope, and is the force that leads objects to move downhill.

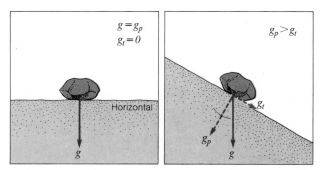

FIGURE 8.1 Effects of gravity on a rock lying on a hillslope. Gravity acts vertically, and can be resolved into two components. One is perpendicular (g_p) and one parallel (g_t) to the surface.

A rock on a hilltop that reached its position because it was lifted, against the pull of gravity, as the hill was created, possesses potential energy. The potential energy (E_p) of a rock of a certain mass (m), raised to a height (h), where g is the acceleration due to gravity, is

$$E_p = mgh$$

Moisture lifted to a cloud by solar energy has potential energy. When the moisture falls as rain, the potential energy is converted to the kinetic energy of falling raindrops. In the same way, when a rock moves downslope, its potential energy is transformed into kinetic energy. The kinetic energy (E_k) can be expressed as

$$E_k = \tfrac{1}{2}\, mv^2$$

where m is the mass of an object moving at velocity v.

Downslope Movement of Debris

A loose particle of regolith possessing potential energy will at some time move downslope. Sooner or later it reaches a stream or some other agent of transport, which will carry it farther. The beginning of its journey, the trip down the nearest slope, can be very slow or very fast, but in either case it is controlled primarily by gravity.

Once regolith begins to move downslope it becomes sediment by definition. Under natural conditions, a slope is likely to evolve toward an angle that allows the quantity of regolith moving from upslope to be balanced by the quantity of material that is moving on downslope. Such a slope is said to be in a balanced, or **steady-state,** condition.

Mass movement is not confined to the land. Sediment transported by mass-wasting is found beneath lakes and it covers vast areas of the seafloor. As on land, movement of sediment in the oceans is controlled mainly by gravity. Therefore, we can think of mass-wasting as universal, and active wherever slopes exist.

MASS-WASTING PROCESSES

Basis of Classification

Mass-wasting processes all share one characteristic: they take place on slopes. Any perceptible downslope movement of a mass of bedrock, regolith, or a mixture of the two is commonly referred to as a **landslide.** However, we can recognize many

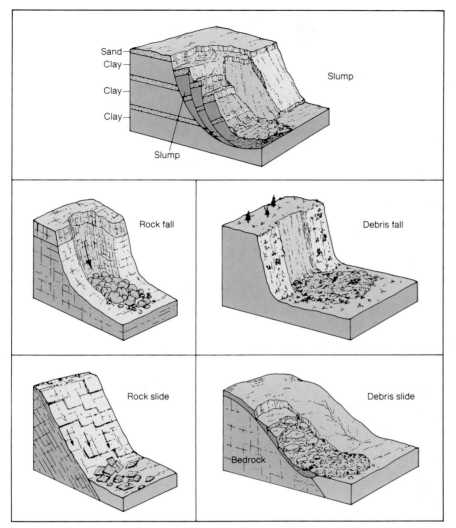

FIGURE 8.2 Examples of slope failures giving rise to slumps, falls, and slides.

different kinds of slope movements, and because they often are gradational into one another, no simple and ideal classification exists. The composition and texture of the sediment involved, the amount of water and air mixed with the sediment, and the steepness of slope all influence the type and velocity of movement. In effect, a progressive transition exists from the flow of clear streamwater, to sediment-laden streamwater, to an array of mass-wasting processes in which water acts as an important lubricant to promote flow. In some processes water plays no direct or significant role.

The approach here will be to separate mass-wasting processes into those involving (1) the sudden failure of a slope that results in the downslope transfer of debris by falling, sliding, or slumping; and (2) the downslope flow of mixtures of sediment, water, and air. In the latter group, processes are distinguished on the basis of their velocity of movement and on the concentration of sediment in the flowing mixture.

Slope Failures

Slumps

A **slump** is a type of slope failure in which a downward and outward rotational movement of rock or regolith occurs along a concave-up slip surface (Fig. 8.2). The top of the displaced block usually is tilted backward, producing a reversed slope. A series of adjacent slump blocks forms a characteristic hummocky topography consisting of depressions more or less concentrically aligned (Fig. 8.3). Slumps frequently are associated with heavy rains or sudden shocks, such as earthquakes (Fig. 8.4).

Slumps are one of the types of mass movement that we are most likely to see, for many result from

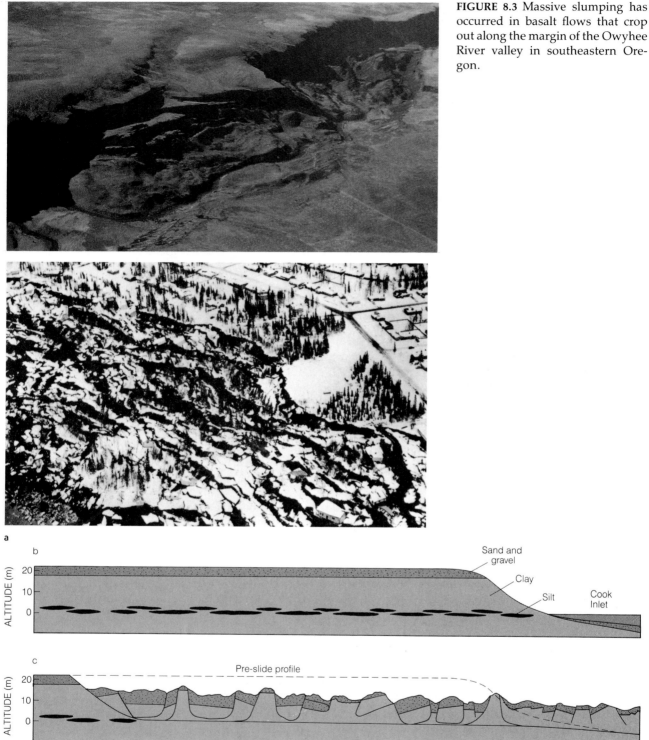

FIGURE 8.3 Massive slumping has occurred in basalt flows that crop out along the margin of the Owyhee River valley in southeastern Oregon.

FIGURE 8.4 Chaotic terrain in Anchorage, Alaska caused by slope failure during the great earthquake of March 27, 1964. (*a*) Slump blocks at Turnagain Heights that have partially destroyed a suburban housing area. (*b*) Cross section through bluffs prior to earthquake. Sandy glacial gravel overlies clay containing lenses of silt. (*Source:* After NAS/NRC, 1984.) (*c*) Cross section after earthquake. Failure has occurred along the weak silt zone. Slumps have produced tilted, chaotic landscape as surface was lowered and blocks of sediment were moved laterally into the bay. (*Source:* After NAS/NRC, 1984.)

human modification of the land. They are numerous along roads and highways where bordering slopes have been oversteepened by construction activity. They also are seen along river banks or seacoasts where currents or waves undercut the base of a slope.

Slumping often is episodic and related to changing climatic conditions. For example, slumping may be seasonal and associated with infiltration of water into the ground during the rainy season. In parts of the American West, increased slumping during recent decades appears to be correlated with an overall increase in average rainfall. An increase in slumping and associated earthflow movements in the northwestern United States about 5000 years ago apparently was related to a shift from a warm, dry climatic regime in the middle Holocene to cooler and wetter conditions since.

Falls and Slides

Rockfalls, the free falling of detached bodies of bedrock from a cliff or steep slope, are common phenomena in mountainous terrain where the products of such events form conspicuous deposits at the base of steep slopes (Fig. 8.2). A rockfall may involve the dislodgment and fall of a single rock fragment, or it may involve the sudden collapse of a huge mass of rock that breaks on impact into a vast number of smaller pieces. These pieces continue to bounce, roll, and slide downslope before

friction and decreasing gradient bring them to a halt. *Debris falls* are similar to rockfalls, but they consist of a mixture of rock and weathered regolith, as well as vegetation (Fig. 8.2).

Rockslides are the sudden downslope movement of newly detached masses of bedrock (or of debris, in the case of **debris slides**) typically across a preexisting inclined bedrock surface such as a bedding surface (Fig. 8.2). Like rockfalls and debris falls, they are common in glaciated mountains where steep slopes abound. When large slides occur, the resulting deposit commonly is a chaotic, jumbled mass of rock or debris, with individual boulders sometimes measuring tens of meters across (Fig. 8.5).

Sliderock and Taluses

Where mechanical weathering is active, accumulations of weathered rock debris mantle the base of steep cliffs. The angular particles range in size from grains of sand to large boulders. The body of rock waste sloping outward from the cliff that supplies it is a *talus* (Fig. 8.6), and the sediment composing it is **sliderock.** From cliff to talus, the movement is chiefly by falling, sliding, and rolling. The rock fragments come to rest at the steepest angle, measured from the horizontal, at which the debris remains stable (Fig. 8.7). This **angle of repose** typically is in the range of 33° to 37°. Few fine particles may be visible, for they tend to settle into the large

FIGURE 8.5 A jumble of giant boulders forms the deposit of a massive rockslide that descended a steeply sloping mountainside in the central Argentine Andes.

open voids between coarser fragments. Because large falling rocks have more momentum than small particles, they tend to move farther down the slope, and some may bound beyond the toe of the talus where they form a scattered array of isolated boulders (Fig. 8.6).

Sediment Flows

Factors Controlling Flow

When a sufficiently large force is applied to a geologic material, it will deform. If the deformation is continuous and irreversible, it is called *flow.* In mass-wasting, the force is gravity and the materials consist of mixtures of regolith or rock debris, water, and air.

Some materials will not flow until a minimum level of stress has been exceeded. Others become increasingly fluid as flow proceeds. At above-freezing temperatures, the way a material flows depends on (1) the relative proportion of solids, water, and air; (2) the grain-size distribution of the solid fraction; and (3) the physical and chemical properties of the sediment.

Mass-wasting processes involving flow can be subdivided according to their relative sediment concentration into **slurry flows** and **granular flows** (Fig. 8.8). These can further be subdivided on the basis of velocity of flow. The more fluid types are transitional, at high velocities, into sediment-laden streamflow. Boundaries between processes are only approximate. In Figure 8.8, they can shift to the right or left, or up or down, depending on the

FIGURE 8.6 Talus at the base of steep cliffs in the Argentine Andes. Large boulders roll and bound beyond the toe of the active talus where they are scattered across the surface of a river terrace.

FIGURE 8.7 Coarse, angular limestone blocks stand at the angle of repose in a talus below steep cliffs in the central Brooks Range, Alaska.

grain-size distribution, sediment concentration, and other factors. The topmost part of the diagram includes combinations of velocity and sediment concentration that probably do not occur under natural conditions.

Slurry Flows

Slurry flows are moving masses of sediment that are saturated with water which is trapped among the grains and transported with the flowing mass.

The mixture is dense enough that large boulders can be suspended in the flow. Boulders too large to remain in suspension can be rolled along by the flow. When flow ceases, fine and coarse particles settle out together, resulting in a nonsorted sediment.

Solifluction. The slow viscous downslope movement of waterlogged soil and surficial debris is known as **solifluction**. Such flowage is especially noticeable in polar latitudes and alpine regions where

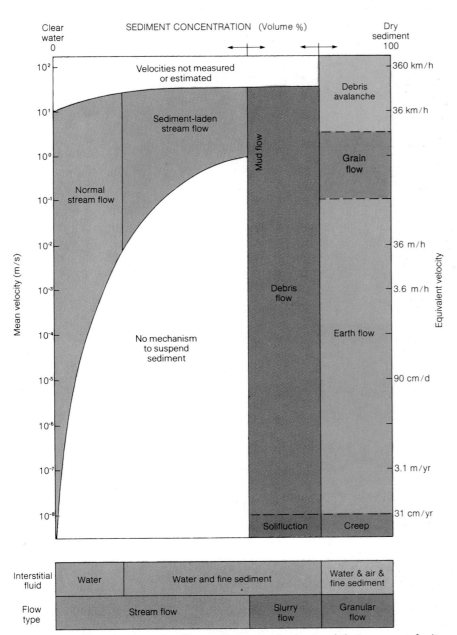

FIGURE 8.8 Classification of sediment flows on the basis of their mean velocity and sediment concentration. (*Source:* After Pierson and Costa, 1987.)

FIGURE 8.9 A meter-thick solifluction lobe has slowly moved downslope and covers glacial deposits on the floor of the Orgère Valley in the Italian Alps.

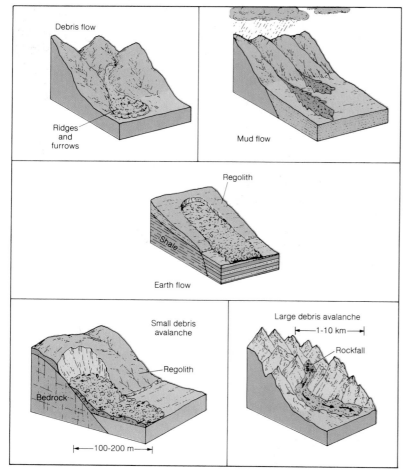

FIGURE 8.10 Examples of slope failures giving rise to debris flows, mudflows, earthflows, and debris avalanches.

shallow surface thawing of frozen ground on hill-slopes results in a layer of saturated debris resting on a frozen base (Fig. 8.9). However, it also occurs on hillslopes in temperate and tropical latitudes where sediment remains moist for long intervals. The movement is so slow, generally no more than a few centimeters a year, that it can be detected only by field measurements made over several seasons. The slow flowage results in distinctive surface features, including lobes and sheets of debris that often override one another. Surface vegetation carried along with the moving debris may be deformed and folded so that it resembles a crumpled carpet.

Debris Flows. **Debris flows** are a conspicuous form of mass-wasting. They involve the downslope movement of unconsolidated regolith, the greater part being coarser than sand. In some cases they begin with a slump, the lower part of which then continues to flow downslope (Figs. 8.10 and 8.11). Some debris flows travel at a rate of no more than 1 m/yr, whereas others travel at velocities measured in km/h.

Commonly, debris flows have an apronlike or tonguelike front. They also possess a very irregular surface marked by concentric ridges and transverse depressions that resemble the deposits of mountain glaciers. Debris flows are frequently associated with intervals of extremely heavy rainfall that lead to oversaturation of the ground.

FIGURE 8.11 Slump and debris flow near Mangaweka on North Island, New Zealand. Movement occurred following heavy rain on a slope that had recently been cleared of forest cover.

FIGURE 8.12 Channel of a mudflow that moved rapidly downslope through mature forest near Lake Chelan, Washington, during a sudden storm.

Mudflows.   A debris flow that has a water content sufficient to make it highly fluid is often called a **mudflow.** Because of their high mobility, mudflows tend to travel along the floors of valleys (Fig. 8.12).

Mudflow sediment ranges in consistency from mud as stiff as freshly poured concrete to a souplike mixture nearly equal to that of very muddy water. In fact, after heavy rains in mountain canyons, a mudflow can start as a muddy stream that continues to pick up loose sediment until its front portion becomes a moving dam of mud and rubble, extending to each steep wall of the canyon and urged along by the force of the flowing water behind it. On reaching open country at the mountain front, the moving dam collapses, floodwater pours around and over it, and mud mixed with boulders is spread as a wide thin sheet (Fig. 8.10). Sediment fans at the foot of mountain slopes in many arid regions, as in the American southwest, the central Argentine Andes, and the Hindu Kush of central Asia, consist largely of superposed sheets of mudflow sediments that are interstratified with stream sediments (Fig. 8.13).

FIGURE 8.13 Three superposed mudflow deposits, separated by buried soils, exposed in an alluvial fan in Arroyo Grande, Argentine Andes.

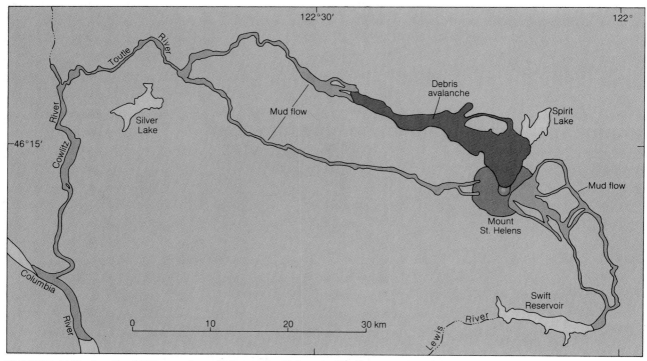

FIGURE 8.14 During the 1980 eruption of Mount St. Helens in Washington State, a massive debris avalanche traveled down the North Fork of the Toutle River, and volcanic mudflows were channeled down valleys west and east of the mountain. Some mudflows reached the Columbia River, having traveled more than 90 km. Flow velocities were as high as 40 m/s and averaged 7 m/s. (*Source:* After Janda et al., 1981.)

In regions of active volcanism, layers of volcanic ash and debris commonly mantle the surface and are especially susceptible to mobilization as mudflows. Highly mobile mudflows can travel great distances and at such high velocities (100 km/h or more) that they constitute one of the major hazards associated with many volcanic eruptions. Mudflows are so widespread in valleys surrounding some volcanoes in the Cascade Range that a substantial percentage of the total eruptive products from a vent actually lies beyond the volcanic cone itself. A mudflow that originated on the slopes of Mount Rainier in the Cascade Range about 5700 years ago traveled as far as 72 km. It spread out beyond the mountain front as a broad lobe where it is as much as 25 m thick. Its volume is estimated at well over a billion cubic meters. Mount St. Helens, an especially active volcano, has produced mudflows throughout much of its 35,000-year history. The most recent occurred during the huge eruption of May 1980 (Fig. 8.14).

Granular Flows

In granular flows, the full weight of the flowing mass of sediment is supported by grain-to-grain contact or collision between grains. The sediment may be largely dry with air filling the pores, or it can be saturated with water but have a grain-size distribution that allows water to escape easily.

Creep. The consistent leaning of old fences, poles, and gravestones, and the fracture and displacement of road surfaces are among the common types of evidence for **creep**, the imperceptibly slow downslope movement of regolith. Natural and artificial exposures of bedrock often show steeply inclined layers of rock bent over in the downslope direction just below the ground surface, a result of slow differential creep (Fig. 8.15).

Loose, incoherent deposits on or at the base of slopes and moving mainly by creep are termed **colluvium.** The particles in a body of colluvium tend to be angular and to lie in a chaotic jumble. These characteristics generally make it possible to distinguish colluvium from sediment deposited by flowing water or air, for such sediments tend to consist of rounded particles, sorted and deposited in layers.

A number of factors contribute to creep (Table 8.1). One factor, common in regions with cold winters, involves the freezing of water in the regolith. As water freezes, it increases in volume. Ice forming in regolith therefore pushes the ground surface up. This lifting of regolith by the freezing of contained water is called **frost heaving.** On a hillside the ground surface is lifted essentially at right angles to the slope. When thawing occurs, each particle tends to drop vertically, pulled downward by gravity. Its net motion, therefore, is a small distance downslope (Fig. 8.16). Long-term movement consists of a complex series of zigzags that can lead to substantial downslope displacement.

FIGURE 8.15 Beds of shale have been overturned by slow downslope creep on a hillslope in the Laramie Basin of Wyoming.

TABLE 8.1 *Factors Contributing to Creep of Regolith*

Frost Heaving	Freezing and thawing, without necessarily saturating the regolith, causing lifting and subsidence of particles
Wetting and Drying	Causes expansion and contraction of clay minerals; creation and disappearance of films of water on mineral particles causes volume changes
Heating and Cooling Without Freezing	Causes volume changes in mineral particles
Growth and Decay of Plants	Causes wedging, moving particles downslope; cavities formed when roots decay are filled from upslope
Activities of Animals	Worms, insects, and other burrowing animals displace particles, as do animals trampling the surface
Dissolution	Dissolution of mineral matter creates voids which tend to be filled from upslope
Activity of Snow	Where a seasonal snow cover is present, it tends to creep downward and drag with it particles from the underlying surface

Although creep occurs at a rate too slow to be seen, careful measurements of the downslope displacement of objects at the surface show the rates involved. As might be expected, rates tend to be higher on steep slopes than on gentle slopes. Measurements in Colorado, for example, document a rate of 1.5 mm/yr on a 19° slope but indicate a rate of 9.5 mm/yr on a slope of 39°. Rates also tend to increase as soil moisture increases. However, in wet climates vegetation cover also increases and the roots of plants, which bind the soil together, may inhibit creep. The creep rate measured on one grassy hillside in England that has a slope of 33° is only 0.02 mm/yr.

Earthflows. An ***earthflow*** is a granular flow with a velocity generally in the range of about 1 m/day to 360 m/h (10^{-5} to 10^{-1} m/s) (Figs. 8.8 and 8.10). Like debris flows, they are often made up of weak or weathered regolith, occur on moderately steep slopes, and frequently are associated with excessive rainfall.

One type of earthflow occurs in highly porous clays, called *quick clays*. Such units will weaken if shaken suddenly, as by an earthquake. This causes them to fluidize and fail abruptly. Any structure built on them or in their path may be quickly demolished.

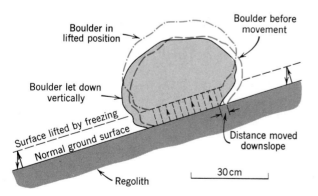

FIGURE 8.16 Stone moved downslope by alternate freezing and thawing of the ground. As freezing occurs, the stone is raised perpendicular to the ground surface which also rises; when the ground thaws, gravity pulls the stone down vertically, giving it a small but significant component of movement downslope.

Debris Avalanches. A granular flow moving at high velocity (>10 m/s; = >36 km/h) is termed a ***debris avalanche.*** Large, rapid debris avalanches are rare but spectacular events. They frequently involve large masses of falling rock and debris that break up, pulverize on impact, and then continue to travel downslope, often for great distances. They move at high velocity, may involve extremely large volumes of broken rocks, and can be extremely destructive (Table 8.2; Fig. 8.17).

Large rockfalls that give rise to rapidly moving debris avalanches have had the greatest human impact in populated mountain regions like the Alps. For example, in September 1717 a large mass of rock and glacier fell from the crest of the Mont Blanc massif along the French–Italian border onto Triolet Glacier (Fig. 8.18). Pulverizing on impact, the frag-

FIGURE 8.17 Deposits of a large debris avalanche that passed along the floor of a valley leading from the upper slopes of Huascaran, a high glacier-clad summit in the Peruvian Andes. A huge mass of rock, snow, and ice was converted into a rapidly moving debris flow that swept down the valley and buried a large town, killing most of its inhabitants.

TABLE 8.2 *Characteristics of Some Large Debris Avalanches*

Locality	Date	Volume (million m³)	Vertical Movement (m)	Horizontal Movement (km)	Calculated Velocity (km/h)
Huascaran, Peru	1971	10	4000	14.5	400
Sherman Glacier, Alaska	1964	30	600	5.0	185
Mt. Rainier, Washington	1963	11	1890	6.9	150
Madison, Wyoming	1959	30	400	1.6	175
Elm, Switzerland	1881	10	560	2.0	160
Triolet Glacier, Italy	1717	20	1860	7.2	≥125
Black Hawk, California	prehistoric	280	1220	8.0	120
Saidmarreh, Iran	prehistoric	2000	1650	14.5	340

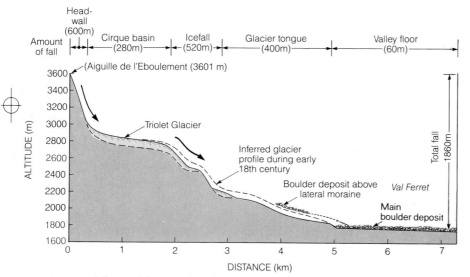

FIGURE 8.18 A large debris avalanche in upper Val d'Aosta, Italy, traveled 7 km from its source high on a mountain spur. Within only a few minutes the debris buried two communities on the valley floor, killing all the inhabitants and livestock. The reconstructed trajectory of the debris avalanche, which occurred in 1717, is based on deposits left along the valley sides. (*Source:* After Porter and Orombelli, 1980.)

mented debris moved rapidly downvalley where it overwhelmed two settlements before its front came to rest some 7 km from, and 1860 m lower than, the site of detachment. An estimate of the velocity can be obtained by equating the kinetic and potential energy of the rock mass

$$\tfrac{1}{2}mv^2 = mgh$$

and then solving for velocity

$$v = \sqrt{2gh}$$

By inserting values for gravitational acceleration (9.8 m/s) and initial distance of fall (400 m), we arrive at a velocity on impact of close to 320 km/h. As the sheet of debris reached the floor of the main valley, its momentum carried it up the opposite valley wall to a height of at least 60 m. Using the same equation, we can calculate that its velocity must have then been at least 125 km/h. The total travel time, from start to finish, over the entire 7 km distance must have been between 2 and 4 min.

Because large debris avalanches are infrequent and extremely difficult to study, there are few observational data about the process. Their extreme mobility has been attributed to the debris riding on a layer of compressed air, somewhat like a commercial hovercraft that moves across a gentle surface on a layer of compressed air. Alternatively, it may be air trapped within the debris mass that causes it to behave in a highly fluid manner.

Subaqueous Mass Movement

Extensive studies of the continental shelves and slopes have shown that submarine mass movement is extremely common and widespread. Mass movements also have been documented in lakes. As on land, wherever slopes occur beneath water, the potential exists for gravity-induced movement of rock and sediment.

Much of our knowledge of submarine mass movements has resulted from systematic investigations for petroleum on the continental shelves and slopes. Off the south coast of Alaska, a region of active tectonism, the continental shelf and slope display three subparallel zones with increasing deformation in the seaward direction. The innermost, marked by incipient slope failures, gives way to a zone marked by deep surface breaks and other signs of extension. Further seaward, on the slope, large rotational slumps are observed, some as thick as 230 m.

As we have seen (chapter 4), submarine slope failures can give rise to turbidity currents, a type of subaqueous sediment flow that travels down submarine canyons and deposits turbidites on the continental rise. In some regions subaqueous mass movements also account for a substantial portion of shelf deposits, especially in areas seaward of large rivers. Most deltaic regions characteristically display surface features and sediments attributable

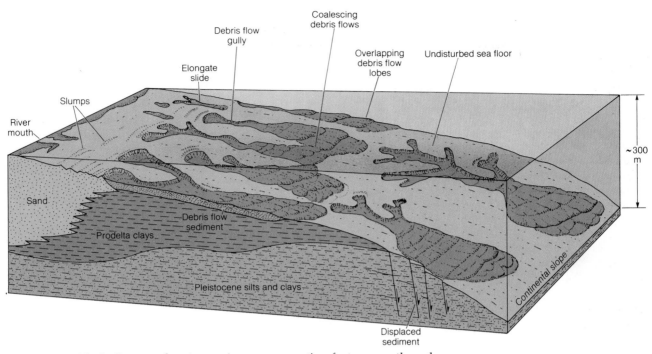

FIGURE 8.19 Block diagram showing various mass-wasting features on the submarine surface of the Mississippi Delta. (*Source:* After Coleman and Prior, 1983.)

to slope failures. In such environments, instability can occur even on slopes of 1° or less.

Slope failures generally display three distinct zones: a source region where subsidence and rotational slumping takes place, a central channel where sediment is transported, and a zone where sediment is deposited, often in the form of overlapping lobes of debris. These and other features are common on the submarine slopes of the Mississippi delta, which is among the best-studied deltas in the world (Fig. 8.19). Slides and sediment flows are extremely active on the delta front, with some movements having documented rates of several hundred meters per year. In places on the delta slope, more than 30 m of sediment has been deposited by sediment flows in the last 100 years.

TRIGGERING OF MASS-WASTING EVENTS

Mass-wasting events sometimes seem to occur at random, with no apparent reason. However, the largest, most disastrous, and most numerous events commonly are related to some extraordinary activity or occurrence.

Shocks

Sudden *shocks,* such as an earthquake, may release so much energy that slope failures of many types and sizes are triggered simultaneously. In 1929 a major earthquake (magnitude 7.7; chapter 14) in northwestern South Island, New Zealand triggered at least 1850 landslides larger than 2500 m² within an area of 1200 km² near the quake's center. An estimated 210,000 m³ of debris was displaced, on average, in each km² of land. Landslides were reported to be most numerous on well-bedded and well-jointed mudstones and fine sandstones. The Alaska earthquake of 1964 triggered many rockfalls, one of which became a huge rock avalanche that swept across the surface of Sherman Glacier, burying it with up to several meters of coarse, angular debris.

Slope Modification

Landslides often result from *slope modification* or *removal of support.* They typically occur where roads or highways have been cut into regolith creating an artificial slope which exceeds the angle of repose (Fig. 8.20). Retaining walls may inhibit landslides, but unless they are very strong, creep of the re-

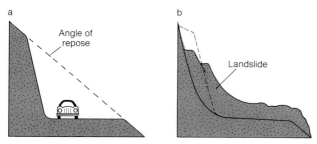

FIGURE 8.20 Modification of a slope in building a road leads to slope failure. (a) In constructing a road, the natural angle of repose of the sediment is exceeded. (b) The oversteepened slope fails, and a landslide buries the road. In the process, the natural angle of repose is re-established.

golith and subsequent failure may render them ineffective.

Undercutting

Slumps and other types of landslides may be triggered by the *undercutting* action of a stream along its bank or by surf action along a coast. Coastal landslides are especially common during large storms which may direct their energy against rocky headlands or along the bases of cliffs of unconsolidated sediments. Seacliffs on many volcanic islands retreat as wave action quarries away thin, well-jointed lava flows, thereby causing overlying rocks to collapse. The resulting debris is then quickly reworked by surf and transported away (Fig. 13.10).

Exceptional Precipitation

Landslides are frequently associated with heavy or prolonged rains that saturate the ground and make it unstable. Such was the case in 1925 when prolonged rains, coupled with melting snow, started a large debris flow in the Gros Ventre River basin of western Wyoming. The water saturated a sandstone that overlies shale and dips toward the valley floor, creating conditions that were ideal for slope failure. An estimated 37 million m^3 of rock, regolith, and organic debris moved rapidly downslope and created a natural dam that ponded the river. Two years later the dam failed, causing a flood that led to several deaths. Today, more than sixty years after the debris flow began, the scar at the head of the slide is still quite obvious, as is the hummocky topography downslope.

Eruptions

Volcanic eruptions are still another means of initiating mass-wasting events. Stratovolcanoes consist of an inherently unstable pile of interstratified lava flows, rubble, and pyroclastic layers. Unconsolidated deposits generally lie at the angle of repose. On glaciated volcanoes, slopes may be oversteepened by glacial erosion. During eruptions, slope failure is common and often widespread. If a volcano supports glaciers or extensive snowfields, melting of snow and ice can release large quantities of water which combine with unconsolidated deposits on the slopes and move rapidly downvalley as mudflows.

Submarine Slope Failures

Factors leading to unstable conditions on continental slopes and delta fronts include the following: (1) high internal pressures resulting from rapid deposition of sediment and the inability of trapped intergrain water to escape; (2) generation of methane gas from organic matter deposited with sediments; (3) local oversteepening of slopes owing to high rates of sedimentation; (4) slope steepening and removal of support at the base of slopes through faulting; and (5) shocks induced by earthquakes.

MASS-WASTING HAZARDS

As the human population increases in number and cities and road systems expand across the landscape, the likelihood that mass-wasting processes will affect people increases (chapter frontispiece). Although it may not always be possible to predict accurately the occurrence of events that will trigger mass movements, a knowledge of the processes and their relationship to local geology can lead to intelligent planning which can help reduce the loss of lives and property.

Hazards Assessments

Assessments of potential hazards resulting from mass-movement events are based mainly on reconstructions of similar past events in order to evaluate the magnitude of their impact on the landscape and the frequency of occurrence. From such information, it is possible to calculate how often an event of a certain magnitude is likely to recur.

Maps showing potential areas of impact of mass-wasting events (such as slumps, debris flows, or

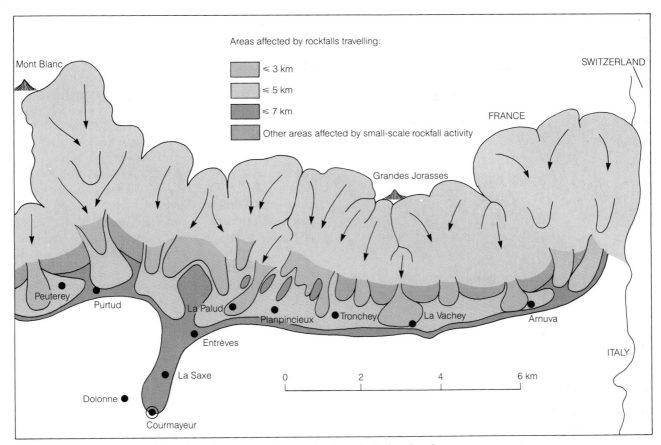

FIGURE 8.21 Map of part of the Mont Blanc massif along the French–Italian border showing zones that could be reached by giant rockfalls like that of Figure 8.18 traveling up to 3, 5, or 7 km from potential source areas high on the mountain slopes. Distribution is based on known pattern of rockfall deposits and on potential trajectories. Several small communities are built on or near large prehistoric rockfall deposits and lie within zones of potential future rockfall activity. (*Source:* After Porter and Orombelli, 1981.)

debris avalanches) are important tools for land-use planners wishing to minimize potential risks from such processes. For example, debris avalanches and smaller rockfalls are ever-present hazards in northern Italy along the south flank of the Mont Blanc massif. Field studies have shown that large debris avalanches similar to that of September 1717 (Fig. 8.18) have blanketed the floors of the major valleys with rocky debris repeatedly during the last 3000 years. From this evidence, a map has been constructed showing the areas that could be affected by future events having various trajectories and distribution patterns (Fig. 8.21). It is apparent that a number of small communities in the valley are at hazard from small- to intermediate-size events (3- to 5-km-long runouts), while several large communities, including Courmayeur with a population of several thousand, could be affected by large debris avalanches (up to 7 km long) like some recorded in deposits on the valley floors.

Valleys draining active volcanoes in the Cascade Range contain deposits of large mudflows that flowed off the volcanoes repeatedly during the last 10,000 years. Based on the number and extent of such deposits, hazards maps have been prepared, like that shown for Mount Rainier and vicinity in Figure 8.22. This map shows that risk from mudflows is high within about 25 km of the volcano's slopes, and that risk exists even at distances of more than 100 km along densely populated valley floors. A similar map prepared prior to the 1980 eruptions of Mount St. Helens proved prophetic, for mudflows generated during that series of eruptive events had distributions closely similar to those predicted on the basis of geologic studies of past events.

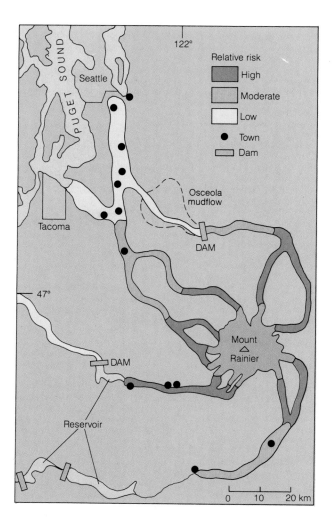

Hazards Mitigation

The impact of mass-wasting processes on human environments can often be reduced or eliminated by advanced planning. Areas subject to slow flowage can be stabilized by draining or pumping water from saturated sediment, while oversteepened hillslopes can be prevented from slumping if they are regraded to angles equal to or less than the natural angle of repose. In mountain valleys subject to mudflows from active volcanoes, man-made reservoirs can be quickly emptied of water so dams can restrain potentially destructive mudflows from reaching population centers (Fig. 8.22). Although large rockfalls and debris avalanches generally can neither be anticipated nor prevented, eliminating or restricting human activities in possible impact zones offers the best means of mitigating such hazards.

FIGURE 8.22 Map of southeastern Puget Lowland, Washington, showing areas of low, moderate, and high risk from mudflows and floods originating at Mount Rainier volcano. The extent of the 5700-year-old Osceola mudflow is shown by a dashed line. (*Source:* After Crandell and Mullineaux, 1975.)

Essay

COLLAPSING VOLCANOES

The mighty eruption of Mount St. Helens in May 1980 provided an awesome example of how an active volcano can literally destroy itself. The gigantic lateral blast of superheated gases and fragmented rock debris that devastated a broad area beyond the mountain's flanks was accompanied by a massive debris avalanche that moved an estimated 2.8 km³ of rock, ice, and regolith downslope at 350 km/h into an adjacent valley (Fig. 8.14). The landslide and blast left a gaping hole in the side of the volcano and lowered its summit by 420 m. The collapse and accompanying blast had not been anticipated by most geologists, and it motivated additional studies which made clear that flank collapse is far more common on volcanoes than once believed.

One problem with recognizing past debris avalanches from volcanoes is often the immensity of their dimensions. The elongated valley that extends some 40 km north of towering Mount Shasta in northern California contains a complex of hillocks and mounds of volcanic rock that many geologists had interpreted as relicts of numerous minor eruptions from isolated vents. Recognition of the similarity between these deposits and those of the huge debris avalanche from Mount St. Helens prompted a reassessment and led to the remarkable conclusion that the entire array of features resulted from a similar gigantic collapse of a flank of the Shasta volcano about 300,000 years ago. In this case, however, the volume of rock involved was at least 26 km³—almost ten times the volume of the landslide on Mount St. Helens. Whereas the St. Helens deposits cover an area of 60 km², the Shasta debris avalanche overwhelmed at least 450 km² (Fig. B8.1).

These and other volcanoes that have suffered collapse of their flanks possess several common attributes. They generally rise to heights of more than 1000 m and their slopes are steeper than 20°. Furthermore, most con-

FIGURE B8.1 A massive prehistoric avalanche deposit (foreground) extends 43 km northwest of Mount Shasta volcano (in the distance) and covers at least 450 km².

sist of piles of jointed lava flows interstratified with loose tephra and colluvium, making them inherently unstable. However, many volcanoes possessing these same characteristics have not suffered collapse. This suggests that the generation of huge debris avalanches may be linked to a triggering mechanism, such as an earthquake or the intrusion of magma into the volcanic edifice, both of which were involved in the Mount St. Helens event. In any case, caution dictates that potential hazard zones around active stratovolcanoes should extend into valleys far beyond the actual flanks of the mountains, for although large flank collapses are not a necessary phase of each volcano's evolution, they nevertheless are relatively common and constitute an ever-present danger.

The largest volcanoes on the Earth rise steeply from the floor of the Pacific Ocean to heights of 7000 m or more. Like their smaller cousins on the continents, they are composed of huge piles of well-jointed lava flows and unstable rubble. Chaotic topography along the submerged lower margins of the Hawaiian Islands has been interpreted as evidence of repeated massive landslides on the volcano flanks. It has been suggested that much of the northern half of East Molokai volcano (Fig. 17.8) lies on the adjacent seafloor below a steep landslide scarp. Anomalous coral-bearing gravels found to altitudes of 60 m or more on the southeastern islands of the chain have recently been attributed to the surge of a giant wave, produced when a huge submarine landslide carried away part of the south side of Lanai about 100,000 years ago.

SUMMARY

1. The pull of gravity causes rock debris to move downslope without a transporting medium. It affects rock debris both on land and beneath the sea.

2. The composition and texture of debris, the amount of air and water mixed with it, and the steepness of slope influence the type and velocity of slope movements.

3. Slumps involve a rotational movement along a concave-up slip surface. They commonly result in hummocky terrain having backward-tilted blocks of rock or regolith and concentrically aligned depressions.

4. Falling and sliding rock and debris masses are common in glaciated mountains where steep slopes abound.

5. Sliderock accumulates at the base of cliffs to produce taluses with slopes that stand at the angle of repose.

6. Slurry flows involve dense moving masses of saturated sediment that form nonsorted deposits when flow ceases. Flow velocities range from very slow (solifluction) to rapid (debris flows).

7. Eroding muddy streams can turn into mudflows that move rapidly down canyons and spread out as thin sheets of nonsorted sediment on fan surfaces at the foot of mountain slopes.

8. Highly mobile volcanic mudflows can travel many tens of kilometers from the slopes of volcanoes where they originate.

9. Although creep is imperceptibly slow, it is widespread and therefore quantitatively important in the downslope transfer of debris.

10. Large, rapidly moving debris avalanches are relatively infrequent but potentially hazardous to humans.

11. Mass-wasting events on hillslopes can be triggered by earthquakes, undercutting by streams, heavy or prolonged rains, or volcanic eruptions. Subaqueous slope failures may further be related to rapid deposition of sediments, oversteepening of slopes, and earthquake shocks.

12. Loss of life and property from mass-movement events can be prevented or mitigated by advanced assessment and planning based on geologic studies of previous occurrences.

IMPORTANT WORDS AND TERMS TO REMEMBER

angle of repose (p.179)

colluvium (p. 185)
creep (p. 185)

debris avalanche (p. 186)
debris fall (p. 179)
debris flow (p. 183)
debris slide (p. 179)

earthflow (p. 186)

flow (p. 180)
frost heaving (p. 185)

granular flow (p. 180)
gravity (p. 176)

landslide (p. 176)

mass-wasting (p. 176)

mudflow (p. 184)

perpendicular component of gravity (p. 176)

quick clay (p. 186)

rockfall (p. 179)
rockslide (p. 179)

shocks (p. 189)
sliderock (p. 179)
slope modification (removal of support) (p. 189)
slump (p. 177)
slurry flow (p. 180)
solifluction (p. 181)
steady-state (p. 176)

talus (p. 179)
tangential component of gravity (p. 176)

undercutting (p. 190)

QUESTIONS FOR REVIEW

1. How does mass-wasting differ from weathering and from stream erosion?
2. Into what two component forces can gravity be resolved on any sloping surface?
3. What distinctive topographic features characterize an area where numerous slumps have occurred?
4. What conspicuous type of deposit generally is found at the base of a cliff subject to frequent rockfalls?
5. Why does the surface of most sliderock deposits stand at a characteristic slope angle?
6. How does a slurry flow differ from a granular flow? What are examples of each?
7. How might one prove that creep is occurring on a slope and how can its rate be measured?
8. What differences in environment might explain why some alluvial fans in mountainous regions consist entirely of alluvium while others contain interstratified layers of alluvium and mudflow deposits?
9. Why are lava flows erupted from stratovolcanoes largely restricted to the volcanic cones, while mudflows originating on the same mountains are often distributed for many tens of kilometers down adjacent valleys?
10. What characteristics might serve to distinguish colluvium from alluvium in natural exposures?
11. Why are large debris avalanches especially dangerous to humans?
12. What distinctive features on the submarine portions of deltas provide evidence of subaqueous mass movement?
13. What are four factors that can trigger mass movement events on hillslopes or on submarine slopes?

CHAPTER 9

Groundwater

Limestone pinnacles rising steeply above the Li River near Guilin, China, are a classic example of tower karst.

WATER IN THE GROUND

Access to water, whether from streams, lakes, springs, direct rainfall, or from underground, is a vital human need. Most early cities and towns were founded close to streams that would provide a reliable source of water. With growth of population, the streams often became insufficient. People then resorted to bringing water from a more distant source through canals, or obtained water from underground supplies by digging wells.

As society has become increasingly industrialized, communities have generated ever-larger amounts of human and industrial wastes, a good deal of which has inevitably found its way into the very water that people must rely on for their existence. In many places water is dwindling both in quantity and quality, creating important questions for the communities involved: Will there be enough clean water to sustain future needs? Is its quality adequate for the uses to which we put it? Is the water being used efficiently, and with a minimum of waste?

Origin of Groundwater

Less than 1 percent of the water on the Earth is *groundwater,* defined simply as all the water contained in spaces within bedrock and regolith. Although the total volume of groundwater is small, it is 35 times larger than the volume of water lying in freshwater lakes or flowing in streams on the Earth's surface.

Groundwater has its origin in rainfall. Most of it is slowly moving on its way back to the ocean, either directly through the ground or by flowing out onto the surface and joining streams (Fig. 1.13).

That groundwater comes from rain was not established on a quantitative basis until the seventeenth century when Pierre Perrault, a French physicist, measured the mean annual rainfall for a part of the drainage basin of the Seine River in eastern France and the mean annual runoff from it in rivers. After allowing for loss by evaporation, he concluded that the difference between the amounts of rainfall and runoff was ample enough, over a period of years, to account for the amount of water in the ground.

Depth of Groundwater

Water is present everywhere beneath the ground surface. More than half of all groundwater, including most of the water that is usable, occurs within about 750 m of the Earth's surface. The volume of water in this zone is estimated to be equivalent to a layer of water approximately 55 m thick spread over the world's land area. Below a depth of about 750 m, gradually though irregularly water decreases in amount. Holes drilled for oil have found water as deep as 9.4 km. A deep hole drilled by Soviet scientists on the Kola Peninsula encountered water at depths of more than 11 km. However, even though water may be present in crustal rocks at such depths, the pressure exerted by overlying rocks is so high and openings in rocks are so small that it is unlikely that much water is present or that it can move freely through the enclosing rocks.

Water Table

Much of our knowledge of where groundwater occurs has been learned from the accumulated experience of generations of people who have dug or drilled millions of wells. This experience tells us that a hole penetrating the ground ordinarily passes first into an **unsaturated zone** (or **zone of aeration**), the zone in which open spaces in regolith or bedrock are filled mainly with air. The hole then enters the **saturated zone,** the zone in which all openings are filled with water. The upper surface of the saturated zone is the **water table,** which normally slopes toward the nearest stream or lake (Fig. 9.1). In moist climatic regions the water table ordinarily lies within a few meters of the surface. Whatever its depth, the water table is a significant surface, for it represents the upper limit of all readily usable groundwater.

MOVEMENT OF GROUNDWATER

Most of the groundwater within a few hundred meters of the surface is in motion. Unlike the swift flow of rivers, which is measurable in kilometers per hour, groundwater moves so slowly that velocities are expressed in centimeters per day or meters per year. To understand why the movement is so slow, we must know something about the porosity and permeability of rocks.

Porosity and Permeability

The amount of water that can be contained within a given volume of rock or sediment depends on the **porosity,** which is the proportion (in percent) of the total volume of a body of bedrock or regolith that consists of open spaces (pore spaces). A very

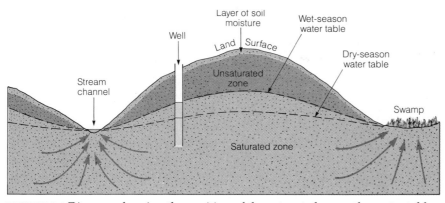

FIGURE 9.1 Diagram showing the position of the saturated zone, the water table, and the unsaturated zone in a typical groundwater system. Arrows show direction of flow in the saturated zone toward low places in the landscape.

porous rock is one containing a comparatively large proportion of pores, regardless of their size. Sediment is ordinarily very porous, ranging from 20 percent in some sands and gravels to as much as 50 percent in some clays. Porosity is affected by the sizes and shapes of the rock particles and the compactness of their arrangement (Fig. 9.2a and b). The degree to which the pores of a sedimentary rock become filled with cementing substances also affects porosity (Fig. 9.2c). The porosity of igneous and metamorphic rocks generally is low. However, if such rocks have many joints and fractures, their porosity will be higher.

Permeability is the capacity for transmitting fluids. A rock of very low porosity is also likely to have low permeability. However, high porosity values do not necessarily mean high permeability values, because size and continuity of the openings influence permeability in an important way. The relationship between the size of openings and the molecular attraction of rock surfaces plays a large part. *Molecular attraction* is the force that makes a thin film of water adhere to a rock surface despite the force of gravity. An example is the wet film on a pebble that has been dipped in water. If the open space between the two adjacent particles in a rock is small enough, the films of water that adhere to the two particles will come into contact with each other. The force of molecular attraction therefore

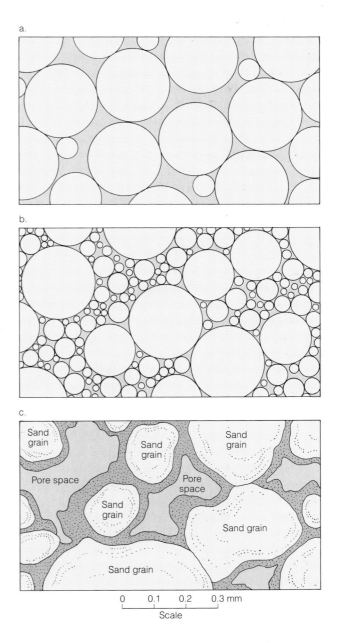

FIGURE 9.2 Contrasts in porosity of sediments. (a) A porosity of 32 percent in a reasonably well-sorted sediment. (b) A porosity of 17 percent in a poorly sorted sediment in which fine grains fill spaces between larger ones. (c) Reduction in porosity of an otherwise porous sediment due to the presence of a cementing agent.

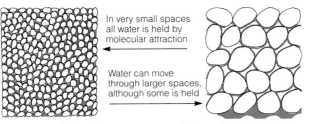

In very small spaces all water is held by molecular attraction

Water can move through larger spaces, although some is held

FIGURE 9.3 The effect on permeability of molecular attraction in the intergranular spaces of a fine sediment (left) and a coarser sediment (right) of equal porosity. The scale is larger than natural size.

extends right across the open space (Fig. 9.3). At ordinary pressures, the water is held firmly in place and permeability is low. That is what happens in a wet sponge before it is squeezed. The same thing happens in clay, the particles of which have diameters of less than 0.005 mm (Table 4.1).

By contrast, in a sediment with grains at least as large as sand (0.06 to 2 mm) the open pores commonly are wider than the films of water adhering to the grains. Therefore, the water in the centers of the openings is free to move (Fig. 9.3). Such sediment is permeable. As the diameters of the openings increase, permeability increases. Gravel, with very large openings, is more permeable than sand and can yield large volumes of water to wells.

Movement in the Unsaturated Zone

Water from a rain shower soaks into the soil, which usually contains clay resulting from the chemical weathering of bedrock. Due to its content of extremely fine clay particles, the soil is generally less permeable than underlying coarser regolith or rock. The low permeability and the fine clay cause part of the water to be retained in the soil by forces of molecular attraction. This is the layer of soil moisture shown in Figure 9.1. Some of this moisture evaporates directly, but much is taken up by plants

which later return it to the atmosphere through transpiration (Fig. 1.13).

Water that molecular attraction cannot hold in the soil seeps downward until it reaches the water table. In fine-grained sediment a narrow fringe as much as 60 cm thick immediately above the water table is kept wet by *capillary attraction,* the same force that draws ink through blotting paper and kerosene through the wick of a lamp. With every rainfall, more water is supplied from above, but apart from soil moisture and the capillary fringe, the unsaturated zone is likely to be nearly dry between rains (Fig. 9.1).

Movement in the Saturated Zone

The movement of groundwater in the saturated zone, called *percolation,* is similar to the flow of water when a saturated sponge is squeezed gently. Water moves slowly by percolation through very small open spaces along parallel, threadlike paths. Movement is easiest through the central parts of the spaces but diminishes to zero immediately adjacent to the sides of each space because the water is held in place there by molecular attraction.

Responding to gravity, water percolates from areas where the water table is high toward areas where it is lowest. In other words, it flows toward surface streams or lakes (Fig. 9.4). Only part of the water travels directly down the slope of the water table by the shortest route. Much of it flows along innumerable long curving paths that go deeper through the ground. Some of the deeper paths turn upward against the force of gravity and enter the stream or lake from beneath. This happens because the water in the saturated zone at any given altitude is under greater pressure beneath a hill than beneath a stream. The water therefore tends to move toward points where pressure is least. However, most of the groundwater entering a stream travels along shallow paths not far beneath the water table.

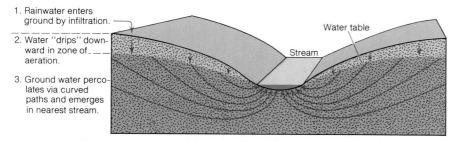

1. Rainwater enters ground by infiltration.

2. Water "drips" downward in zone of aeration.

3. Ground water percolates via curved paths and emerges in nearest stream.

Water table

Stream

FIGURE 9.4 Movement of groundwater in a humid region in uniformly permeable rock. Long curved arrows represent only a few of many possible paths.

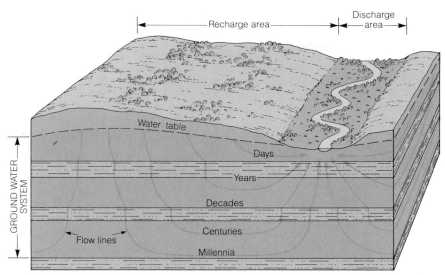

FIGURE 9.5 Recharge and discharge areas in a humid landscape. The time required for groundwater to reach the discharge area from the recharge area depends on the route and distance of travel. Downward and upward percolation is faster and more direct in the most porous strata. (*Source:* After Heath, 1983.)

THE NATURE OF THE GROUNDWATER SYSTEM

The shallower part of the Earth's groundwater system operates continually as a small but integral part of the hydrologic cycle (chapter 1). It acts both as reservoir and conduit. Water entering the system moves through the ground and escapes into stream valleys on its way toward the ocean.

Recharge and Discharge Areas

Water enters the groundwater system as precipitation falling on *recharge areas,* which are areas where water is added to the saturated zone (Fig. 9.5). It moves through the system to *discharge areas,* which are areas where subsurface water is discharged to streams or to bodies of surface water. The areal extent of recharge areas is invariably larger than that of discharge areas. In humid regions, recharge areas encompass nearly all areas except streams and their adjacent floodplains (Fig. 9.5). In more arid regions, recharge occurs mainly in mountains and in the alluvial fans that border them. It also occurs along channels of major streams that are underlain by permeable alluvium through which water leaks downward and recharges the groundwater (Fig. 9.6). The time it takes for water to move through the ground from the recharge area to the nearest discharge area depends on rates of flow and the travel distance. It may take only a few days, or possibly thousands of years in cases where water moves through the deeper parts of the groundwater system (Fig. 9.5).

Fluctuations of the Water Table

In humid regions the water table is a subdued imitation of the ground surface above it. It is high beneath hills and low at valleys because water tends to move toward low points in the topography where the pressure on it is least. If all rainfall were to cease, the water table would slowly flatten and gradually approach the levels of the valleys. Percolation would diminish and cease, and the streams

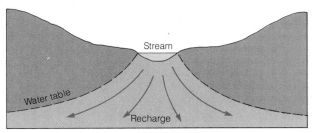

FIGURE 9.6 In arid regions, direct recharge is minimal, and the water table lies below the riverbed. Large through-flowing streams and intermittent streams during times of flow lose water, which percolates downward to resupply groundwater at depth. Compare with Figure 9.5 showing recharge in a humid region.

in the valleys would dry up. In times of drought, when rain may not fall for several weeks or even months, we can sense the flattening of the water table in the drying up of wells. When that occurs we know that the water table has fallen to a level below the bottoms of the wells. It is repeated rainfall, dousing the ground with fresh supplies of water, that maintains the water table at a normal level.

Discharge and Velocity

What, then, determines the steepness of the slopes of a water table and the rate of flow of the percolating groundwater? It was discovered by experiment that the rate of flow of groundwater increases as the slope of the water table increases, as long as the permeability of the ground remains uniform. In 1856, Henri Darcy, a French engineer, showed that the rate of groundwater flow through permeable materials is directly proportional to the product of the cross-sectional area through which flow can occur, the permeability, and the slope of the water table (the *hydraulic gradient*).

Because of the large amount of friction involved in percolation, flow rates are slow. Normally velocities range between half a meter a day and several meters a year. The largest rate yet measured in the United States, in exceptionally permeable material, is only about 250 m/yr.

AQUIFERS

Water-Table Aquifers

When we look for a good supply of groundwater we search for an *aquifer* (from the Latin for water carrier), a body of permeable rock or regolith saturated with water through which groundwater moves. Bodies of gravel and sand are commonly good aquifers, and so are many sandstones. However, the presence of a cementing agent between grains of a sandstone reduces the diameter of the openings and so reduces the effectiveness of these rocks as aquifers.

It might seem that claystones, igneous rocks, and metamorphic rocks would not be aquifers, because the spaces between their mineral grains are extremely small. However, what is true for small samples does not necessarily apply to large bodies of the same rock. Many such rock bodies contain fissures, spaces between layers, and other openings such as joints that permit free circulation of groundwater.

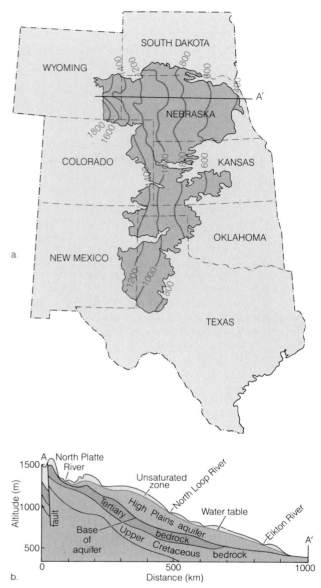

FIGURE 9.7 High Plains aquifer. (a) Map showing the distribution of aquifer and contours (in m) on the water table. Water flow is generally east, perpendicular to the contour lines. (*Source:* After Gutentag et al., 1984.) (b) Cross section along profile A–A' showing the relation of High Plains aquifer to underlying bedrock units and the position of the water table.

High Plains Aquifer

About 30 percent of the groundwater used for irrigation in the United States is obtained from an extensive aquifer that lies at shallow depths beneath the High Plains (Fig. 9.7). The aquifer is tapped by about 170,000 wells and supplies water for more than 20 percent of the irrigated land of the country. It averages about 65 m thick, and consists of a number of young sandy and gravelly rock units that lie

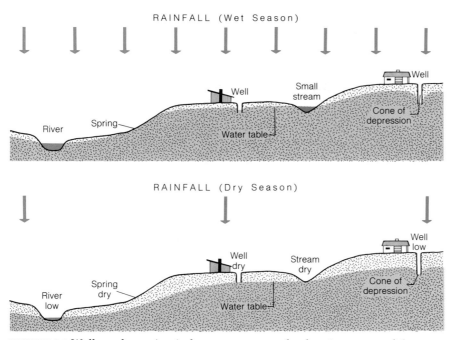

FIGURE 9.8 Wells and a spring in homogeneous rock, showing cones of depression and the effect of seasonal fluctuation of the water table. The slopes of the water table are steeper in the wet season, when the input of water into the system is greatest, than in the dry season, when the input is least.

at depths of less than 350 m. The water table slopes gently from west to east and water flows through the aquifer at an average rate of about 30 cm/day.

Development of groundwater for irrigation in the High Plains was spurred by severe regional drought in the 1930s and again in the 1950s. Annual recharge of the High Plains aquifer from precipitation is much less than the amount of water being withdrawn, so the inevitable result is a long-term fall of the water table. A dramatic increase in pumping rates has led to serious declines in water level. In parts of Kansas, New Mexico, and Texas, the saturated thickness has declined by more than 50 percent. The resulting decreased water yield and increased pumping costs have led to major concern about the future of irrigated farming on the High Plains.

Springs

A spring is a flow of groundwater emerging naturally at the ground surface. The simplest spring is one that issues from a place where the land surface intersects the water table (Fig. 9.8). A vertical or horizontal change in permeability is a common reason for the localization of springs (Fig. 9.9). Often this involves the presence of an *aquiclude*, a body of impermeable or distinctly less permeable rock

adjacent to an aquifer. If a porous sand overlies a relatively impermeable clay, water percolating downward through the sand will flow laterally when it reaches the underlying clay and may emerge as a spring where the stratigraphic contact crops out, as along the side of a valley or a coastal cliff. Springs may also be localized along structural features, such as faults. In fact, one way in which a fault can sometimes be identified at the land surface is by a series of springs aligned along its trace.

Small springs are found in all kinds of rocks. Almost all large springs issue from lava, limestone, or gravel aquifers.

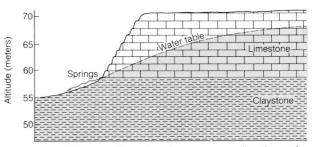

FIGURE 9.9 Cross section showing an unconfined aquifer overlying an impermeable bed. At the foot of the scarp is a zone of seeps or small springs.

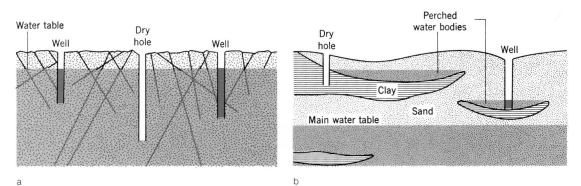

a   b

FIGURE 9.10 Ordinary wells and adjacent dry holes in rock that is not homogeneous. (a) In fractured massive rocks. (b) In bodies of permeable sand containing discontinuous bodies of impermeable clay. Two perched water bodies are shown.

Wells

An ordinary well fills with water simply because it intersects the water table (Fig. 9.8). If inflow of groundwater to the well cannot replenish water being removed, the water level in the well lowers. This creates a conical depression in the water table immediately surrounding a well called a *cone of depression* (Fig. 9.8). In most small domestic wells the cone of depression may be hardly discernible. Wells pumped for irrigation and industrial uses, however, withdraw so much water that the cone can become very wide and steep, and can lower the water table in all wells of a district. Figure 9.8 shows that a shallow well can become dry at times, whereas a nearby deeper well may yield water throughout the year.

If rocks are not homogeneous, the yields of wells may vary considerably within short distances. Massive igneous and metamorphic rocks (Fig. 9.10a), for example, are not likely to be very permeable except where they are cut by fractures. A hole that does not intersect fractures is therefore likely to be dry. Because fractures generally die out downward, the yield of water to a shallow well can be greater than to a deep one. Discontinuous bodies of permeable and impermeable rock or sediment (Fig. 9.10b) result in very different yields to wells. They also create *perched water bodies* (water bodies perched in positions higher than the main water table above an aquiclude). The impermeable layer catches and holds the water reaching it from above.

Confined Aquifers

In some regions the geometry of inclined rock layers permits groundwater to circulate through an aquifer that is confined between impermeable beds. Three essentials of the pattern are (1) a series of inclined strata that include a permeable layer sandwiched between impermeable ones; (2) rainfall to feed water into the permeable layer where the layer intersects the land surface; and (3) a fissure or well situated so that water from the permeable layer can escape upward through the impermeable rock above (Fig. 9.11).

Artesian Systems

When these essentials are present, we have an *artesian system.* The input consists of precipitation, which enters the permeable layer (now an aquifer by definition) and percolates through it. The output consists of water forced upward through fissures or wells that penetrate the capping rock. The upward flow is caused by the pressure of water flowing through the inclined aquifer from the area of recharge.

Figure 9.11 illustrates such a system. Except for a zone near the surface that lies above the water table, the whole series of strata is saturated with water. In the impermeable rock above and below the aquifer the water is motionless because it is held in place by capillary attraction in tiny spaces between mineral grains. However, within the aquifer water flows slowly downward, escaping from the system through fissures or wells that reach the land surface.

If rainfall reaching the area of recharge has a greater volume than the water discharged through fissures or wells, only enough water to balance output can enter the system. The excess flows away over the land surface. On the other hand, if wells

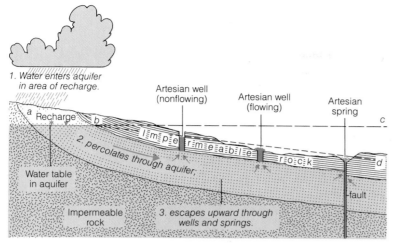

FIGURE 9.11 Artesian wells and springs. Three essential conditions are an aquifer, an impermeable roof, and water pressure sufficient to make the water in any well rise above the aquifer. The water rises, in any well, to the height (*bc*) of the water table in the recharge area (*ab*), minus an amount determined by the loss of energy in friction of percolation. Thus the water rises only to the line *bd*, which slopes downward away from the recharge area.

are drawing from the system more water than can enter through recharge, the flow from wells will decline until a balance is achieved.

Artesian Wells and Springs

A well in which water rises above the aquifer is an **artesian well.** Under unusually favorable conditions, water pressure can be great enough to create fountains that rise as much as 60 m above ground level. Natural springs that draw their supply of water from a confined aquifer in the same manner are called **artesian springs.**

Tertiary Limestone Artesian System

A major artesian aquifer confined to Tertiary limestone strata in Florida is full of caves and smaller openings that have been dissolved in the rock. In the central and northwestern parts of the peninsula these rocks are exposed at the surface and form the recharge area. To the northwest, northeast, and south they are covered by younger strata and slope downward toward the Gulf and Atlantic coasts (Fig. 9.12).

The age of water at various places within the system has been determined by measuring the radiocarbon content of carbonate molecules dis-

solved in the water. Samples were collected from wells along a line 133 km long and approximately parallel to the slope of the aquifer. Most of the radiocarbon enters the ground as precipitation falling on the recharge area and moves through the aquifer with the groundwater. The age was found to increase systematically away from the recharge area. Water in the well farthest from the recharge area is calculated to have been in the ground for at least 19,000 years.

Contamination of Groundwater Supplies

Pollution by Sewage

The *quality* of a body of water refers to its temperature and the amount and character of its content of mineral particles, dissolved substances, and organic matter (chiefly bacteria) in relation to its intended use. The most common source of water pollution in wells and springs is sewage. The infectious disease most often communicated by polluted waters is typhoid. Drainage from septic tanks, broken sewers, privies, and barnyards contaminates groundwater. If water contaminated with sewage bacteria passes through sediment or rock with large openings, such as very coarse gravel or

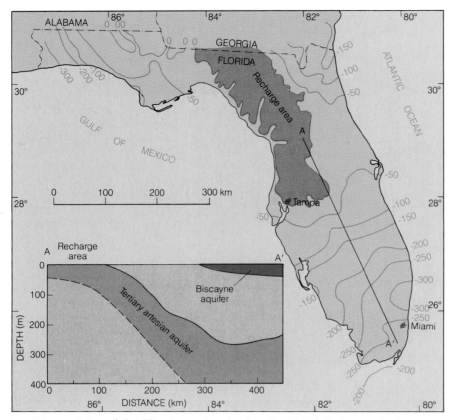

FIGURE 9.12 Map of the Florida peninsula showing depth to top of the Tertiary limestone artesian aquifer (in meters) and the area of recharge. Inset diagram shows a stratigraphic section along line A–A'. In the southern part of the state, the Tertiary aquifer lies below a surface aquifer in younger rocks. (*Source:* After Cederstrom et al., 1979.)

cavernous limestone, it can travel long distances without much change (Fig. 9.13). If, on the other hand, it percolates through sand or permeable sandstone, it can become purified within short distances, in some cases less than about 30 m (Fig. 9.14). Sand is especially ideal, for it promotes purification by (1) mechanically filtering out bacteria (water gets through but most of the bacteria do not); (2) destruction of bacteria by oxidation; and (3) destruction of bacteria by other organisms, which consume them. For this reason, purification plants that treat municipal water supplies and sewage percolate these fluids through sand. Unfortunately, dissolved chemicals (including pesticides and heavy metals) are not filtered out by this technique.

Contamination by Seawater

Along coasts, fresh groundwater is separated from seawater along a narrow transition zone (Fig. 9.15). Any pumping from an aquifer near the coast will

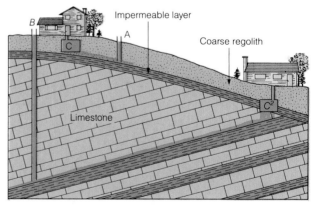

FIGURE 9.13 Pollution of wells. The shallow well, A, was unwisely located a short distance downslope from a septic tank, C, and received polluted drainage from it. The owner then drilled a deeper well, B. This well tapped layers of cavernous limestone dipping toward it from the lower septic tank, C^2. The water flowed through openings in the limestone and reached the bottom of well B unpurified by percolation. The well owner must relocate his septic tank or else dig a shallow well located upslope from C.

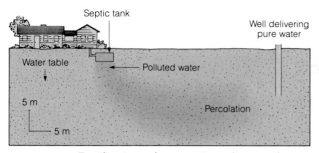

FIGURE 9.14 Purification of contaminated groundwater in sand and gravel during percolation through a short distance.

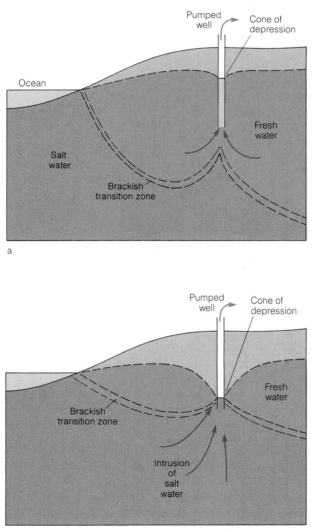

a.

b.

FIGURE 9.15 Seawater contamination of wells. (*Source: After T. Dunne and L. B. Leopold, 1978.*) (a) Near the coast, groundwater occurs as a lens over salty marine water. (b) Heavy pumping of groundwater forms a large cone of depression both at the top and at the base of the groundwater lens, eventually permitting saline water to enter the well.

reduce the flow of fresh groundwater toward the sea. In turn, this may allow saltwater to move landward through permeable strata. Excessive pumping that exceeds the natural flow of fresh groundwater toward the sea may eventually permit saline water to encroach far inland and reach major pumping centers. Such seawater intrusion can then contaminate water supplies (Fig. 9.15). Once intrusion occurs, it is very difficult to reverse.

GEOLOGIC ACTIVITY OF GROUNDWATER

Dissolution

As soon as rainwater infiltrates the ground, it begins to react with minerals in regolith and bedrock and weathers them by chemical means. This chemical weathering process, whereby minerals and rock material pass directly into solution, is known as *dissolution.* The dissolved matter contained in the groundwater reappears in streams and is carried to the sea, where it joins other substances in solution and eventually enters into the building of limestone and other marine sedimentary rocks.

Carbonate rocks, such as limestone and marble, are especially susceptible to dissolution (Fig. 9.16). By measuring over a period of time the amount of dissolution observed on small, precisely weighed limestone tablets placed at open sites in various areas, it is possible to calculate the average rate at which limestone landscapes are being lowered by dissolution. In temperate regions with high rainfall, a high water table, and a nearly continuous cover of vegetation, carbonate landscapes are being lowered at rates of up to about 10 mm/1000 yr. In dry regions, with scanty rainfall, low water tables, and discontinuous vegetation, rates are much lower. Measured rates of dissolution by groundwater in carbonate terrains of the United States show that the erosion rate can be greater than the average erosional reduction of the surface by mass-wasting, sheet erosion, and streams.

Chemical Content of Groundwater

Analyses of many wells and springs show that the elements and compounds dissolved in groundwater consist mainly of chlorides, sulfates, and bicarbonates of calcium, magnesium, sodium, potassium, and iron. We can trace these substances to the common minerals in the rocks from which they were weathered. As might be expected, the

FIGURE 9.16 Marble balustrade of the Forbidden City in Beijing, China shows the effects of more than 300 years of dissolution. The original sharply carved design has become smooth and indistinct.

composition of groundwater varies from place to place according to the kind of rock in which it occurs. In much of the central United States the water is "hard," that is, rich in calcium and magnesium bicarbonates, because the bedrock includes abundant limestones and dolostones that consist of those carbonates. In some places within arid regions the concentration of dissolved sulfates and chlorides is so great that the groundwater is unfit for human consumption. In particularly dry regions, evaporation of water in the unsaturated zone leads to deposition of substances that make soils unsuitable for agriculture. Such conditions may result from slow circulation of groundwater through sedimentary rock, from which salts are dissolved.

Chemical Cementation and Replacement

The conversion of sediment into sedimentary rock is primarily the work of groundwater. A body of sediment lying beneath the sea is generally satu-

rated with water, as is sediment lying in the saturated zone beneath the land. Substances in solution in the water are precipitated as cement in the spaces between rock particles that form the sediment. This activity transforms the loose sediment into firm rock. Calcite, silica, and iron compounds (mainly hydroxides) are, in that order, the chief cementing substances.

Less common than the deposition of cement between the grains of a sediment is *replacement*, the process by which a fluid dissolves matter already present and at the same time deposits from solution an equal volume of a different substance. Evidently replacement takes place on a volume-for-volume basis because the new material preserves the most minute textures of the material replaced. Both mineral and organic substances can be replaced. Petrified wood is a common example of the replacement of organic matter (Fig. 9.17).

Carbonate Caves and Caverns

Limestone, dolostone, and marble are carbonate rocks that consist of the minerals calcite and dolomite in various proportions. These rocks underlie millions of square kilometers of the Earth's surface. Although carbonate minerals are nearly insoluble in pure water, they are readily dissolved by the carbonic acid formed by the interaction of carbon dioxide and rainwater (Table 7.1, reaction 1), which percolates into the groundwater reservoir. As a result, the groundwater becomes charged with calcium and bicarbonate ions (Table 7.1, reaction 5).

The weathering attack occurs mainly along joints and other partings in the carbonate bedrock. The result is impressive. When granite is weathered chemically, quartz and other resistant minerals remain. However, when limestone weathers, nearly all its volume can be carried away in solution in slowly moving groundwater.

The process of dissolution creates cavities of many sizes and shapes. A natural underground opening, generally connected to the surface and large enough for a person to enter is called a *cave*. A very large cave or system of interconnected cave chambers is often called a *cavern*. Although most caves are small, some are of exceptional size. The Carlsbad Caverns in southeastern New Mexico include one chamber 1200 m long, 190 m wide, and 100 m high. Mammoth Cave, Kentucky, consists of interconnected caverns with an aggregate length of at least 48 km. The recently discovered Good Luck Cave on the tropical island of Borneo has one chamber so large that into it could be fitted not only the world's

FIGURE 9.17 Logs of petrified wood weathering out of mudstone layers in Petrified Forest National Park, Arizona.

largest previously known chamber (in Carlsbad Cavern, New Mexico), but also the largest chamber in Europe (in Gouffre St. Pierre Martin, France), and the largest chamber in Britain (Gaping Ghyll).

Cave Deposits

Some caves have been partly filled with insoluble clay and silt, originally present as impurities in limestone and gradually concentrated by dissolution. Others contain partial fillings of *dripstone* and *flowstone*, deposits chemically precipitated from dripping and flowing water, respectively, in the open air or in an air-filled cavity. Both are commonly composed of calcium carbonate. The precipitates take on many curious forms, which are among the chief attractions to cave visitors. The most common shapes are **stalactites** (iciclelike forms of dripstone and flowstone, hanging from ceilings); **stalagmites** (blunt "icicles" of dripstone projecting upward from cave floors); and **columns** (stalactites joined with stalagmites, forming connections between the floor and roof of a cave) (Fig. 9.18).

As its name implies, dripstone is deposited by successive drops of water. As each drop forms on the ceiling of a cave, it loses a tiny amount of carbon dioxide gas and precipitates a particle of calcium carbonate (Fig. 9.19). This chemical reaction is simply the reverse of the one by which calcium carbonate is dissolved by carbonic acid.

Dripstone can be deposited only in caves that are filled with air and therefore lie above the water table. Yet many, perhaps most, caves are believed to have formed below the water table, as is sug-

FIGURE 9.18 Dripstone formations adorn a large limestone cavern in Carlsbad Caverns National Park, New Mexico. In places, stalactites have merged with stalagmites to form columns, the sides of which are sometimes ornately fluted.

FIGURE 9.19 Drop of water collects at the end of a growing stalactite in Carlsbad Caverns. As the water loses carbon dioxide, a tiny amount of calcium carbonate precipitates from the solution and is added to the end of the dripstone formation.

gested by their shapes and by the fact that some caverns are lined with crystals, which can only form in an aqueous environment. How can we reconcile these apparently conflicting observations? The answer probably lies in a depression of the water table. This may be due to uplift of the land that causes a stream to cut downward into the landscape, or to a change of climate which causes a lowering of the regional water table. If caves formed during a period when the water table was high, they would pass into the zone of aeration as the water table falls. Dissolution could then give way to deposition of dripstone (Fig. 9.20).

Sinkholes

In contrast to a cave, a **sinkhole** is a large solution cavity open to the sky. Some sinkholes are caves whose roofs have collapsed; others are formed at the surface, where rainwater is freshly charged with carbon dioxide and is most effective as a solvent. Many sinkholes that are located at the intersections of joints, where downward movement of water is most rapid, have funnel-like shapes.

Sinkholes of the Yucatan Peninsula in Mexico, which are locally called *cenotes*, have high, vertical sides and contain water because they extend below the water table (Fig. 9.21). The cenotes were the primary source of water for the ancient Maya and

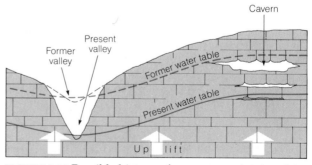

FIGURE 9.20 Possible history of a cavern containing dripstone. The cavern was excavated below the water table. When streams deepened their valleys, the water table was lowered as it adjusted to the deepened valleys. This left the cavern above the water table.

FIGURE 9.21 Sacred well at Chichen Itza, a ruined Mayan city on the Yucatan Peninsula. This cenote, formed in flat-lying limestone beds, contained a rich store of archeological treasures that were cast into the water with human sacrifices.

FIGURE 9.22 The Arecibo Radio Telescope was constructed in a natural depression within surrounding hills of tower karst terrain in Puerto Rico.

formerly supported a considerable population in Yucatan. A large cenote at the ruined city of Chichen Itza was sacred and dedicated to the rain gods. Remains of more than 40 human sacrifices, mostly young children, have been recovered from the cenote, together with huge quantities of jade, gold, and copper offerings.

In the carbonate landscape of the Florida Peninsula new sinkholes are forming constantly. In one small area of about 25 km², more than 1000 collapses have occurred in recent years. In this case the cause may be lowering of the water table by drought and excessive pumping of local wells, and to the subsequent collapse of cave roofs.

Karst Topography

In some regions of exceptionally soluble rocks, sinks and caverns are so numerous that they combine to form a peculiar topography characterized by many small closed basins (Fig. 9.22). In this kind of land-scape the drainage pattern is irregular. Streams disappear abruptly into the ground, leaving their valleys dry, and then reappear elsewhere as large springs. Such terrain is called *karst topography* after the Karst region of Yugoslavia where it is strikingly developed. It is defined as an assemblage of topographic forms resulting from dissolution of the bedrock and consisting primarily of closely spaced sinkholes. While most typical of carbonate rock landscapes, karst can also develop in areas underlain by gypsum.

In the United States, karst topography is developed over wide areas in Kentucky, Tennessee, southern Indiana, northern Florida, and northern Puerto Rico (Fig. 9.22). One of the most famous and distinctive karst regions lies near Guilin in southeastern China where towerlike peaks of limestone rise up to 200 m high. The dramatic landscape of this region has inspired both classical Chinese painters and present-day photographers (chapter frontispiece).

Essay

UNDERGROUND STORAGE OF HAZARDOUS WASTES

One of the leading environmental concerns of industrialized countries is the necessity of dealing with dangerous waste products, especially those which are highly toxic or radioactive. Experience has demonstrated that surface dumping quickly leads to contamination of surface and subsurface water supplies, and can result in serious health problems and even death. Countries with nuclear capacity have the special problem of disposing of high-level radioactive waste products, substances so highly toxic that even minute quantities can prove fatal if released to the surface environment.

Most studies concerning disposal of toxic and nuclear wastes have concluded that underground storage is appropriate, provided safe sites can be found. In the case of high-level nuclear wastes that can remain dangerous for tens or hundreds of thousands of years due to the long half-lives of the radioactive isotopes, a primary requirement is a site that will be stable over such long intervals of time. The only completely safe sites would be ones where waste products and their containers would not be affected chemically by groundwater, physically by natural deformation such as earthquakes, or accidentally by people.

The placement of toxic-waste products underground, even far underground, immediately raises concerns about groundwater. Water is a nearly universal solvent, and the weakly acid character of most groundwater means that any toxic substance put in contact with it, as well as the container that holds the substance, is likely to corrode, dissolve, and be transported away from the site of storage. Water is present in crustal rocks to depths of many kilometers and in many of those rocks it is circulating at rates of 1 to 50 m/yr. Over tens or hundreds of thousands of years, even such slow rates can move dissolved substances over great distances and introduce them to more rapidly flowing parts of the hydrologic system.

Geologists are in general agreement that the ideal underground storage site should possess the following characteristics:

1. The enclosing rock should have few fractures and low permeability.
2. The enclosing rock should have no present or future economic mineral potential.
3. Local groundwater flow should be away from the biosphere.
4. Only very long paths of groundwater flow should be directed toward places accessible to humans.

conductive

SUMMARY

1. Groundwater is derived from rainfall and occurs everywhere beneath the land surface.
2. The water table is the top of the saturated zone. In humid regions its form is a subdued imitation of the ground surface above it.
3. Groundwater flows chiefly by percolation, at rates far slower than those of surface streams. With constant permeability, velocity of flow of groundwater increases as the slope of the water table increases.
4. In moist regions groundwater percolates away from hills and emerges in valleys. In dry regions it is likely to percolate away from beneath

large surface streams thereby recharging the ground. *what?*
5. Major supplies of groundwater are found in aquifers, among the most productive of which are porous sand, gravel, and sandstone.
6. Groundwater flows into most wells directly by gravity, but into artesian wells under hydrostatic pressure. Withdrawal of water through wells creates cones of depression in the water table.
7. Groundwater dissolves mineral matter from rock. It also deposits substances as cement between grains of sediment, thereby reducing

FIGURE B9.1 Toxic wastes leak from rusting containers and soak into the groundwater system beneath this unsupervised waste disposal dump.

5. The area should have low rainfall.
6. The unsaturated zone should be thick.
7. The rate of erosion should be very low.
8. The probability of earthquakes or volcanic activity should be very low.
9. Future change of climate in the region should be unlikely to affect groundwater conditions substantially.

The safe long-term storage and eventual disposal of toxic and nuclear wastes at underground sites therefore requires considerable knowledge of local and regional groundwater systems. It also requires that we gain an understanding of how these systems are likely to change in the future as a result of crustal movements, local and global climatic change, and other natural factors that can affect the stability of a storage site.

porosity and converting the sediments to sedimentary rock.
8. In carbonate rocks, groundwater not only creates caves and sinkholes by dissolution but also, in some caves, deposits calcium carbonate as flowstone and dripstone.
9. Polluted water percolating through permeable sand or sandstone can often become purified within short distances.

IMPORTANT WORDS AND TERMS TO REMEMBER

aquiclude (p. 203)

aquifer (p. 202)

artesian spring (p. 205)

artesian system (p. 204)

artesian well (p. 205)

capillary attraction (p. 200)
cave (p. 208)
cavern (p. 208)
column (p. 209)
cone of depression (p. 204)
confined aquifer (p. 204)

discharge area (p. 201)
dissolution (p. 207)
dripstone (p. 209)

flowstone (p. 209)

groundwater (p. 198)

hard water (p. 208)

hydraulic gradient (p. 202)

karst topography (p. 211)
molecular attraction (p. 199)

perched water body (p. 204)
percolation (p. 200)
permeability (p. 199)
petrified wood (p. 208)
porosity (p. 198)

recharge (p. 201)
recharge area (p. 201)
replacement (p. 208)

saturated zone (p. 198)
seawater intrusion (p. 207)
sinkhole (p. 210)
spring (p. 203)
stalactite (p. 209)
stalagmite (p. 209)

unsaturated zone (p. 198)

water quality (p. 205)
water table (p. 198)
well (p. 204)

zone of aeration (p. 198)

QUESTIONS FOR REVIEW

1. What is the ultimate source of groundwater?
2. Explain the difference between porosity and permeability.
3. Why does a thin zone immediately above the water table remain continuously moist?
4. Describe the kind of flow that occurs as water percolates through the ground.
5. Why do the flow paths of groundwater moving beneath a hill tend to turn upward toward a stream in an adjacent valley?
6. What factors determine how long it takes water to move from a recharge area to a discharge area?
7. What is the hydraulic gradient and what importance does it have in determining the rate of flow of groundwater?
8. Why are sandstones generally better aquifers than siltstones or shales?
9. What features in igneous and metamorphic rocks promote the flow of groundwater through them?
10. How are springs related to the water table?
11. What causes a cone of depression near a producing well?
12. Explain what causes water to rise to or above the ground surface in an artesian well.
13. Why is sand especially effective in promoting purification of water flowing through it?
14. What is the origin of "hard" water in regions of carbonate bedrock?
15. Explain the origin of petrified wood.
16. What is the origin of most limestone caves that are now festooned with dripstone and flowstone formations?
17. What geologic and climatic factors lead to the development of karst landscapes?
18. What characteristics should the ideal underground disposal site for hazardous waste possess?

C H A P T E R 10

Streams and Drainage Systems

*In this oblique satellite view, the muddy Ganges River flows across the green
plains of northern India, gathering water from tributaries that drain the steep
southern flank of the Himalaya in the distance.*

STREAMS IN THE LANDSCAPE

Almost anywhere we travel over the Earth's land surface we can see evidence of the work of running water. Even in places where no rivers flow today, we are likely to find deposits and landforms that tell us water has been instrumental in shaping the landscape. Most of these features can be related to the activity of streams that are part of complex drainage systems. A *stream* is a body of water that carries rock particles and dissolved substances, and flows down a slope along a clearly defined path. The path is the stream's *channel,* and the rock particles constitute the bulk of its *load,* the material that the stream moves or carries.

Streams play an important role in our lives. They are an important source of water for human and industrial consumption. Many rivers are avenues of transportation. They also have great scenic and recreational value. The floors of stream valleys are generally fertile, and building on them is easy because the terrain is gentle. As a result, they tend to invite large populations which then must face the danger of damage by floods as well as the necessity of controlling stream pollution from the discharge of wastes.

In addition to their immediate practical and esthetic importance, streams are vital geologic agents for the following reasons:

1. Streams carry most of the water that goes from land to sea, and so are an essential part of the hydrologic cycle.

2. Every year streams transport billions of tons of sediment to the oceans, where it is deposited and can ultimately become part of the rock record.

3. Streams shape the surface of the continental crust. Most of the Earth's landscapes consist of stream valleys separated by higher ground and are the result of weathering, mass-wasting, and stream erosion working in combination (Fig. 10.1).

EROSION BY RUNNING WATER

Erosion by water begins even before a distinct stream has been formed. It occurs in two ways: by impact as raindrops hit the ground, and by sheets of water flowing over the ground during heavy rains. As raindrops strike bare ground they dislodge small particles of loose soil, spattering them in all directions. On a slope the result is net displacement

FIGURE 10.1 Master streams flowing down the steep southern slope of the Himalaya in Nepal are joined by tributaries that follow belts of weak rock and fault zones. Rock debris transported to the mountain front is deposited across a broad alluvial plain as the streams flow south to join the Ganges River in northern India. The distance across this satellite view is about 100 km and total relief is about 8000 m.

downhill. One raindrop has little effect, but the number of raindrops is so great that together they can accomplish a large amount of erosion.

The average annual rainfall on the area of the United States is equivalent to a layer of water 76 cm thick covering the entire land surface. Of this layer, 45 cm returns to the atmosphere by evaporation and transpiration (Fig. 1.13), and 1 cm infiltrates the ground, recharging the groundwater; the remaining 30 cm forms runoff. We can subdivide the runoff into *overland flow,* the movement of runoff in broad sheets or groups of small, interconnecting rills, and *streamflow,* the flow of surface water in a well-defined channel. While streamflow is very obvious, overland flow is less so. Usually overland flow occurs only through short distances before it concentrates into channels and forms streams. Such flow takes place wherever rainfall is greater than the capacity of the ground to absorb it. The erosion performed by overland flow is called *sheet erosion.*

The effectiveness of raindrops and overland flow in eroding the land is greatly diminished by a protective cover of vegetation. The leaves and branches of trees break the force of falling raindrops and cushion their impact upon the ground. More importantly, the intricate network of roots forms a tight mesh that holds soil in place, greatly reducing erosion. The root network also holds water, letting it percolate slowly down through the soil beneath. As a consequence, in vegetated areas less water runs off over the surface.

In places where crops are grown, the surface is ordinarily bare during part of each year. On unvegetated, sloping fields, pastures that are too closely grazed, and areas planted with widely spaced crops such as corn, rates of erosion can be high. Wise farmers therefore reduce areas of bare soil to a minimum and prevent the grass cover on pastures from being weakened by overgrazing (Fig. 10.2). If crops such as corn, tobacco, and cotton must be planted on a slope, strips of such crops are often alternated with strips of grass or similar plants that resist sheet erosion.

Under certain natural conditions the splash of raindrops and the work of sheet erosion are so effective that they combine to remove large volumes of fine rock particles. In some subtropical grasslands all the rainfall is concentrated within a single rainy season. During the long dry season, evaporation so depletes soil moisture that grass

FIGURE 10.3 Effects of accelerated sheet erosion on a silty soil in the Samburu District of northern Kenya. The ground surface has been lowered about 60 cm in the last 50 years, as indicated by the exposed tree roots below remnants of the original ground surface.

becomes sparse, covering no more than 40 to 60 percent of the ground. Although kept bare by natural causes, the soil is as vulnerable to erosion as soil laid bare by farming (Fig. 10.3).

GEOMETRY OF STREAMS

Stream Channel

A stream's channel is designed as an efficient conduit for carrying water. The **discharge**, or quantity of water that passes a given point in the channel per unit time, varies both along the channel and through time. In response, most channels are self-adjusting, continually modifying their shape to changing conditions. The size of any particular channel cross section reflects the typical stream conditions at that place. However, it may not be large enough to carry exceptional flows which overtop the stream banks and spread across the adjoining *floodplain*—the part of any stream valley that is inundated during floods.

Cross-Sectional Shape

A stream channel typically is rounded in cross section. Very small streams may be as deep as they are wide, whereas very large streams usually have widths many times greater than their depths (Fig. 10.4). It follows that along any stream channel the ratio of width to depth is likely to change downstream as the volume of water increases. It also changes as the volume and character of the load change.

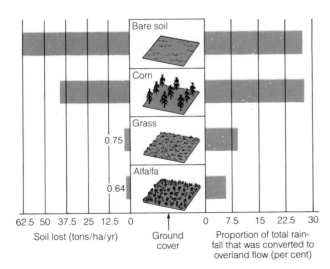

Soil lost (tons/ha/yr)					Ground cover	Proportion of total rainfall that was converted to overland flow (per cent)			
62.5	50	37.5	25	12.5	0	0	7.5	15	22.5	30

Bare soil

Corn

Grass 0.75

Alfalfa 0.64

FIGURE 10.2 Effect of plant cover on rate of sheet erosion measured over four years at Bethany, Missouri. The soil is silty, slope is 4.5°, and annual rainfall is 1000 mm. The measurements show that grass and alfalfa, with their continuous network of roots and stems, are nearly 300 times as effective as "row crops," such as corn, in holding the soil in place. Erosional loss from bare soil in this area occurs at a rate of about 45 cm/100 yr.

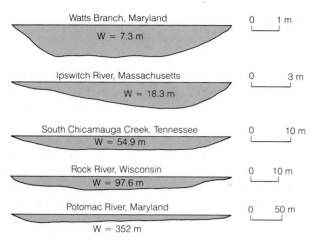

FIGURE 10.4 Cross sections of some natural streams, drawn so that their widths (W) are at the same scale. In general, the wider the channel, the larger the ratio of width to depth. (*Source:* After Leopold et al., 1964.)

Long Profile

The **gradient** (or **slope**), which is a measure of the vertical drop over a given horizontal distance, may be 60 m/km or even more for a mountain stream, whereas near the mouth of a large river it may be 0.1 m/km or even less. The average gradient of a river decreases downstream, and so its *long profile* (a line drawn along the surface of a stream from its source to its mouth) is generally a curve that decreases in gradient downstream (Fig. 10.5). However, it is not a perfectly smooth curve because of irregularities along the channel. A local change in gradient may occur, for example, where a channel passes from a belt of resistant rock into one that is more erodible, or where a landslide or lava flow forms a temporary dam. A man-made dam similarly introduces an irregularity in the long profile of a stream channel, usually creating an extensive reservoir upstream from the obstruction. Where abrupt changes in the long profile occur, water tends to flow more rapidly and turbulently through a stretch of rapids, or it may plunge over a steep drop as a waterfall.

Velocity

The flow of water within a channel is not uniform. Velocity decreases toward the bed and the channel walls because of increasing frictional resistance to flow. In large sinuous rivers, the maximum velocity in a straight segment of channel is at or near the surface in midchannel. However, as the stream bends, the zone of highest velocity swings toward the outside of the channel (Fig. 10.6).

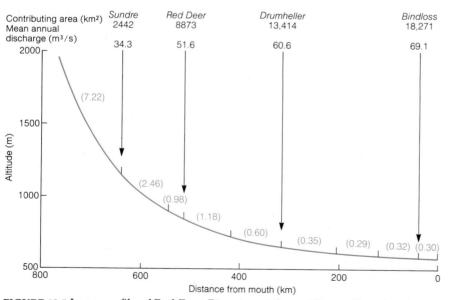

FIGURE 10.5 Long profile of Red Deer River in southern Alberta, Canada, showing typical downstream change in gradient. The size of the land area that contributes water to the stream increases downstream, as does discharge. The slope of the channel (in parentheses) is measured in meters per kilometer (m/km) for nine segments along the course and decreases rather steadily downstream (*Source:* After Campbell, 1977.)

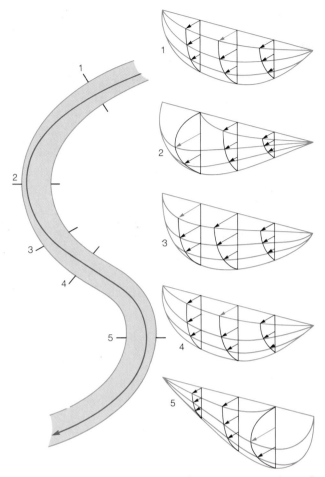

FIGURE 10.6 Velocity distribution in cross sections through a sinuous channel (lengths of arrows indicate relative flow velocities). The zone of highest velocity (red arrow) lies near the surface and toward the middle of the stream where the channel is relatively straight (sections 1 and 4). At bends, the maximum velocity swings toward the outer bank and lies below the surface (sections 2 and 5). (*Source:* After Leopold, 1964.)

Channel Patterns

Straight Channels

Except where the landscape is strongly influenced by joints or faults, straight channel segments are rare. Generally they occur for only brief stretches before they turn sinuous. Close examination of a straight segment of channel in the field shows that it has many of the features of sinuous channels. A line connecting the deepest parts of the channel, called the ***thalweg,*** typically does not follow a straight path equidistant from the banks, but wanders back and forth across the channel (Fig. 10.7a). Where the thalweg swings to one side of a channel, a deposit of alluvium (a ***bar***) tends to accumulate on the opposite side where velocity is lower. Because the thalweg is sinuous, bars alternate on opposite sides of the channel. Clearly, a straight channel implies neither a symmetrical, unchanging streambed nor a straight thalweg.

Meandering Channels

Meandering Pattern. The pattern of most streams is a series of smooth, looplike bends that are similar in size. Such a looplike bend of a stream channel is called a *meander* (from the Menderes River in southwestern Turkey [Latin = Meander], noted for its winding course). Meanders are not accidental. They occur most commonly in channels having gentle gradients in fine-grained alluvium, and they occur even in streams having no load at all. The meandering pattern reflects the way in which a river minimizes resistance to flow and dissipates energy as uniformly as possible along its course. It is a pattern of equilibrium.

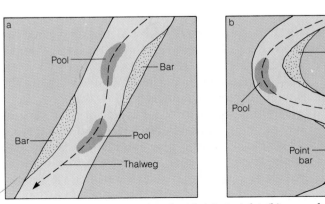

FIGURE 10.7 Features associated with (a) straight; (b) meandering; and (c) braided streams. The arrows indicate the direction of stream flow. (*Source:* In part after Ritter, 1978.)

Migration of Meanders. The nearly continuous shift or migration of a meander is accomplished by erosion on the outer banks of meander bends. Along the inner side of each meander loop, where water is shallowest and velocity is lowest, coarse sediments accumulate to form a distinctive ***point bar*** (Fig. 10.7b).

Slumping of the stream banks occurs most frequently along the downstream side of a meander bend. This is where the thalweg, and the highest current velocity, impinges on the channel side, causing erosion and undercutting of the banks. As a result, meanders tend to migrate slowly down the valley, subtracting from, and adding to, various pieces of real estate along the banks. This constant shifting of land from one bank to the other can lead to legal disputes over property lines and even over the boundaries between countries and states.

Meander Cutoffs. The behavior of streams in laboratory channels shows that if the bank sediment is uniform, meanders are symmetrical and migrate downvalley at the same rate. However, the material of natural banks generally is not uniform. Wherever the downstream limb of a meander encounters less erodible sediment, such as clay, its migration can be slowed. Meanwhile the upstream limb, migrating more rapidly, may intersect and cut into the slower-moving limb (Fig. 10.8a). Thus, the channel bypasses the loop between the two limbs and the cutoff loop is converted into a curved ***oxbow lake***. The new course is shorter than the older course, so the channel gradient is steeper there and the stream length is shortened (Fig. 10.8b).

Nearly 600 km of the Mississippi River channel has been abandoned through cutoffs since 1776. However, the river has not been shortened appreciably because the segments lost through cutoffs have been balanced by lengthening as other meanders were enlarged.

Braided Channels

Water in a braided stream flows through two or more adjacent but interconnected channels separated by bars or islands (Fig. 10.7c). If a stream is unable to transport all the available load, it may deposit the coarsest sediment as a bar which locally divides the flow and concentrates it in the deeper stretches to either side. As the bar builds up, it may emerge above the surface as an island and become stabilized by vegetation that anchors the sediment and inhibits erosion.

Large braided rivers typically have numerous constantly shifting shallow channels and abundant

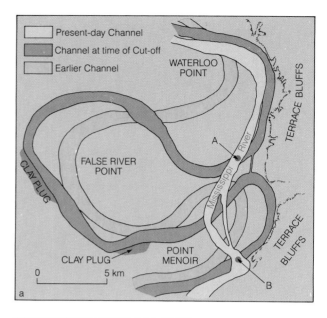

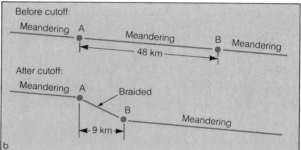

FIGURE 10.8 Cutoff of a meander loop of the Mississippi River in Louisiana. (*Source:* After Russell, 1967.) (a) The downvalley migration of the river was halted when the channel encountered a body of clay in the floodplain sediments. This allowed the next meander loop to advance and finally cut off the river segment surrounding False River Point. (b) The new, shorter channel had a steeper gradient than the abandoned course, and a braided pattern developed.

sediment load (Fig. 10.9). Although at any moment the active channels of such a stream may cover no more than 10 percent of the width of the entire channel system, within a single season all or most of the surface sediment may be reworked by the laterally shifting channels.

The braided pattern tends to form in streams having highly variable discharge and easily erodible banks that can supply abundant sediment load to the channel system. If a meandering stream is unable to move all the load it is transporting through a certain channel segment, a change to a braided pattern will apparently increase the ability of the same discharge to move a greater load. This is possible because the cumulative width of the channel

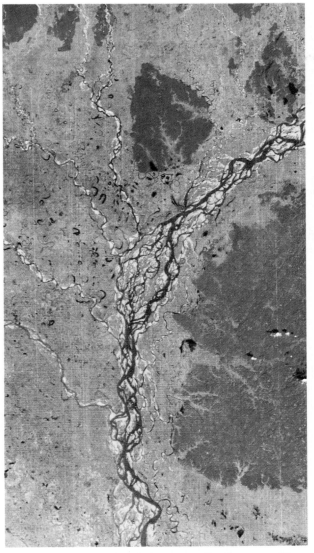

FIGURE 10.9 Intricate braided pattern of Brahmaputra River where it flows out of the Himalayas en route to the Ganges delta. Noted for its huge sediment load, the river is as wide as 8 km during the rainy monsoon season.

system becomes greater and the slope locally increases (Fig. 10.8). The braided pattern, therefore, may represent an adjustment by which a stream is able to transport a larger load more efficiently.

DYNAMICS OF STREAMFLOW

The channels of most natural streams consist partly or wholly of sediment. As streams move this material from place to place, their channels are continually being altered. Because stream and channel are closely related and are ever-changing, they should be examined together as an interrelated system.

Factors in Streamflow

Five basic factors control the manner in which a particular stream behaves: (1) the *channel dimensions* (width and depth), expressed in meters (m); (2) the *gradient*, expressed in meters per kilometer (m/km); (3) the *average velocity*, expressed in meters per second (m/s); (4) the *discharge*, expressed in cubic meters per second (m³/s); and (5) the *load*, consisting of rock particles plus matter in solution, expressed in metric tons per cubic meter (m.t./m³). Unlike rock particles that constitute the mechanical load, dissolved matter generally makes little difference to the behavior of the stream.

A stream system is characterized by a continual interplay among these factors. Measurements of natural streams show that as discharge changes, velocity and channel shape also change. The relationship can be expressed by the formula:

$$\underset{\substack{\text{Discharge} \\ (m^3/s)}}{Q} = \underset{\substack{\text{Width} \\ (m)}}{w} \times \underset{\substack{\text{Depth} \\ (m)}}{d} \times \underset{\substack{\text{Average velocity} \\ (m/s)}}{v}$$

Depth varies continuously across the stream, and velocity differs at every point in any cross section. Therefore, these values are expressed as averages, which are difficult to obtain accurately. When discharge changes, as it does continually, one or more of the other three terms must change for balance to be maintained. With increased discharge, the velocity also increases. The stream erodes and enlarges its channel, rapidly if it flows on alluvium, much more slowly if it flows on bedrock. The increased load is carried away. This continues until the increased discharge can be accommodated in a larger channel and by faster flow. In contrast, when discharge decreases, some of the load is dropped, decreasing the channel depth and width, and the velocity is reduced by increased friction. In these ways width, depth, and velocity are continually readjusted to changing discharge.

A dramatic example of changes in stream factors can be seen when floods occur. During 1956, the channel of the Colorado River at Lees Ferry, Arizona, experienced a major change in dimensions as discharge increased and then declined (Fig. 10.10). Prior to the flood, the channel averaged about 2 m deep and 100 m wide. As discharge increased in late spring, the water rose in the channel and erosion scoured the bed until at peak flow the channel was about 7 m deep and 125 m wide. Together with an increase in velocity, the enlarged channel was now able to accommodate the increased flood discharge and carry a greater load. As discharge fell, the stream was unable to transport as much

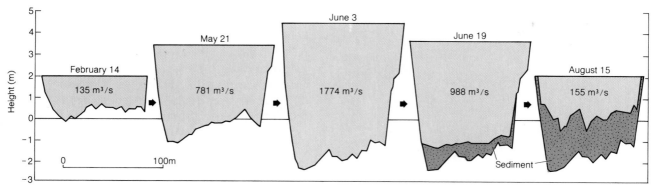

FIGURE 10.10 Changes in the cross-sectional area of the Colorado River at Lees Ferry, Arizona, during 1956. As discharge increased from February to June, the channel floor was scoured and deepened and the water level rose higher against the banks. During the falling-water phase, the river level fell and sediment was deposited in the channel, decreasing its depth. (*Source:* After Leopold et al., 1964.)

sediment, and the excess load was dropped in the channel, causing its floor to rise. At the same time, the water level fell, thereby returning the cross-sectional area to its preflood dimensions.

Thus, a stream and its channel are related intimately. We can think of them as a single system. The channel is so responsive to changes in discharge that the system, at any point along the stream, is continually close to a balanced condition.

Changes Downstream

Traveling down a river from its head to its mouth, one can see that orderly adjustments occur along it. For example, (1) discharge increases (Fig. 10.5); (2) width and depth of the channel increase (Fig. 10.11); (3) velocity increases slightly (Fig. 10.11); and (4) gradient decreases (Figs. 10.5 and 10.11).

The fact that velocity increases downstream seems to contradict the common observation that water rushes turbulently down steep mountain slopes and flows smoothly over nearly flat lowlands. However, the physical appearance of a stream is not a true measure of its velocity. Discharge is low in the headward reaches of a stream, so average velocity is also low. Discharge increases downstream as tributaries introduce more water. To accommodate the greater volume of water, velocity increases accordingly, together with the cross-sectional area of the channel.

Base Level

As a stream flows downslope and enters the sea, its potential energy falls to zero and it no longer has the ability to deepen its channel. The limiting level below which a stream cannot erode the land is called its **base level**. The **ultimate base level** for most streams is global sea level, projected inland as an imaginary surface (Fig. 10.12). Exceptions are streams that drain into closed interior basins having no outlet to the sea. Where the floor of a tectonically downfaulted basin lies below sea level (for example, Death Valley, California, or the basin of the Dead Sea in the Middle East), such streams can erode their channels well below the level of the world ocean.

For a stream ending in a lake, the level of the lake acts as **local base level**, for the stream cannot erode below it (Fig. 10.12). Such local base levels put a halt to the stream's ability to erode downward. But if the lake is destroyed by erosion at its outlet, the stream, having acquired additional potential energy, would then be able to deepen its channel. By doing so it would adjust its long profile to the changed conditions.

Artifical Dams

The practice of building artifical dams across rivers, thereby creating a reservoir upstream from the dam (a local base level), has been widespread in many countries. Some of the dams make possible the generation of hydroelectric power, and they also reduce seasonal floods by allowing lowered reservoirs to fill during times of flood. Because the reservoirs are artificial lakes, they trap nearly all the sediment that the river formerly carried uninterruptedly to the ocean. The accumulating sediment will eventually fill the reservoirs, making them useless. However, the sedimentation process is slow,

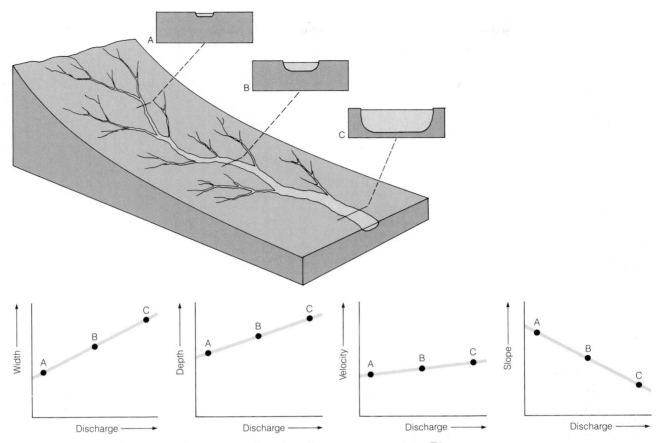

FIGURE 10.11 Changes in the downstream direction along a stream system. Discharge increases as new tributaries join the main stream. Width and depth of the channel are shown by cross sections A, B, and C. Graphs show the relationship of discharge to channel width and depth, to velocity, and to slope at the same three cross sections.

and in most cases it will be at least several hundred years before reservoirs are filled with sediment.

The Graded Stream

In adjusting to changes in discharge and erodibility of its bed a stream modifies its channel so that irregularities are minimized and the least energy is expended in the movement of water and sediment along its course. Frequently, this involves enlargement of the channel cross section by erosion, as in the example of the Colorado River flood, or reductions of its dimensions through deposition. The overall tendency is toward a smooth long profile—a profile of equilibrium in which all factors are in a state of balance. Geologists have long used the

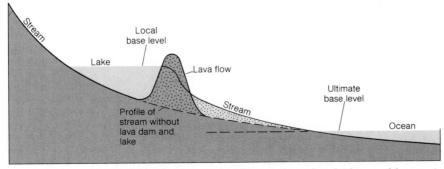

FIGURE 10.12 Relationship of a stream to ultimate base level (the world ocean) and to a local base level (a lake) along its course.

term *grade* in referring to a stream that has achieved such a balance. A **graded stream** is one in which the slope has become so adjusted, under conditions of available discharge and prevailing channel characteristics, that the stream is just able to transport the sediment load available to it. If any of the controlling factors change, then the stream will adjust to absorb the change and restore an equilibrium condition.

In actuality, it is unlikely that a condition of perfect equilibrium is ever achieved in natural stream systems. In the drainage basin of a typical river, changes are constantly taking place that upset the balance. A passing rain cloud may suddenly increase the discharge in one tributary, collapse of a bank may locally introduce an excess of sediment to the channel, or the stream may abruptly encounter a less erodible rock along its course. Adjustments take place, but each event perturbs the system and leads it away from a state of perfect balance. For this reason, it is more appropriate to think of a stream as reaching a condition of approximate equilibrium in which adjustments are continually taking place.

Large-Scale Changes in Equilibrium

Significant changes in the equilibrium condition of a stream system may require much longer periods of adjustment. For example, a major storm can introduce a sudden increase in discharge requiring substantial channel adjustment through a large part of the stream system. A long-term change in climate can also have dramatic effects on a stream if it results in a change in mean annual discharge, in the seasonal distribution of precipitation, or in the vegetation cover and related rates of runoff. Fluctuations of world sea level may lead to major changes in river systems as streams adjust their long profiles to a rise or fall of base level. Tectonic activity can also cause streams to change course, to become ponded, or to erode downward or build up their channels in response to local or regional crustal movements. In each of these cases, both long-term adjustments of drainage systems and shorter-term adjustments related to seasonal, local, or small-scale events are likely.

Floods

The seasonal distribution of rainfall causes many streams to rise seasonally in flood. This is discharge great enough to cause a stream to overflow its banks (Fig. 10.13). People affected by floods are fre-

FIGURE 10.13 Mekong River in Kampuchea in Cambodia, swollen by monsoon rains and colored brown with its load of silt. Villages, built on poles and perched on the highest parts of the levees, are surrounded by water during times of flood.

quently surprised and even outraged at what the rampaging stream has done to them, but geologists tend to view floods as normal and expectable events.

As discharge increases during a flood, so does velocity. This has the double effect of enabling a stream to carry not only a greater load, but also larger particles. The collapse of the large St. Francis Dam in southern California in 1928 provides an extreme example of the exceptional force of floodwaters. As the dam gave way, the water behind it rushed down the valley, moving blocks of concrete weighing as much as 9000 metric tons through distances of more than 750 m. Because natural floods are also capable of moving very large objects as well as great volumes of sediment, they are able to accomplish considerable geologic work.

Major floods can be disastrous events, causing both loss of life and extensive property damage (Table 10.1), so it would be highly desirable to be able to predict their occurrence. By analyzing the frequency of occurrence of past floods of different

TABLE 10.1 *Fatalities From Some Disastrous Floods*

River	Date	Fatalities (est.)	Remarks
Huang He, China	1887	900,000	Flood inundated 129,500 km² (50,000 mi²). Many villages swept away
Johnstown, Pennsylvania	1889	2,100	South Fork Dam failed. Wall of water 10 to 12 m high rushed down valley
Yangtze, China	1911	100,000	Formed lake 130 km long and 50 km wide
Yangtze, China	1931	200,000	Flood extended from Shanghai to Hankow
Vaiont, Italy	1963	2,000	Landslide into lake produced wave that overtopped dam and inundated villages below

Sources: NOAA, U.S. Geological Survey, and Encyclopedia Americana, 1983.

size, we can establish the probable interval, in years, between floods of a given magnitude. This interval (termed the *recurrence interval*) is found by plotting on a probability graph the frequency of floods of different magnitudes that a stream has experienced during a period of record. The resulting curve is then used to estimate how frequently a flood of a certain magnitude is likely to recur. In Figure 10.14, for example, a flood having a discharge of 1750 m³/s is likely to recur once in every 10 years, whereas a larger flood of 2500 m³/s will probably recur only once in every 50 years.

Exceptional floods—well outside the stream's normal range—occur infrequently, only once in many decades or even centuries. A large flood that recurs, for example, only once in every 100 years is referred to as a 100-year flood. During such floods the geologic work accomplished may be prodigious. Nevertheless, their long-term effects on the landscape may be less than the cumulative effect of annual floods over the same interval.

THE STREAM'S LOAD

A stream's load of solid particles consists largely of coarse particles that move along or close to the streambed (the *bed load*) and fine particles that are suspended in the stream (the *suspended load*). Where these solid particles are deposited, they constitute alluvium. In addition to solid particles, streams carry dissolved substances (the *dissolved load*) that are chiefly a product of chemical weathering.

Bed Load

The average rate of movement of bed-load particles is less than that of the water, for the particles are

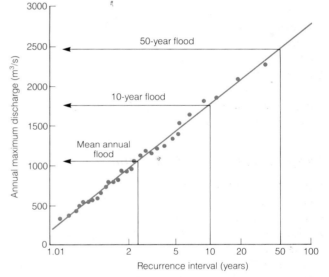

FIGURE 10.14 Curve of flood frequency for Skykomish River at Gold Bar, Washington, plotted on a probability graph. Once in every 10 years a flood of close to 1750 m³/s can be anticipated, whereas only once in every 50 years is there likely to be a flood of 2500 m³/s. (*Source:* Data from U.S. Geological Survey; after Dunne and Leopold, 1978.)

not in constant motion. Instead they move discontinuously by rolling or sliding. Where forces are sufficient to lift a particle, it may move short distances by *saltation,* a motion that is intermediate between suspension and rolling or sliding. It involves the progressive, forward movement of a particle in a series of short intermittent jumps along arcuate paths. Saltation will continue as long as currents are turbulent enough to lift particles and permit them to travel some distance downstream.

The bed load generally amounts to between 5 and 50 percent of the total load of most streams, but reliable measurements are rare.

Suspended Load

The muddy character of many streams is due to the presence of fine particles of silt and clay moving in suspension (Fig. 10.13). The Yellow River (Huang He) of China is yellow because of the great load of yellowish silt it erodes and transports seaward from widespread deposits of eolian sediment that underlie much of its basin.

Because upward-moving currents within a turbulent stream exceed the velocity at which particles of silt and clay can settle toward the bed under the pull of gravity, such particles tend to remain in suspension longer than they would in nonturbulent waters. They can settle and be deposited only where velocity decreases and turbulence ceases, as on a floodplain, in a lake, or in the sea.

Most of the suspended load in streams is derived from fine-grained regolith washed from areas unprotected by vegetation and from sediment eroded and reworked by the stream from its own banks.

Dissolved Load

Even the clearest streamwater contains dissolved chemical substances that constitute part of its load. Only seven ions comprise the bulk of the dissolved content of most rivers. These are bicarbonate, calcium, sulfate, chloride, sodium, magnesium, and potassium. Although in some streams the dissolved load may represent only a few percent of the total load, in others it amounts to more than half. Streams that receive large contributions of groundwater generally have higher dissolved loads than those whose water comes mainly from surface runoff.

Competence and Capacity

When a clear, gently flowing stream is transformed into a torrent of muddy water during flood season, its ability to transport sediment can increase dramatically. Normally this ability is referred to in terms of a stream's *competence* and *capacity*.

Competence

The size of particles a stream can transport under a given set of hydrologic conditions is a measure of its **competence,** expressed as the diameter of the largest sediment that can be moved as bed load. Competence depends mainly on velocity. This is why a major flood can move very large boulders, or even objects like railroad locomotives, while un-

der average flow conditions the same stream may be able to transport only fine gravel.

Capacity

A stream's **capacity** is the potential load it can carry, measured as the volume of sediment passing a given point in the channel per unit of time. It depends especially on discharge, channel gradient, and the character of the load. Numerous field measurements of natural streams show that suspended sediment load tends to increase with increasing discharge (Fig. 10.15).

Downstream Changes in Grain Size

Because competence is mainly related to velocity, one might expect the average size of sediment to increase in the downstream direction as velocity increases. In fact, the opposite is true; sediment normally decreases in coarseness downstream. In

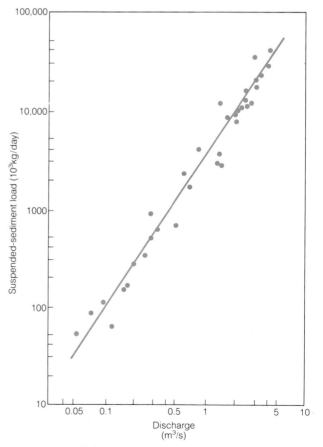

FIGURE 10.15 Relationship between suspended sediment load of the Rio Grande near Bernalillo, New Mexico, and the discharge for 1952. (*Source:* After Leopold et al., 1964.)

mountainous headwaters of large rivers, tributary streams mostly flow through channels floored with coarse gravel that may include boulders as large as a meter or more in diameter. Because fine sediment is easily moved, even by streams having low discharge, it is readily carried away by small mountain streams, leaving the coarser sediment behind. Through time, the coarse bed load is gradually reduced in size by abrasion and impact as it moves slowly along. When the stream eventually reaches the sea, its bed load may consist mainly of alluvium no coarser than sand.

Factors Influencing Sediment Yield

Some streams run clear nearly all the time, whereas others are constantly muddy and are quite obviously transporting a sizable sediment load. Such contrasts suggest that some land areas are being eroded more rapidly than others. The differences are related to a combination of geologic, climatic, and topographic factors, including rock type and structure, local climate, and relief and slope. These factors control the sediment yield from a drainage basin.

Climate

Climate influences erosion in several ways. One might expect that the greater the precipitation, the greater the erosion. However, in humid regions, the nearly continuous vegetation cover tends to inhibit erosion. In such areas with high average precipitation, therefore, erosion rates may actually be less than in some dry regions that lack a continuous cover of vegetation. Field measurements suggest that the greatest sediment yields are from landscapes transitional between full-desert conditions and grassland (Fig. 10.16). With increasing humidity and the development of forest cover, sediment yields tend to decline.

Topography and Geology

The highest measured sediment yields are from basins that drain mountainous regions having steep slopes and high relief. Rates are also higher in areas underlain by erodible clastic sediments or sedimentary rocks, or by low-grade metamorphic rocks, than in areas where crystalline or highly permeable carbonate rocks crop out. Structural factors also play a role, for rocks that are more highly jointed or fractured are more susceptible to erosion than massive ones.

Human Activity

Human activity, especially the clearing of forests, development of cultivated land, damming of streams, and construction of cities, also affects erosion rates and sediment yields in the drainage basins where they occur. Sometimes the results are dramatic. In parts of the eastern United States, areas cleared for construction produce between 10 and 100 times more sediment than comparable rural or natural areas that are vegetated. On the other hand, in urbanized areas sediment yield tends to be low because the land is almost completely covered by buildings, sidewalks, and roads that protect the underlying rocks and sediments from erosion.

DEPOSITIONAL FEATURES OF STREAMS

Floodplains and Levees

When a stream rises in flood, it may overflow its banks and inundate the floodplain (Fig. 10.17). Many streams are bordered by natural levees—broad, low ridges of fine alluvium built along both sides of a stream channel by water that spreads out of the channel during floods. The alluvium of which levees are composed becomes still finer away from the river and grades into a thin cover of silt and clay over the rest of the floodplain.

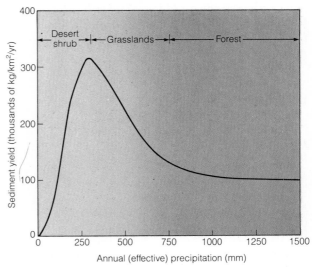

FIGURE 10.16 Relationship between sediment yield and precipitation. As precipitation increases, so does sediment yield, so long as vegetation cover remains restricted. Once the moisture level is sufficient to support continuous vegetation, erosion is reduced and sediment yield declines. (*Source:* After Langbein and Schumm, 1958.)

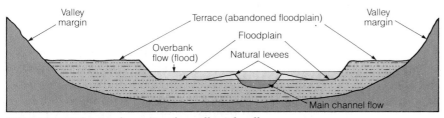

FIGURE 10.17 Main elements of an alluvial valley.

Natural levees are built, and are continually added to, only during floods so high that the floodplain is converted essentially into a lake deep enough to submerge the levees. As the water flows out of the submerged channel during a flood and across the adjacent submerged floodplain, depth, velocity, and turbulence decrease abruptly at the channel margins. The decrease results in sudden, rapid deposition of the coarser part of the suspended load (usually fine sand and silt) along the margins of the channel. Farther away from the channel, finer silt and clay settle out in the quiet water.

Terraces

Most stream valleys contain **terraces,** which are abandoned floodplains formed when a stream flowed at a level above that of its present channel and floodplain (Fig. 10.17), Their presence indicates that a change in the equilibrium condition of the stream has occurred. Terraces form as a stream erodes downward through its deposits to a new level. A terrace may be underlain by sediments, by bedrock, or by both. In other words, a terrace is a landform and is distinct from the materials that compose it.

Paired Terraces

Terraces may occur at many levels, implying a complex history of stream events. Terrace remnants on opposite sides of a valley that lie at the same level are termed **paired terraces** (Fig. 10.18 a). A set of paired terraces implies an interval of deposition (also called **aggradation**) to produce a sedimentary fill, followed by an episode of downcutting (also called **degradation**). A series of such terraces therefore points to a succession of aggradational and degradational events.

Nonpaired Terraces

Terraces on opposite sides of a stream that do not match are called **nonpaired terraces.** (Fig. 10.18 b) and must differ in age, for a stream cannot flow at different levels simultaneously. They can reflect a single episode of degradation as a stream shifts laterally from one side of its valley to the other while cutting downward. Many terrace sequences include examples of both paired and nonpaired types, making interpretation of stream history difficult.

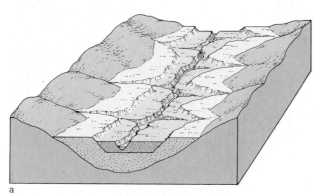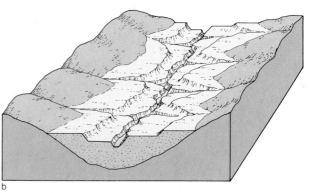

a   b

FIGURE 10.18 Examples of (a) paired and (b) nonpaired alluvial terraces. A succession of two paired terraces implies three changes of stream condition (from aggradation to degradation to aggradation to degradation), whereas the nonpaired terrace sequence illustrated required a single change (from aggradation to degradation) as the stream shifted laterally while cutting into the alluvial fill.

Alluvial Fans

When a stream flowing through a steep highland valley comes out suddenly onto a nearly level valley floor or plain, it experiences an abrupt decrease of slope and a corresponding decrease in competence. Therefore, it deposits the part of its load that cannot be transported on the gentler slope.

As sediments are deposited, the stream channel shifts laterally toward lower ground. Through constant shifting of the channel, the deposit takes the form of an *alluvial fan,* defined as a fan-shaped body of alluvium typically built where a stream leaves a steep mountain valley (Fig. 10.19). The profile of the fan, from top to base in any direction, has the same curved form characteristic of the long profiles of streams. The exact form of the profile depends chiefly on discharge and on the diameters of particles in the bed load. Hence, no two fans are exactly alike. A small stream carrying a load of coarse particles builds a shorter, steeper fan than a larger stream carrying a load of finer particles.

The area of a fan generally is closely related to the size of the area upstream from which its sediments are derived.

A fan is originally localized by a decrease of slope. As soon as its long profile has become smooth, further deposition occurs as water is redistributed through a network of spreading channels, resulting in net reduction in discharge, velocity, and competence in each channel. Deposition may also occur as water percolates down into the underlying porous fan sediments, thereby reducing surface discharge and competence. In some cases, such percolation will cause a stream to disappear near the top of a large fan, only to reappear again near the fan's base.

Deltas

As the water of a stream enters the standing water of the sea or a lake, its speed drops rapidly, it loses both competence and capacity, and it deposits its load. The sedimentary deposit that results may de-

FIGURE 10.19 Alluvial fan built into Death Valley, California, a down-faulted desert basin with white, salt-encrusted playas on its floor.

velop a crudely triangular shape like the Greek letter delta (Δ), from which the deposit derives its name (Fig. 10.20).

Coarse-Grained Deltas

Delta Types. Deltas built by streams transporting coarse sediments are of two types. Gravel-rich deltas formed where an alluvial fan is building outward into a standing body of water are called *fan deltas* (Fig. 10.21a). They typically occupy basins adjacent to a mountain front. Stream-channel, sheet-flood, and debris-flow sediments characteristic of alluvial fans form the upper parts of these deposits. Such sediments display evidence of highly variable currents and abrupt changes of facies. *Braid deltas* are coarse-grained deltas constructed by braided streams that prograde (build outward) into a standing body of water (Fig. 10.21a). Their upper parts display features characteristic of braided streams. Such deltas are especially common, for example, where braided glacial meltwater streams flow into lakes or the sea.

Delta Sediments. As the stream enters standing water, particles of the bed load are deposited first, in order of decreasing weight. Then the suspended sediments settle out. A layer representing one depositional event (such as a single flood) is sorted, grading from coarse sediment at the stream mouth to finer sediment offshore. The accumulation of many successive layers creates an embankment that grows progressively outward (Fig. 10.21a). A coarse, thick, steeply sloping part of a depositional unit in a delta is a *foreset layer.* Traced away from the shore, the same unit becomes rapidly thinner and finer, covering the bottom over a wide area. This gently sloping, fine, thin part of a depositional unit in a delta is a *bottomset layer.*

As deposition proceeds, the coarse foreset layers progressively overlap the bottomset layers. The stream gradually extends outward over the growing delta, so that its course is lengthened. Coarse channel deposits and finer interchannel deposits together form *topset layers,* defined as the layers of stream sediment that overlie the foreset layers in a delta.

Fine-Grained Deltas

Some of the world's greatest rivers, among them the Ganges-Brahmaputra, the Huang He, the Amazon, and the Mississippi, have built massive deltas at their mouths. Each delta has its own pecul-

FIGURE 10.20 Delta of the Nile River, as seen from an orbiting spacecraft. Meandering distributary channels cross the flat, vegetated delta surface and build lobes of sediment where they enter the Mediterranean.

iarities. These are determined by such factors as the stream's discharge, the character and volume of its load, the shape of the bedrock coastline at the delta, the offshore topography, and the intensity and direction of longshore currents and waves. Most major rivers transport large quantities of fine suspended sediment, the bulk of which is carried seaward as the fresh river water overrides denser salt water at the coast. The fine sediment then settles out to form the gently sloping front of a marine delta. Where currents and wave action are so strong that most sediment reaching the coast is reworked, delta formation may be inhibited. However, if the rate of sediment supply exceeds the rate of removal by erosion along the coast, then a delta will be built seaward. The Mississippi River delivers such a large load to the Gulf of Mexico that long fingerlike bodies of sediment are deposited along and around distributary channels to form a complex delta front (Fig. 1.15). The coarsest sediment lies along the channel, while finer sediment reaches the front of

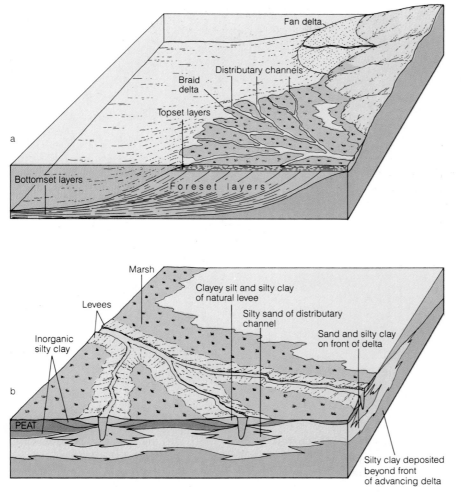

FIGURE 10.21 Main features of deltas. (a) A braid delta built into a lake displays topset, foreset, and bottomset layers. A nearby fan delta is an alluvial fan that is building out into the body of water. (b) Part of a large fine-grained delta built into the sea shows the intertonguing relationship of coarse channel deposits and finer sediments deposited on the delta front and beyond. (*Source:* Partly after Pettijohn et al., 1972.)

the delta and also accumulates between distributary channels during times of overbank flooding. The result is a complex intertonguing of facies (Fig. 10.21b).

DRAINAGE SYSTEMS

Drainage Basins and Divides

Every stream or segment of a stream is surrounded by its ***drainage basin,*** the total area that contributes water to the stream. The line that separates adjacent drainage basins is a ***divide.*** Drainage basins range in size from less than a square kilometer to vast areas of subcontinental dimension (Fig. 10.22).

In North America, the huge drainage basin of the Mississippi River encompasses an area that exceeds 40 percent of the area of the contiguous United States. Not surprisingly, the areas of drainage basins bear a close relationship to both the length and mean annual discharge of the streams that drain them.

Stream Order

The arrangement and dimensions of streams in a drainage basin tend to be orderly. This can be verified by examining a stream system on a map and numbering the observed stream segments according to their position, or *order,* in the system. The smallest segments lack tributaries and are classified

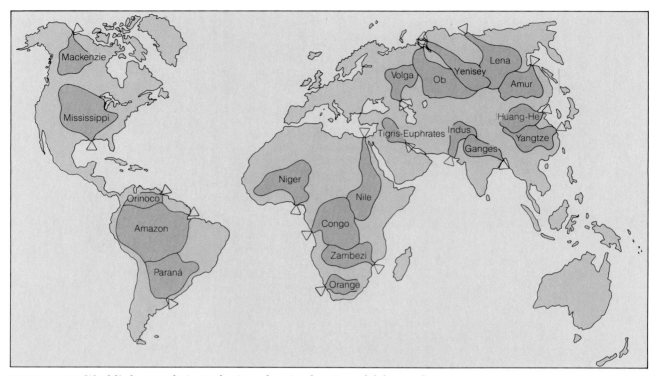

FIGURE 10.22 World's largest drainage basins, showing location of deltas at their mouths. (*Source:* After Evans, 1981.)

as first-order streams. Second-order streams form where two first-order streams join; they have first-order tributaries. Third-order streams are formed by the joining of two second-order streams and can have first-order and second-order tributaries, and so forth for successively higher stream orders (Fig. 10.23). If the number of segments of each order is tabluated, it can quickly be seen that for any stream system the number of segments increases with decreasing stream order. In other words, a stream system is somewhat like a tree, with a trunk (the main stream) and numerous branches (the tributaries). This orderliness is like that inherent in a stream's long profile, in which gradient decreases systematically from head to mouth, while discharge, velocity, and channel dimensions increase. All these relationships imply that in response to a given quantity of runoff, stream systems develop with just the size and spacing required to move the water off each part of the land with greatest efficiency.

Evolution of Drainage

A system of streams does not necessarily require much time to develop, as indicated by the following example. In August 1959, an earthquake oc-

curred at Hebgen Lake, near West Yellowstone, Montana. The movement tilted the terrain in such a way that a large area of silt and sand, formerly part of the lakebed, emerged and was subjected to runoff. Small drainage systems began to develop immediately. Sample areas were surveyed and mapped one and two years after the earthquake occurred. The results showed the same basic geometry that characterizes much larger and older systems. The small, newly formed valleys, together with the areas between them, were disposing of the available runoff in a highly systematic way, and all within a period of two years after the surface had emerged from beneath the lake.

As a drainage system develops, details of its pattern change. New tributaries are added, and some old tributaries are lost due to **stream capture** (or **piracy**), the interception and diversion of one stream by another stream that is expanding its basin by erosion in the headward direction. In the process some stream segments are lengthened and others are shortened. Just as the hydraulic factors within a stream are constantly adjusting to changes, so too is the drainage system constantly changing and adjusting as it grows. Like the stream channel, it is a dynamic system tending toward a condition of equilibrium.

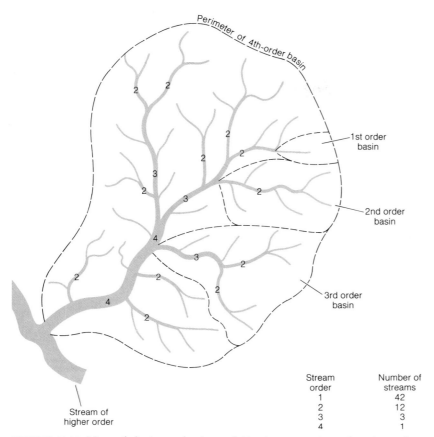

FIGURE 10.23 Map of drainage basin and its stream system showing tributary channel segments numbered according to stream order. Both the number of tributaries and their length are related to stream order. Basins are classified according to the order of the largest stream they contain. All unnumbered channels are first-order stream segments.

Stream order	Number of streams
1	42
2	12
3	3
4	1

Stream History

Streams can be classified into several categories that reflect distinctive histories. By analyzing the relationship of streams to the rock units across which they flow, an experienced geologist may learn a great deal about the structure and geologic history of an area.

Consequent Streams

A **consequent stream** is one whose pattern is determined solely by the direction of slope of the land (Fig. 10.24 a). Consequent streams often are found in massive or gently sloping rocks.

Subsequent Streams

A **subsequent stream** is a stream whose course has become adjusted so that it occupies belts of weak rock or other geologic structures (Fig. 10.24 b). When such belts are long and straight, subsequent streams following them are also straight.

Antecedent Streams

An **antecedent stream** is one that has maintained its course across an area of the crust that was raised across its path by folding or faulting (Fig. 10.24 c). The name comes from the fact that the stream is antecedent to (older than) the uplifted structure. Such streams are often recognizable because they cross topographically high ridges through deep gorges, rather than taking a more obvious path around the end of the uplift.

Superposed Streams

A **superposed stream** was let down, or superposed, from overlying strata onto buried bedrock having a lithology or structure unlike that of the covering strata (Fig. 10.24 d). Most superposed streams began as consequent streams on the surface of covering strata. Their initial paths, therefore, were not controlled by the rocks on which they are now flowing.

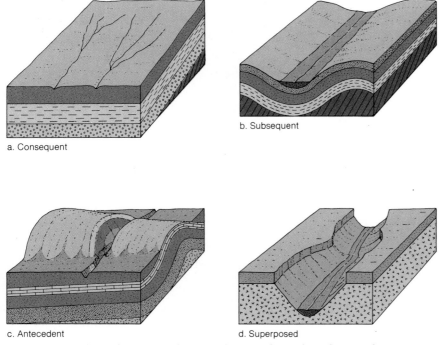

a. Consequent

b. Subsequent

c. Antecedent

d. Superposed

FIGURE 10.24 Three-dimensional views showing how the relation of streams to underlying rock structure provide information about stream history. (a) A consequent stream flows down the slope of recently tilted strata. (b) The course of a subsequent stream has become adjusted to the folded structure of underlying rocks by flowing along a valley between adjacent parallel folds. (c) An antecedent stream cuts across folded strata that have been uplifted across its path. (d) A superposed stream has cut downward through a sedimentary stratum and its channel now lies in the underlying igneous rocks.

Essay

CATASTROPHIC FLOODS AND THE CHANNELED SCABLAND

Most of the Columbia River Basin in the northwestern United States is underlain by extensive basalt flows that collectively form the Columbia Plateau. A rich topsoil overlying the basalt makes the plateau a prime wheat-producing region—but not all of it. A glance at a satellite image of the plateau in eastern Washington shows an array of dark channel-like features where lava lies at the surface, stripped of fertile topsoil (Fig. B10.1). Although they resemble a large braided stream system, the channels are now mostly dry.

In the 1920s, geologist J Harlen Bretz began a study of this curious landscape, which he named the Channeled Scabland. Bretz carefully documented the character and distribution of an assemblage of landscape features that provided evidence about the origin of the Scabland. These included dry coulees (canyons) with abrupt cliffs marking former huge cataract systems, plunge pools, potholes, deep rock basins carved in the basalt, massive gravel bars containing enormous boulders, deposits of gravel in the form of gigantic current ripples, and upper limits of water-eroded land that lie hundreds of meters above valley floors.

Although Bretz considered alternative hypotheses to explain this array of features, he

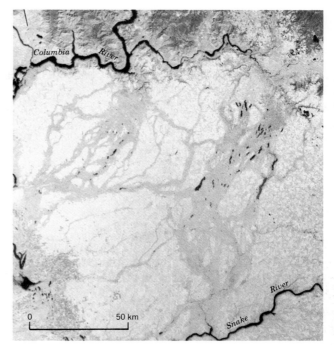

FIGURE B10.1 Vertical satellite view of the Channeled Scabland in eastern Washington where glacial floodwaters stripped away a cover of light-colored eolian silt to expose the underlying dark-gray basaltic lava flows. The floodwaters traveled from the upper right to the lower left.

was led inescapably to conclude that they could only be accounted for by a catastrophic event—a truly gigantic flood, far larger than any historic flood. A simple calculation pointed to an enormous volume of water, the source of which was resolved with the discovery that the continental ice sheet covering western Canada during the last glaciation had advanced across the Clark Fork River and dammed a huge lake in the vicinity of Missoula, Montana. Glacial Lake Missoula contained between 2000 and 2500 km^3 of water when it was filled, and remained in existence only so long as the ice dam was stable. Whenever the glacier retreated or began to float in the rising lake, the dam failed, and water was released rapidly from the basin, as through a plug had been pulled from a gigantic bathtub. The only possible exit route lay across the Columbia Plateau and down the Columbia River to the sea.

Initially there was much resistance to the flood hypothesis, but alternative ideas were discarded one by one as evidence steadily accumulated. Further study throughout the Columbia River Basin has generated new information and further controversy. It is now generally agreed that many floods crossed the plateau, but the exact number, timing, and geologic effects of these events have yet to be worked out in detail.

The landforms produced by the raging waters provide clues about the nature of the floods. The giant ripples, for example, have been carefully measured. Their height and width has been related to former water depth and water-surface slope, as reconstructed from field measurements of the highest evidence of flood activity. Standard equations used by hydraulic engineers then permitted estimates to be made of flow velocities during flood events. The results indicate that the giant ripples formed in water depths of 12 to 150 m where the water surface sloped 1.3 to 5.6 m/km, and under average flow velocities of 9 to 18 m/s (32 to 65 km/h).

Such investigations hold promise for gaining a more complete understanding of the dynamics of catastrophic flooding and the manner in which giant floods can radically alter a landscape. Although the prehistoric Scabland floods were of exceptional magnitude, many other examples of flood-sculptured terrain make it apparent that such catastrophic geologic events have played a locally important role in landscape evolution.

SUMMARY

1. Streams are part of the hydrologic cycle and the chief means by which water returns from the land to the sea. They shape the continental crust and transport sediment to the oceans.

2. Raindrop impact and sheet erosion are effective in dislodging and moving regolith on bare, unprotected slopes.

3. The average gradients of streams decrease from head to mouth, and their long profiles tend to be smooth concave-upward curves.

4. Straight channels are rare. Meandering channels form on gentle slopes and where sediment load is small to moderate. Braided patterns develop on steeper slopes and where bed load is large.

5. Discharge, velocity, and cross-sectional area of a channel are interrelated such that when discharge changes, the product of the other two factors also changes to restore equilibrium.

6. As discharge increases downstream, channel width and depth increase, and velocity increases slightly.

7. World sea level constitutes the base level for most streams. A local base level, such as a lake, may temporarily halt downward erosion upstream.

8. Streams tend toward a graded condition in which slope is adjusted so that the available sediment load can just be transported. Changes continually occur that upset a stream's balance, requiring adjustments to be made in the channel factors.

9. Streams experiencing large floods have increased competence and capacity, and so are capable of transporting great loads of sediment as well as very large boulders. Exceptional floods, however, have a low recurrence interval.

10. Although bed load in some streams may amount to as much as 50 percent of the total load, it is very difficult to measure accurately. Most suspended load is derived from erosion of fine-grained regolith or from stream banks. Streams that receive large contributions of groundwater commonly have higher dissolved loads than those deriving their discharge principally from surface runoff.

11. Sediment yield is influenced by lithology, structure, climate, and topography. The greatest sediment yields are recorded in small basins that are transitional from desert to grassland conditions, and in mountainous terrain with steep slopes and high relief. Under moist climates vegetation anchors the surface, thereby reducing erosion.

12. During floods, streams overflow their banks and construct natural levees that grade laterally into silt and clay deposited on the floodplain. Terraces are due to the abandonment of a floodplain as a stream erodes downward.

13. Alluvial fans are constructed where a stream experiences a sudden decrease in gradient, as where it leaves a steep mountain valley. It thereby loses competence and deposits the coarser fraction of its load. The area of a fan is closely related to the size of the drainage basin upstream from which its sediments originated.

14. A delta forms where a stream enters a body of standing water and loses its ability to transport sediment. The shape of a delta reflects the balance between sedimentation and erosion along the shore.

15. A drainage basin encompasses the area supplying water to the stream system that drains it. Its area is closely related to the stream's length and annual discharge.

16. Stream systems possess an inherent orderliness, with the number of stream segments increasing with decreasing stream order.

IMPORTANT WORDS AND TERMS TO REMEMBER

aggradation (p. 230)
alluvial fan (p. 231)
antecedent stream (p. 235)

bar (p. 221)

base level (p. 224)
bed load (p. 227)
bottomset layer (p. 232)
braid delta (p. 232)
braided channel (p. 222)

capacity (p. 228)
channel (p. 218)
competence (p. 228)
consequent stream (p. 235)

degradation (p. 230)
discharge (p. 219)
dissolved load (p. 227)
divide (p. 233)
drainage basin (p. 233)

fan delta (p. 232)
flood (p. 226)
floodplain (p. 219,229)
foreset layer (p. 232)

graded stream (p. 226)
gradient (p. 220)

load (p. 218)
local base level (p. 224)
long profile (p. 220)

meander (p. 221)
meandering channel (p. 221)

natural levee (p. 230)
nonpaired terraces (p. 230)

oxbow lake (p. 222)
overland flow (p. 218)

paired terraces (p. 230)
piracy (p. 234)
point bar (p. 222)
prograde (p. 232)

recurrence interval (p. 227)
runoff (p. 218)

saltation (p. 227)
sheet erosion (p. 218)
stream (p. 218)
stream capture (p. 234)
stream order (p. 233)
streamflow (p. 218)
subsequent stream (p. 235)
superposed stream (p. 235)
suspended load (p. 227)

terrace (p. 230)
thalweg (p. 221)
topset layer (p. 232)

ultimate base level (p. 224)

QUESTIONS FOR REVIEW

1. What evidence leads us to think that streams are a major force in shaping the Earth's landscapes?

2. Describe how overland flow differs from streamflow.

3. Why does vegetation decrease the effectiveness of sheet erosion?

4. How do a stream's channel dimensions (depth, width) and velocity adjust to changes in discharge?

5. What controlling factors would have to change to cause a braided channel system to change to a meandering system?

6. Why does stream velocity generally increase downstream, despite a decrease in stream gradient?

7. Why would a stream flowing into the ocean be likely to experience changes in its erosional or depositional regime if significant changes in sea level occur?

8. What are the principle characteristics of a graded stream?

9. What is meant by a "200-year flood"? What is the likely effect of such a flood on landscape evolution compared to the cumulative effect of annual floods?

10. What factors in a stream would have to change to cause fine gravel being moved as bedload to be transported as suspended load? In what way would the forward movement of such particles also be likely to change?

11. Why, if velocity increases downstream, does sediment in transport typically decrease in size in that direction?

12. How does increasing urbanization affect the amount of sediment eroded from a drainage basin, and why?

13. How might internal stratification and sedimentary characteristics permit you to distinguish between a delta and an alluvial fan that are preserved in the stratigraphic record?

14. What is a reasonable explanation for the orderly arrangement of tributaries that is typical of most drainage systems?

CHAPTER 11

Deserts and Wind Action

A sea of sand dunes has accumulated at the margin of a semiarid basin along the foot of the Sangre de Cristo Mountains in Colorado's Great Sand Dunes National Monument.

GEOGRAPHY OF DESERTS

Although the word **desert** means literally a deserted, uninhabited, or uncultivated area, the modern development of artificial water supplies has changed the meaning of this word by making many dry regions livable and suitable for agriculture. As a result, the term *desert* is now generally used as a synonym for arid land, whether "deserted" or not, in which annual rainfall is less than 250 mm (10 in.) or in which the evaporation exceeds the precipitation rate. Aridity, then, remains the chief characteristic of any desert.

Origins of Deserts

Arid lands of various kinds total about 25 percent of the land area of the world outside the polar regions. In addition, a smaller though still large percentage of semiarid land exists in which the annual rainfall ranges between about 250 and 500 mm (10 and 20 in.). These dry and semidry areas form a distinctive pattern on the world map (Fig. 11.1).

The meaning of the pattern becomes clear if we examine the general plan of circulation of the atmosphere (Fig. 1.14). The most extensive arid lands are associated with the two circumglobal belts of dry, descending subtropical air centered between latitudes 15° and 30°; examples include the Sahara and Kalahari deserts of Africa, the Rub-al-Khali Desert of Saudi Arabia, and the Great Australian Desert. Together with other subtropical deserts, these are examples of one of five recognized types of deserts.

A second type of desert is found in continental interiors far from sources of moisture where warm summers and dry cold winters prevail. The Gobi and Takla Makan deserts of central Asia fall into this category.

A third, more local kind of desert is found on the lee side of mountain ranges. The mountains create a barrier to the flow of moist air producing a **rainshadow** effect. As the air rises against the windward slope of a range it cools, thereby enabling it to retain less and less moisture. The bulk of the moisture is then lost through precipitation. Air reaching the lee side of the mountain range is deficient in moisture, resulting in a dry climate over the country beyond. The lofty Sierra Nevada in eastern California forms such a barrier and is largely responsible for the arid climate of the desert basin immediately east of it.

Coastal deserts constitute a fourth category. They occur locally along the margins of continents where cold upwelling seawater cools maritime air flowing onshore, thereby decreasing its ability to hold

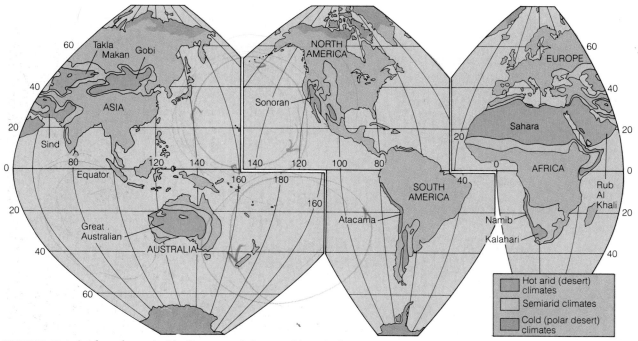

FIGURE 11.1 Arid and semiarid climates of the world and the major deserts associated with them. Very dry areas of the polar regions include areas known as polar deserts.

moisture. As the air encounters the warm land, its limited moisture condenses, giving rise to coastal fogs. However, the air contains too little moisture to generate much precipitation. Coastal deserts of this type in Chile, Peru, and southwest Africa, are among the driest places on Earth.

The four kinds of deserts mentioned thus far are hot deserts, where rainfall is low and temperatures are high. A fifth category consists of vast deserts that occur in the polar regions where precipitation is also extremely low due to the sinking of cold, dry air. However, the surfaces of these *polar deserts,* unlike those of more tropical latitudes, are often underlain by abundant water, but nearly all is in the form of ice. Even in midsummer, with the sun above the horizon for 24 hours, the temperature may remain below freezing. Examples of polar deserts are found in northern Greenland, arctic Canada, and in the ice-free valleys of Antarctica (Fig. 7.5). Such deserts are considered to be the closest earthly analogs to the surface of Mars, where temperatures also remain below freezing and the rarefied atmosphere is extremely dry.

Desert Climate

The arid climate of a hot desert results from the combination of high temperature, low precipitation, and high evaporation rate. The world's record temperature of 57.7°C (135.9°F) was measured in the Libyan Desert in northern Africa. In the Atacama Desert of northern Chile intervals of a decade or more have passed without measurable rainfall. The higher the temperature, the greater the rate of evaporation. In parts of the southwestern United States, evaporation from lakes and reservoirs is 10 to 20 times more than the annual precipitation.

During daylight hours, the air over hot desert regions is heated and expands in volume causing it to rise. Strong winds are produced as surface air moves in rapidly to take the place of the rising hot air.

SURFACE PROCESSES IN DESERTS

No major geologic process is restricted entirely to desert regions. Rather, the same processes operate with different intensities in moist and arid landscapes. In a desert, the landforms, soils, and surface sediments show some distinctive differences from those of humid regions.

Weathering and Mass-Wasting

In a moist region, regolith covers the ground almost universally and is comparatively fine-textured because it usually contains clay, a product of chemical weathering. The regolith moves downslope, mainly by creep, and it is covered with almost continuous vegetation. As a result of creep, hillslope form can usually be described as a series of curves.

By contrast, the regolith in a desert is thinner, less continuous, and coarser in texture. Much of it is the product of mechanical weathering. Although chemical weathering takes place, its intensity is greatly diminished because of reduced soil moisture. Slope angles developed by downslope creep become adjusted to the average particle size of the regolith; the coarser the particles, the steeper the slope required to move them. Because the particles created by mechanical weathering tend to be coarse, slopes are generally steeper than in a moist region.

Mechanically weathered fragments of rock tend to break off along joints, leaving steep, rugged cliffs. Hills with cliffy slopes are common in dry regions, particularly where layers of exposed bedrock are nearly horizontal (Fig. 11.2).

FIGURE 11.2 Jointed sandstone at Dead Horse Point, Utah, breaks into coarse blocks that litter the steep talus at base of the cliff.

FIGURE 11.3 Aboriginal drawings have been etched in a dark coating of desert varnish on a rock outcrop in Owens Valley, California. Removal of the oxide coating has exposed unweathered rock of lighter color.

FIGURE 11.4 A steep-walled canyon in the Canyon de Chelly region of Arizona has a flat floor underlain by sandy alluvium.

In many desert areas, the light hues of recently deposited sediments contrast with the darker hues of older deposits. The darker color can generally be attributed to *desert varnish,* a thin, dark shiny coating, consisting mainly of manganese and iron oxides, formed on the surface of stones and rock outcrops in desert regions after long exposure (Fig. 11.3). Concentration of manganese in the varnish is thought to be due either to release of the element from desert dust (which settles on the ground and is weathered by summer rainstorms) or, alternatively, to the manganese-concentrating activity of microorganisms that live on rock surfaces.

Streams and Lakes

Most desert streams never reach the sea, for they soon disappear as evaporation takes place and water soaks into the ground. Exceptions are long rivers like the Nile in Egypt and the Colorado in the southwestern United States, both of which originate in high mountains that receive abundant precipitation. Such rivers carry so much water that they keep flowing to the ocean despite great losses where they cross a desert.

The sparse vegetation cover in deserts is no great impediment to surface runoff, and the loose dry regolith is easily eroded. Major rainstorms are likely to be accompanied by *flash floods,* sudden, swift floods that transport large quantities of sediment. The debris forms fans at the base of mountain slopes and on the floors of wide valleys and basins.

Often streams in flood pass rapidly through desert canyons where they erode preexisting alluvium and undercut valley sideslopes causing the slopes to cave. As a flood subsides, the load is deposited rapidly, creating a flat alluvial surface (Fig. 11.4). The stratigraphy of such alluvial fills often discloses a complex history of cutting and filling.

In arid regions, runoff is rarely abundant enough to sustain permanent lakes. Instead, one often encounters a **playa** (Spanish for beach), which is a dry lake bed in a desert basin (Fig. 11.5). White or grayish salts at the surface, left by the repeated formation and evaporation of temporary lakes, can reach thicknesses of tens of meters and constitute an important source of industrial chemicals. Following a large rainstorm, runoff may be sufficient to form a temporary **playa lake** that may persist for several weeks.

FIGURE 11.5 A playa about 240 km north of Broken Hill, Australia, has whitish deposits of salt at its surface. Inactive sand dunes in the foreground suggest that in the past the climate was drier and vegetation cover was less, thereby permitting dunes to form.

Wind

Wind is an effective geologic agent in dry regions where landforms resulting from erosion and deposition by wind are often widespread. However, contrary to popular belief, most deserts are not characterized mainly by sand dunes. Only a third of Arabia, the sandiest of all dry regions, and only a ninth of the Sahara are covered with sand. Much of the nonsandy area of deserts is crossed by systems of stream valleys or is covered by alluvial fans and alluvial plains (Fig. 11.6). Thus, in most deserts it is apparent that more geologic work is done by streams than by wind.

FLUVIAL LANDFORMS IN DESERTS

Fans and Bajadas

Alluvial fans can develop under a wide range of climatic conditions, but they are especially common in arid and semiarid lands. In such environments fans typically are composed both of alluvium and debris-flow deposits. They are a characteristic landform of deserts where they can be the major source of groundwater for irrigation. In some places entire cities have been built on alluvial fans or fan complexes (for example, San Bernardino, California and Teheran, Iran). Many fans in Iran, Afghanistan, and Pakistan are dotted with mounds

FIGURE 11.6 Dry stream channels that cross the desert landscape of the Sinai Peninsula just west of the Gulf of Aqaba point to former intervals of greater surface runoff.

of debris that mark the sites of deep shafts connecting horizontal tunnel systems. These were designed to collect water within the upper reaches of fans for use in surface irrigation. Some such systems date back a thousand years or more.

In desert basins of the American Southwest, the Middle East, and central Asia, alluvial fans form a prominent part of the landscape. In these regions they border highlands, with the top of each fan lying at the mouth of a mountain canyon. Where a mountain front is straight and the canyons are widely spaced, each fan will encompass an arc of about 180° (Fig. 10.19). More typically one finds a *bajada* (Spanish for slope) along the base of a mountain range (Fig. 11.7). This is a broad alluvial apron composed of coalescing adjacent fans that

FIGURE 11.7 A broad bajada of coalescing alluvial fans slopes upward from the strip of fertile land along the floor of Owens Valley, California, toward the snow-capped Sierra Nevada in this vertical false-color satellite image.

has an undulating surface due to the convexities of the component fans.

Pediments

In some arid and semiarid landscapes a sloping surface at the base of mountain slopes resembles a bajada, but the surface is not underlain by thick alluvium. Instead it is erosional and is cut across bedrock. Scattered over it are rock fragments, some brought by running water from adjacent mountains and some derived by weathering from the rock immediately beneath. Downslope, the rock particles gradually thicken to form a continuous cover of alluvium. The eroded bedrock surface may extend for many kilometers along the mountain front and it passes downslope beneath the margin of the basin fill.

Such a bedrock landform is a *pediment*, defined as a sloping surface, cut across bedrock and thinly or discontinuously veneered with alluvium, that slopes away from the base of a highland (Fig. 11.8). Pediments are especially characteristic of arid and semiarid environments.

The long profile of a pediment, like that of a fan, is concave-up; it becomes progressively steeper toward the mountain front. Such a form is typically associated with the work of running water. Faint, shallow channelways on pediment surfaces show that water is indeed involved in their formation.

A pediment meets the mountain slope at its head at a distinct angle. This suggests that desert mountain slopes do not become gentler with time, as they would in a wet region where chemical weath-

ering and creep of regolith are dominant. Instead, they seem to adopt an angle determined by the resistance of the bedrock and maintain that angle as they gradually retreat under the attack of weathering and mass-wasting. In this way, retreat of the mountain slope will lengthen a pediment at its upslope edge. The growth of the pediment, at the expense of the mountain, may continue until the entire mountain has been consumed.

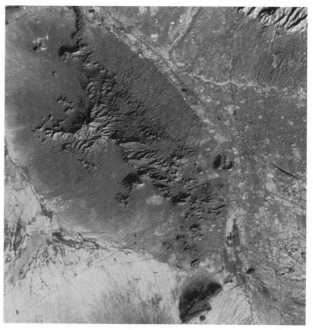

FIGURE 11.8 An elongate pediment surrounds residual rock highlands of the Tucson Mountains near Tucson, Arizona, which lies along the right side of this satellite image.

FIGURE 11.9 Ayres Rock, a massive inselberg, rises about 360 m above the surrounding flat plain in central Australia.

Eyewitness accounts of sheetfloods and of lateral erosion by stream floods (accompanying intense desert storms) have led both processes to be proposed as agents responsible for pediment formation. It is generally agreed that pediments are slopes across which sediment is transported. However, the exact manner in which pediments form is still not established.

Inselbergs

Steep-sided mountains, ridges, or isolated hills rise abruptly from adjoining monotonously flat plains in many arid and semiarid regions. These features are *inselbergs* (German for island mountains). They resemble rocky islands standing above the surface of a broad, flat sea (Fig. 11.9). Although they have been reported from many environmental settings, ranging from coastal to interior and arid to humid, they are especially common and well developed in semiarid grasslands in the middle of stable continents. Numerous examples can be found in southern and central Africa, northwestern Brazil, and central Australia.

Field evidence suggests that inselbergs form in areas of massive or resistant rock (most commonly granite or gneiss, but also sedimentary rocks) that contrasts with surrounding rocks which are more susceptible to weathering. Differential weathering over long intervals lowers adjacent terrain, leaving these rock masses standing high. Once formed, the bare rock hills tend to shed water, whereas surrounding debris-mantled plains absorb water and therefore weather more rapidly.

Inselbergs remain as stable parts of a landscape and may persist for tens of millions of years or more. Some are believed to date back to the Mesozoic Era, in which case they have remained prominent landforms since the time of the dinosaurs.

WIND AS A GEOLOGIC AGENT

The effects of wind action are visible in most deserts. In many local areas landforms resulting from wind erosion or deposition predominate. Although typical of arid and semiarid regions, eolian deposits are found along many seacoasts. They are also widespread in temperate zones where they provide evidence of former eolian activity under drier and windier conditions.

Wind Erosion

Flowing air erodes in two ways. **Deflation** (from the Latin word meaning to blow away) is the picking up and removal of loose rock and soil particles by wind. This process provides most of the wind's load. The second process, **abrasion**, results when rock is impacted by wind-driven grains of fine sediment.

Deflation

Deflation on a large scale takes place only where little or no vegetation exists and where loose rock particles are fine enough to be picked up by the wind. Areas of significant deflation are found mainly in deserts. Other nondesert places include ocean beaches, the shores of large lakes, and floodplains of large meltwater streams (Fig. 11.10). Of greatest economic importance, however, is the deflation of bare plowed fields in farmland during times of drought, when no moisture is present to hold soil particles together.

In most areas the results of deflation are not easily seen because the whole surface tends to be lowered in an irregular fashion. In places, however, measurement is possible. In the dry 1930s, deflation in parts of the western United States amounted to 1 m or more within only a few years. This is a tremendous rate compared with the long-term average erosion rate (only a few centimeters per thousand years).

Deflation Basins.   Basins excavated by wind are among the most conspicuous evidence of deflation.

FIGURE 11.10 Active deflation of meltwater sediments downstream from Tasman Glacier in the Southern Alps of New Zealand produces clouds of dust. Loess is accumulating on vegetation-covered glacial deposits in the foreground.

Tens of thousands occur in the semiarid Great Plains region of North America from Canada to Texas. Most are less than 2 km long and only a meter or two deep. In wet years they are carpeted with grass and even contain shallow lakes. However, in dry years soil moisture evaporates, grass dies away, and wind deflates the bare soil.

Where sediments are particularly prone to deflation, depths of deflation basins can reach 50 m or more. The immense Qattara Depression in the Libyan Desert of western Egypt, the floor of which lies more than 100 m below sea level, has been attributed to intense deflation. In any basin, the depth to which deflation can reach is limited finally only by the water table. As deflation approaches the level of the water table, the surface soil is moistened, thereby encouraging the growth of vegetation which inhibits further wind erosion.

Desert Pavement.   As sand and silt are blown away from a deposit of alluvium, or locally removed by sheet erosion, stones too large to be moved are concentrated at the surface (Fig. 11.11). Eventually a continuous cover of stones forms a **desert pavement**. Coarse particles at the surface are concentrated chiefly by deflation and sheet wash and fit together almost like the blocks of a cobblestone pavement (Fig. 11.12).

Ventifacts

Where bedrock and loose stones are abraded by wind-driven sand and silt, they acquire a distinctive shape and surface polish. Any bedrock surface or stone that has been abraded and shaped by wind-blown sediment is a **ventifact**. It is recognized by its smooth, polished surfaces which typically are separated from each other by sharp keel-like edges (Fig. 11.13).

Yardangs

Among the common eolian landforms of some desert regions are elongate, streamlined, wind-eroded

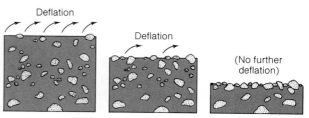

FIGURE 11.11 Progressive deflation of a poorly sorted sediment leads to the development of a desert pavement.

ridges called *yardangs* (from the Turkic word *yar*, meaning steep bank). Typically they are carved from indurated sediments or from highly weathered crystalline rocks and have a shape similar to that of an inverted ship's hull (Fig. 11.14). Individual yardangs range up to a few tens of kilometers long and up to 100 m high. Generally they occur in groups. Yardangs probably form by differential erosion along irregular surface depressions that parallel the wind direction. They then increase in size as wind abrasion deepens and broadens the depressions, often leaving sharp intervening ridges.

Wind-Transported Sediments

As sediment is moved by the wind, the sand grains forming the bed load move rather slowly and are

FIGURE 11.12 A desert pavement on the floor of Searles Valley, California, consists of a layer of gravel, too coarse to be moved by the wind, that covers finer sediment and inhibits further deflation.

FIGURE 11.13 Ventifacts litter the ground surface near Lake Vida in Victoria Valley, Antarctica. The most intensely abraded surfaces are inclined to the right, in the direction from which strong winds blow off the East Antarctic Ice Sheet.

deposited quickly when wind velocity subsides. However, the finer particles of silt and clay that form the suspended load travel faster, longer, and much farther before they settle to the ground. The wind typically deposits sand in heaps or small hills, whereas the finer sediment tends to be deposited as a smooth blanket across the landscape.

Wind-Blown Sand

Experiments with sand blown artificially through glass-sided wind tunnels show that sand grains move by saltation along arclike paths, much as they do in a stream of water (Fig. 11.15). In this mode of sediment transport, the particles follow a motion that is intermediate between suspension and rolling or sliding. The jumps involve elastic bounces that are similar to those of a Ping-Pong ball.

Most sand grains probably enter an airstream by bouncing or being knocked into the air by the impact of another grain. If a wind is strong enough, it can start a grain rolling along the surface where it may impact another grain and knock it into the air. When this second particle hits the ground it will impact other grains, some of which are projected upward into the airstream. Within a very short time the air close to the ground may contain a very large number of sand grains that move along with the wind in arclike trajectories as long as the wind velocity is great enough to keep them moving.

FIGURE 11.14 Aligned yardangs, resembling inverted ship's hulls, parallel the direction of prevailing wind in the coastal desert of Peru. Cut in erodible Tertiary sedimentary rocks, the largest reach lengths of up to 500 m.

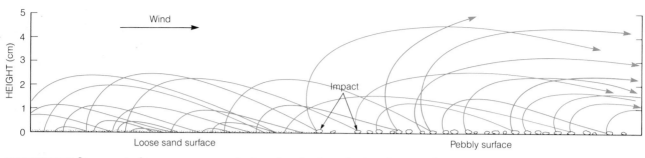

FIGURE 11.15 Strong wind causes movement of sand grains by saltation. Impacted grains bounce into the air and are carried along by the wind as gravity pulls them back to the land surface where they impact other particles, repeating the process.

Saltating sand grains seldom rise far off the ground. This is shown by abrasion marks on utility poles and fence posts that are sandblasted to a height of up to a meter, but no higher. Because blowing sand always moves close to the land surface, obstacles in its path tend to halt its forward motion, thereby promoting the formation of dunes.

Sand Ripples

Sheets of well-sorted sand that have accumulated at the surface are inherently unstable, even under gentle winds. As the wind passes across such an accumulation, saltation begins to move the smaller, most easily transported grains. Sand grains too large to be moved are left behind. The saltating finer grains impact the surface at some average distance downwind that depends on wind speed, particle size, and particle density. There more fine particles are set in motion and another accumulation of coarse grains develops as the fine sand moves onward. Through this process, the coarse grains form small linear ridges of sand with their long axes oriented perpendicular to the wind direction (Fig. 11.16). Such **sand ripples** are often seen on the windward sides of dunes. Under very strong winds the ripples disappear, for then all grains can be moved and sorting is less likely to occur.

Dunes

A **dune** is a mound or ridge of sand deposited by wind. Generally a dune forms where an obstacle distorts the flow of air. Velocity within a meter or two of the ground varies with the slightest irregularity of the surface. On encountering an obstacle, wind sweeps over and around it, but leaves a pocket of slower-moving air immediately behind the obstacle. In these pockets of low velocity, moving sand grains drop out and form mounds. The grow-

ing mounds in turn influence the flow of air. As more sand piles up, the mounds join together to form a single dune.

Dune Form.   A dune is asymmetrical. It has a steep, straight lee (downwind) slope and a gentler windward slope (Fig. 11.17). Sand grains move up the windward slope by saltation to reach the crest of the dune. As most saltation jumps are much shorter than the length of the lee face, grains making it past the brink of the dune generally fall onto the lee surface near its top. The bulge thus created through grain-by-grain accumulation eventually reaches an unstable angle. The sand then ava-

FIGURE 11.16 Sand ripples cross the surface of a desert sand sheet on the floor of Monument Valley, Arizona.

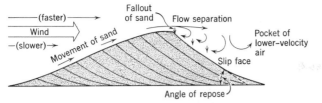

FIGURE 11.17 A cross section through a dune shows the development of typical windward and leeward slopes and internal stratification.

lanches (or "slips") downward, spreading the grains in the bulge down the lee face. For this reason, the straight lee slope of an active dune is known as the *slip face.* The avalanching keeps the slip face at the angle of repose, which is typically 30° to 34°. The accumulation and avalanching of sand grains on the slip face of a dune produces cross strata much like the foreset layers in a delta (Fig. 11.17).

The angle of slope of the windward flank varies with wind velocity and grain size, but it is always much less than that of the slip face, which it meets

at a sharp angle. The asymmetry of a dune provides a means of telling the direction of the wind that shaped it, for the slip face always lies on the side toward which the prevailing wind is blowing.

Dune Height. Many dunes grow to heights of 30 to 100 m, and some massive desert dunes in the western Alashan Plain of China are reported to reach heights of more than 500 m (Fig. 11.18). The height to which any dune can grow probably is determined by the upward increase in wind velocity. At some level this will become great enough to carry the sand grains up into suspension off the top of a dune as fast as they arrive there by saltation up the windward slope.

Dune Migration. Transfer of sand from the windward to the lee side of an active dune causes the whole dune to migrate slowly downwind. Measurements of desert dunes of the barchan type (Table 11.1) show rates of migration as great as 25 m/yr. The migration of dunes, particularly along

FIGURE 11.18 Huge sand dune in the Alashan Plain of western China reaches a height of several hundred meters.

FIGURE 11.19 The ancient temple of Mama Cuna (A.D. 1440–1530) at Pachacamac south of Lima, Peru, has been buried by shifting coastal sand dunes.

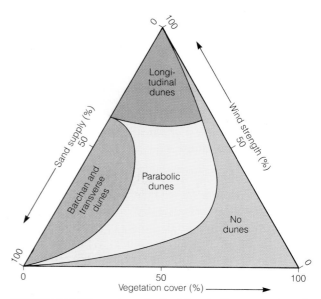

FIGURE 11.20 Dune type in the Navajo Country of north-eastern Arizona is related to vegetation cover, amount of available sand, and wind strength. (*Source*: After Hack, 1941.)

coasts just inland from sandy beaches, has been known to bury houses, fill in irrigation canals, and even threaten the existence of towns (Fig. 11.19). In such places, sand encroachment is countered most effectively by planting vegetation that can survive in the very dry sandy soil of the dunes. Continuous plant cover inhibits dune migration for the same reason that it inhibits deflation; if the wind cannot move sand grains across it, a dune cannot migrate.

Factors Affecting Dune Type. Dune type is controlled by the degree of vegetation cover, as well as by the strength of the wind and the amount of available sand (Fig. 11.20). Where sand is plentiful and lack of moisture inhibits growth of vegetation, strong winds from one direction build **barchans** and **transverse dunes** (Table 11.1). If less sand is available, **linear dunes** tend to form. As moisture increases and vegetation begins to encroach, **parabolic dunes** predominate. With a further increase in vegetation and with declining wind strength, dune formation may cease.

Sand Seas

Some large deserts contain vast tracks of shifting sand known as **sand seas**. Some of the best examples are found in northern and western Africa and in the vast desert regions of the Arabian Peninsula (Fig. 11.21). They contain a variety of dune forms, ranging from low mounds of sand and barchans to **star dunes** (Table 11.1). In some deserts huge dune complexes form a seemingly endless and monotonous landscape.

Wind-Blown Dust

Mobilization of Dust. Particles of silt and clay below a very thin layer of motionless air at the ground surface may be so small and closely packed that they present a smooth surface to the wind which cannot lift them directly off the ground. Instead, they commonly are moved into the airstream by the impact of saltating sand grains or through physical disruption of the smooth surface.

We can see how this happens by looking at a dusty desert road covered by dry silt on a windy day. The wind blowing across the road generates little or no dust, but a vehicle driving over the road creates a choking cloud, which is blown a short distance before settling once more to the ground. The passing wheels have broken up the crusted surface of powdery silt that was too smooth to be disturbed by the wind.

Transport of Dust. Once in the air, fine particles constitute the wind's suspended load. They are continually tossed about by eddies, like particles in a stream of turbulent water, while gravity tends to pull them toward the ground. Meanwhile they are carried forward. Although in most cases suspended sediment is deposited not far from its place of origin, strong winds are known to carry very fine dust thousands of kilometers. Fine particles of

TABLE 11.1 *Principal Types of Dunes Based on Form*

Dune Type	Definition and Occurrence	
Barchan dune	*A crescent-shaped dune with horns pointing downwind.* Occurs on hard, flat floors of deserts. Constant wind and limited sand supply. Height 1 m to more than 30 m	
Transverse dune	*A dune forming an asymmetrical ridge transverse to wind direction.* Occurs in areas with abundant sand and little vegetation. In places grades into barchans	
Parabolic dune	*A dune of U-shape with the open end of the U facing upwind.* Some form by piling of sand along leeward and lateral margins of areas of deflation in older dunes	
Linear dune	*A long, straight, ridge-shaped dune parallel with wind direction.* As much as 100 m high and 100 km long. Occurs in deserts with scanty sand supply and strong winds varying within one general direction. Slip faces vary as wind shifts direction	
Star dune	*An isolated hill of sand having a base that resembles a star in plan.* Ridges converge from basal points to central peak as high as 100 m. Tends to remain fixed in place in area where wind blows from all directions	

(*Source:* After McKee, 1979)

quartz deflated from Asian deserts have been found in the Hawaiian Islands where they form a foreign component of surface soils. Reddish dust deflated from the Sahara has settled on the decks of ships in the Atlantic Ocean and also is deposited on glaciers in the Alps. The amount of sediment actually moved by the atmosphere, year in and year out, is probably only a fraction of a percent of its potential capacity, for the air is rarely if ever fully loaded.

Dust Storms. During a succession of dry years in the 1930s, large dust storms swept the Great Plains of the western United States. In a particu-larly large storm on March 20, 1935, a cloud of suspended sediment rose 3.6 km above the ground. Enough sediment was carried eastward on March 21 to bring temporary twilight conditions at midday over New York and New England, 3000 km beyond the principle source area in Colorado. The distance and travel time imply wind velocities of about 80 km/h.

Loess

Most regolith contains a small proportion of fine sediment deposited from suspension in the air, but so thoroughly mixed with other materials as to be

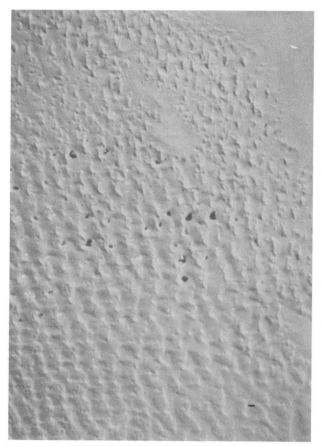

FIGURE 11.21 Satellite image of a vast sand sea that covers a portion of the Gobi Desert in central Asia. Small interdune lakes occur locally where the water table intersects the ground surface.

FIGURE 11.22 Silty loess exposed in a steep bluff near Xian, China, was derived by deflation of desert regions to the northwest. The dark reddish-brown zone at the level of the figure is a paleosol.

indistinguishable from them. However, in some regions wind-deposited sediment is so thick and uniform that it constitutes a distinctive deposit and may control the primary landscape characteristics. It is known as *loess* (German for loose) and is defined as wind-deposited silt, commonly accompanied by some fine sand and clay.

Characteristics. Loess possesses two characteristics that indicate it was deposited by the wind: It forms a rather uniform blanket, mantling hills and valleys alike through a wide range of altitudes, and it contains fossils of land plants and animals. It is massive, typically lacks stratification, and generally stands at such a steep angle that it forms cliffs (Fig. 11.22), just as though it were firmly cemented rock. This is the result of the fine grain size of loess, in which molecular attraction is strong enough to make the particles very cohesive. Porosity is extremely high, commonly exceeding 50 percent. Loess,

therefore, absorbs and holds water, and typically supports productive soils.

Origin. The distribution of loess shows that its principal sources are deserts and the floodplains of glacial meltwater streams. The loess that covers some 800,000 km² in central China, and reaches a thickness of more than 300 m in some places (Fig. 11.23), was blown there from the floors of the great desert basins of central Asia. It is likely that these extensive and thick loess deposits resulted from the breakdown of rocks by frost action and glacial processes in the high mountains of northwestern China and the subsequent deflation of resulting fine particles from large fans of alluvium in adjacent desert basins.

Glacial Loess. Loess of glacial origin is abundant in the middle part of North America (especially Nebraska, South Dakota, Iowa, Missouri, and Illinois) and in east-central Europe (especially Austria, Hungary, and Czechoslovakia). It has two distinctive features: First, the shapes and mineral composition of its particles resemble the fine sediment produced by the grinding action of glaciers. Second, glacial loess is thickest immediately downwind from large sediment-laden rivers, such as the

FIGURE 11.23 Heimugou Gulch exposes nearly 135 m of loess that underlies the southern Loess Plateau near Luochuan in Shaanxi Province, China. The loess was deposited over the last 2.4 million years, mainly during intervals when glacial climates prevailed.

Mississippi and Missouri. Areas just outside the margins of large ice-age glaciers were very cold and windy, and the floodplains remained largely bare and were easily deflated. The wind-blown sediment settled out, forming deposits 8 to 30 m thick adjacent to the source valleys (Fig. 11.24). Downwind the sediment decreases progressively in thickness and average grain size.

ADVANCING DESERTS

African Sahel

In the region south of the Sahara lies a belt of very dry grassland known as the Sahel (Arabic for border). There the annual rainfall is normally only 100 to 300 mm and most of it falls during a single short season.

In the early 1970s the Sahel experienced the worst drought of this century (Fig. 11.25). For several years in succession the rains failed, causing adjacent desert to spread southward—according to one estimate as much as 150 km. The effects of the drought extended from the Atlantic to the Indian Ocean, a distance of 6000 km. It affected a population of at least 20 million people, many of them seminomadic herders of cattle, camels, sheep, and goats. The results of the drought were intensified by the fact that between about 1935 and 1970 the human population had doubled, and with it that of the livestock. This increase of people and animals led to severe overgrazing, so that with the coming of the drought the grass cover almost completely failed. Some 40 percent of the cattle—a great many millions—died. Millions of people suffered from thirst and starvation. Many succumbed as vast numbers migrated southward in search of food and water. By 1975 the rains returned briefly. Then in the 1980s, the continuing drought, felt especially in Ethiopia and the Sudan, led to widespread famine. Mass starvation was only alleviated by worldwide relief efforts.

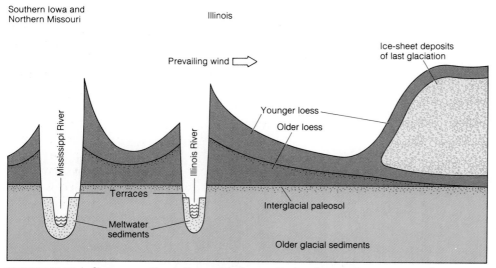

FIGURE 11.24 A diagrammatic section, with the vertical scale greatly exaggerated, shows the distribution and relative thickness of loess deposits in the upper Mississippi drainage basin. Loess is thickest and coarsest adjacent to major river valleys, especially on their downwind sides, indicating that the valleys were the primary source areas for the wind-blown dust. (*Source*: After Willman et al., 1968.)

Desertification

Such invasion of desert into nondesert areas, referred to as ***desertification***, can result from natural environmental changes as well as from human activities. The major symptoms are declining groundwater tables, increasing saltiness of water and topsoil, reduction in supplies of surface water, unnaturally high rates of soil erosion, and destruction of native vegetation. Although we can find evidence of natural desertification events in the geologic record, there is increasing concern that human activities, regardless of natural climatic trends, can in themselves help promote widespread desertification.

FIGURE 11.25 Overgrazing during years of drought killed most of the vegetation around wells in the Azaouak Valley, Mali. Without vegetation, soil blows away and the desert advances.

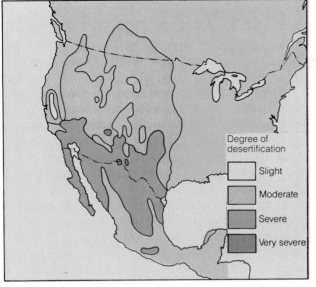

FIGURE 11.26 Vast areas of the American West show evidence of recent moderate to severe desertification. (*Source*: After Dregne, 1977.)

Impact of Desertification

The impact of desertification on human life in North America is less severe than in more densely populated regions of the world, but it nevertheless has important and far-reaching implications for the continent's food, water, and energy supplies, as well as its natural environment. Nearly 37 percent

of the arid lands of the continent have experienced "severe" desertification (Fig. 11.26). Within the arid southwestern United States, about 10 percent of the land area—an area approximately the size of the original 13 states—has been severely affected by desertification in the last century. Large shifting sand dunes have formed, erosion has largely denuded the landscape of vegetation, numerous gullies have developed, and salt crusts have accumulated on nearly impermeable irrigated soils. Mostly this has been brought about by overgrazing, by excessive withdrawal of groundwater, and by unsound water-use practices, in part allied with population increase and expanded agricultural production.

Countermeasures

How can these detrimental effects be halted and even reversed? The answer lies largely in understanding the geologic principles involved, and in intelligent application of measures designed to reestablish a natural balance in the affected areas. Reduction of incentives to exploit arid lands beyond their natural capacity, coupled with long-range planning aimed at minimizing the negative effects of human activity, should help in reaching the desired goal. Because arid lands of the western United States supply some 20 percent of the total agricultural output of the nation, the long-term benefits could be substantial.

Essay

THE WINDS OF MARS

When the Mariner 9 spacecraft went into orbit around Mars in 1971 and began sending photographs back to anxious scientists on Earth, the initial results were disappointing. They gave a picture of a big fuzzy ball, lacking distinctive surface features. Eventually it was realized that the spacecraft had arrived in the midst of a huge planetwide dust storm, so dense that it obscured such remarkable surface features as spectacular deep valleys, gigantic volcanoes, and a vast array of impact craters. Dust particles were observed as high as 55 km above the land surface, winds reached velocities of 300 km/h, and the storm raged unabated for several months.

As the winds died down and the dust settled, the returning images of the planetary surface disclosed a variety of features related to wind erosion and deposition. The recognition of eolian landforms like those found on the Earth has enabled geologists to interpret the role of wind in shaping the Martian surface.

Wind streaks, the most common eolian feature, have been used to infer seasonal patterns of atmospheric circulation. Resembling the tails of comets, these temporary bright or dark patterns resulting from erosion and deposition of surface dust change shape and size with each major storm (Fig. B11.1).

The effectiveness of wind abrasion is seen in extensive fields of yardangs, which are indicators of strong prevailing winds (Fig. B11.2). Some of these tapered ridges are tens of kilometers long and separated by valleys nearly 1 km wide. Probably they have been cut in erodible pyroclastic flows, mudflows, and consolidated regolith. Craters formed by meteorite impacts, standing like pedestals above surrounding plains, probably were preserved by blocky ejecta that armored the nearby surface while surrounding nonarmored terrain was lowered by deflation.

Fields of sand dunes are also widespread on Mars and are comparable in size to sand seas

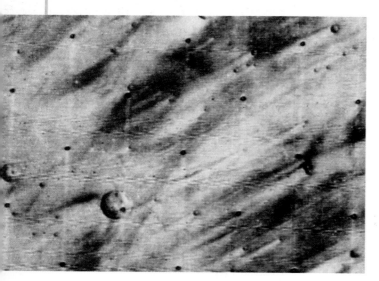

FIGURE B11.1 Wind streaks that form distinctive patterns beyond high-standing rims of impact craters on Mars are used to infer prevailing wind directions.

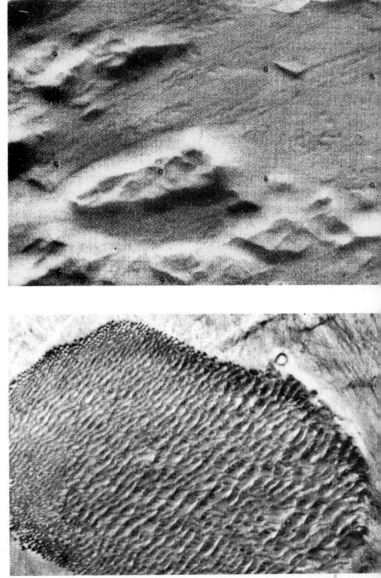

FIGURE B11.2 Linear yardangs aligned in the direction of prevailing winds are common features on images of the Martian surface.

FIGURE B11.3 Dark-colored area of sand dunes on Mars, near Hellas. The distance from crest to crest of adjacent dunes is about 1.5 km.

on the Earth (Fig. B11-3). Small dune fields are often associated with craters. The pictures reveal both barchans and transverse dunes, but no linear or star dunes. Their observed asymmetry provides a reliable basis for determining the directions of strong winds. The low-density atmosphere of Mars requires wind velocities as high as 320 km/h to move the sand-size particles.

Probable dust storms like that of 1971 were earlier observed through telescopes by Earth-bound astronomers in 1892, 1924, 1941, and 1956. Such gigantic storms are probably common occurrences on the planet and may make the Martian surface the windiest and dustiest place in our solar system.

SUMMARY

1. Hot deserts constitute about a quarter of the world's nonpolar land area and are regions of slight rainfall, high temperature, excessive evaporation, relatively strong winds, sparse vegetation, and interior drainage. Polar deserts occur at high latitudes where descending cold, dry air creates arid conditions.

2. Although no major geologic process is confined to deserts, mechanical weathering, flash floods, and winds are especially effective geologic agents. The water table generally is low.

3. Fans, bajadas, and pediments are conspicuous features of many deserts. Pediments are mainly shaped by running water and are surfaces across which sediment is transported.

4. Inselbergs form in massive or resistant rocks and may remain as persistent landforms for millions of years.

5. Wind carries a bed load of saltating sand grains close to the ground and a suspended load of fine particles at higher levels. Sorting of sediment results.

6. Through deflation and abrasion, winds create deflation basins, desert pavement, ventifacts, and yardangs.

7. Dunes often originate where obstacles distort the flow of air. Bare dunes have steep slip faces and gentler windward slopes. They migrate in the direction of wind flow, forming cross strata that slope downwind.

8. Loess is deposited chiefly downwind from deserts and from floodplains of active glacial meltwater streams. Once deposited, it is stable and is little affected by further wind action.

9. Recurring natural droughts can lower the water table, cause high rates of soil erosion, and destroy vegetation, thereby leading to the invasion of deserts into nondesert areas. Overgrazing, excessive withdrawal of groundwater, and other human activities can promote desertification. It can be halted or reversed by measures that restore the natural balance.

IMPORTANT WORDS AND TERMS TO REMEMBER

abrasion (p. 248)

bajada (p. 245)
barchan dune (p. 252)

deflation (p. 248)
desert (p. 242)
desert pavement (p. 248)
desert varnish (p. 244)

desertification (p. 256)
dune (p. 250)

flash flood (p. 244)

inselberg (p. 247)

linear dune (p. 252)
loess (p. 254)

parabolic dune (p. 252)
pediment (p. 246)
playa (p. 244)
playa lake (p. 244)
polar desert (p. 243)

rainshadow (p. 242)

sand ripples (p. 250)

sand sea (p. 252)
slip face (p. 251)
star dune (p. 252)

transverse dune (p. 252)

ventifact (p. 248)

yardang (p. 248)

QUESTIONS FOR REVIEW

1. What atmospheric factors cause most of the world's large hot deserts to be concentrated in belts lying between 30° and 15° from the equator?

2. Why do slope angles in arid landscapes tend to be steeper and sharper than those in humid landscapes?

3. What evidence can you cite that points to streams as being effective agencies of erosion and sediment transport in desert regions?

4. Why do playas on the floors of desert basins often have a distinctive deposit of salts at their surface?

5. How would you tell a bajada from a pediment in the field? What processes are involved in the formation of each?

6. How do inselbergs form, and why are they persistent features of a landscape?

7. Explain what controls the depth to which deflation is effective in arid regions.

8. Why are the erosional effects of blowing sand generally confined to a zone within about a meter of the ground surface?

9. Explain the origin of the internal stratification of a sand dune. How do sand dunes migrate downwind?

10. How might you tell the former direction of the prevailing wind from the form and internal stratification of an ancient inactive sand dune?

11. What helpful advice could you give a farmer in the American southwest who wished to halt the imminent encroachment of migrating sand dunes across his productive fields?

12. Why are smooth surfaces of fine dust difficult for the wind to erode unless the surface is disturbed?

13. How might you tell a deposit of loess from an alluvial silt having a similar range of particle sizes?

14. What are some of the obvious symptoms of desertification, and how might they be retarded or reversed by human intervention?

CHAPTER 12

Glaciers and Glaciation

Glaciers in eastern Greenland spill from cirques on mountain flanks and fill deep ice-eroded valleys that channel them toward the sea.

GLACIERS

Glaciers as Part of the Cryosphere

Glaciers constitute an important part of the *cryosphere,* that portion of the planet where temperatures are so low that water exists primarily in the frozen state. Other components of the cryosphere include snow, perennially frozen ground, and sea, lake, and river ice. Most glacier ice on the Earth resides in the polar regions, above the Arctic and Antarctic circles. In these regions sea ice forms a vast sheet over the polar seas, but its extent fluctuates seasonally. Because it is so thin (generally 3 m or less), it is not so large, volumetrically, as the ice contained in large ice sheets, but it has an important effect on global climate because of its highly reflective surface.

Large fluctuations also occur in the areal extent and volume of glaciers, but on a longer time scale. Widespread deposits and glacially eroded terrain beyond existing glaciers point to changing climatic conditions on the Earth, for glaciers and other forms of ice are very sensitive to climate. Therefore, a study of their past distribution can provide important information about global changes of climate over millions of years.

Forms of Glaciers

Defined simply, a *glacier* is a permanent body of ice, consisting largely of recrystallized snow, that shows evidence of downslope or outward movement due to the pull of gravity. On the basis of form and extent, several classes of glacier can be distinguished (Table 12.1).

Mountain Glaciers and Ice Caps

The smaller types are confined by surrounding topography that determines their shape and direction of movement. The smallest glaciers mostly occupy protected hollows or depressions on the sides of mountains (Fig. 12.1a). Larger glaciers spread downward onto valley floors where their shapes are controlled by the bedrock landscape on which they lie (frontispiece and Fig. 12.1b).

Still larger glaciers spread beyond the confining walls of mountain valleys and onto gentle slopes where topography exerts little control on their form (Fig. 12.1c). Ice caps of various sizes cover mountain highlands or lower-lying lands at high latitude, and display generally radial outward flow (Fig. 12.1d).

TABLE 12.1 *Principal Types of Glaciers, Classified According to Form*

Glacier Type	Characteristics
Cirque glacier	Occupies bowl-shaped depression on the side of a mountain (Fig. 12.1a)
Valley glacier	Flows from cirque(s) onto and along floor of valley (Fig. 12.1b)
Fjord glacier	Occupies a submerged coastal valley and its base lies below sea level. May have steep terminus that recedes rapidly by frontal calving (Fig. 12.10)
Piedmont glacier	Terminates on open slopes beyond confining mountain valleys and is fed by one or more large valley glaciers (Fig. 12.1c)
Ice cap	Dome-shaped body of ice and snow that covers mountain highlands, or lower-lying lands at high latitudes, and displays generally radial outward flow (Fig. 12.1d)
Ice field	Extensive area of ice in a mountainous region that consists of many interconnected alpine glaciers. Lacks domal shape of ice caps. Its flow is strongly controlled by underlying topography
Ice sheet	Continent-sized masses of ice that overwhelm nearly all land within their margins (Fig. 12.2)
Ice shelf	Thick glacier ice that floats on the sea and commonly is located in coastal embayments (Fig. 12.2)

Ice Sheets

Ice sheets are the largest glaciers on the Earth. These continent-sized masses of ice overwhelm nearly all the land surface within their margins. Modern ice sheets are confined to Greenland and Antarctica, and collectively comprise about 95 percent of all glacier ice on our planet. Their combined estimated volume of close to 24 million km^3 would be sufficient to raise world sea level by nearly 66 m if all were to melt.

The Greenland Ice Sheet, which has an area approximately equal to that of the United States west of the Rocky Mountains, reaches such a great thickness (some 3000 m) that the crust of the Earth beneath much of it has been depressed below sea level by its weight. Antarctica is covered by two large ice sheets that meet along the lofty Transantarctic Mountains (Fig. 12.2). The East Antarctic Ice Sheet is the larger of the two and covers the continent of Antarctica. It is the only truly polar ice sheet on Earth, for the North Pole lies at the center of the deep Arctic Ocean which is covered only by a thin layer of sea ice. Because of its ice sheet, Antarctica has the highest average altitude and the

(a)

(b)

(c)

(d)

FIGURE 12.1 Glaciers of mountainous regions. (a) A small cirque glacier on the crest of the Cascade Range in central Washington State; (b) valley glacier, Barnard Glacier, Alaska; (c) piedmont glacier, Malaspina Glacier, Alaska; (d) ice cap, South Patagonian Ice Cap, Chile and Argentina.

lowest average temperature of all the continents. The smaller West Antarctic Ice Sheet overlies numerous islands of the Antarctic archipelago. Parts of it rest on land that rises above sea level, while extensive portions cover land lying below sea level.

Ice Shelves

Ice shelves occur at several places along the margins of the Antarctic ice sheets, as well as locally in the Canadian Arctic islands. They are mainly located in large coastal embayments (Fig. 12.2), are attached to land on one side, and their seaward margin generally forms a steep ice cliff rising as much as 50 m above sea level. The largest ice shelves extend hundreds of kilometers seaward from the coastline and can reach a thickness of at least 1000 m. They are nourished by ice streams flowing off the land, as well as by direct snowfall on their surface.

Temperatures in Glaciers

Except for a thin surface layer that is chilled below freezing each winter, the ice throughout many gla-

ciers is at the *pressure melting point,* the temperature at which ice can melt at a particular pressure (Fig. 12.3). Under these conditions meltwater and ice can exist together in equilibrium. In such *temperate* (or *warm*) *glaciers,* which are found mainly in low and middle latitudes, ice is at the pressure melting point throughout. At high latitudes and high altitudes, where the mean annual air temperature lies below freezing, ice temperature drops below the pressure melting point and little or no seasonal melting occurs. Glaciers whose ice remains below the pressure melting point are termed *polar* (or *cold*) *glaciers.* In *subpolar glaciers,* an intermediate type, a thin surface zone reaches the melting point in summer, but temperatures beneath the upper several meters of ice remain below freezing.

Snowline

Glaciers can only form at or above the *snowline.* The snowline is a line that marks the intersection of the land with an irregular surface representing the lower limit of perennial snow. Because its altitude is controlled mainly by temperature and pre-

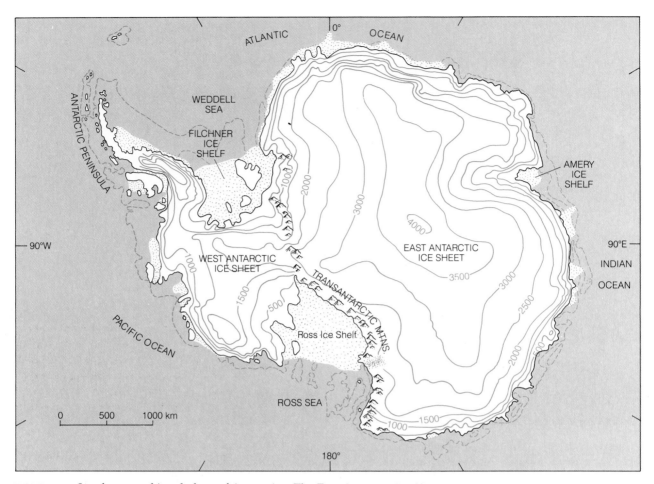

FIGURE 12.2 Ice sheets and ice shelves of Antarctica. The East Antarctic Ice Sheet overlies the continent of Antarctic whereas the much smaller West Antarctic Ice Sheet overlies a volcanic island arc and adjacent seafloor. Three major ice shelves occupy large embayments. The ice-covered regions of Antarctica nearly equal the combined area of Canada and the conterminous United States. (*Source*: After Denton et al., 1984.)

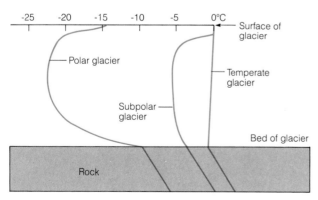

FIGURE 12.3 Temperature profiles for polar, subpolar, and temperate glaciers. Ice in temperate glaciers is at the pressure melting point from surface to bed, whereas in polar glaciers the temperature remains below freezing and the ice is frozen to its bed. In subpolar glaciers only a thin surface zone may seasonally reach the melting point. (*Source*: After Meier, 1964.)

cipitation, the snowline rises from near sea level in polar latitudes to altitudes of about 5000 to 6000 m in the tropics, and it also rises inland from moist coastal regions toward the drier interiors of large islands and continents. Where high coastal peaks intercept moist air traveling onshore, resulting in strong climatic contrasts on opposite sides of mountain ranges, the snowline rises inland with a steep gradient.

Conversion of Snow to Glacier Ice

Glacier ice is essentially a metamorphic rock that consists of interlocking crystals of the mineral ice that has been deformed by flow owing to the weight of overlying snow and ice. Newly fallen snow is very porous, having a density less than a tenth that of water. Air easily penetrates the pore spaces where the delicate points of snowflakes gradually disappear through evaporation. The resulting water va-

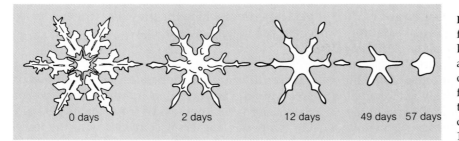

FIGURE 12.4 Conversion of a snowflake into a granule of old snow. Delicate points of a snowflake disappear through melting and evaporation. The resulting water refreezes and vapor condenses near the center of the crystal, making it denser. (*Source*: After Bader et al., 1939.)

por then condenses, mainly in constricted places near the centers of ice crystals. In this way, snowflakes gradually become smaller, rounder, and thicker, and the pore spaces between them disappear (Fig. 12.4). Snow that survives a year or more and achieves a density that is transitional between snow and glacier ice is called *firn.* Ultimately, firn passes into true glacier ice when it becomes so dense that it is no longer permeable to air. Although now a rock, such ice has a far lower melting point than any other naturally occurring rock, and its density of about 0.9 g/cm³ means that it will float in water.

Changes in Glacier Mass

Mass Balance

The mass of a glacier is constantly changing as the weather varies from season to season and, on longer time scales, as local and global climates change. These ongoing environmental changes cause fluctuations in the amount of snow added to the glacier surface, and in the amount of snow and ice lost by melting. These, in turn, determine the *mass balance* of the glacier, which is a measure of the change in total mass during a year.

Accumulation and Ablation

Mass balance is measured in terms of *accumulation,* the addition of mass to the glacier, and of *ablation,* which is the loss of mass from the glacier. Accumulation occurs mainly as snowfall, whereas ablation takes place mainly through melting or the breaking off of bergs into the sea or into lakes marginal to the ice.

If, during a year, more mass is added to a glacier than is lost, the result is a positive mass balance (Fig. 12.5). By contrast, if more mass is lost than gained, the glacier experiences a negative mass balance. If the balance is mainly positive over a period of years, it means that the glacier is increasing in

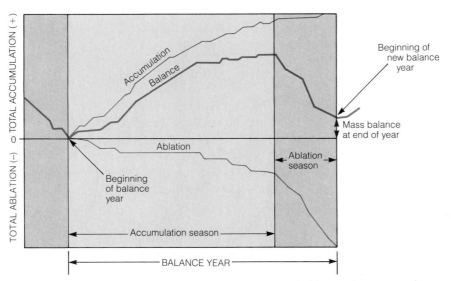

FIGURE 12.5 Diagram showing how accumulation and ablation determine glacier mass balance (heavy line) over the course of a balance year. The balance curve rises during the accumulation season as mass is added to the glacier, then falls during the ablation season as mass is lost. Mass balance at the end of the balance year reflects the difference between mass gain and mass loss. (*Source*: After International Commission of Snow and Ice, 1969.)

mass. Accordingly, the front, or **terminus,** of the glacier is likely to advance as the glacier grows. A succession of predominantly negative years normally leads to retreat of the terminus. Alternatively, if no net change in mass occurs, the glacier is in a balanced state. If this condition persists, the terminus is likely to remain relatively stationary.

If a mountain glacier is viewed at the end of the summer ablation season, two zones are generally visible on its surface (Fig. 12.6). An upper zone, the **accumulation area,** is the part of the glacier covered by remnants of the previous winter's snowfall and is an area of net gain in mass. Below it lies the **ablation area,** a region of net loss char-

acterized by a dark-toned surface of bare ice and old snow from which the previous winter's snow-cover has largely melted away.

Equilibrium Line

The **equilibrium line** separates the accumulation area from the ablation area and marks the level on the glacier where net loss of mass equals net gain. On temperate glaciers it coincides with the lower limit of fresh snow at the end of the summer (the snowline). Being very sensitive to climate, the equilibrium line fluctuates in altitude from year to year and is higher in warm dry years than in cold wet years (Fig. 12.7).

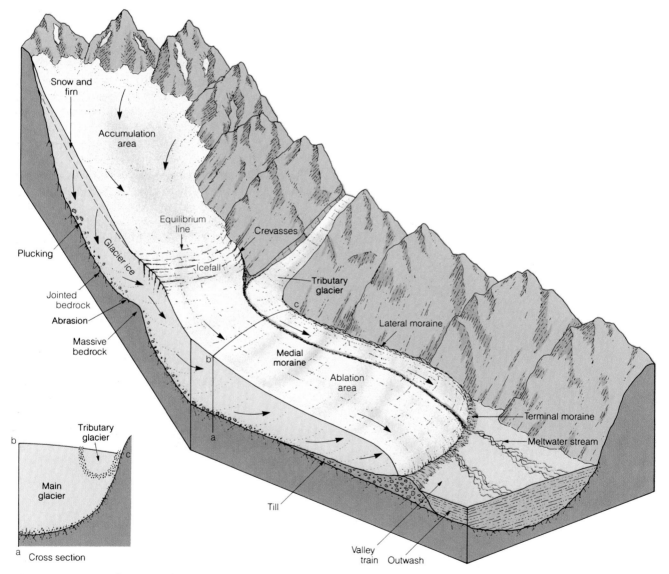

FIGURE 12.6 Main features of a valley glacier and its deposits. The glacier has been cut away along its center line; only half is shown. Crevasses form where the glacier passes over a steeper slope at its bed. Arrows show directions of ice flow.

Response of the Glacier Terminus

Measurement of the mass balance of a glacier can provide an excellent indication of its current "state of health." Observations of a glacier's marginal fluctuations are not as good an indicator, for a lag occurs between a change in climate and the response of the glacier terminus to the change. The lag reflects the time it takes for the effects of a change in accumulation above the equilibrium line to be transferred through ice flow to the glacier terminus. The length of the lag depends on the

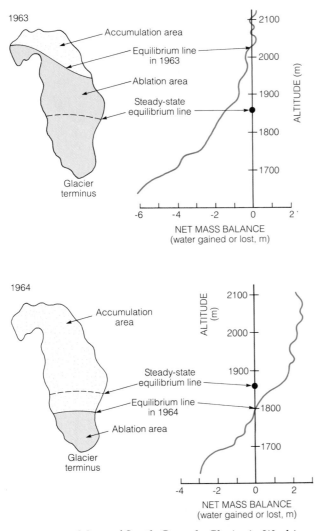

size and flow characteristics of a glacier, and will be longer for large glaciers than for small ones, and longer for cold glaciers than for warm ones. For glaciers of modest size in temperate latitudes (like those in the European Alps), response lags may range from several years to a decade or more. This partly explains why in any area having glaciers of different sizes, fluctuations of glacier margins may not be in phase.

Movement of Glaciers

Internal Flow

At some point, depending on the steepness of the surface on which it is lying and on the surrounding temperature, a mass of compacted snow and ice begins to deform and flow downslope under the pull of gravity. The flow takes place mainly through movement within individual ice crystals which are subjected to higher and higher stress as the weight of the overlying snow and ice increases. Under this stress, deformation (*creep*) takes place along internal planes in an ice crystal in much the same way that playing cards in a deck slide past one another if the deck is pushed from one end. As movement proceeds, differential stresses between crystals cause some to grow at the expense of others, and the resulting larger crystals end up having a similar orientation. This leads to increased efficiency of flow, for the internal creep planes of all crystals now are approximately parallel.

Basal Sliding

Ice temperature is very important in controlling the way glaciers move and their rate of movement. Meltwater at the base of temperate glaciers acts as a lubricant and permits the ice to slide across its bed. In some glaciers such sliding accounts for up to 90 percent of the rate of flow (Fig. 12.8b). By contrast, polar glaciers are so cold they are frozen to their bed. Their motion does not involve basal sliding, and their rate of movement is greatly reduced.

FIGURE 12.7 Maps of South Cascade Glacier in Washington State at end of the 1963 and 1964 balance years, showing the position of the equilibrium line relative to the position it would have under a balanced (steady-state) condition. The curves show values of mass balance as a function of altitude. During 1963, a negative balance year, the glacier lost mass and the equilibrium line was high (2025 m). In 1964, a positive balance year, the glacier gained mass and the equilibrium line was low (1800 m). (*Source*: After Meier and Tangborn, 1965).

Crevasses

The surface portion of a glacier, having little weight upon it, is brittle. Where a glacier flows over an abruptly steepened slope, such as a bedrock cliff, the surface ice is subjected to tension and it cracks. The cracks open up and form *crevasses*, which are deep, gaping fissures in the upper surface of a gla-

FIGURE 12.8 Flow of ice within a glacier. (a) Snow accumulating above the equilibrium line is compacted and flows downward and toward the terminus. The flow lines emerge at the surface below the equilibrium line in the ablation area. (b) Three-dimensional view through half of a glacier showing horizontal and vertical velocity profiles. A portion of the total observed movement is due to internal flow within the ice, whereas part is due to sliding of the glacier along its bed, lubricated by a film of meltwater.

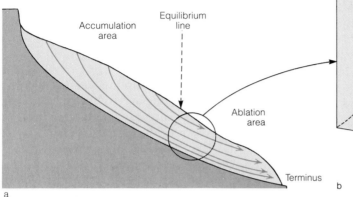

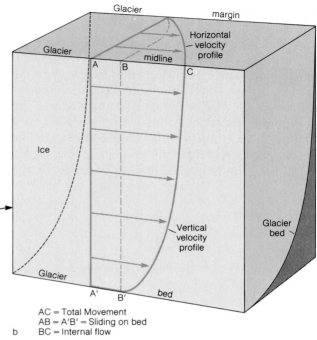

AC = Total Movement
AB = A'B' = Sliding on bed
BC = Internal flow

cier (Fig. 12.6). Crevasses are rarely as much as 50 m deep. At greater depths internal flow prevents crevasses from forming. Because it cracks at the surface, yet flows at depth, a glacier is analogous to the upper layers of the Earth itself, which consist of a surface zone that cracks and fractures (the lithosphere) and a deeper zone (the asthenosphere) that can flow very slowly.

Velocities and Directions of Flow

Measurements of surface velocity across a valley glacier show that surface ice in the central part of the glacier moves faster than ice at the sides, similar to the velocity distribution in a river (Figs. 12.8b and 10.6). The reduced rates of flow toward the margins are due to frictional drag against the valley walls. A similar reduction in flow rate toward the bed is observed in a vertical profile of velocity.

Although snow continues to pile up in the accumulation area each year, while melting removes snow and ice from the ablation area, the surface profile of a glacier does not change much because ice is transferred from the accumulation area to the ablation area. The dominant flow is downward in the accumulation area, where mass is being added, and upward in the ablation area, where mass is being lost (Fig. 12.8a). Crystals of ice that enter the glacier near its head, therefore, have a long path

to follow before they emerge near the terminus. Those falling closer to the equilibrium line, however, travel only a short distance through the glacier before reaching the surface again.

Flow velocities in most glaciers range from only a few centimeters to a few meters a day, or about the same slow rate that groundwater percolates through crustal rocks. Hundreds of years have elapsed since ice now exposed at the terminus of a very long glacier fell as snow near the top of its accumulation area.

Glacier Surges

Most glaciers slowly expand or contract in size as the climate fluctuates, but from time to time some glaciers change dimensions rapidly and in a way that is either unrelated, or only secondarily related, to climatic change. Such events, called *surges*, are unusually rapid rates of movement marked by dramatic changes in glacier flow and form. When a surge occurs a glacier seems to go berserk. Before a surge, the lower part of the glacier often consists of immobile (stagnant) ice. As the surge begins, the boundary between moving ice in the upper valley and stagnant ice below moves rapidly downglacier and a chaos of crevasses and ice pinnacles forms (Fig. 12.9). Some glaciers have advanced up to several kilometers during surges. Rates of move-

FIGURE 12.9 Moraines of Tweedsmuir Glacier, Alaska, are contorted by periodic surges of tributary ice streams.

FIGURE 12.10 Icebergs break away from the calving front of a fjord glacier in southwest Greenland. This fjord was filled by an arm of the sea as the glacier retreated back rapidly by frontal calving.

high pressures are generated within the water that lead to widespread separation of the ice from its bed. The resulting effect is similar to what happens when a rapidly moving automobile encounters a wet street during a rainstorm. The weight of the car places the water layer beneath the tires under so much pressure that they are floated off the wet pavement and the vehicle slides along out of control.

Calving Glaciers

During the last century and a half many Alaskan fjord glaciers have receded at rates far in excess of typical glacier retreat rates on land. Their dramatic recession is due to frontal *calving,* a process that involves the progressive breaking off of icebergs from a glacier that terminates in deep water (Fig. 12.10). Although the base of a fjord glacier may lie far below sea level along much of its length, its terminus can remain stable as long as it is resting (or "grounded") against a shoal. However, if the front recedes into deep water, calving can commence and may continue rapidly and irreversibly until the glacier front once again becomes grounded, generally near the head of the fjord.

GLACIATION

Glaciation, the modification of the land surface by the action of glacier ice, has occurred so recently over large parts of North America and Europe that weathering, mass-wasting, and erosion by running water have not had time to alter the landscape ap-

ment as great as 100 times those of normal glaciers and averaging as much as 6 km a year have been measured.

The causes of surges is still imperfectly understood. It is generally believed that as water accumulates beneath a glacier over a period of years, preciably. Except for a cover of vegetation, the appearance of these glaciated landscapes has remained nearly unchanged since they emerged from beneath the ice. Like the geologic work of other surface processes, glaciation involves erosion, transport, and deposition of sediment.

Glacial Erosion and Sculpture

In changing the surface of the land over which it moves, a glacier acts collectively like a plow, a file, and a sled. As a plow it scrapes up weathered rock and soil and plucks out blocks of bedrock; as a file it rasps away firm rock; and as a sled it carries away the load of sediment acquired by plowing and filing, along with additional rock debris fallen onto it from adjacent slopes.

Small-Scale Erosional Features

Glacial Striations. The base of a temperate glacier is studded with rock particles of various sizes. The fragments move with the flowing ice as it flows across bedrock and produces long subparallel scratches called **glacial striations** (Fig. 12.11). Grooves aligned in the direction of ice flow are produced by larger stones that are dragged across the subglacial bed. Fine particles of sand and silt in the basal ice act like sandpaper, and can polish the rock until it has a smooth reflective surface.

Asymmetrical Landforms. At the same time the basal ice drags at bedrock, breaking off blocks (usually along joints or fractures) and quarrying them out. Blocks are mainly removed on the downglacier sides of hillocks, whereas on the upglacier sides abrasion and polishing of the rock is dominant. The asymmetry of the resulting landforms clearly indicates the direction in which the glacier was moving (Fig. 12.12).

Landforms of Glaciated Mountians

Cirques. Among the most characteristic landforms of glaciated mountains is the **cirque,** a bowl-shaped hollow on a mountainside, open downstream and bounded upstream by a steep slope (the **headwall**). Cirques are excavated mainly by frost-wedging, glacial plucking, and abrasion (Fig. 12.13). The floors of many cirques are rock basins. Some contain small lakes, called **tarns,** ponded behind a bedrock threshold at the edge of the cirque.

A cirque probably begins to form beneath a large snowbank or snowfield just above the snowline.

FIGURE 12.11 Recently deglaciated bedrock surface beyond Findelen Glacier, Swiss Alps. Debris carried at the base of the glacier produced grooves, striations, and polish on bedrock as the ice flowed forward in the direction of the Matterhorn.

As meltwater infiltrates rock openings beneath the snow, it refreezes and expands, disrupting the rock and dislodging fragments. Small rock particles are then carried away by snowmelt runoff during periods of thaw. This activity gradually creates a depression in the land and enlarges it. As the snowbank turns into a glacier, plucking helps to enlarge the cirque still more and abrasion at the bed will further deepen it.

Headward growth of cirques on opposite sides of a mountain crest can produce a narrow serrated ridge called an **arête.** Where three or more cirques have sculptured a mountain mass, the result can be a high sharp-pointed peak, or **horn,** the classic example of which is the Matterhorn in the European Alps (Fig. 12.11 and 12.13).

Glacial Valleys. Glaciated valleys differ from ordinary stream valleys in several ways. Their chief characteristics include a cross profile that is trough-

FIGURE 12.12 Asymmetrical glacially sculptured bedforms in front of Franz Josef Glacier in New Zealand's Southern Alps. The glacier flowed from right to left. Upglacier slopes are smooth and polished. Scarps facing downvalley result from the plucking of bedrock blocks by flowing ice.

like (U-shaped) and a floor that lies below the floors of smaller tributary valleys from which streams often descend as waterfalls or cascades (Fig. 12.13). The long profile of a glaciated valley floor may possess steplike irregularities and shallow basins. These are related to the spacing of joints in the rock, which influences the ease of glacial plucking, or to changes in rock type along the valley. Finally, the valley typically heads in a cirque or group of cirques.

Fjords. A *fjord* is a segment of a deep glaciated valley partly filled by an arm of the sea. Fjords are common features along the mountainous west-facing coasts of Norway, Alaska, British Columbia,

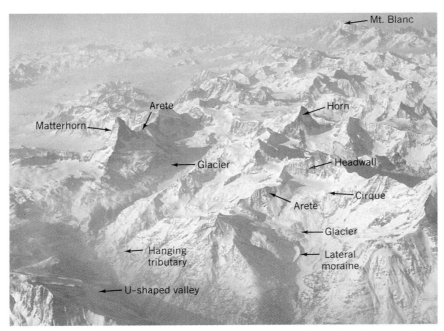

FIGURE 12.13 Photograph of central Swiss Alps near Zermatt showing alpine glacial landforms. Major U-shaped valleys head in cirques and have hanging tributary valleys. The sharp-pointed Matterhorn (middle distance) is a classic *horn*.

FIGURE 12.14 Fjords along the northeastern coast of Baffin Island in the Canadian Arctic.

at the pressure melting point. In peripheral zones, evidence of glacial erosion is often less obvious, leading to the conclusion that the thinner ice there may have been very cold and largely frozen to its bed.

Where ice sheets overwhelmed mountainous terrain, as in the cordillera of northwestern North America, the upper limit of glaciation can frequently be identified where smooth, abraded slopes pass abruptly upward into rugged frost-shattered peaks and mountain crests that stood above the ice surface. Some divides between adjacent drainages are broad and smooth and show evidence of glacial abrasion and plucking where they were overridden by ice.

Streamlined Forms. In many areas inside the limits of former ice sheets, the land surface has been molded into smooth, nearly parallel ridges that range up to many kilometers long (Fig. 12.15). These forms resemble the streamlined bodies of supersonic airplanes and offer minimum resistance to glacier ice flowing over and around them. Among the most distinctive is the *drumlin,* a streamlined hill consisting largely of glacially deposited sediment and elongated parallel with the direction of ice flow. Glacially molded drumlin-shaped hills of bedrock, called *rock drumlins,* also owe their streamlined shape to erosion by flowing ice. Like a true drumlin, the long axis of a rock drumlin lies parallel to the flow direction of the overriding glacier.

Chile, and New Zealand, and in northern Canada (Fig. 12.14). The floors of many fjords contain elongate basins that reach depths of 300 m or more and are evidence of deep glacial erosion. As a result, fjords often are shallow at their seaward end and deepen inland. For example, Sognefjord in Norway reaches a depth of 1300 m, yet near its seaward end it shallows to only about 150 m.

Landforms Associated with Ice Caps and Ice Sheets

Abrasional Features. Landscapes overridden by ice sheets display the same small-scale erosional features typical of valley glaciation. Striations, especially, have been helpful to geologists in reconstructing the flow lines of long-vanished northern ice sheets. They also demonstrate that basal sliding occurred in the thick central zones of former ice sheets and that ice resting on the glacier bed was

Glacial Transport

A glacier differs from a stream in the way in which it carries its load of rock particles. Part of its load can be carried at its sides and even on its surface. A glacier can carry much larger pieces of rock and it can transport large and small pieces side by side without segregating them according to size and density into a bed load and a suspended load. Because of these differences, deposits made directly from a glacier are neither sorted nor stratified.

The load of a glacier typically is concentrated at its base and sides because these are the areas where glacier and bedrock are in contact and where abrasion and plucking are effective. Much of the rock debris on the surface of valley glaciers arrived there by rockfalls from adjacent cliffs. Where two glaciers join, rock debris at their margins merges to form a *medial moraine,* visible as a dark linear ridge of sediment on the ice surface (Figs. 12.1b and 12.6).

A good deal of the load in the base of a glacier consists of *rock flour.* These very fine sand and silt

FIGURE 12.16 Bouldery till in an end moraine of an alpine glacier in the eastern Cascade Range displays a wide range in grain size and lack of sorting.

FIGURE 12.15 Drumlin topography in Canada west of Hudson Bay produced by erosion and deposition at the base of the Laurentide Ice Sheet during the last glaciation.

particles have sharp, angular surfaces that are produced by crushing and grinding at the base of the glacier (Fig. 4.18).

Glacial Deposits

Glacial Drift

Sediment deposited directly by glaciers or indirectly by meltwater in streams, in lakes, and in the sea together constitute *glacial drift.* The term *drift* dates from the early nineteenth century when it was vaguely conjectured that all such deposits had been "drifted" to their resting places during the biblical flood of Noah or in some other ancient body of water. Included within drift are several kinds of sediment that form a gradational series ranging from nonsorted to sorted types.

Till. At one end of the range is *till,* which is nonsorted drift deposited directly from ice. The name was given by Scottish farmers long before the origin of the sediment was understood. The constituent rock particles in till lie just as they were released from the ice (Fig. 12.16). Most tills are a random mixture of rock fragments. A matrix of fine-grained sediment surrounds larger stones of various sizes. The *till matrix* consists largely of sand and silt particles derived by abrasion of the glacier bed and from reworking of preexisting fine-grained sediments. Pebbles and larger rock fragments in till often have faceted and abraded surfaces, and some are striated. Both the stones and the coarser matrix grains in till tend to have their longest axis aligned in the direction of ice flow.

Glacial-Marine Drift. Closely resembling till, *glacial–marine drift* is sediment deposited on the seafloor from floating ice shelves or bergs. As an iceberg or the base of an ice shelf melts, the contained sediment is released and settles to the seafloor. Stones dropped from passing bergs plunge into marine sediments, deforming any laminated structure they possess. Such *dropstones* are diagnostic features of glacial–marine and ice–marginal lake environments.

FIGURE 12.17 Massive end moraines of Miage Glacier in the Italian Alps rise abruptly from the valley floor. Glacier ice behind the moraines is thickly mantled with debris generated by frequent rockfalls from steep valley walls upglacier.

Stratified Drift. By contrast, **stratified drift** is both sorted and stratified. It is not deposited by glacier ice, but by meltwater emanating from the ice. Stratified drift ranges from coarse, very poorly sorted sandy gravels that are transitional into till, to fine-grained, well-sorted silts and clays deposited in quiet-water environments.

Deposits of Active Ice

Moraines. In actively flowing glaciers, sediment transported by the ice is plastered onto the ground as till or is released by melting at the glacier margin where it either accumulates as a moraine or is reworked by meltwater and transported beyond the terminus. Widespread drift with a relatively smooth surface topography consisting of gently undulating knolls and shallow, closed depressions is known as *ground moraine*. Commonly it consists of till that blankets the landscape and may reach a thickness of 10 m or more. An *end moraine*, on the other hand, is a ridgelike accumulation of drift deposited along the margin of a glacier. An accumulation near the glacier terminus is a *terminal moraine*, whereas a *lateral moraine* forms along the lateral margin of a glacier below the equilibrium line. Both types normally form a single, continuous landform (Fig. 12.6).

End moraines can form by a bulldozing action of the glacier front, by slumping of loose surface debris off the glacier margin as the ice melts, by repeated plastering of drift from basal ice onto the ground, or by streams of meltwater that build up deposits of stratified drift at the glacier margin. They range in height from a few meters to hundreds of meters (Fig. 12.17). The great height and thickness of some lateral moraines are due to the repeated addition of drift during successive ice advances. Buried land surfaces, marked by paleosols and organic remains, are sometimes exposed in lateral moraines and provide evidence of their composite character.

Erratics and Boulder Trains. Some of the boulders and smaller rock fragments in till are the same kind of rock as the bedrock on which the till was deposited, but many are of other kinds, having been brought from greater distances. A glacially deposited rock or rock fragment with a lithology different from that of the underlying bedrock is an *erratic* (Latin for wanderer). Some huge erratics have estimated weights of thousands of tons.

In areas of ice-sheet glaciation, erratics derived from distinctive bedrock sources may have a fanlike distribution, spreading out from the area of outcrop and reflecting the diverging pattern of ice flow. A

FIGURE 12.18 Outwash terraces rise above meandering Cave Stream on South Island, New Zealand.

group of erratics spread out fanwise is a *boulder train,* so named in the nineteenth century when rock particles of all sizes were called boulders.

Outwash. Stratified sediment deposited by meltwater streams as they flow away from a glacier margin is called *outwash* ("washed out," beyond the ice). Such streams typically have a braided pattern because of the large sediment load they are moving. If the streams are free to swing back and forth widely beyond the glacier terminus, they deposit outwash to form a broad *outwash plain.* Meltwater streams confined by valley walls will build a *valley train* (Fig. 12.6).

During glacier retreat, a stream's sediment load is reduced and the underloaded stream cuts down into its outwash deposits to produce *outwash terraces.* Series of terraces are common in valleys that have experienced repeated glaciations (Fig. 12.18). Generally each major terrace can be traced upstream to an end moraine or former ice limit.

Deposits of Stagnant Ice

When rapid ablation greatly reduces ice thickness in the terminal zone of a glacier, movement may virtually cease. Sediment carried by meltwater flowing over or beside the nearly motionless stagnant ice is deposited as stratified drift, which slumps and collapses as the supporting ice slowly melts away. Such sediment is called *ice-contact stratified drift.* It is recognized by abrupt changes in grain size, by distorted, offset, and irregular stratification, and by extremely uneven surface form.

Bodies of ice-contact stratified drift are classified according to their shape (Fig. 12.19). Among the most common forms is a *kame,* a moundlike hill of stratified drift. Some consist of sediment deposited in crevasses or on the surface of stagnant ice, whereas others are deltas or fans built on or against ice. A *kame terrace* is a terracelike body of sediment along the side of a valley that was deposited against glacier ice that occupied the valley floor. A basin within a body of glacial sediment created by the melting out of a mass of underlying ice is a *kettle.* Some extensive end moraine systems of former ice sheets consist of broad belts of *kettle-and-kame topography,* extremely uneven terrain consisting of a complex of adjacent kames and kettles; most of the multitude of lakes that dot the land surface in the states of Michigan, Minnesota, and Wisconsin occupy kettles in such terrain. Long, narrow ice-contact ridges of stratified drift, often having a sinuous shape, are called *eskers.* Most are deposited by meltwater streams flowing through tunnels in stagnant ice.

THE GLACIAL AGES

History of the Concept

As early as 1821 European scientists began to recognize features characteristic of glaciation in places far from any existing glaciers. They drew the then

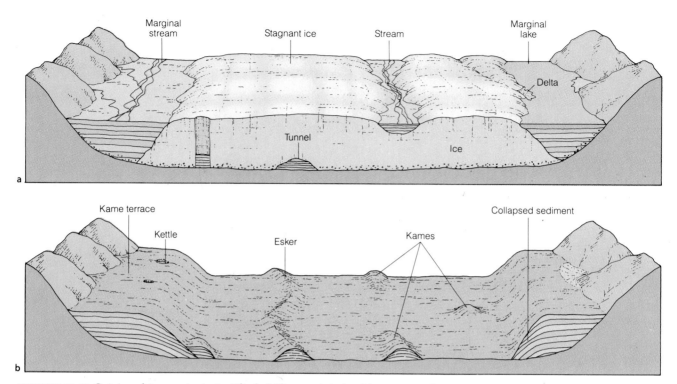

FIGURE 12.19 Origin of ice-contact stratified drift associated with stagnant-ice terrain. (a) Nearly motionless melting ice furnished temporary retaining walls for bodies of sediment deposited chiefly by meltwater streams and in meltwater lakes. (b) As ice melts, bodies of sediment slump, creating kettle-and-kame topography.

remarkable conclusion that glaciers must once have covered wide regions. The concept of a glacial age with widespread effects was first proposed in 1837 by Louis Agassiz, a Swiss scientist who achieved considerable fame through his hypothesis. Although at first many regarded the idea as outrageous, gradually, through the work of many geologists, the concept gained widespread acceptance.

The Last Glaciation

Ice Cover

Before 10,000 years ago, the Earth for many millennia was in the last glacial age. It was a world that was very different from the one with which we are familiar. As the climate cooled about 25,000 years ago, an extensive ice sheet that had formed over north-central Canada began to expand south toward the United States and west toward the Rocky Mountains. Simultaneously in northern Europe another great ice sheet spread southward and overwhelmed the landscape (Fig. 12.20). Other large ice sheets expanded over arctic regions of North America and Eurasia, including some areas now

submerged by shallow polar seas, and over the mountainous cordillera of western Canada (Fig. 12.20). The ice sheets in Greenland and Antarctica expanded as falling sea level exposed areas of the surrounding continental shelves. Glacier systems also grew in the world's major mountain ranges, including the Alps, Andes, Himalaya, and Rockies, as well as in numerous smaller ranges and on isolated peaks scattered widely through all latitudes.

On a global scale, the areas of former glaciation add up to an impressive total of more than 44 million km², or about 29 percent of the entire land area of the Earth. Today, for comparison, only about 10 percent of the world's land area is covered with glacier ice; of this area, 84 percent lies in the Antarctic.

Deformation of the Crust

The weight of the massive ice sheets caused the crust of the Earth to subside beneath them, a process described further in chapter 14. The contrast between the density of crustal rocks (about 2.7 g/cm) and glacier ice (about 0.9 g/cm) means that an ice sheet 3 km thick could cause the crust to

FIGURE 12.20 Areas of the Northern Hemisphere that were covered by glaciers during the last glacial age. Arrows show the general direction of ice flow. Coastlines are shown as they were at that time, when world sea level was about 100 m lower than present. Sea ice, shown covering the Arctic Ocean, extended south into the North Atlantic. Some scientists postulate that thick ice shelves, rather than sea ice, covered these portions of the ocean. The extent of former glacier ice over the Barents and Kara seas, as well as in parts of northern North America, is controversial.

subside by as much as 1 km. The Hudson Bay region of Canada, which lay near the center of the vast Laurentide Ice Sheet (Fig. 12.20), is still rising as the crust adjusts to the removal of this ice load.

Lowering of Sea Level

Whenever large glaciers formed on the land, the moisture needed to produce and sustain them was derived primarily from the oceans. As a result, sea level was lowered in proportion to the volume of ice on land. During the last glacial age, world sea level fell at least 100 m, thereby causing large expanses of the shallow continental shelves to emerge as dry land (Fig. B4.1). At that time the Atlantic coast of the United States between New York and Florida lay about 150 km east of its present position. At the same time, lowering of sea level joined Britain to France where the English Channel now lies, and North America and Asia formed a continuous landmass across what is now the Bering Strait (Fig. 12.20). These and other land connections allowed plants and animals, including humans, to pass freely between land areas that now are separated by ocean waters.

Earlier Glaciations

Until rather recently, it was thought that the Earth had experienced four glacial ages during the Pleistocene Epoch. This assumption was based on studies of ice-sheet and mountain-glacier deposits and had its roots in early studies of the Alps where geologists identified stream terraces they thought were related to four ice advances. This traditional view was discarded when studies of deep-sea sediments disclosed a long succession of glaciations, the most recent of which was shown by radiocarbon dating to equate with the youngest glacial drift on land. Paleomagnetic dating of the cores (chapter 6) showed that glacial–interglacial cycles average about 100,000 years long and that during the last million years alone there had been about 10 such episodes. For the Pleistocene Epoch as a whole, at least 15 to 20 glacial ages are recorded, rather than the traditional four.

Seafloor Evidence

With increasing depth in a core, the biologic component of seafloor sediments shows repeated shifts of surface-water animal and plant populations, from warm interglacial forms to cold glacial forms. The ratio of the isotopes ^{18}O to ^{16}O in calcareous ooze also fluctuates in a similar pattern. The isotopic variations in Pleistocene marine sediments are thought primarily to represent changes in global ice volume. During glacial ages, when water is evaporated from the oceans and precipitated on land to form glaciers, water containing the light isotope ^{16}O is more easily evaporated than water containing the heavier ^{18}O. As a result, Pleistocene glaciers contained more of the light isotope, while the oceans became enriched in the heavy isotope. Isotope curves derived from the sediments therefore give us a continuous reading of changing ice volume on the planet (Fig. 12.21). Because glaciers wax and wane in response to climatic changes, the isotopes also give a generalized view of global climatic change.

Pre-Pleistocene Glaciations

Ancient glaciations, identified mainly by tillites and striated rock surfaces, are known from still older parts of the geologic column (Fig. 4.10). The earliest recorded glaciation dates to about 2.3 billion years ago in the early Proterozoic. Evidence of others has been found in rocks of late Proterozoic, early Paleozoic, and late Paleozoic age. During the latest of these intervals 50 or more glaciations are believed to have occurred. Evidence from such low-latitude regions as South America, Africa, and India, as well as from Antarctica, suggest that the Earth's land areas must have had a very different relationship to one another during the late Paleozoic glaciation than they do today.

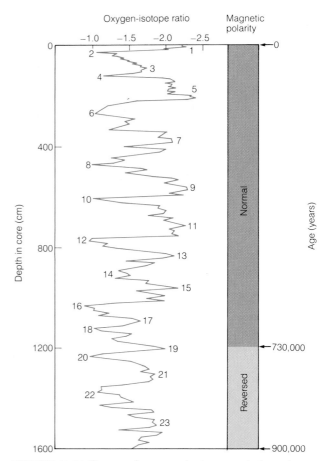

FIGURE 12.21 Curve of oxygen-isotope variations in an equatorial Pacific deep-sea core representing global changes in ice volume during the last 900,000 years. Each glacial–interglacial cycle is about 100,000 years long. Even-numbered peaks are glacial maxima; odd-numbered peaks are interglacial maxima.

The Little Ice Age

Old lithographs and paintings of Alpine valleys from the eighteenth and early nineteenth centuries show glaciers terminating far beyond their present positions (Fig. 12.22), sometimes close to farmland and buildings that date to earlier centuries. Documents from the sixteenth and seventeenth centuries describe glaciers that advanced over small villages and became larger than at any time in human memory. Similar ice advances took place in other parts of the world and in many cases led to the greatest expansion of glaciers since the end of the last glacial age, some 10,000 years earlier. The climatic change that heralded these conditions apparently occurred in the thirteenth century, bringing an end to an era of relatively mild climate during the Middle Ages. This interval of generally cool climate between the middle thirteenth and middle nineteenth centuries, during which mountain gla-

FIGURE 12.22 Lithograph showing Lower Grindelwald Glacier in the Swiss Alps in 1826 at the culmination of the last phase of the Little Ice Age.

ciers expanded worldwide, is commonly referred to as the *Little Ice Age.*

Events similar to the Little Ice Age have occurred repeatedly during the last 50,000 years and show that minor climatic changes of short duration are superimposed on much longer glacial–interglacial cycles.

WHAT CAUSES GLACIAL AGES?

Glacial Eras and Shifting Continents

Successions of glacial ages, each lasting tens of millions of years, can be identified in the geologic record of the last 2.5 billion years (Fig. 12.23). They were separated by long periods of mild climate,

that of the Mesozoic Era being the most recent example. What seems to be the only reasonable explanation for their pattern is suggested by slow but important geographic changes that affect the crust of the planet. These changes include (1) the movement of continents as they are carried along with shifting plates of lithosphere; (2) the large-scale uplift of continental crust where one plate overrides another; (3) the creation of high mountain chains where continents collide; and (4) the opening or closing of ocean basins and seaways between moving landmasses.

The effect of such earth movements on climate is illustrated by the fact that low temperatures are found, and glaciers tend to form and persist, in two kinds of situations: high latitudes and high altitudes—especially in places where winds can supply abundant moisture evaporated from a nearby ocean. The Earth's largest existing glacier is centered on the South Pole where temperatures are constantly below freezing and the land is surrounded by ocean. The only glaciers found at or close to the equator lie at extremely high altitudes.

Abundant evidence now leads us to conclude that the positions, shapes, and altitudes of landmasses have changed with time (chapter 16), in the process altering the paths of ocean currents and of atmospheric circulation. Where evidence of ancient ice-sheet glaciation is now found in low latitudes, we infer that such lands were formerly located in higher latitudes where large glaciers could be sustained. Although this conclusion appears to explain the pattern of glaciation during and since the late Paleozoic, information about earlier glacial eras is very fragmentary and less easy to evaluate.

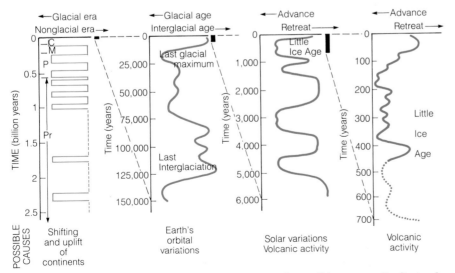

FIGURE 12.23 Time scales of climatic variations and possible causes. In first column: Pr = Proterozoic, P = Paleozoic, M = Mesozoic, and C = Cenozoic.

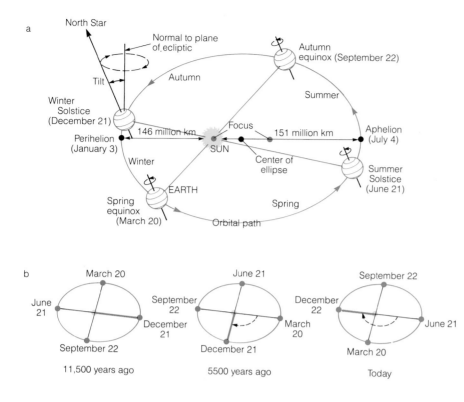

FIGURE 12.24 (a) Geometry of the Sun–Earth system. The Earth's orbit is an ellipse with the Sun at one focus. The Earth moves around its orbit in the direction of the arrows. Simultaneously it spins about its axis, which is tilted to the plane of the equinoxes and which causes the position of the equinoxes and the solstices to shift slowly around the Earth's orbit. (b) About 11,500 years ago the winter solstice occurred near the aphelion (the most distant point on the orbit from the Sun), whereas today it occurs at the opposite end of the orbit near the perihelion (the closest approach to the Sun).

Ice Ages and the Astronomical Theory

As initially discovered through studies of glacial deposits, and later verified by studies of deep-sea cores, glacial and interglacial ages have alternated for more than a million years (Fig. 12.21). The cause of this basic climatic pattern, which produced worldwide geographic changes and major relocations of plants and animals, has long been a fundamental challenge to the development of a comprehensive theory of climate. A preliminary answer was provided by Scottish geologist John Croll, in the middle nineteenth century, and was later elaborated by Milutin Milankovitch, a Serbian astronomer of the early twentieth century.

Croll and Milankovitch recognized that minor variations in the path of the Earth in its orbit around the Sun and in the inclination, or tilt, of the Earth's axis cause slight but important variations in the amount of radiant energy reaching the top of the atmosphere at any given latitude. Three movements are involved (Fig. 12.24). First, the axis of rotation, which now points to the North Star, wobbles like a spinning top, moving slowly in a circular path. Simultaneously, the elliptical orbit of the Earth is also rotating, but much more slowly. These two motions together cause a progressive change in the Earth–Sun distance for a given date (Fig. 12.24b). At the spring and autumn equinoxes day and night are of equal length. At the summer and winter

solstices, days are longest and shortest, respectively. The equinoxes move slowly around the orbital path, completing one full cycle in about 23,000 years, a motion referred to as the *precession of the equinoxes* (or simply *precession*). Second, the *tilt* of the axis, which now averages 23.5°, shifts about 1.5° to either side during a span of about 41,000 years (Fig. 12.24a). Finally, the shape, or *eccentricity*, of the orbit changes over an average period of nearly 100,000 years. About 100,000 years ago the orbit was highly eccentric (less circular) compared to what it has been for the last 10,000 years.

Each of these astronomical factors varies on a different time scale, so when they interact the resulting pattern is complicated (Fig. 12.25). Mathematical analysis of climatically controlled biologic and isotopic records in deep-sea cores shows clearly that they fluctuate on the same time scales as those of axial tilt and precession. This persuasive evidence has provided considerable support for the theory that astronomical factors control the timing of the glacial–interglacial cycles.

Solar Variations, Volcanic Activity, and Little Ice Ages

Climatic fluctuations measured in centuries or decades are too short to be influenced either by movements of continents or by the three primary astronomical variations, and require us to seek other

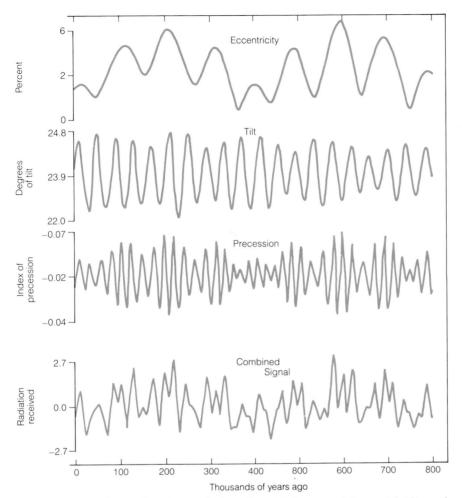

FIGURE 12.25 Curves showing variations in orbital eccentricity, axial tilt, and precession during the last 800,000 years. Adding these factors produces a combined signal that can be used to compute the amount of radiation reaching the Earth at any latitude through time. (*Source*: After Imbrie et al., 1984.)

explanations for their cause. Two have received special attention (Fig. 12.23).

One hypothesis is based on the concept that the energy output of the Sun fluctuates through time. The idea is appealing because it might explain climatic variations on several different time scales. However, although correlations have been found between weather patterns and rhythmic fluctuations in the number of sunspots appearing on the surface of the Sun, as yet there has been no clear demonstration that solar variations are responsible for climatic changes on the scale of the Little Ice Age.

Large explosive volcanic eruptions can eject large quantities of fine ash into the atmosphere to create a veil of fine dust that will circle the globe. The fine

particles tend to scatter incoming solar radiation, resulting in a slight cooling at the Earth's surface. Although the dust settles out rather quickly, generally within a few months to a year, tiny droplets of sulfuric acid, produced by the interaction of volcanically emitted SO_2 gas and water vapor, also scatter the Sun's rays and remain concentrated in the upper atmosphere for several years. During a large eruption, the average surface air temperature can be lowered 0.5° to 1.0°C, sufficient to influence the mass balance of glaciers. Close association of intervals of glacier advance and periods of increased volcanic activity during the last several centuries lend support to the hypothesis that volcanic emissions can produce detectable changes of climate on the scale of decades.

Essay

THE EARTH'S FUTURE CLIMATES

As our expanding society becomes increasingly vulnerable to changes of climate, even of small magnitude, the need for reliable prediction of future climatic changes becomes obvious. At present, prediction of future climatic trends is based largely on extrapolation, that is, on estimation of the probable future continuation of a trend shown by past events of a similar kind. However, this approach requires that we understand the basic causes of fluctuations in the climatic record, or our predictions could prove unreliable.

What can be said about probable future climates? Because the Croll–Milankovitch astronomical variations apparently control the timing of the glacial ages, the climatic curve based on them can be extended into the future by calculating the interacting effects of the Earth's predictable movements. The results show that although we may be nearing the end of the present interglaciation, the next full-scale glaciation will not culminate for about another 23,000 years (Fig. B12.1). Obviously, it will be a long time before this prediction is proved right or wrong!

The near future is clearly of greater interest to us. Can we say for certain that the climate of the Little Ice Age is behind us and that we can look forward to equable climates in the coming centuries before we descend into a future glacial age? Some scientists have suggested that a cooling trend which began in the middle 1940s may herald a return to Little Ice Age conditions, or even mark the start of a downward trend leading to the next glaciation. Others consider it but a minor reversal in an otherwise longer-term trend toward still warmer climates that are likely to result from the ever-increasing release of CO_2 and other gases into the atmosphere.

The progressive increase of atmospheric CO_2, totaling about 25 percent since 1850, has been taking place since large-scale burning of fossil fuels began at the start of the industrial revolution. The buildup of CO_2, together with other industrial gases that contribute to the backscattering of incoming solar radiation, is expected to cause significant worldwide warming within the next century. Computer simulations suggest that average global surface temperatures could increase about 4°C. Even if fossil-fuel consumption and related CO_2 buildup were to cease abruptly (a highly unlikely scenario), the surface air temperature would continue to rise by nearly 1°C as the atmosphere moves toward a new equilibrium condition.

These results have led some scientists to predict that the Earth is about to enter a "super interglaciation" in which temperatures will

SUMMARY

1. Glaciers consist of ice which has been transformed from snow by compaction, recrystallization, and flow. They form part of the cryosphere, the area of the planet where water exists primarily in the frozen state.

2. The major types of glaciers, based on their geometry, are cirque glaciers, valley glaciers, ice caps, ice sheets, and ice shelves.

3. Ice in temperate glaciers is at the pressure melting point and water exists at the glacier bed. Polar glaciers consist of ice that is below the pressure melting point. In subpolar glaciers, a thin surface zone reaches the melting point in summer, but the ice beneath is below freezing.

4. Glaciers depend for their survival on low temperature and adequate precipitation. They bear a close relationship to the snowline, which is low in polar regions and rises to high altitudes in the tropics.

5. The mass balance of a glacier is measured in terms of accumulation and ablation. The equilibrium line separates the accumulation area

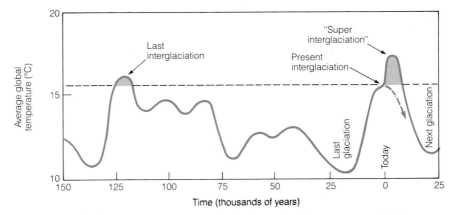

FIGURE B12.1 Course of climate during the last 150,000 years and 25,000 years into the future. The natural course of future climate (dashed line) would be one of declining temperatures until the next glacial maximum, about 23,000 years from now. With the CO_2-induced greenhouse effect, continued warming may lead to a super interglaciation within the next several hundred years. During such an interval, temperature may rise above that of the last interglaciation. The decline toward the next glaciation would thereby be delayed by several millennia. (*Source*: After Imbrie and Imbrie, 1979.)

reach levels not experienced since the last interglaciation 120,000 years ago, or even since the warm Cretaceous Period (Fig. B12.1). The ultimate results of this huge man-made "experiment" we have embarked upon are uncertain, but they might be dramatic in some regions. A warmer Earth could lead to extensive reduction in polar ice bodies, with an accompanying rise in world sea level of many meters. The slow inundation of coastal lands, where a high percentage of the world's inhabitants dwell, would cause large-scale dislocations of populations. Agriculture, forestry, fisheries, water supplies, energy production, and transportation, would also be impacted, in some cases severely, producing social and political repercussions that are difficult to foresee. The uncertainties involved pose a major challenge to Earth scientists who have at their disposal an array of geologic information about past climates and surface conditions that can offer important clues about environmental changes that the future may hold in store for us.

from the ablation area and marks the level on the glacier where net gain is balanced by net loss.

6. The motion of temperate glaciers includes both internal flow and sliding along the bed. In polar glaciers, which are frozen to their bed, motion is much slower and involves only internal flow.

7. Surges involve extremely rapid flow, probably related to excess water at the glacier bed. Frontal calving can lead to rapid recession of glacier margins that recede into deep water.

8. Glaciers erode rock by quarrying and abrasion. They transport the waste and deposit it as drift.

9. Mountain glaciers convert stream valleys into U-shaped troughs with hanging tributaries and with cirques at their heads. Fjords are excavated far below sea level by thick ice streams in high-latitude coastal regions.

10. Flow directions of ice sheets are inferred from striations, grooves, and drumlins.

11. The load, carried chiefly in the base and sides of a glacier, includes rock fragments of all sizes, from fine rock flour to large boulders.

12. Till is deposited directly by glaciers, while glacial–marine drift is deposited on the seafloor from floating glacier ice. Stratified drift is deposited by meltwater. It includes outwash deposited as outwash plains or valley trains beyond the ice margin, and ice-contact stratified drift deposited upon or against stagnant ice.

13. Ground moraine is built up beneath a glacier, whereas end moraines (both terminal and lateral) form at the glacier margins.

14. During glacial ages huge ice sheets repeatedly covered northern North America and Europe, eroding bedrock and spreading drift over the outer parts of the glaciated regions. As the ice sheets grew, world sea level fell and the crustal rocks beneath them subsided due to the added weight.

15. Glacial ages are discerned in many parts of the geologic column; they extend back at least 2.3 billion years.

16. Glacial ages have alternated with interglacial ages in which temperatures approximated those of today. Studies of marine cores indicate that glacial–interglacial cycles were about 100,000 years long, and that during the Pleistocene there have been 15 or more such cycles. In most glaciated regions we see evidence only of the three or four most extensive glaciations.

17. The occurrence of glacial ages probably is related to favorable positioning of continents and ocean basins, brought about by movements of lithospheric plates. The timing of the glacial–interglacial cycles appears to be closely controlled by changes in the orbital path and axial tilt of the Earth, which affect the distribution of solar radiation received at the surface.

18. Climatic variations on the scale of centuries and decades may result from fluctuations in energy output from the Sun, or from injections of volcanic dust and gases into the atmosphere.

IMPORTANT WORDS AND TERMS TO REMEMBER

ablation (p. 267)
ablation area (p. 268)
accumulation (p. 267)
accumulation area (p. 268)
arête (p. 272)

boulder train (p. 277)

calving (p. 271)
calving glacier (p. 271)
cirque (p. 272)
cirque glacier (p. 264)
col (p. 273)
creep (of glacier ice) (p. 269)
crevasse (p. 269)
cryosphere (p. 264)

dropstones (p. 275)
drumlin (p. 274)

eccentricity (of the orbit) (p. 282)
end moraine (p. 276)
equilibrium line (p. 268)

erratic (p. 276)
esker (p. 277)

firn (p. 267)
fjord (p. 273)
fjord glacier (p. 264)

glacial drift (p. 275)
glacial–marine drift (p. 275)
glacial striations (p. 272)
glaciation (p. 271)
glacier (p. 264)
ground moraine (p. 276)

hanging valley (p. 273)
headwall (p. 272)
horn (p. 272)

ice cap (p. 264)
ice-contact stratified drift (p. 277)
ice field (p. 264)
ice sheet (p. 264)
ice shelf (p. 264)

kame (p. 277)
kame terrace (p. 277)
kettle (p. 277)
kettle-and-kame topography (p. 277)

lateral moraine (p. 276)
Little Ice Age (p. 281)

mass balance (p. 267)
medial moraine (p. 274)

outwash (p. 277)
outwash plain (p. 277)
outwash terrace (p. 277)

piedmont glacier (p. 264)
polar (cold) glacier (p. 265)
precession of the equinoxes (p. 282)
pressure melting point (p. 265)

rock drumlin (p. 274)
rock flour (p. 274)

snowline (p. 265)
stratified drift (p. 276)
subpolar glacier (p. 265)
surge (p. 270)

tarn (p. 272)

temperate (warm) glacier (p. 265)
terminal moraine (p. 276)
terminus (of a glacier) (p. 268)
till (p. 275)
till matrix (p. 275)

tilt (of axis) (p. 282)

valley glacier (p. 264)
valley train (p. 277)

QUESTIONS FOR REVIEW

1. What distinguishes temperate glaciers from polar glaciers?

2. What is the snowline and how are glaciers related to it?

3. Describe the steps in the conversion of snow to glacier ice.

4. Why does the position of the equilibrium line provide a rough estimate of the glacier's mass balance?

5. Why is there a time lag between a change of climate and the response of a glacier's terminus to the change?

6. In what ways does ice temperature influence the way a glacier moves?

7. Describe the unique motions of surging and calving glaciers.

8. What small-scale and large-scale erosional features can be used to infer directions of flow of former glaciers?

9. How do glaciated valleys differ from nonglaciated valleys?

10. How might one distinguish till from stratified drift in a roadside outcrop?

11. In what different ways are moraines built at a glacier margin?

12. Under what conditions does kettle-and-kame topography develop?

13. Why, and by how much, does world sea level fall and rise during glacial–interglacial cycles?

14. What evidence obtained from deep-sea cores indicates that glacial–interglacial cycles have occurred repeatedly during the Pleistocene Epoch?

15. What natural factors explain the recurrence of glacial events on time scales of about 23,000, 41,000 and 100,000 years?

16. How can large volcanic eruptions influence climate and cause glaciers to expand or contract in size?

C H A P T E R 13

The Ocean Margins

Waves rework a gravelly beach near the mouth of the Rakaia River along the margin of the Canterbury Plain on the coast of South Island, New Zealand.

CURRENTS, TIDES, AND WAVES

The shoreline, where land and ocean meet, is one of the most dynamic places on our planet. An annual visit to almost any shore will disclose changes that have taken place since the previous visit; sometimes they are small, but often they are substantial. The energy driving these changes comes from ocean currents and waves, which derive their energy from winds, and from tides that are generated by the gravitational attraction between the Earth and the Moon.

Surface Ocean Currents

Origin

Surface ocean currents are broad, slow drifts of surface water. They are set in motion by the prevailing surface winds. Air that flows across a water surface drags the water slowly foward, creating a current of water as broad as the current of air, but rarely more than 50 to 100 m deep. In low latitudes surface seawater moves westward with the trade winds. The general westerly direction of the North and South Equatorial currents (Fig. 13.1) is reinforced by the Earth's rotation. Both north and south of the equator, the westerly moving currents eventually are deflected where a current encounters a coast, and by the Coriolis effect (chapter 1). On reaching higher latitudes, the currents travel eastward, moved by the prevailing westerly winds. The result is a circular motion of water in each of the major ocean basins, both north and south of the equator.

Warm and Cold Currents

Along most continental margins, the predominant flow of water roughly parallels the coast. Warm surface water originating in the equatorial region moves along the eastern margins of continents between latitudes 45° N and S, while cold currents originating in higher latitudes encounter, and are deflected along, the western margins (Fig. 13.1). At high latitudes, cold currents generally prevail, except in a few cases where warm currents are able to sweep north into subpolar latitudes (e.g., North Atlantic current, Alaska current).

Tides

Tides are the rhythmic rise and fall of ocean waters resulting from the gravitational attraction of the Moon and (to a lesser degree) the Sun acting on the Earth.

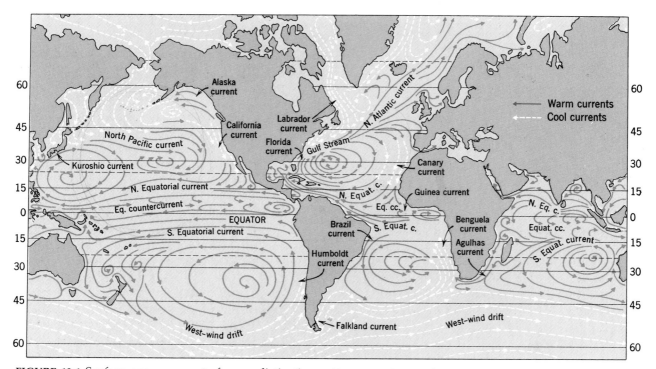

FIGURE 13.1 Surface ocean currents form a distinctive pattern, curving to the right in the Northern Hemisphere and to the left in the Southern.

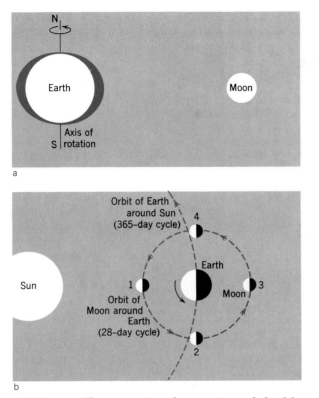

a

b

FIGURE 13.2 The gravitational attractions of the Moon and Sun on the Earth raise tidal bulges in the ocean. (a) Idealized diagram of the tidal bulges, relative to the Earth's axis of rotation, and the position of the Moon. (b) When the Moon and Sun attract in the same direction (Moon positions 1 and 3), highest tides are observed. When the Moon and Sun have opposition positions (Moon positions 2 and 4), lowest tides are experienced.

Tidal Bulges

Gravitational attraction of the Moon creates **tidal bulges** on opposite sides of the Earth that appear to move continually around the Earth as it rotates (Fig. 13.2a). In fact, the bulges remain stationary while the Earth rotates beneath them. They are caused chiefly by the difference in the Moon's attraction for the ocean and for the solid Earth, which in turn is related to differences in their distance from the Moon. On the side toward the Moon, ocean water is attracted more strongly by the Moon than is the solid Earth beneath; on the far side, the water is pulled less strongly than the underlying Earth. The result is two tidal bulges on opposite sides of the planet.

At most places on the ocean margins two high tides and two low tides are observed each day. However, the Sun also affects the tides, sometimes opposing the Moon by pulling at right angles, and sometimes aiding by pulling in the same direction (Fig. 13.2b). The Sun is only about half as effective as the Moon in producing tides, so the two tidal effects never entirely cancel each other.

Tidal bulges cannot move around the Earth unhindered because continents get in the way. Therefore, water piles up against the continental margins whenever a tidal bulge arrives. In effect, a mass of water runs into the coastline at every high tide. Water from a high tide flows back to the ocean basin as the Earth rotates and the tidal bulge passes.

Tidal Currents

Tidal currents are caused mainly by the twice-daily tidal bulges that pass around the Earth. In the open sea, tidal effects are small; however, in bays, straits, estuaries, and other narrow places, tidal fluctuations can generate rapid currents. Tidal flows can approach 25 km/h in places, and tidal heights of more than 16 m are known (Fig. 13.3). Such fast-moving currents, although restricted in extent, readily move sediment around. Large linear sand ridges can be built paralleling such currents, as well as sand waves, which are large sand ripples oriented perpendicular to the current direction.

Waves

Wave Motion

Ocean waves are generated by winds that blow across the surface. Figure 13.4 shows the significant dimensions of a wave traveling in deep water where it is unaffected by the bottom far below. The motion of a wave is very different from the motion of any parcel of water within it. As wind sweeps across a field of grain or tall grass, the individual stalks bend forward and return to their positions, creating a wavelike effect. In similar fashion the *form* of a wave in water moves continuously forward, but each water parcel revolves in a loop, returning, as the wave passes, very nearly to its former position. This looplike or *oscillating* motion of the water, first predicted theoretically, was later proved by injecting droplets of colored water into waves in a glass tank and then photographing their paths with a movie camera. Waves receive their energy from wind, and so can receive it only at the surface where air and water meet. Because the wave form is created by a looplike motion of water parcels, the diameters of the loops at the water surface exactly equal wave height (*H* in Fig. 13.4). Downward from

a

b

FIGURE 13.3 Tidal fluctuations in coastal zones. (a) Coastline near Nelson, New Zealand at high tide. (b) Same coast about 12 hours later at low tide. (c) Inrushing tide in an estuary at Moncton, New Brunswick, Canada, forms a turbulent wall-like wave of water (a *tidal bore*) that moves rapidly (10 to 15 km/h) against the river flowing gently from right to left.

c

the surface a progressive loss of energy occurs, expressed in diminished diameters of the loops. At a depth equal to only half the *wavelength,* the distance between successive wave crests or troughs (*L* in Fig. 13.4), the diameters of the loops have become so small that motion of the water is negligible.

Erosion by Waves

Wave Base.   The depth *L*/2 is the effective lower limit of wave motion (Fig. 13.4). Therefore, it must be the lower limit of erosion of the bottom by waves. This depth is generally referred to as the *wave base.* In the Pacific Ocean, wavelengths as great as 600 m

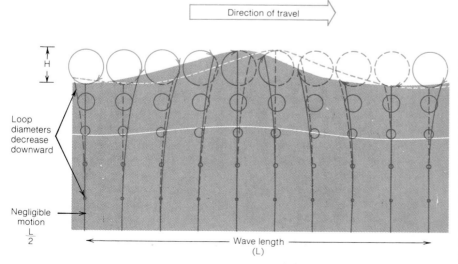

FIGURE 13.4 Looplike motion of water parcels in a wave in deep water. To follow the successive position of a water parcel at the surface, follow the arrows in the largest loops from right to left. This is the same as watching the wave crest travel from left to right. Parcels in smaller loops underneath have corresponding positions, marked by continuous, nearly vertical lines. Dashed lines represent waveform and parcel positions one-eighth period later.

have been measured. For them, $L/2$ equals 300 m, a depth half again as great as the outer edge of the average continental shelf. Although the wavelengths of most ocean waves are far less than 600 m, it is nevertheless possible for very large waves approaching these dimensions to affect even the outer parts of continental shelves, which average 200 m in depth. Landward of depth $L/2$ the wave motion continuously lifts and drops fine particles of bottom sediment, very slowly moving them seaward along the gently sloping bottom. Such erosion is very slow and not at all spectacular, but over a million years or more the cumulative result is great.

Breaking Waves. What happens in the shallow water at the shore is more rapid, sometimes spectacular, and always different in style. When a wave moving toward shore reaches depth $L/2$ its base meets the bottom and the form of the wave begins to change. The looplike paths of water parcels gradually become elliptical, and velocities of the parcels increase. Interference of the bottom with wave motion distorts the wave by increasing its height and shortening wavelength. Often the height is doubled. In other words, the wave grows steeper. The front of the wave is in shallower water than the rear part and is steeper. Eventually the steep front is unable to support the advancing wave, and as the rear part continues to move forward, the wave collapses or *breaks* (Fig. 13.5).

Surf. When a wave breaks, the motion of its water instantly becomes turbulent, like that of a swift river. Such "broken water" is called **surf**, defined as wave activity between the line of breakers and the shore. In turbulent surf, each wave finally dashes against rock or rushes up a sloping beach until its energy is expended; then it flows back. Water piled against the shore returns seaward in an irregular and complex way, partly as a broad sheet along the bottom and partly in localized narrow channels as **rip currents**, which are responsible for dangerous "undertows" that can sweep unwary swimmers out to sea.

Surf possesses most of the original energy of each wave that created it. This energy is quickly consumed in turbulence, in friction at the bottom, and in moving the sediment that is thrown violently into suspension from the bottom. Although fine sediment is transported seaward from the surf zone, most of the geologic work of waves is accomplished by surf shoreward of the line of breakers.

Depth of Erosion by Surf. How deep below sea level can surf erode rock and move sediment? The answer depends on the depth at which waves break. Most ocean waves break at depths that range between wave height and 1.5 times wave height. Because waves are seldom more than 6 m high, the depth of vigorous erosion by surf should be limited to 6 m times 1.5, or 9 m below sea level. This theoretical limit is confirmed by observation of breakwaters and other structures, which are found to be only rarely affected by surf at depths of more than about 7 m.

Erosion Above Sea Level. During great storms, surf can strike effective blows well above sea level. The west coast of Scotland is exposed to the full

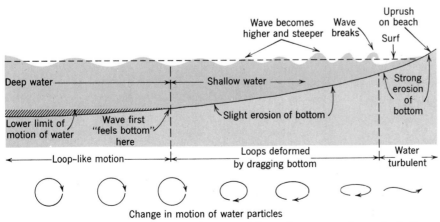

FIGURE 13.5 Waves change form as they travel from deep water through shallow water to shore. Circles and ellipses are not drawn to scale with the waves shown above. Compare with Figure 13.4.

force of Atlantic waves. During a great storm on that coast a solid mass of stone, iron, and concrete weighing 1200 metric tons was ripped from the end of a breakwater and moved inshore. The damage was repaired with a block weighing more than 2300 metric tons, but five years later storm waves broke off and moved that one too. The pressures involved in such erosion were about 27 metric tons/m². Even waves having much smaller force break loose and move blocks of bedrock from sea cliffs, partly by compressing the air in fissures, which then pushes out blocks of rock.

The vertical distance through which water can be flung against the shore would surprise anyone whose experience of coasts is limited to periods of calm weather. During a winter storm in 1952, again on the west coast of Scotland, the bow half of a small steamship was thrown against a cliff and left there, wedged in a big crevice, 45 m above sea level.

Abrasion in the Surf Zone.   Another important kind of erosion in the surf zone is the wearing down of rock by wave-carried rock particles. By continuous rubbing and grinding with these tools, the surf wears down and deepens the bottom and eats into the land, at the same time smoothing, rounding, and making smaller the tools them-

selves. As we have seen, this activity is limited to a depth of only a few meters below sea level. The surf, therefore, is like an erosional knife-edge or saw, cutting horizontally into the land (Fig. 13.6).

Wave Refraction

A wave approaching a coast generally does not encounter the bottom simultaneously all along its length. As any segment of the wave touches the seafloor, that part slows down. Accordingly, wavelength then begins to decrease and wave height increases. Gradually the wave swings around to parallel the bottom contours and is said to be refracted (Fig. 13.7). **Wave refraction** is the process by which the direction of a series of waves, moving in shallow water at an angle to the shoreline, is changed. Thus, waves approaching the margin of a deep-water bay at an angle of 40° or 50° may, after refraction, reach the shore at an angle of 5° or less. Waves passing over a submerged ridge off a headland will converge on the headland. Convergence, plus the increased wave height that accompanies it, concentrates wave energy on the headland. Conversely, refraction of waves approaching a bay will make them diverge, diffusing their energy at the shore. Because of refraction, headlands are eroded more vigorously than are bays,

FIGURE 13.6 A nearly horizontal wave-cut bench has formed along the coast at Bolinas Point, California, as the surf, acting like an erosional saw, has cut into the exposed tilted sedimentary rocks.

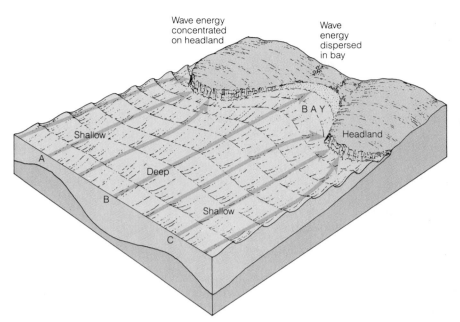

FIGURE 13.7 Refraction of waves concentrates wave energy on headlands and disperses it along shores of bays. Oblique view shows how waves become progressively distorted as they approach the shore over a bottom that is deepest opposite the bay.

so that in the course of time irregular coasts tend to become smoother and less indented.

Transport of Sediment

Transport Along the Shore

Longshore Currents. Despite refraction, most waves reach the shore at an angle. The oblique approach of waves sets up a longshore current within the surf zone that flows parallel to the shore (Fig. 13.8a). Such currents easily move fine sand suspended in the turbulent surf. The direction of longshore currents may change seasonally if the prevailing wind directions change, thereby causing changes in the direction of the arriving waves.

Beach Drift. Meanwhile, on the exposed beach, a second kind of movement alongshore is occurring. Because waves generally strike the beach at an angle, the *swash* of each wave travels obliquely up the beach, but the *backwash* flows straight down the slope of the beach. The result is **beach drift**, a zigzag movement of sand and pebbles, with net progress along the shore (Fig. 13.8b). The greater the angle of waves to shore, the greater the longshore movement. Pebbles tagged and timed have been observed to drift along a beach at a rate of more than 800 m/day. When the volume of sand

moved by beach drift is added to that moved by longshore currents, the total can be very large.

Offshore Transport and Sorting

Seaward of the surf zone, in deeper water, bottom sediment is shifted by unusually large waves during storms and by currents, with net movement seaward. Each particle is picked up again and again, whenever the energy of waves or currents is great enough to move it. As the particle gets into ever-deeper water, it is picked up less and less frequently. With increasing depth, energy related to wave motion decreases, so only finer-sized grains can be moved. As a result, the sediment becomes sorted according to diameter, from coarse in the surf zone to finer offshore.

As sediments accumulate on a continental shelf, they normally grade seaward from sand into mud. This gradation is true not only of the particles eroded from the shore by surf, but also of the particles contributed by rivers, whose currents carry suspended loads to the sea.

THE SHAPING OF COASTS

The coast is a boundary between two realms, land and water. At a coast, ocean waves that may have

a

b

FIGURE 13.8 Longshore currents and beach drift. (a) Waves arriving obliquely onshore along a coast near Oceanside, California, generate a longshore current that moves sediment down the coast. (b) Surf swashes obliquely onto a Brazilian beach and forms a series of arcuate cusps as the water loses momentum and flows back down the sandy slope.

traveled unimpeded across thousands of kilometers of open ocean encounter an obstruction to further progress. They dash against firm rock, erode it, and move the eroded rock particles. Over the long term, the net effect is substantial.

Waves and the currents created by waves are the agents responsible for most of the erosion of coasts, as well as for most of the transport and deposition of the sediment created by wave erosion or washed into the sea by rivers. As sediment moves offshore and is deposited, it contributes to the progressive growth of continental shelves, and thus to the growth of continents.

The Shore Profile

To understand the changes along a coast, we must first look at what happens to the *shore profile,* a vertical section along a line perpendicular to the shore. If we combine this with what we know about the forces that act along (parallel to) the shore, we will then have a three-dimensional picture of coastal activities.

Elements of the Profile

Seen in profile, the usual elements of a cliffed coast (Fig. 13.9) are a wave-cut cliff and wave-cut bench, both the work of erosion, and a beach, the result of deposition. On noncliffed coasts, the beach constitutes the primary shore environment.

Wave-Cut Cliff.   A *wave-cut cliff* is a coastal cliff cut by surf. Acting like a horizontal saw, the surf cuts most actively at the base of the cliff. As the upper part of the cliff is undermined, it collapses and the resulting debris is redistributed by surf. An undercut which has not yet collapsed may have a well-developed *notch* at its base (Fig. 13.10). The notch is a concave part of the shore profile, overhung by the part above. Other erosional features associated with cliffed coasts include *sea caves, sea arches,* and *stacks* (Fig. 13.11). Each is the result of differential erosion as surf attacks a cliff, causing its gradual retreat.

Wave-Cut Bench.   A *wave-cut bench* is a bench or platform cut across bedrock by surf. It slopes gently seaward and is extended progressively landward as the cliff retreats. Some benches are bare or partly bare, but most are covered with sediment that is in transit from shore to deeper water. The shoreward parts of some benches are exposed at low tide (Fig. 13.6). If the coast has been raised by

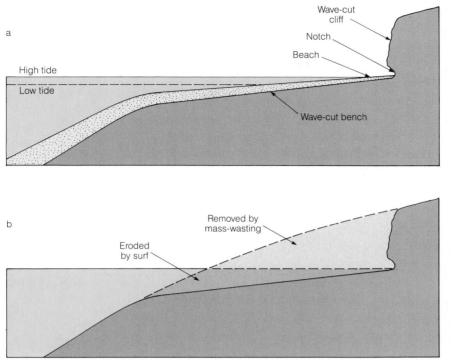

a

Wave-cut cliff

Notch

Beach

High tide

Low tide

Wave-cut bench

b

Removed by mass-wasting

Eroded by surf

FIGURE 13.9 (a) Principal features of a shore profile along a cliffed coast. (b) The comparatively large proportion of material removed by mass-wasting relative to that eroded by surf. Notching of the cliff by surf action undermines the rock or sediment which collapses and is reworked by surf.

FIGURE 13.10 A notch is excavated by waves striking the base of a cliff of jointed basalt along the windward coast of Hawaii. Progressive undercutting ultimately causes collapse of the overlying rocks and retreat of the sea cliff.

recent faulting, a wave-cut bench can be wholly exposed (Fig. 13.12).

Beach. A beach is regarded by most people as the sandy surface above the water along a shore. Actually, it is more than this. We define a ***beach*** as wave-washed sediment along a coast, extending throughout the surf zone. In this zone, sediment is in very active movement. The sediment of a beach is derived in part from erosion of adjacent cliffs or from cliffs elsewhere along the shore. However, along most coasts a much higher percentage of it comes from alluvium contributed by rivers.

On low, open shores a typical beach may consist of several distinct elements (Fig. 13.13). The first is a rather gently sloping ***foreshore,*** a zone extending from the level of lowest tide to the average high-tide level. Next is a ***berm,*** which is a nearly horizontal or landward-sloping bench formed of sediment deposited by waves. Beyond this lies the ***backshore,*** a zone extending inland from a berm to the farthest point reached by waves. On some

FIGURE 13.11 Stack and sea arch along the French shore of the English Channel near Étretat carved in horizontally bedded white chalk. The surf hollows out sea caves in the most erodible part of the bedrock. A cave cut through a headland is transformed into an arch. Isolated remnants of the cliff stand as stacks on a wave-cut bench.

FIGURE 13.12 Emerged wave-cut bench at Tongue Point, southwest of Wellington, New Zealand. Crustal uplift along this coast has raised the former seafloor to form a broad emergent platform. A wave-cut cliff marks the seaward edge of the platform, below which a younger wave-cut bench is forming.

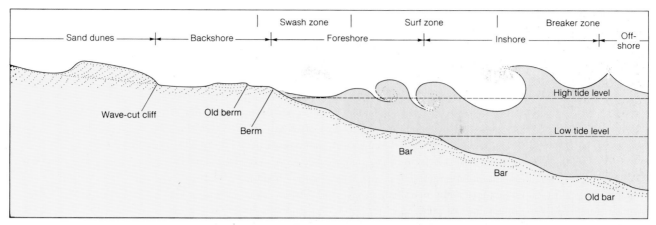

FIGURE 13.13 Typical profile across a beach showing foreshore, berm, and backshore elements. Length of profile about 100 m. Vertical scale exaggerated about twice.

beaches the backshore ends in a wave-cut scarp or in a line of sand dunes.

Steady State Along a Coast

Water meets land in a zone of dynamic activity marked by erosion and the creation, transport, and deposition of sediment. Through these activities, the form of the land slowly changes, and the water in motion moves and shapes the sediment derived from the land. The forces that fashion the shore profile—cliff, bench, and beach—tend to reach and maintain a condition of equilibrium or *steady state*, a compromise in the water–land conflict.

Beach Profile. The compromise is reached in several different ways. On a beach, for instance, the swash of a wave running up the beach as a thin sheet of water moves sediment upslope, while gravity pulls it back again (Fig. 13.13). More energy is needed to move pebbles than to move sand grains downslope. Therefore, the pebbles moved by the swash remain until the slope becomes steep enough for them to be carried back. This partly explains why gravel beaches are generally steeper than beaches built of sand. Another factor is permeability. Much of the water swashing up a beach quickly moves downward into the beach sediment, thereby reducing the volume and the transporting capability of the backwash.

Seasonal Changes. During storms, the increased energy in the surf erodes the exposed part of the beach and makes it narrower. In calm weather, the exposed beach is likely to receive more sediment than it loses and consequently becomes wider.

Storminess may be seasonal, resulting in seasonal changes in beach profiles. Along parts of the Pacific coast of the United States, winter storms tend to carry away fine sediment, and the remaining coarse fraction assumes a steep profile. In calm summer weather, fine sediment drifts in and the beach assumes a more gentle profile. At any time, however, the beach profile represents an average steady-state condition among the forces that are shaping it.

Depositional Features Along Coasts

Up to this point we have been describing the erosional effects of surf and the shaping of the shore profile. The deposits made by surf and by the currents it sets up are equally important, for they result in large part from longshore movement of sediment and occur as recognizable landforms.

Marine Deltas

The position of the outer limit of a marine delta—the extent to which it projects seaward from the land—is also a compromise between the rate at which a river delivers sediment at its mouth and the ability of currents and waves to erode the sediment and move it elsewhere along the coast (Fig. 13.14). The great size of the Mississippi delta (Fig. 13.15) testifies to the huge volume of sediment carried by the river, and to the relative ineffectiveness of waves in destroying it. The Columbia River of the northwestern United States formerly transported a large load of sediment to the Pacific coast (much of the sediment is now trapped behind numerous hydroelectric and irrigation dams built along its course). Yet there is no large delta at its mouth.

FIGURE 13.14 Mouth of Yangtze River at Shanghai, China. Light-colored silty sediment is carried seaward, settles to the seafloor, and is added to the prograding delta.

Winter storms and the persistent action of waves moving shoreward from the North Pacific erode sediment as quickly as it arrives. The sediment is moved laterally along the shore and is added to extensive spits built across major coastal embayments.

The Mississippi delta, with an area of 31,000 km², is really a complex of several coalescing subdeltas built successively during the last several thousand years (Fig. 13.15). Each subdelta probably began when a major flood created a new distributary. The present active delta has a very irregular front. The fronts of the adjacent inactive subdeltas have been extensively modified by coastal erosion and by gradual submergence as the crust slowly subsides under the weight of the accumulating sedimentary pile.

Spits, Bay Barriers, and Other Forms

Common among other conspicuous forms on many coasts is the **spit,** an elongate ridge of sand or gravel that projects from land and ends in open water (Fig. 13.16). Well-known large examples are Sandy Hook on the southern side of the entrance to New York Harbor, and Cape Cod, Massachusetts (Fig. 13.17). Most spits are merely continuations of beaches, built by beach-drifted sediment dumped at a place where the water deepens, as at the mouth

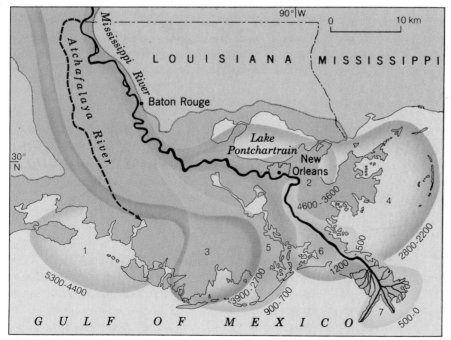

FIGURE 13.15 The Mississippi River has built a series of overlapping subdeltas (numbered 1 to 7) while occupying successive distributary channels. The ages of subdeltas are given in radiocarbon years before the present. (*Source*: After Morgan, 1970.)

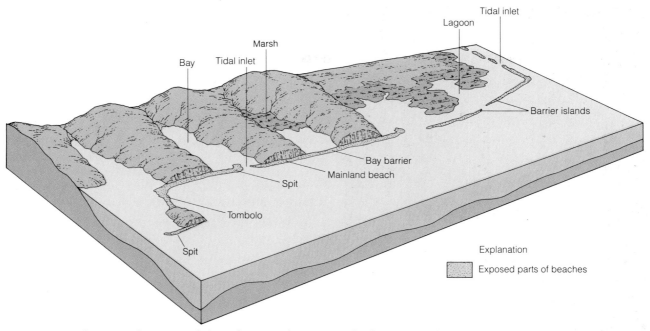

FIGURE 13.16 Common depositional shore features along a stretch of coast. Local direction of beach drift is toward the free end of spits.

of a bay. When the spit has been built up to sea level, waves act on it just as they would act on a beach. Much of a spit is likely to be above sea level, although the tip of it cannot be. The free end curves landward in response to currents created by surf.

FIGURE 13.17 Spit forming on Cape Cod, Massachusetts. Sediment derived by the reworking of glacial deposits is transported northward and southward by longshore drift. An eddy carries sediment around the north point of the spit and into Cape Cod Bay.

A *tombolo* is a ridge of sand or gravel that connects an island to the mainland or to another island (Fig. 13.16). It forms in much the same way as a spit does.

Along an embayed coast with abundant sediment supply, a **bay barrier** may form as a ridge of sand or gravel that completely blocks the mouth of a bay (Fig. 13.16). It develops as beach drift lengthens a spit across a bay in which tidal or river currents are too weak to scour away the spit as fast as it is built.

Barrier Islands

A **barrier island** (Figs. 13.16 and 13.18a) is a long island built of sand, lying offshore and parallel to the coast. Such islands are found along most of the world's lowland coasts. Well-known examples are Coney Island and Jones Beach (New York City's coastal playground areas), the long chain of islands centered at Cape Hatteras on the North Carolina coast, and Padre Island, Texas, some 130 km long.

Many barrier islands off the southeastern United States probably were built as the rapid rise of sea level at the end of the last glacial age began to slow in the middle Holocene (Fig. 13.18b). Shells collected at depths of 5 to 10 m from the basal deposits of some barriers have radiocarbon ages of about 5000 years, thereby supporting such a history. As the sea then rose more slowly across the very gentle continental shelf and the shoreline moved pro-

(a)

FIGURE 13.18 (a) Chandeleur Island, an elongate barrier island off the coast of Louisiana, displays the effects of waves that surge across the low ridge of sand during major storms. (b) Cross section through Galveston Island, a barrier island off the coast of Texas. Dashed lines show the former position of the seaward side of the island, based on radiocarbon dating. Since 3500 years ago, the island has grown southeastward toward the Gulf of Mexico (*Source*: After Kraft and Chrzastowski, 1985.)

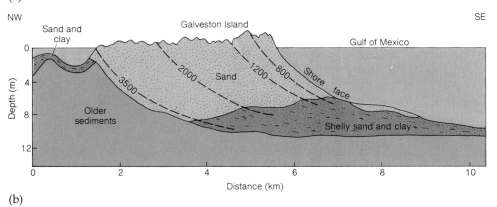

(b)

gressively inland, waves breaking in shallow water offshore eroded the bottom and piled up sand to form long bars. Gradually built up above sea level, the bars became barrier beaches, and ultimately barrier islands.

A barrier island generally consists of one or more ridges of dune sand related to successive shorelines occupied as the island formed. During great storms, surf washes across low places in the barrier and erodes it, cutting inlets that may remain permanently open. At the same time, fine sediment is washed into the lagoon between barrier and mainland. Sediment is thereby transferred landward by overwash, as well as laterally by longshore drift, as an island evolves. It is apparent that the development of barrier islands must be closely related to such factors as sediment supply, direction and intensity of waves and nearshore currents, the shape of the offshore profile, and the relative stability of sea level.

Constructional Features Along Coasts

While all coasts are subjected to the erosional effects of waves and currents, in some places con-

structional processes may ***prograde*** (build out) the coastline more rapidly than it can be destroyed by surf. We have already seen that the seaward expansion of a delta depends largely on such a relationship. However, other constructional activities can be observed along some coasts.

Lava Flows

Coastal areas subject to volcanic activity may expand seaward and change form as lava flows enter the sea. Substantial tracts of new land have thus been created along the coasts of Iceland and Hawaii, for example (Fig. 13.19). When hot lava enters the sea, it is quickly quenched and breaks apart. The resulting fragments accumulate to build a lava delta. The shoreline will continue to advance seaward as fluid lava moves across the top of the expanding delta surface.

Organic Reefs

Many of the world's tropical coastlines consist of limestone reefs built by vast numbers of tiny colonial organisms that secrete calcium carbonate. Such

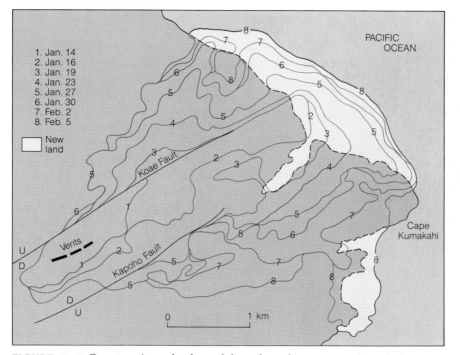

FIGURE 13.19 Construction of a lava delta where lava enters the sea from an eruption along the east rift zone of Kilauea volcano, Hawaii. As the fresh lava meets the water, it is quickly quenched and breaks up, building a delta of lava debris over which the flow continues to move, thereby extending the new landform progressively seaward. (*Source*: After Richter et al., 1970.)

organisms require shallow, clear water of near-normal salinity in which the temperature remains above about 18°C. Reefs, therefore, are built only at or close to sea level and are characteristic of tropical latitudes.

Fringing and Barrier Reefs.   Three principal reef types are recognized. A *fringing reef* is a reef directly attached to or bordering the adjacent land (Fig. 13.20a). It therefore lacks a *lagoon,* which is a bay inshore from an enclosing reef or piece of land and paralleling the coast. Typically, a fringing reef has a tablelike upper surface as much as 1 km wide, and its seaward edge plunges steeply into deeper water.

A *barrier reef* is a reef separated from the land by a lagoon that may be of considerable length and width (Fig. 13.20b). Such a reef may lie far off the coast of a continent, as is the case with the Great Barrier Reef off Queensland, Australia.

Atolls.   When a tropical volcanic island with a fringing reef slowly subsides, the reef will have to grow upward if its organisms are to survive near sea level. With continued subsidence, the area of exposed volcanic land will become progressively

smaller and the fringing reef will become an off-shore barrier reef. If the volcanic island subsides still farther, it will disappear, and its surrounding reef will be transformed into an *atoll,* a roughly circular coral reef enclosing a shallow lagoon (Fig. 13.20c). Atolls generally lie in deep water of the open ocean and range in diameter from as little as 1 to as much as 130 km.

Coastal Evolution

The world's coasts do not all fall into easily identifiable classes. Their variety is great because their configurations depend largely on the structure and erodibility of coastal rocks, the active geologic processes, the time over which they have operated, and the history of world sea-level fluctuations.

Structural Control

Some coasts, like most of the Pacific coast of North America, are steep and rocky and consist of mountains or hills separated by deep valleys. Others, like the Atlantic and Gulf coasts from New York City to Florida and onward into northern Mexico,

a

b

c

FIGURE 13.20 Chief kinds of tropical coral reefs. (a) Fringing reef on the island of Oahu in the Hawaiian Islands. (b) Barrier reef enclosing island of Moorea in the Society Islands. A narrow lagoon separates the high island, the remnant of a formerly active volcano, from a shallow reef flat. (c) A small atoll in the Society Islands is surmounted by low, vegetated sand islets that lie inside a line of breakers along the reef margin.

cut across a broad coastal plain that slopes gently seaward, and are festooned with barrier islands. These coasts represent two extremes, between which are many intermediate kinds. Each owes its general character to its structural setting. The rugged and mountainous Pacific coast lies along the margin of the American lithospheric plate which is continually being deformed where it interacts with adjacent plates to the west. Uplifted and faulted marine terraces are common features along parts of this coast and similar coasts that are emerging from the sea. By contrast, the eastern continental margin lies within the same lithospheric plate as the adjacent ocean floor, but in a zone that is tectonically passive. The old bedrock has low relief and much of the coastal zone borders young sedimentary deposits of the Atlantic and Gulf coastal plains.

Where rocks of contrasting erodibility are exposed along a coast, marine erosion will commonly produce a shoreline that is strongly controlled by rock type and structure. Such control is especially impressive in folded sedimentary or metamorphic belts that have been partially submerged (Fig. 13.21). It also is responsible for some of the world's deeply embayed fjord coasts, the pattern of which reflects deep glacial erosion along regional fracture systems and subsequent drowning as glaciers retreated to the heads of their fjords (Fig. 12.14).

Coastal Processes

The effectiveness of erosional and depositional processes is not equal along all coasts. Coasts lying at latitudes between about 45° and 60° are subject to higher-than-average storm waves, whereas subtropical east-facing coasts are subjected to infrequent but often disastrous hurricanes (called *typhoons* west of the 180th meridian). In the polar regions sea ice becomes an effective agent of coastal erosion. These and other factors influence the amount of energy expended in erosion along the shore and, together with structural and compositional properties of the exposed rocks, contribute to the variety of coastal landforms.

Changing Sea Level

Submergence. Whatever their nature, nearly all coasts have experienced **submergence** due to the worldwide rise of sea level that has occurred during the last 15,000 years as the Earth emerged from a glacial age. Over this interval, world sea level rose at least 100 m, leading to widespread submergence of most coastal zones.

Because of recent submergence, evidence of lower, glacial-age sea levels is almost universally found only beyond the present coastlines and to depths of 100 m or more. Former beaches, coastal sand dunes, and similar features on the inner continental shelves mark shorelines built by the rising sea at the end of the glacial age, and later drowned.

Emergence. Evidence of higher sea levels is related mainly to past interglacial ages. Inland from the Atlantic coast of the United States from Virginia to Florida are many marine beaches, spits, and barriers, the highest of which reach an altitude of more than 50 m. These landforms owe their present altitude to a combination of broad upward arching of the crust and submergence during times when climates were warmer than now, glaciers were smaller, and sea level was therefore higher. The position of such features above present sea level points to *emergence* of the land following their formation.

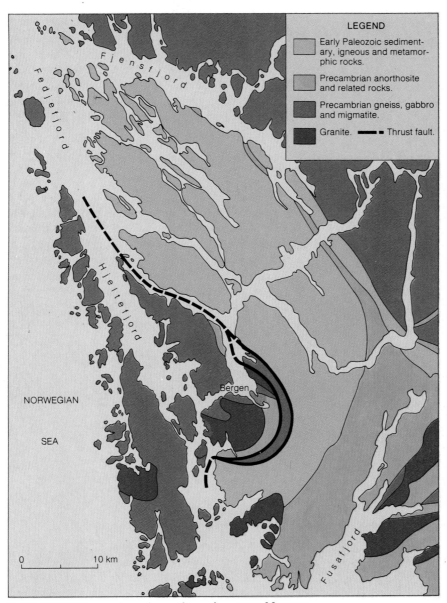

FIGURE 13.21 The embayed coastline of western Norway is controlled by the folded structure of the bedrock which has been differentially eroded by ice-age glaciers.

FIGURE 13.22 Uplifted coral reefs along the margin of Huon Peninsula in eastern Papua, New Guinea. The reefs, formed at sea level during the last several hundred thousand years, have been progressively uplifted to altitudes as high as several hundred meters.

Sea-level Cycles. Varied evidence demonstrates that many coastal and offshore features are relics of times when sea level was either higher or lower than now. The youngest deposits along a coast often form a thin blanket over older, similar units that date to earlier times. Repeated emergence and submergence over many glacial–interglacial cycles, each accompanied by erosion and redeposition of shoreline deposits, has resulted in complex coastal landform assemblages that test the ability and experience of geologists to interpret them correctly.

Relative Movements of Land and Sea. Rise and fall of sea level are universal movements, affecting all parts of the world ocean at the same time. Uplift and subsidence of the land, also causing emergence or submergence along a coast, are piecemeal movements, generally involving only parts of landmasses. Geologically rapid *relative* changes of sea level may characterize such regions; for example, some coastal stretches of North America and Europe that were covered by large ice sheets as recently as 10,000 years ago are still rising, at rates of nearly 1 cm/yr, in response to the removal of the thick ice load (chapter 14). This uplift has been in progress ever since the ice sheets began to melt away, and has raised some former ocean beaches hundreds of meters above present sea level (Fig. 14.25). Similarly, vertical tectonic movements at the boundary of converging plates of lithosphere have elevated beaches and tropical reefs to positions far above sea level (Figs. 13.12 and 13.22). Because changes of land and sea may occur simultaneously, either in the same or opposite directions, unraveling the history of sea-level fluctuations along a coast can be a difficult and challenging exercise.

Protection Against Shoreline Erosion

Erosion During Great Storms

The approximate equilibrium among the forces that operate on coasts is occasionally interrupted by exceptional storms that erode cliffs and beaches at rates far greater than the long-term average (Fig. 13.23). During a single storm in 1944, cliffs of compact sediment on Cape Cod in eastern Massachusetts retreated up to 5 m, or more than 50 times the normal annual rate of retreat. Such infrequent

FIGURE 13.23 Strong surf generated by Hurricane Gloria in September, 1986 cuts vigorously into the sandy south shore of Nantucket Island off the south coast of Massachusetts.

bursts of rapid erosion not only can be quantitatively important in the natural evolution of a coast, but can also have a significant impact on coastal inhabitants. More than 75 percent of the population of the United States now lives in coastal belts that include only 5 percent of the land area of the nation. The concentration of such large numbers of people in coastal areas means that infrequent large storms not only can be hazardous to life, but can cause extraordinary damage to property.

Protection of Seacliffs

A strip of shore that consists of comparatively erodible rock or sediment, such as that on Cape Cod, can be protected from erosion in several ways. A cliff can be clad with an armor consisting of tightly packed boulders so large that they can withstand the onslaught of storm waves. It can also be defended by a strong *seawall* built parallel to the shore on foundations deep enough to prevent undermining by surf during storms. Both structures offer cliffs some protection against ordinary storms, but both are expensive.

Protection of Beaches

Because of their great recreational value, beaches in densely populated regions justify greater expense for maintenance than most headlands do. A beach, however, presents a special sort of problem. As a result of beach drift (Fig. 13.8), what happens on one part of a beach affects all other parts that lie in the downdrift direction. For example, a seawall, dock, or other structure built at the updrift end of a beach reduces the amount of sand available for beach drift. The surf becomes underloaded and makes good the loss by eroding sand from along the beach. Small beaches have been completely destroyed by this process in only a few years.

Groins. Such erosion can be checked, at least to some extent by building groins at short intervals along the beach. A *groin* is a low wall, built on a beach, that crosses the shoreline at a right angle (Fig. 13.24). Many groins contain openings that permit some water to pass through them. Groins act as a check on the rate of beach drift and so cause sand to accumulate against their updrift sides. Some erosion, however, occurs beyond them on their downdrift sides.

Artificial Nourishment. Another way of protecting a beach that is being eroded is to bring in sand

FIGURE 13.24 Groins built along the shoreline of Miami Beach, Florida, to prevent excessive loss of sand by longshore drift at this popular resort area. Sand piles up on the upcurrent side of each groin.

artificially and pile it on the beach at the updrift end. Surf then erodes the pile and drifts the new sand down the length of the beach. Using this method, however, the sand that artificially nourishes a beach must be continuously replenished. As can be imagined, both the feeding of a beach and the construction and maintenance of groins are expensive.

Effects of Human Interference

Beaches in southern California are deteriorating because of human interference. Most of the sand on those beaches is supplied, not by erosion of wave-cut cliffs, but by alluvium dumped into the sea by streams at times of flood. The floods themselves cause damage to man-made structures along the stream courses, so dams have been built across the streamways to control flooding. Of course, the dams also trap the sand and gravel carried by the streams, thus preventing the sediment from reaching the sea. This in turn has affected the balance among the factors involved in longshore currents and beach drift, and has resulted in significant erosion of some beaches.

A similar situation has developed along the Black Sea coast of the Soviet Union. Of the sand and pebbles that form the natural beaches there, 90 percent was supplied by rivers as they entered the sea. During the 1940s and 1950s, three things occurred: large resort developments including high-rise hotels were built at the beaches; by construction of breakwaters, two major harbors were extended into the sea; and dams were built across some rivers inland from the coast. All this construction interfered with the steady state that had existed among the supply of sediment to the coast, longshore currents and beach drift, and deposition of sediment on beaches. By 1960, it was estimated that the combined area of all beaches along the coast had decreased by 50 percent. Then beachfront buildings began to sag or collapse as the surf undermined and ate away at their foundations. An ironic twist to the chain of events lies in the fact that large volumes of sand and gravel were removed from beaches for use as concrete aggregate, not only to construct buildings but also to build the dams that cut off the supply of sediment to the coast.

Essay

HOW TO MODIFY AN ESTUARY

San Francisco Bay is a grand example of an estuary: It is a drowned river mouth where freshwater encounters seawater, and where tidal effects are evident. When first discovered by Europeans, its shores and productive waters supported as many as 20,000 Native Americans. The population boom that followed the discovery of gold in California in 1848 and which has continued unabated to the present day has led to major changes in the river systems that enter the bay and thereby has greatly affected the estuary environment (Fig. B13.1).

The need for water is a continuing concern of Californians who need it not only for human consumption and industry, but also to support huge agricultural enterprises that require large quantities of fresh water. Diversion of water from natural streams has reduced freshwater inflow to the estuary to less than 40 percent of the historic (A.D. 1850) flow; by the year 2000, the projected flow will decline still further to only 30 percent of the historic average. At the same time, increasing urbanization has led to a loss of 95 percent of the bordering tidal marshes as a result of filling and diking. Hydraulic mining for gold in the foothills of the Sierra Nevada produced vast quantities of fine sediment that choked streams, destroyed fish spawning grounds, obstructed navigation, and ultimately reached San Francisco Bay where it reduced the area and volume of the estuary and modified tidal circulation.

The unforeseen consequences of this human tampering with a natural stream and estuarine system have been many. Most commercial fisheries have disappeared. Reduced freshwater flushing of the bay has concentrated agricultural, domestic, and industrial waste products in estuarine sediments, contaminating them and the organisms that feed on them, and raising increasing concerns about human health. Natural habitats of migrating birds have been extensively destroyed, with presumed major effects on bird populations.

Such changes are not unique to San Francisco Bay. Many other large estuaries, which are favored sites of urban development and industrialization, have similar problems (the Rhine River in the Netherlands, the Thames River in England, and the Susquehanna and Potomac rivers in the United States). All are sensitive to human-induced changes and susceptible to steady deterioration. The best hope for reversing their decline is through an improved understanding of how human activities affect the physical, chemical, and biological processes in the river–estuarine systems and of what measures must be taken to curtail potentially destructive actions.

FIGURE B13.1 Vertical satellite image of the shrinking San Francisco Bay estuary. Filling and diking of tidal marsh to create farmland, evaporation ponds, and residential and industrial developments has reduced 2200 km² of marsh land that existed before 1850 to less than 130 km² today.

SUMMARY

1. Surface seawater circulates as currents in a number of huge subcircular cells that rotate clockwise in the Northern Hemisphere and counterclockwise in the Southern Hemisphere. Surface ocean currents are driven by winds and move warm equatorial water toward the polar regions.

2. Rapid tidal currents, generated by twice-daily passage of tidal bulges that sweep around the Earth, produce movement of sediment in bays, straits, estuaries, and other restricted places along coasts.

3. In deep water, waves have little or no effect on the bottom. At a depth equal to half the wavelength waves can begin to stir the bottom sediments.

4. Most of the geologic work of waves is performed by surf at depths of 9 m or less.

5. Wave refraction tends to concentrate wave erosion on headlands and to diminish it along the shores of bays.

6. Longshore currents and beach drift transport great quantities of sand along coasts.

7. On rocky coasts the shore profile includes a wave-cut cliff, a wave-cut bench, and a beach. On gentle, sandy coasts the beach typically consists of a foreshore, berm, and backshore.

8. Depositional shore features include beaches, spits, bay barriers, barrier islands, and tombolos. Most barrier islands form offshore, in areas where a rising sea advances over a gently sloping coastal plain.

9. Constructional landforms along coasts include marine deltas, tropical organic reefs that may be separated from the shore by a lagoon, and lava flows that build lava deltas where they enter the sea.

10. Nearly all coasts have experienced recent submergence due to postglacial rise of sea level. Some have experienced more complicated histories of emergence and submergence due to tectonic and isostatic movements on which are superimposed the worldwide sea-level rise.

11. The shape of coasts partly reflects the amount of energy available to erode and deposit sediment. Rock structure and degree of erodibility help dictate the form of rocky coasts.

12. A shore cliff can be protected for a time, at least, by a seawall or an armor of boulders. A beach can be temporarily protected by a series of groins or by importation of sand.

IMPORTANT WORDS AND TERMS TO REMEMBER

atoll (p. 303)

backshore (p. 297)
backwash (p. 295)
barrier island (p. 301)
barrier reef (p. 303)
bay (p. 301)
bay barrier (p. 301)
beach (p. 297)
beach drift (p. 295)
berm (p. 297)
breaker (p. 293)

emergence (of a coast) (p. 305)

foreshore (p. 297)
fringing reef (p. 303)

groin (p. 307)

hurricane (p. 304)

lagoon (p. 303)

longshore current (p. 295)

notch (coastal) (p. 296)

prograde (p. 302)

rip current (p. 293)

sea arch (p. 296)
sea cave (p. 296)
seawall (p. 307)
shore profile (p. 296)
spit (p. 300)
stack (p. 296)
submergence (of a coast) (p. 304)
surf (p. 293)

swash (of wave) (p. 295)

tidal bore (p. 292)
tidal bulge (p. 291)
tidal current (p. 291)
tidal inlet (p. 301)
tide (p. 290)
tombolo (p. 301)
typhoon (p. 304)

wave (p. 291)
wave base (p. 292)
wave-cut bench (p. 296)
wave-cut cliff (p. 296)
wave-cut notch (p. 296)
wavelength (p. 292)
wave refraction (p. 294)

QUESTIONS FOR REVIEW

1. Where are tidal currents most rapid, and therefore most effective in moving sediment?

2. Describe the motion of a parcel of water below the ocean surface as a wave passes and how it differs from the motion of the wave form at the surface.

3. How is wave base related to wavelength?

4. What causes a wave to "break" near the shore?

5. What is the effective depth of erosion by surf, and what determines it?

6. How does wave refraction explain why rocky headlands are more vigorously eroded by surf than bays?

7. What causes longshore currents to develop and, in some cases, to shift direction seasonally?

8. Describe the typical elements of a shore profile on a cliffed coast; on a low sandy coast.

9. How do waves and currents affect the development of (a) a marine delta? (b) a spit? (c) a barrier island?

10. Why are the sediments in a beach generally well sorted and well stratified?

11. How are limestone atolls related to volcanic islands?

12. What features would provide evidence of coastal emergence due to tectonic uplift?

13. Describe measures that can be taken to reduce the impact of erosion (a) along a cliffed coast and (b) along a sandy beach.

14. How might the construction of a dam along the lower part of a large river affect the coast where the river enters the ocean?

QUESTIONS FOR REVIEW OF PART 2

1. How has submergence caused by the postglacial rise of world sea level affected the evolution of coasts?

2. Why do minerals in rocks at the Earth's surface vary in their susceptibility to chemical weathering?

3. What subsurface geologic factors need to be evaluated in selecting a site for underground disposal of toxic waste substances? Why are they significant?

4. If the long-term average precipitation of a temperate mountain region were to increase, how might it affect (a) the competence of a stream draining the mountains; (b) the rate of chemical weathering of exposed granitic rocks; (c) the rate of downslope creep on mountain slopes; and (d) the dimensions of small-valley glaciers?

5. Why are we justified in regarding soils as a nonrenewable resource?

6. How do joints affect the weathering, mass-wasting, and erosion of rocks at the Earth's surface?

7. Landforms resulting from different surface processes sometimes can be easily misinterpreted. How might you distinguish between the following on the basis of characteristics other than form: (a) a linear dune and an esker; (b) an end-moraine complex and a debris avalanche deposit; (c) the terrace of a meandering stream and a kame terrace?

8. How might the Earth be different if water had a density of 1.1 instead of 0.9?

9. In what ways is the flow of water in a stream similar to the flow of ice in a glacier? In what ways is it different, and why?

10. People who live on valley floors seem to be especially susceptible to geologic hazards related to various mass-wasting and fluvial processes. Explain why.

11. Describe some of the potentially adverse consequences of excessive withdrawal of groundwater for human consumption.

12. Explain the ways in which the construction of a large dam can affect the natural channel system of a river.

13. How can alluvial terraces and the relationship of a stream to bedrock structures across which it flows provide clues about the history of a stream?

14. Why is wind a more effective agent of erosion and transport of fine sediment in desert lands than in more humid regions? How can the adverse effects of wind erosion be curtailed or reversed?

15. What causes glacial and interglacial ages to recur in a cyclic pattern? How might one explain the fact that glaciation has been more widespread during the last 2 million years (the Quaternary Period) than in the preceding geologic period?

The Lost River Range, Idaho is part of the Basin and Range Province. The range is a horst, and the valley in the foreground is a graben.

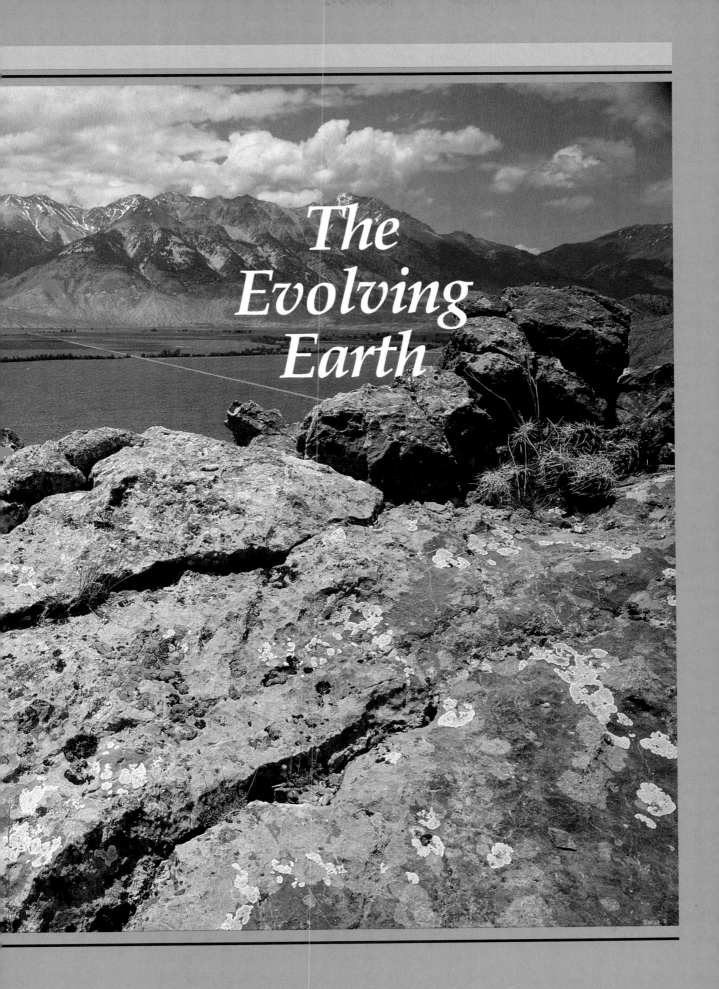

The
Evolving
Earth

Earthquakes, Isostasy, and the Earth's Internal Properties

Gaping fissures opened in a residential area of Anchorage, Alaska, during the earthquake of March 27, 1964.

EARTHQUAKES

Preceding chapters have dealt largely with activities that can be seen and studied on the Earth's surface as they happen—activities that are driven by the Sun's heat energy. The next three chapters discuss internal activities, such as earthquakes, bending of rock, uplifting of mountains, and movement of tectonic plates, that cannot be seen while they are in progress. Such activities are driven by the Earth's internal heat energy. Surely no manifestation of that energy is more dramatic than an earthquake.

Earthquake Risk

No locality on the Earth's surface is free from earthquakes, but in many places the quakes that do occur are weak and not dangerous to people or dwellings. For example, scientists believe that in southern Florida, southern Texas, and parts of Alabama and Mississippi, the probability of damaging earthquakes is almost zero. All other parts of the United States have experienced damaging quakes in the past, and more can be expected to occur in the future (Fig. 14.1).

Most Americans think immediately of California when earthquakes are mentioned. However, the most intense earthquake to jolt North America in the past 200 years was centered near New Madrid, Missouri. Three earthquakes of great size occurred on December 16, 1811, and January 23 and February 7, 1812. The actual sizes of these earthquakes are unknown because instruments to record them did not exist at the time. However, judging from the local damage caused, plus the fact that tremors were felt and minor damage occurred as far away as New York and Charleston, South Carolina, it is estimated that the largest of the quakes was larger than the one that leveled San Francisco in 1906.

Earthquake Disasters

Every year the Earth experiences many hundreds of thousands of earthquakes. Fortunately, only one or two are large enough, or close enough to major centers of population, to cause loss of life. Certain areas are known to be earthquake prone, and special building codes in such places require structures to be as resistant as possible to earthquake damage. However, all too often an unexpected earthquake will devastate an area where buildings are not ad-

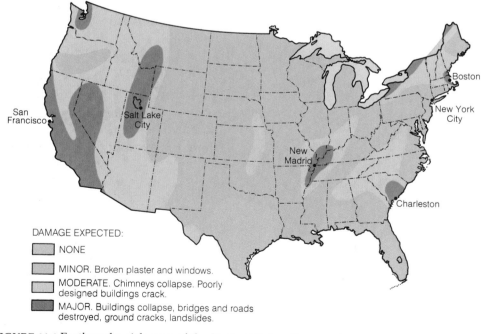

DAMAGE EXPECTED:

☐ NONE

☐ MINOR. Broken plaster and windows.

☐ MODERATE. Chimneys collapse. Poorly designed buildings crack.

☐ MAJOR. Buildings collapse, bridges and roads destroyed, ground cracks, landslides.

FIGURE 14.1 Earthquake-risk map of the United States. Zones refer to maximum earthquake intensity and, therefore, to maximum destruction that can occur. The map does not indicate frequency of earthquakes. For example, frequency in southern California is high, but in eastern Massachusetts it is low. Nevertheless, when earthquakes occur in eastern Massachusetts, they can be as severe as the more frequent quakes in southern California.

FIGURE 14.2. The Hotel DeCarlo was one of the buildings that collapsed during the great earthquake that struck Mexico City in 1985. Proper building design can minimize damage. Nearby buildings of sturdier construction withstood the shaking.

TABLE 14.1 *Earthquakes During the Past 800 Years That Have Caused 50,000 or More Deaths*

Place	Year	Estimated Number of Deaths
Silicia, Turkey	1268	60,000
Chihli, China	1290	100,000
Shaanxi, China	1556	830,000
Shemaka, U.S.S.R.	1667	80,000
Naples, Italy	1693	93,000
Catania, Italy	1693	60,000
Beijing, China	1731	100,000
Calcutta, India	1737	300,000
Lisbon, Portugal	1755	60,000
Calabria, Italy	1783	50,000
Messina, Italy	1908	160,000
Gansu, China	1920	180,000
Tokyo and Yokohama, Japan	1923	143,000
Gansu, China	1932	70,000
Quetta, Pakistan	1935	60,000
T'ang-shan, China	1976	700,000

equately constructed. One such example was the earthquake that destroyed parts of the center of Mexico City in 1985 (Fig. 14.2).

Sixteen earthquake disasters are known to have caused 50,000 or more deaths apiece (Table 14.1). The most disastrous earthquake on record occurred in 1556, in Shaanxi Province, China, where an estimated 830,000 people died. Many of those people lived in cave dwellings excavated in loess (chapter 11), which collapsed as a result of the quake. The second most disastrous earthquake also occurred in China, at T'ang-shan, in 1976. The town was completely destroyed and approximately 700,000 people are believed to have died. Since 1900 there have been 39 earthquakes, worldwide, in which 500 or more people have died.

Origin of Earthquakes

Earthquakes and the vibrations they cause happen when the Earth is suddenly jolted, as if struck by a giant hammer. Make an experiment yourself. Have a friend hit one end of a wooden plank or the top of a wooden table with a hammer while you press your hand on the other end. You will feel vibrations set up in the plank or tabletop by the energy of the hammer blow. The harder the blow, the stronger the vibrations. The reason you can feel those vibrations is that some of the energy imparted by the hammer blow is transferred by elastic vibrations through the solid wood. Fortunately, giant hammers don't hit the Earth, but a bomb blast or a violent volcanic explosion will serve as an energy source just as well. So too will the sudden slipping of rock masses along a fracture, causing two hard, rocky surfaces to slide suddenly past each other.

Faults. A fracture through a body of rock along which the opposite sides have been displaced relative to each other is called a ***fault***. Sudden movement along faults is the cause of most earthquakes. But it cannot be that simple. Sliding occurs every time pressure is applied to a fault; some earthquakes are millions of times stronger than others. The same energy that in one case will be released by thousands of tiny slips and tiny earthquakes will in another case be stored and released in a single giant earthquake. The answer seems to be that if fault surfaces do not slip easily, energy can be stored in elastically deformed bodies of rock, just as in a steel spring that is compressed.

Evidence supporting the idea of energy being stored in elastically deformed rocks came first from studies of the San Andreas Fault. During long-term

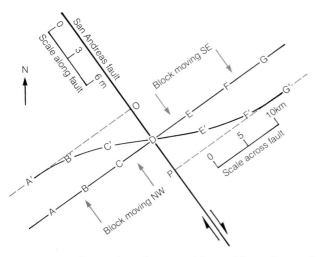

FIGURE 14.3 An earthquake caused by sudden release of energy. Sketch based on detailed surveys near the San Andreas Fault, California, before and after the abrupt movement that caused the earthquake of 1906. The seven survey points, A to G, were originally aligned. Slowly, movement of the two fault blocks bent the crust and displaced the points to new positions, A' to G'. Friction between the two sides of the fault prevented steady slippage. Suddenly, the frictional lock was broken and the rocks on either side of the fault rebounded. The surveyed points lay along the lines A'O and PG'. The sudden offset along the fault, distance OP, was 7 m.

field observations in central California, beginning in 1874, scientists from the U.S. Coast and Geodetic Survey determined the precise positions of many points both adjacent to and distant from the fault (Fig. 14.3). As time passed, movement of the points revealed that the crust was slowly being bent. For some reason, in the area of measurement near San Francisco, the fault was locked and did not slip. On April 18, 1906, the two sides of the fault shifted abruptly. The stored energy was released as the bent crust broke and snapped back, thereby creating a violent earthquake. Repetition of the survey then revealed that the bending had disappeared.

Most earthquakes occur in the brittle rock of the lithosphere. As discussed in chapter 15, brittleness is the tendency for a solid to fracture when the deforming force exceeds the limits of elasticity. At great depth, temperatures and pressures are too high for brittle fracture to happen. Rocks can neither fracture nor store elastic energy under such conditions. Rather, they are like putty and undergo permanent changes of shape, even after the deforming forces have been removed. Earthquakes, then, are phenomena of the outer, cooler portion of the Earth.

How Earthquakes Are Studied

The name given to the study of earthquakes is *seismology,* a word that comes directly from the ancient Greek term for earthquakes, *seismos.*

Seismographs

The device used to record the shocks and vibrations caused by earthquakes is a *seismograph.* The ideal place to record the vibrations and motions would be from a stable platform that is not connected to the Earth. As the observer on the platform would not be influenced by the vibrations, accurate measurements of the shaking surface below could be made. An observer who must stand on the Earth's vibrating surface will move with the surface, making the act of measurement difficult because there is no fixed frame of reference against which to make measurements.

Inertial Seismographs. To overcome the frame of reference problem, most seismographs make use of *inertia,* which is the resistance a large mass has to sudden movement. If a heavy mass, such as a block of iron, is suspended from a light spring (Fig. 14.4), the iron block has so much inertia it will remain almost stationary when the spring is suddenly extended. If the spring is connected to the ground, and the ground vibrates, the spring will expand and contract but the iron block will stay almost stationary. Then the distance between the ground and the iron mass can be used to sense vertical displacement of the ground surface. Horizontal displacement can be similarly measured by suspending a large mass from a string to make a pendulum (Fig. 14.4a). Because of its inertia, the mass does not keep up with the horizontal ground motion, and the difference between the pendulum and ground movement records the horizontal ground motions. Seismographs with inertial masses are commonly used in groups so that motions can be measured simultaneously in up–down, east–west, and north–south directions.

Strain Seismographs. Another kind of device, called the Benioff strain seismograph, employs two concrete piers in the ground spaced at a distance of about 35 m. Attached to one pier is a long, rigid, silica-glass tube (Fig. 14.4b). The other pier carries a very sensitive detector to measure even the slightest movement in the end of the silica-glass tube. Strain seismographs are commonly installed in mines, tunnels, and other places where a constant

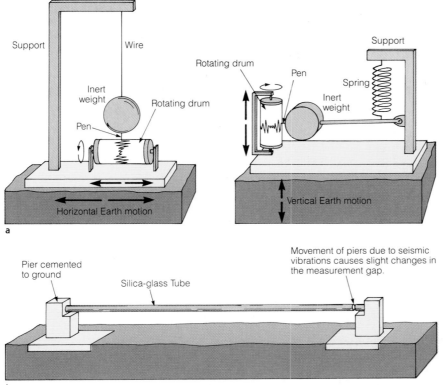

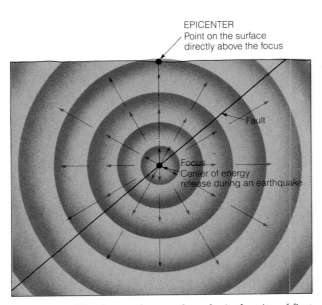

FIGURE 14.4 Seismographs measure vibrations sent out by earthquakes. (a) Two kinds of inertial seismographs. The pendulum device measures horizontal motions, the spring-supported device measures vertical motions. (b) A strain seismograph. A rigid silica-glass tube, as long as 35 m, is supported on a solid concrete pier anchored to the ground. A second pier carries a sensitive electronic measuring device to record any movement in the end of the rod. The distance between the two piers changes when a seismic vibration disturbs the surface.

FIGURE 14.5 The focus of an earthquake is the site of first movement on a fault and the center of energy release.

temperature is maintained, and where wind disturbance is minimal.

Modern seismographs are incredibly sensitive. Vibrational movements as tiny as one hundred millionth (10^{-8}) of a centimeter can be detected. Indeed, many instruments are so sensitive they can detect ground depression caused by a moving automobile several blocks away, and even ocean waves and tides several kilometers from the seashore.

Earthquake Focus and Epicenter

The point where energy is first released to cause an earthquake is called the *earthquake focus.* The focus generally lies at some depth below the surface. It is more convenient to identify the *epicenter,* which is the point on the Earth's surface that lies vertically above the focus of an earthquake (Fig. 14.5). A good way to describe the location of an

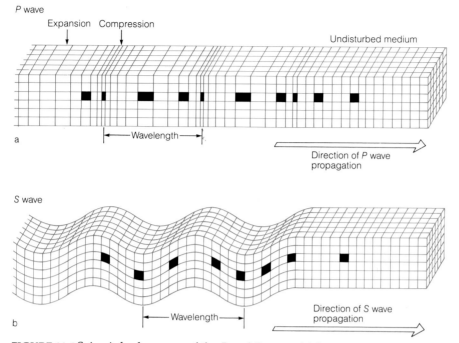

FIGURE 14.6 Seismic body waves of the *P* and *S* types. (a) *P* waves cause alternate compressions and expansions. An individual point in a rock will move back and forth parallel to the direction of *P*-wave propagation. As wave after wave passes through, a square will repeatedly change its shape to a rectangle, then back to a square. (b) *S* waves cause a shearing motion. An individual point in a rock will move up and down, perpendicular to the direction of *S*-wave propagation. A square will repeatedly change to a parallelogram, then back to a square again.

earthquake focus is to state the location of its epicenter and its depth.

Seismic Waves

How is the energy of an earthquake transmitted from the focus to other parts of the Earth? As with any vibrating body, waves (vibrations) spread outward from the focus. The waves, called *seismic waves,* spread out in all directions from the focus, just as sound waves spread in all directions when a gun is fired. Seismic waves are elastic disturbances, so unless the elastic limit is exceeded, the rocks through which they pass return to their original shapes after passage of the waves. Seismic waves must be measured and recorded while the rock is still vibrating. For this reason, many continuously recording seismograph stations are installed around the world.

Body Waves

Seismic waves are of two kinds. *Body waves* travel outward from the focus, passing entirely through the Earth. *Surface waves,* on the other hand, are guided by the Earth's surface, with only a loose constraint imposed by the atmosphere and the ocean. Body waves are analogous to rays of light, or sound waves, which travel outward in all directions from their points of origin. Surface waves arise from the movement of the surface caused by vibration of the entire mass of the Earth.

P Waves.   Body waves are of two kinds. **Compressional** waves deform materials by change of volume in the same way that sound waves do, and consist of alternating pulses of compression and expansion acting in the direction of travel (Fig. 14.6a). Compression and expansion produce changes in the volume and density of a medium. Compressional waves can pass through solids, liquids, or gases because each can sustain changes in density. When a compressional wave passes through a medium, the compression pushes atoms closer together. Expansion, on the other hand, is an elastic response to compression and it causes the distance between atoms to be increased. Movement in a solid, subjected to compressional waves, is back and forth in the line of the wave motion. Compressional waves have the greatest velocity of all

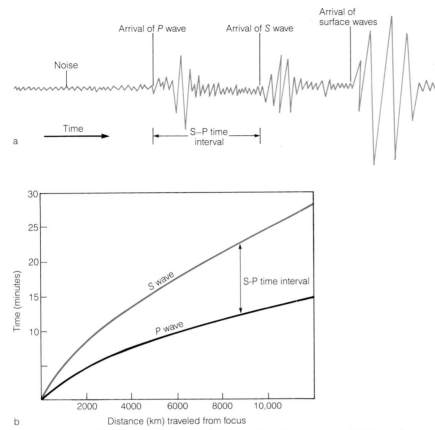

FIGURE 14.7 Differing travel times of *P*,*S*, and surface waves. (a) Typical record made by a seismograph. The *P* and *S* waves leave the earthquake focus at the same instant. The fast-moving *P* waves reach the seismograph first, and some time later the slower-moving *S* waves arrive. The delay in arrival times is proportional to the distance traveled by the waves. The surface waves travel more slowly than either *P* or *S* waves. (b) Average travel–time curves for *P* and *S* waves in the Earth.

seismic waves—6 km/s is a typical value for the uppermost portion of the crust—and they are the first waves to be recorded by a seismograph after an earthquake. They are therefore called *P* (for *primary*) *waves.*

S Waves. The second kind of body waves are **shear waves.** They deform materials by change of shape but not change of volume. Because gases and liquids do not have the elasticity to rebound to their original shapes shear waves can only be transmitted by solids. Shear waves consist of an alternating series of sidewise movements, each particle in the deformed solid being displaced perpendicular to the direction of wave travel (Fig. 14.6b). A typical velocity for a shear wave in the upper crust is 3.5 km/s. Shear waves are slower than *P* waves, and reach a seismograph some time after a *P* wave arrives, so they are called *S* (for *secondary*) *waves* (Fig. 14.7a).

Surface Waves

Surface waves are the result of the whole Earth vibrating like a tub of Jell-O. Such motions cause the shape and/or size of the whole Earth to change, just as a mass of Jell-O changes in shape and size as it vibrates. Such motions are whole-body oscillations, but to an observer on the surface of the Earth, the oscillations are detected as movements of the surface. For this reason they are called surface waves.

To an observer at the surface, surface waves appear very similar to ordinary *P* and *S* waves. However, they travel more slowly than *P* and *S* waves, and in addition they must pass around the Earth rather than through it. Thus, surface waves are the last to be detected by a seismograph (Fig. 14.7a).

Location of Epicenter

The location of an earthquake's epicenter can be determined from the arrival times of the *P* and *S*

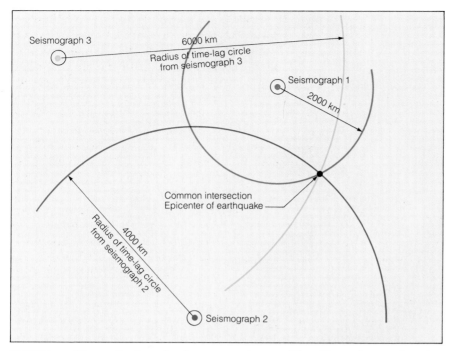

FIGURE 14.8 The method used to locate an epicenter. The effects of an earthquake are felt at three different seismograph stations. The time differences between the first arrival of the *P* and *S* waves depends on the distance of a station from the epicenter. The following distances are calculated by using the curves in Figure 14.7

	Time Difference	Calculated Distance
Seismograph 1	1.0 min	2000 km
Seismograph 2	5.6 min	4000 km
Seismograph 3	7.5 min	6000 km

On a map, a circle of appropriate radius is drawn around each of the stations. The epicenter is where the three circles intersect.

waves. The further a seismograph is away from an epicenter, the greater the time difference between the arrival of the *P* and *S* waves (Fig. 14.7). After determining how far an epicenter lies from a seismograph, the seismologist draws a circle on a map around the station with a radius equal to the calculated distance. The exact position of the epicenter can be determined when data from three or more seismographs are available—it lies where the circles intersect (Fig. 14.8).

Magnitudes of Earthquakes

Very large earthquakes (of the kind that destroyed San Francisco in 1906, T'ang-shan, China, in 1976, and parts of Mexico City in 1985) are, fortunately, relatively infrequent. In earthquake-prone regions, such as San Francisco and the surrounding area, very large earthquakes occur about once a century. In some areas they occur more frequently, in others

less frequently, but a century is an approximate average. This means that the time needed to build up elastic energy to a point where the frictional locking of a fault is overcome, is about 100 years. Small earthquakes may occur along a fault during this time due to local slippage, but even so elastic energy is accumulating because most of the fault remains locked. When the lock is broken and an earthquake occurs, the elastic energy is released during a few terrible minutes. By careful measurement of elastically strained rocks along the San Andreas Fault, seismologists have found that about 100 J of elastic energy can be accumulated in 1 m^3 of deformed rock. This is not very much—it is only equivalent to about 25 calories of heat energy—but when billions or trillions of cubic meters of rock are strained, the total amount of stored energy can be enormous. The volume of deformed rock that released stored energy at the time of the 1906 San Francisco earthquake exceeded 10^{10} m^3. The elastic

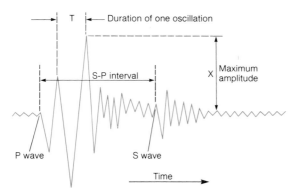

FIGURE 14.9 Determination of Richter magnitude (M) from a seismograph. Divide the maximum amplitude, X, measured in steps of 10^{-4} cm on a suitably adjusted seismograph, by the duration of one oscillation, T, in seconds. Then add a correction factor, Y, determined from the S–P wave arrival time interval. X/T is a measure of the maximum energy reaching the seismograph and the epicenter. The formula is $M = \log X/T + Y$.

energy stored in 10^{10} m³ is 10^{12} J. The energy released by the atom bomb that destroyed Hiroshima during World War II was 4.0×10^{12} J!

Richter Magnitude Scale

Measurements of elastically deformed rocks before an earthquake, and of undeformed rocks after an earthquake, can provide an accurate measure of the amount of energy released. The task is very time consuming, and all too frequently the pre-earthquake measurements are simply not available. Therefore, seismologists have developed a way to estimate the energy released by measuring the amplitudes of the seismic waves recorded on seismographs. The **Richter magnitude scale,** named after the seismologist who developed it, is defined by the amplitudes of the P and S waves at a distance of 100 km from the epicenter of an earthquake (Fig. 14.9). Because wave signals vary in strength by factors of a hundred million or more, the Richter scale is logarithmic, which means it is divided into steps called magnitudes, starting with magnitude 1 and increasing upward. Each unit increase in magnitude corresponds to a tenfold increase in the amplitude of the wave signal. Thus, a magnitude 2 signal has an amplitude that is ten times larger than a magnitude 1 signal, and a magnitude 3 is a hundred times larger.

The energy in a wave is a function of both amplitude (A) and frequency (ω) which is the number of waves that pass a given point each second. Energy is proportional to $\omega^2 A^2$. Thus, if one Richter scale unit corresponds to a tenfold increase in A,

the increase in energy should be proportional to A^2, which is to say a hundredfold. However, the range of frequencies differs from one earthquake to another (in particular, the most energetic earthquakes have higher proportions of low-frequency waves). As a result, the energy increase corresponding to one Richter scale unit increase, when summed over the whole range of frequencies in a wave record, is only a thirtyfold increase. Thus, the difference in energy released between an earthquake of magnitude 4 and one of magnitude 7 is $30 \times 30 \times 30 = 27{,}000$ times!

How big can earthquakes get? The largest recorded to date have Richter magnitudes of about 9.5, which means they release about as much energy as 10,000 atom bombs of the kind that destroyed Hiroshima at the end of World War II. It is possible that earthquakes do not get any larger than this because rocks cannot store more elastic energy. If they are deformed further, they fracture and so release the energy.

Earthquake Damage

The dangers of earthquakes are profound and the havoc they can cause is often catastrophic. Their effects are of six principal kinds: the first two are primary effects, ground motion and faulting, and they cause damage directly; the other four effects are secondary and cause damage indirectly as a result of processes set in motion by the earthquake.

1. Ground motion results from the movement of seismic waves, and especially surface waves, through surface-rock layers and regolith. The motions can damage and sometimes completely destroy buildings. Proper design of buildings (including such features as steel framework and foundations tied to bedrock) can do much to prevent such damage, but in a very strong earthquake even the best buildings may suffer some damage.

2. Where a fault breaks the ground surface, buildings can be split, roads disrupted, and any feature that crosses or sits on the fault can be broken apart.

3. A secondary effect, but one that is sometimes a greater hazard than moving ground, is fire. Ground movement displaces stoves, breaks gas lines, and loosens electrical wires, thereby starting fires. In the earthquakes that struck San Francisco in 1906, and Tokyo and Yokohama in 1923, more than 90 percent of the damage to buildings was caused by fire.

FIGURE 14.10 Ground motion during a 1968 earthquake, Amori Prefecture, Japan, caused the sediment underneath this fertile farmland to lose coherence and develop fluidlike properties. Once weakened, the ridge and embankment slumped and flowed outward.

4. In regions of hills and steep slopes, earthquake vibrations may cause regolith to slip, cliffs to collapse, and other rapid mass-wasting movements to start (chapter 8). This is particularly true in Alaska, parts of southern California, China, and hilly places such as Iran and Turkey. Houses, roads, and other structures are destroyed by rapidly moving regolith.

5. The sudden shaking and disturbance of water-saturated sediment and regolith can turn seemingly solid material to a liquidlike mass such as quicksand (Fig. 14.10). This is called liquefaction and it was one of the major causes of damage during the earthquake that destroyed much of Anchorage, Alaska on March 27, 1964, and that caused apartment houses to sink and collapse in Niigata, Japan that same year.

6. Finally, there are *seismic sea waves* (commonly called by their Japanese name, *tsunami*) that occur following violent movement of the seafloor.

Seismic sea waves, often incorrectly called tidal waves, have been particularly destructive in the Pacific Ocean. About 4.5 h after a severe submarine earthquake near Unimak Island, Alaska in 1946, such a wave struck Hawaii. The wave traveled at a velocity of 800 km/h. Although the height of the wave in the open ocean was less than 1 m, the height increased dramatically as the wave approached land. When it hit Hawaii the wave had a crest 18 m higher than normal high tide. This destructive wave demolished nearly 500 houses, damaged a thousand more, and killed 159 people.

Modified Mercalli Scale

Because damage to the land surface and to human property is so important, a scale of earthquake-damage intensity (called the *Modified Mercalli Scale*) has been developed based on observed damage.

TABLE 14.2 *Earthquake Magnitudes and Frequencies for the Entire Earth and Damaging Effects*

Richter Magnitude	Number per Year	Modified Mercalli Intensity Scale[a]	Characteristic Effects of Shocks in Populated Areas
<3.4	800,000	I	Recorded only by seismographs
3.5–4.2	30,000	II and III	Felt by some people
4.3–4.8	4,800	IV	Felt by many people
4.9–5.4	1,400	V	Felt by everyone
5.5–6.1	500	VI and VII	Slight building damage
6.2–6.9	100	VIII and IX	Much building damage
7.0–7.3	15	X	Serious damage, bridges twisted, walls fractured
7.4–7.9	4	XI	Great damage, buildings collapse
>8.0	One every 5–10 yr	XII	Total damage, waves seen on ground surface, objects thrown in the air

Source: After B. Gutenberg, 1950.
[a]Mercalli numbers are determined by the amount of damage to structures and the degree to which ground motions are felt. These depend on the magnitude of the earthquake, the distance of the observer from the epicenter, and whether an observer is in or out of doors.

The correspondence among damage caused, Richter magnitudes, and the estimated number of earthquakes is listed in Table 14.2.

THE EARTH'S INTERNAL STRUCTURE

We have seen that *P* and *S* waves travel through rock at different velocities. These waves respond to changing rock properties in differing degrees. The arrival times of *P* and *S* waves at seismographs stationed around the world provide records of waves that have traveled along many different paths. From such records it is possible to calculate how the rock properties change and where distinct boundaries occur between layers having sharply different properties.

Layers of Differing Composition

If the Earth's composition were uniform, and if no polymorphic changes occurred in the minerals present, the velocities of *P* and *S* waves should increase smoothly with depth. This is so because higher pressure leads to an increase in the density and the rigidity of a solid and these control the wave velocities. For an Earth of uniform composition it should be possible to predict how long seismic waves would take to pass through the Earth. However, observed travel times differ greatly from such predictions. These differences can best be ac-

counted for by supposing that velocities do not change smoothly with depth, and that composition is not constant throughout.

Reflection and Refraction

To find out where the compositional changes take place, we must use additional wave properties. Seismic body waves behave like light waves and sound waves, which is to say they can be both transmitted through a medium and also *reflected* and *refracted.* Reflection is the familiar phenomenon seen as light bouncing off the surface of a mirror or a glass of water. Seismic body waves are reflected by numerous surfaces in the Earth.

Refraction is a less familiar phenomenon. It occurs whenever a wave velocity changes. The velocity change can be either gradual or abrupt. An abrupt change is seen when a ray of light strikes a surface of water. Some of the light is reflected, but some also crosses the surface and travels through the water. The velocity of light is different in water than in air, and the ray path is sharply bent at the surface. The ray is said to have been refracted. Similarly, a seismic wave can be both reflected and refracted by a surface in the Earth (Fig. 14.11). Refraction occurs whether velocity changes are sudden or gradual. Sudden changes occur at several depths in the Earth. Because seismic-wave velocities also increase gradually with depth between sudden velocity changes, wave paths are curved by refraction (Fig. 14.11).

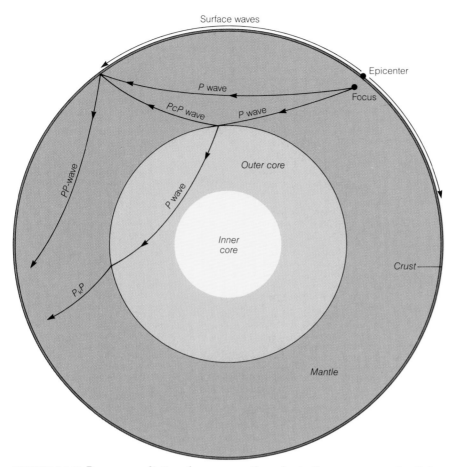

FIGURE 14.11 *P* waves radiating from an earthquake in the upper mantle. Seismographs at some places receive both direct *P* waves as well as reflected and refracted *P* waves. A *P* wave reflected off the surface is called a *PP* wave; one reflected off the core–mantle boundary is a *PcP* wave; one refracted through the liquid outer core is a P_kP wave. *S* waves also show reflection and refraction effects.

Both *P* and *S* waves are strongly influenced by a pronounced boundary at a depth of 2900 km. When *P* waves reach that boundary, they are reflected and refracted so strongly that the boundary actually casts a *P*-wave shadow over part of the Earth (Fig. 14.12). Because the boundary is so pronounced, it can be inferred that it is the place where the comparatively light silicate material of the mantle meets the dense metallic iron of the core. The same boundary casts an even more pronounced *S*-wave shadow, but the reason is not reflection or refraction. Shear waves cannot traverse liquids. Therefore, the huge *S*-wave shadow lets us infer that the outer core is liquid.

Refraction and reflection of seismic waves define three very pronounced boundaries separating four fundamental zones within the Earth. The boundaries correspond to those separating the crust from the mantle, the mantle from the outer core, and the outer core from the inner core.

The Crust

Early in the twentieth century the boundary between the Earth's crust and mantle was demonstrated by a scientist named Mohorovičić (Mo-ho-ro-vitch-ick), who lived in what today is Yugoslavia. He noticed that in measurements of seismic waves arriving from an earthquake whose focus lay within 40 km of the surface, the seismographs about 800 km from the epicenter recorded two distinct sets of *P* and *S* waves. He concluded that one pair of waves must have traveled from the focus to the station by a direct path through the crust, whereas the other pair represented waves that had arrived slightly earlier because they had been re-

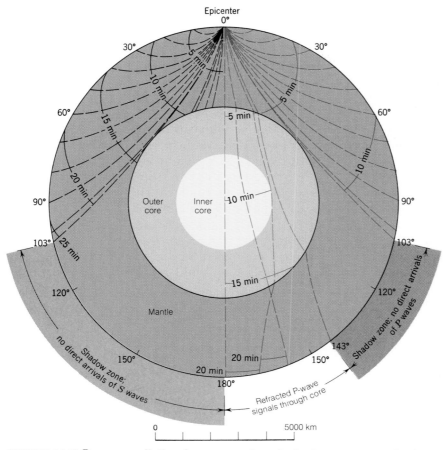

FIGURE 14.12 *P* waves radiating from an earthquake in the upper mantle. Seismographs at some places receive both direct *P* waves as well as reflected and refracted *P* waves. A *P* wave reflected off the surface is called a *PP* wave; one reflected off the core–mantle boundary is a *PcP* wave; one refracted through the liquid outer core is a *PkP* wave. *S* waves also show reflection and refraction effects.

fracted by a boundary at some depth in the Earth. Evidently, the refracted waves had penetrated a deeper zone of higher velocity below the crust, had traveled within that zone, and then had been again refracted upward to the surface (Fig. 14.13). Mohorovičić hypothesized that a distinct compositional boundary separates the crust from the underlying zone of differing composition. Scientists now refer to this boundary as the ***Mohorovičić discontinuity*** and recognize it as the seismic discontinuity that marks the base of the crust. The feature is commonly called the ***M-discontinuity,*** and in conversation is shortened still further to ***moho.***

Thickness and Composition of the Crust

By seismic methods it is possible to determine the thickness of the crust. Seismic-wave velocities can

be measured for different rock types in the laboratory and in the field. When the velocities of waves received at a number of seismographs are calculated, laboratory measurements can be used to determine the depth of the moho, and to estimate

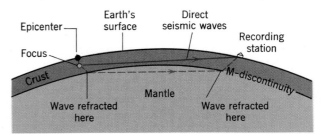

FIGURE 14.13 Travel paths of direct and refracted seismic waves from shallow-focus earthquake to nearby seismograph station.

the probable composition of the crust. Beneath ocean basins the crust is generally less than 10 km thick. Elastic properties of the oceanic crust are those characteristic of basalt and gabbro. But in the continental crust both thickness and composition are very different. The continental crust ranges in thickness from 20 to nearly 60 km, and tends to be thickest beneath major mountain masses (Fig. 14.14), a fact discussed later in this chapter. Velocities in the continental crust are distinctly different from those in the oceanic crust. They indicate elastic properties like those of rock such as granite and diorite, although at some places just above the M-discontinuity, velocities close to those of oceanic crust are often observed. These conclusions agree well with what is known about the composition of the crust from other lines of evidence such as geo-

logical mapping and deep drilling. The agreement gives geologists confidence in drawing conclusions about the mantle, where these other lines of evidence are scarce.

Seismic Exploration of the Crust

Artificial earthquakes provide a powerful way of mapping buried structures in the crust. An artificial earthquake is created by an explosion, or by striking the ground surface sharply and repeatedly with a powerful air hammer. The waves generated by the explosion travel down through the rocks, they are reflected from the upper surfaces of buried strata, from faults, unconformities, and other discontinuities, and travel back to seismographs located on the surface (Fig. B6.1). The exact focus of the ar-

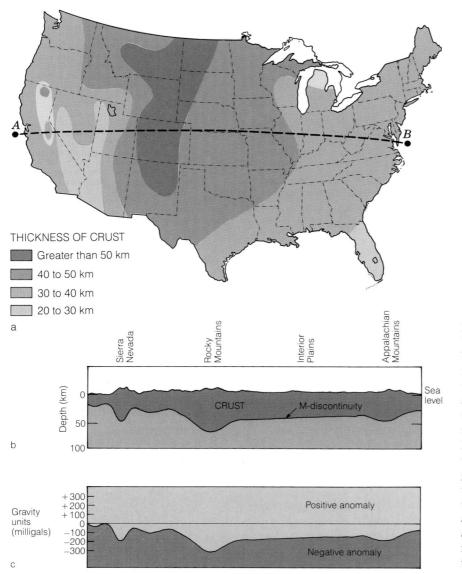

FIGURE 14.14 Crust beneath the United States. (a) Thickness of crust beneath United States, determined from measurements of seismic waves. (b) Section through the crust along the line *A–B* (above). The crust tends to thicken beneath major mountain masses such as the Sierra Nevada, the Rocky Mountains, and the Appalachians. (c) Profile of gravity traverse, adjusted for latitude, free air, and Bouguer corrections. The negative gravity anomalies over the Sierra, the Rockies, and the Appalachians are due to the roots of low-density rocks beneath these topographic highs.

tificial earthquake is known, as are the exact positions of the seismographs, so a very accurate and informative picture can be prepared of the unseen structures below.

Seismic exploration using man-made earthquakes can be used equally well at sea. Vibrations are created in the water by explosions from compressed-air guns. *P* waves pass through water as well as rock. In the oceanic crust the waves are reflected by any discontinuities present, and the reflections are recorded by seismographs towed behind a ship. In many instances, seismic records are even better at sea than they are on land.

The Mantle

The mantle is something of an enigma. It is huge and it controls much of what happens in the crust, but it cannot be seen. *P*-wave velocities in the crust range between 6 and 7 km/s. Beneath the M-discontinuity, velocities are greater than 8 km/s. Laboratory tests show that rocks common in the crust, such as granite, gabbro, and basalt, all have *P*-wave velocities of 6 to 7 km/s. But rocks like peridotite that are rich in dense minerals, such as olivine and

pyroxene, have velocities greater than 8 km/s. We infer, therefore, that such rock must be among the principal materials of the mantle. This inference is consistent with what little direct evidence is available concerning the composition of the upper part of the mantle; for example, evidence from rare samples of mantle rocks found in **kimberlite pipes,** narrow pipelike masses of igneous rock, sometimes containing diamonds, that intrude the crust but originate deep in the mantle (Fig. 14.15).

The Low-Velocity Layer

When *P*- and *S*-wave velocities are calculated for different regions of the mantle (using the arrival times of direct, refracted, and reflected waves), the way velocities increase with depth is far from regular (Fig. 14.16).

The first of the major velocity changes starts at a depth of about 100 km below the surface. The *P*-wave velocity at the top of the mantle is about 8 km/s and it increases to 14 km/s at the core/mantle boundary. From the base of the crust to a depth of about 100 km, the velocity rises slowly to about 8.3 km/s. However, the velocity then starts to drop slowly to a value just below 8 km/s and it remains low to a depth of about 350 km. The zone of reduced velocity is not sharply defined and it is better developed beneath the oceans than beneath the continents. Although the upper boundary of the layer is reasonably sharp and well defined, the lower boundary is rather diffuse. The low-velocity layer can be seen as a small trough, or blip, in both the *P*- and *S*- wave velocity curves in Figure 14.16. No evidence exists to suggest that the low-velocity layer is a zone where density decreases or, for that matter, where the composition changes. To account for the velocity changes, therefore, it is inferred that the zone has the same composition as the mantle immediately above and below, but is less rigid, less elastic, and more plastic than the regions above and below.

A possible explanation for the low-velocity layer is that between 100 and 350 km the geothermal gradient reaches temperatures close to the onset of partial melting of mantle rock. If the explanation is correct, either the rigidity drops sharply close to the melting curve, or melting actually starts and a small amount of liquid develops and forms very thin films around the mineral grains, thus serving as a lubricant. The amount of melting, if it occurs at all, must be very small, because the low-velocity zone does transmit *S* waves, and *S* waves cannot pass through liquids. Any liquid, like a thick film

FIGURE 14.15 Kimberlite, Monarch Pipe, South Africa. Fragments of rock from deep in the mantle are carried upward by the forceful intrusion of kimberlite magma. Rounded fragments are the transported blocks; fragmental, grayish background material is the kimberlite.

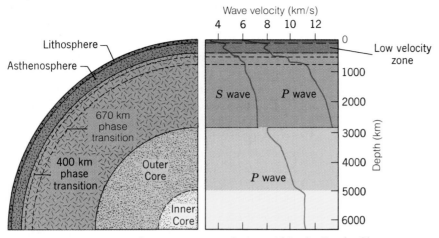

FIGURE 14.16 Variation of seismic-wave velocity within the Earth. Changes occur at the boundaries between the crust and mantle and between the mantle and core, owing to change in composition. Another change occurs at a depth of 100 km corresponding to the lithosphere–asthenosphere boundary. Change also occurs at 400 km below the surface owing to phase transitions, and at 670 km from unknown causes.

of oil, merely serves to lubricate the grains, and at the same time reduce wave velocities by reducing the elastic properties.

The Asthenosphere

An integral part of the theory of plate tectonics is the idea that plates of lithosphere slide over a somewhat plastic zone in the mantle. The importance of the low-velocity layer for the theory is that it proves the existence of the asthenosphere. The top of the low-velocity layer coincides with the base of the lithosphere. Thus, the low velocity zone coincides with the asthenosphere.

The asthenosphere is not a region of uniform thickness, lying everywhere at a constant depth of 100 km. Beneath the oceans, the top of the asthenosphere rises, in places to depths as shallow as 20 km; it is closest to the surface near an ocean ridge but progressively deeper away from the ridge. Beneath the continents the top of the asthenosphere is closer to 100 km, but it sinks to much greater depths beneath the thickest parts of the crust and beneath seafloor trenches. The schematic section through the upper portions of the Earth, shown in Figure 14.17, illustrates the present understanding of the variability of the asthenosphere.

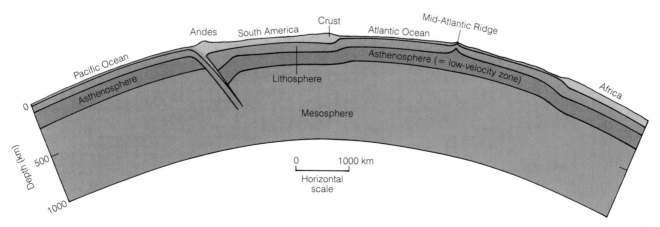

FIGURE 14.17 The asthenosphere varies in thickness and depth. A section through crust and upper mantle shows that the low-velocity zone corresponding to the asthenosphere is deeper beneath continents than beneath ocean basins and dips sharply down beneath the Andes. The section appears distorted because the vertical scale is twice the horizontal scale.

The 400-km Seismic Discontinuity

From the *P*- and *S*-wave curves in Figure 14.16 it is apparent that the velocity increases sharply at about 400 km. Sharp though it seems, the increase is not sharp enough to be accounted for by a change in composition; the cause must be something else. A probable explanation is suggested by laboratory experiment. When olivine is squeezed at a pressure equal to that at a depth of 400 km, the atoms rearrange themselves into a denser polymorph (chapter 2). This process of atomic repacking caused by changes in pressure and temperature is called a *polymorphic transition.* In the case of olivine, the repacking involves a change to a structure resembling that found in a family of minerals called the spinels, of which magnetite is a well-known example. The structural repacking involves a 10 percent increase in density. It is likely that the increase in seismic-wave velocities at 400 km is caused by the olivine–spinel polymorphic transition rather than by compositional changes. The density increase determined from seismic-wave velocities is almost exactly 10 percent.

The 670-km Seismic Discontinuity

An increase in seismic wave velocities—particularly the *P*-wave velocity—also occurs at a depth of 670 km. It is difficult to explain. The observed increase in density is about 10 percent, but the boundary is diffuse. It is not possible to determine from seismic evidence and laboratory experiment whether the boundary is solely due to a polymorphic transition, a compositional change, or both. Some scientists suggest that the increase at 670 km results solely from a polymorphic transition involving the repacking of atoms in the pyroxene minerals present in mantle rocks. Others suggest that the diffuse boundary indicates a polymorphic

change affecting all silicate minerals present. One idea involves the rearrangement of silicon and oxygen to create more dense anions in which each silicon atom is surrounded by six oxygen atoms rather than four. Still other suggestions center on compositional changes. One of the intriguing pieces of evidence comes from earthquakes. The deepest earthquakes have foci of 670 km. Deep earthquakes are associated with sinking slabs of cool, oceanic lithosphere.

If a compositional boundary exists at 670 km, it has been suggested that subducting lithosphere can sink no deeper. Opposing this point of view is evidence from a research technique called *seismic tomography.* The method is similar to that used in medicine in which a three-dimensional picture of the interior of the human body is developed from slight differences in the intensities of X rays passing through in different directions (*CAT scan* is the common name for X-ray tomography). In a similar manner, inhomogeneities in the mantle can be revealed by measuring slight differences in the frequencies and velocities of seismic waves (Fig. 14.18). Seismic tomography does not reveal the presence of a compositional boundary at 670 km. For the present it must be concluded that the cause of the 670-km seismic discontinuity is unknown.

The Core

Seismic waves indicate the way in which density increases with depth. Aided by both increasing pressure and phase transitions, density increases slowly from about 3.3 g/cm^3 at the top of the mantle to about 5.5 g/cm^3 at the base of the mantle. The mean density of the whole Earth is 5.5 g/cm^3. Therefore, to balance the less dense crust and mantle, the core must be composed of material with a density of at least 10 to 11 g/cm^3. The only common

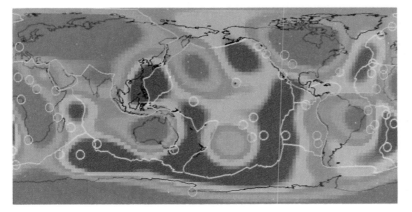

FIGURE 14.18 Lateral heterogeneity in the upper mantle revealed through seismic tomography. Seismic waves travel faster through cooler, more rigid material (shown in blue), more slowly in hotter, less rigid material (red). White lines show plate boundaries; white circles are centers of volcanic activity. Note that the red-colored, low-velocity zones lie beneath spreading edges.

substance that comes close to fitting this requirement is iron. Iron meteorites are samples of material believed to have come from the core of an ancient, tiny planet, now disintegrated. All iron meteorites contain a little nickel and the Earth's core presumably does too. Because S waves do not travel beyond the core–mantle boundary, it is inferred that the core adjacent is molten and that its composition is mostly iron but that nickel and small amounts of other elements are present too.

The Inner Core

P-wave reflections indicate the presence of a solid inner core enclosed within the molten outer core. The two cores (outer and inner) appear to be essentially identical in composition. The reason for the change from a liquid to a solid probably relates to the effect of pressure on the melting temperature of iron. As the center of the Earth is approached, pressure rises to values millions of times greater than atmospheric pressure. Temperature rises also, but not steeply enough to offset the effect of pressure. From the base of the mantle (at a depth of 2900 km) to a depth of 5350 km, temperature and pressure are so balanced that iron is molten. But at a depth of 5350 km another strong reflecting and refracting boundary occurs and the boundary has properties consistent with a change from a liquid to a solid. Apparently, from 5350 km to the center of the Earth, rising pressure overcomes rising temperature, and iron is solid, creating the solid core.

WORLD DISTRIBUTION OF EARTHQUAKES

Now that the concept has been developed to show the way in which seismic waves are used to build an understanding of the Earth's interior, let us turn to the pattern of occurrence of earthquakes. This pattern suggests a great deal about the shapes and motions of the plates of lithosphere.

Although no part of the Earth's surface is exempt from earthquakes, several well-defined *seismic belts* are subject to frequent earthquake shocks (Fig. 14.19). Of these the most obvious is the Circum-Pacific belt, for it is here that about 80 percent of all recorded earthquakes originate. The belt follows the mountain chains in the western Americas from Cape Horn to Alaska, crosses to Asia where it extends southward down the coast, through Japan, the Philippines, New Guinea, and Fiji, where it finally loops far southward to New Zealand. Next in prominence, giving rise to 15 percent of all earthquakes, is the Mediterranean–Asiatic belt, extending from Gibraltar to Southeast Asia. Lesser belts follow the midocean ridges.

Seismic belts are places where a lot of the Earth's internal energy is released. Therefore, it might be expected that other manifestations of internal energy would also appear in these belts, and indeed, some of them do. Midocean ridges, deep-sea trenches, andesitic volcanoes, and many other features that outline the margins of plates of lithosphere, coincide with, or closely parallel, these margins. Compare Figure 14.19 with Figures 1.10 and 3.11 to see that earthquake belts outline the plate boundaries.

The depths of earthquake foci around the edges of the plates are also informative. Most foci are no deeper than 100 km, because, as already mentioned, earthquakes occur in brittle rocks and the brittle lithosphere is only 100 km thick. However, a few earthquakes do originate at greater depths. The epicenters of deep earthquakes, with foci deeper than 100 km, are plotted in Figure 14.19b. It is noteworthy that the deep earthquakes are not associated with oceanic ridges. Rather, they are related to seafloor trenches. Those trenches mark the places where cool, brittle lithosphere sinks down into the mantle.

Benioff Zone

Detailed study of deep-earthquake foci beneath a seafloor trench (Fig. 14.20) shows that the foci follow a well-defined zone called a **Benioff zone.** This important observation strongly suggests that deep earthquakes originate within the relatively cold, downward-moving plate of lithosphere. Because some earthquake foci can be as deep as 670 km, it must be concluded that rapidly descending lithosphere can retain at least some brittle properties to that depth. The reason that earthquakes do not seem to originate at depths below 670 km remains an unsolved problem, but as previously mentioned, may be due to a physical barrier presented by the 670-km seismic discontinuity. It may also be that no physical barrier exists and that even a rapidly sinking slab of lithosphere is sufficiently hot by the time it reaches a depth of 670 km that it has become ductile rather than brittle.

Locations of earthquakes reveal a great deal about the structure of the moving lithosphere. But they provide a static picture, a sort of snapshot of the way things are at the moment. In order to discover

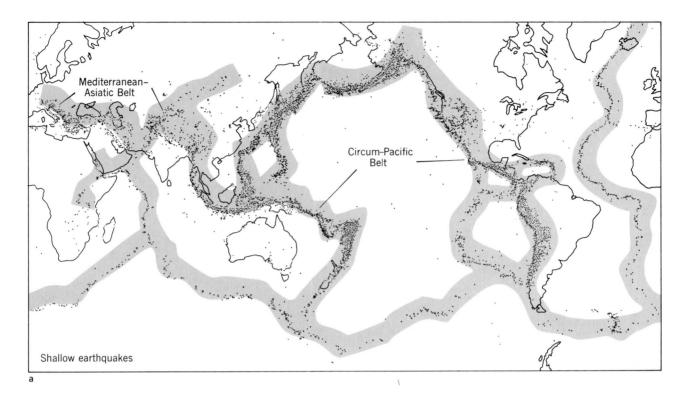

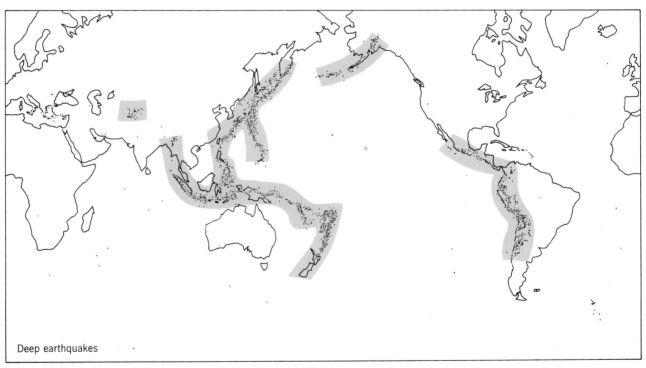

FIGURE 14.19 Epicenters of earthquakes recorded by the U.S. Coast and Geodetic Survey between 1961 and 1967. Each dot represents a single earthquake. (a) Earthquakes of all depths are plotted. Most are shallow, with foci within 100 km of the surface. The epicenters fall into well-defined seismic belts (blue shading) that coincide closely with the margins of plates of the lithosphere. (b) Epicenters of earthquakes having foci deeper than 100 km. They form belts that coincide closely with the seafloor trenches.

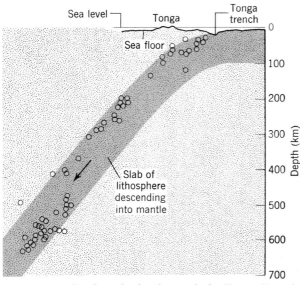

FIGURE 14.20 Earthquake foci beneath the Tonga Trench, Pacific Ocean, during several months in 1965. Each circle represents a single earthquake. The earthquakes define the Benioff zone and are generated by downward movement of a comparatively cold slab of lithosphere. (*Source:* After Isacks et al., 1968.)

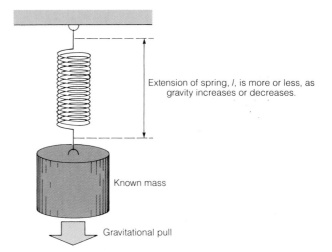

FIGURE 14.21 A gravimeter is a mass of metal suspended on a sensitive spring. The mass exerts a greater or lesser pull on the spring as gravity changes from place to place, extending the spring more or less. The mass of metal and the spring are contained in a vacuum together with exceedingly sensitive measuring devices.

how the Earth changes over time as it responds to forces that make materials move and flow, it is helpful to include other observations besides those from seismology. The most revealing is information derived from measurements of the Earth's gravitational attraction.

GRAVITY ANOMALIES AND ISOSTASY

The Earth is not a perfect sphere; careful measurement reveals that it is actually an ellipsoid that is slightly flattened at the poles and bulged at the equator.

The radius at the equator is 21 km larger than it is at the poles. Therefore, the pull exerted by the Earth's gravitational attraction is slightly greater at the poles than it is at the equator. Thus, a man who weighs 90.5 kg (199 lb) at the North Pole would observe his weight decreasing slowly and steadily to 90 kg (198 lb) by simply traveling to the equator. If the weight-conscious traveler made very exact measurements as he traveled, he would observe that his weight changed irregularly, rather than smoothly. From this he could conclude that the pull of gravity must change irregularly. If the traveler went one step further, and carried a sensitive device called a *gravimeter* (or *gravity meter*) for mea-

suring the pull of gravity at any locality, he would indeed find an irregular variation.

Gravity Measurements

Gravimeters consist of a heavy mass suspended by a sensitive spring (Fig. 14.21). When the ground is stable and free from vibrations due to earthquakes, the pull exerted on the spring by the heavy mass provides an accurate measure of the gravitational pull. Modern gravity meters are incredibly sensitive. The most accurate devices in operation can measure variations in the force of gravity as tiny as one part in a hundred million (10^{-8}).

In order to compare the pull of gravity from point to point on the Earth, three corrections must be applied to gravimeter measurements.

1. A correction must be applied for the latitude of the place of measurement. This correction takes care of the departure of the Earth's shape from a perfect sphere.

2. Topographic variations on the Earth's surface mean that measurements on mountains are made further from the center of the Earth than are measurements made in valleys. A gravimeter can detect the increase in gravitational pull when it is moved as little as a meter from the top of a table to the floor. Therefore, all measurements must be adjusted for elevation. In order to do

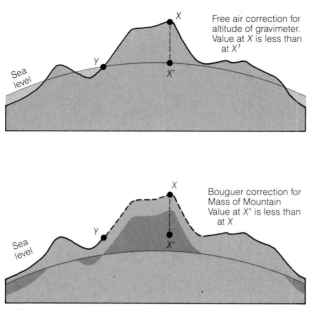

FIGURE 14.22 In order to compare gravitational force measured on a mountain top, *X*, with that at sea level, *Y*, the value at *X* must be corrected for altitude, increasing it to the value it would have at *X'*. The gravitational attraction of the mountain mass between *X* and *X'* must then be corrected by subtracting the attraction corresponding to the amount of rock between *X* and *X'*.

so, gravimeter readings are adjusted as if they were made at the surface of a reference ellipsoid that lacks topographic variations. The Earth's reference ellipsoid corresponds approximately with sea level. The topographic correction is called the *free-air correction* (Fig. 14.22).

3. The final correction that must be made is an adjustment to account for the free-air correction.

When a gravimeter reading is adjusted to the reference ellipsoid value, account must be taken of the gravitational attraction exerted by the mass of rock lying between the gravimeter and the ellipsoid. The correction that completes the adjustment for topography is called the **Bouguer correction**; it has an effect opposite to that of the free-air correction.

Gravity Anomalies

After corrections are made for latitude, free air, and Bouguer, the adjusted figures for the force of gravity might be expected to be the same everywhere on the Earth. In fact the adjusted figures reveal large and significant variations called **gravity anomalies.** The anomalies are due to bodies of rock having differing densities. A simple example of an anomaly is shown in Figure 14.23. From the anomalies, a great deal of important information can be derived.

The thickness of the crust beneath the United States, as determined from seismic measurements of the M-discontinuity, is shown in Figure 14.14. Beneath the three major mountain systems (the Appalachians, the Rockies, and the Sierra Nevada) the crust is thickened. In profile (Fig. 14.14b), the crust beneath the mountains resembles icebergs with high peaks, but with massive roots below the waterline. The accuracy of this analogy is demonstrated by the gravity profile across the United States, shown in Figure 14.14c. Negative gravity anomalies are observed where the crust is thickest. The anomalies are caused by the roots of low-density rock beneath the mountains, just as the basin

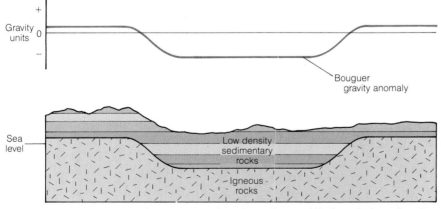

FIGURE 14.23 Example of a gravity anomaly; a basin filled with low-density sedimentary rocks sitting on a basement of dense igneous rocks. Gravity measurements, corrected for latitude, free-air, and Bouguer effects, reveal a pronounced gravity low. The magnitude of the Bouguer anomaly can be used to calculate the thickness of rocks of the basin.

FIGURE 14.24 Beach ridges raised by postglacial isostatic uplift of the land parallel to the shore of James Bay in central Canada.

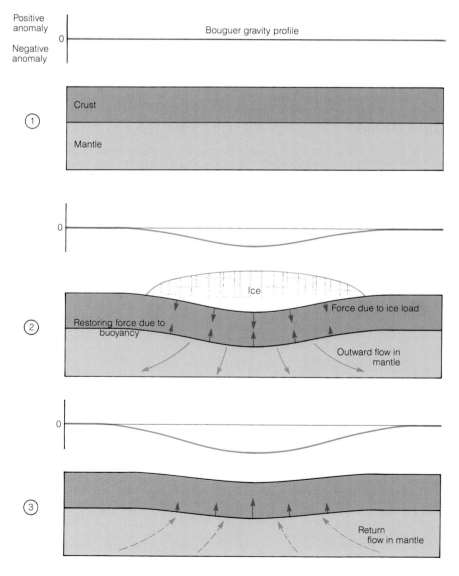

FIGURE 14.25 Depression of the crust by a continental ice sheet. (1) Prior to formation of the ice sheet there is no gravity anomaly. (2) When the ice sheet forms it depresses the crust. At some depth in the mantle, material must slowly flow outward to accommodate the sagging crust. (3) When the ice melts, buoyancy slowly restores the crust to its original level. A negative Bouguer gravity anomaly continues until the depression is removed. The viscosity of the mantle controls the rate of flow and, therefore, the slowness of recovery.

of low-density sediment produces the gravity anomaly shown in Figure 14.23.

The reason a root of low-density rock forms in the first place provides some interesting insights into the Earth's physical properties. Mountains stand high and have roots beneath them because they are comprised of low-density rocks and are supported by the buoyancy of weak, easily deformed but more dense rocks below. Mountains are, in a sense, floating. Even rock can flow and be deformed when pressure is applied slowly and steadily for a very long time. But it is not the crust that is floating on the mantle. Rather it is the lithosphere, capped by a mass of thickened crust, that floats on the asthenosphere. Strange as it may seem, the topographic variations observed at the surface of the Earth do not arise from the strength of the lithosphere, but rather from weakness and the buoyancy of the asthenosphere.

Isostasy

The property of flotational balance among segments of the lithosphere is referred to as ***isostasy.*** The great ice sheets of the last glaciation provide an impressive demonstration of isostasy. The weight of a large continental ice sheet, which may be 3 to 4 km thick, will depress the crust. When the ice melts, the land surface slowly rises again. A spectacular example of glacial depression and rebound is shown in Figure 14.24. The effect is very much like pushing a block of wood into a bucket of thick, viscous oil. When the wood is released, it slowly rises again to an equilibrium position determined by its density. The speed of its rising is controlled by the viscosity of the oil. Just like the block of

wood, glacial depression and rebound mean that somewhere in the mantle, rock must flow laterally when the ice depresses the crust, and then must flow back again when the deforming force is removed (Fig. 14.25). The flow must be slow because parts of northeastern Canada and Scandinavia are still rising, even though most of the thick ice sheets that covered them during the last glaciation had melted away by 7000 years ago.

Continents and mountains are composed of low-density rock, and they stand high because they are thick and light; ocean basins are topographically low because the thin oceanic crust is composed of dense rock. Isostasy and differences in the density of rocks beneath the continents and the oceans, therefore, are the reasons that the Earth has two pronounced topographic levels, as shown in Figure 1.4. The important point to be drawn from this discussion of isostasy is that the lithosphere acts as if it were "floating" on the asthenosphere. (*Floating* is not exactly the correct word, because the Earth is solid, but the system is buoyant and acts as though it were floating.) Sometimes gravity measurements suggest that a mountain seems to be top heavy and has too little root of low-density rock to counterbalance its upper mass. Sometimes, as in the seafloor trenches, it is observed that low-density crust has been dragged down to form a root without a mountain mass above it. These and many other situations lead to local gravity anomalies. The anomalies do not seem to become very large. This suggests that the Earth is always moving toward an isostatic balance. Indeed, isostasy is the principal explanation for vertical motions of the Earth's surface, just as plate tectonics is the principle explanation for lateral motions.

Isostatic Rebound?

Essay

A GIANT EARTHQUAKE IN THE PACIFIC NORTHWEST?

Subduction zones are places of intense seismic activity. Many of the greatest earthquakes ever recorded occurred in subduction zones: examples are the Chilean earthquake of 1960, Richter magnitude 9.5, and the Alaskan earthquake of 1964, magnitude 9.2. Such high-magnitude earthquakes are totally destructive.

Most earthquakes in subduction zones originate within either the sinking or overriding plate. Such earthquakes tend to be no larger than about Richter magnitude 7.5. Presumably they are caused by the bending and stretching of rock within the plates. Really big quakes have foci right at the interface between two plates. Presumably this happens when the downgoing plate sticks to the bottom of the overriding plate. When the lock is broken, a giant earthquake occurs.

From northern California to central British Columbia the Juan de Fuca plate has been slipping beneath the North American plate at a rate of 3 to 4 cm/yr for the past million years (Fig. B14.1), but for the past 200 years no giant earthquakes have occurred on the Cascadia subduction zone. Furthermore, seismographs cannot detect any current seismic activity along the interface between the two plates. Is the Pacific Northwest a likely spot for a giant quake sometime in the future, or is the Juan de Fuca plate sliding smoothly downward?

Geologists of the U.S. Geological Survey and other institutions are attempting to answer this question by looking at the stratigraphic record. Great earthquakes at subduction zones tend to cause pronounced elevation changes along the nearby coastline. Coastal subsidence up to 2 m or more can change well-vegetated coastal lowlands into intertidal mud flats. Careful mapping along the coastline of Washington reveals a number of places where marshes and swamps have suddenly become barren tide flats in the geologically recent past.

Radiocarbon dating of organic matter buried in sediment when lush coastal lowlands were suddenly flooded shows that a catastrophic change in the coastline occurred about 300 years ago in southwestern Washington. Evidence also suggests that catastrophic coastline changes occurred about 1700, 2700, 3100, and 3400 years ago in that area.

Although the evidence found as of 1988 does not prove that giant earthquakes have occurred in the Pacific Northwest, it is likely that they have, and that they will happen again.

SUMMARY

1. Abrupt movement on faults results in earthquakes, many of which cause destructive damage to dwellings and other man-made structures.

2. Ninety-five percent of all earthquakes originate in the Circum-Pacific belt and the Mediterranean–Asiatic belt. The remaining 5 percent are widely distributed along the midocean ridges and elsewhere.

3. Energy released at an earthquake's focus radiates outward as *P* (compressional) waves, and as *S* (shear) waves. Earthquake energy also makes the whole Earth vibrate. Whole Earth vibrations are sensed as surface waves.

4. From the study of seismic waves, scientists infer the internal structure of the Earth by locating boundaries or discontinuities in its composition. Pronounced compositional boundaries occur between the crust and mantle and between the mantle and the outer core.

5. Within the mantle there are two zones, at depths of 400 and 670 km, where sudden density changes produce seismic-wave discontinuities. The change at 400 km is probably produced by a polymorphic transition of olivine. The 670-km change might be due to either a polymorphic transition, a compositional change, or a combination of both.

At this point in the research earth scientists have come to think that the probability of a future giant earthquake is low but real. What they cannot do yet is say exactly when it will happen.

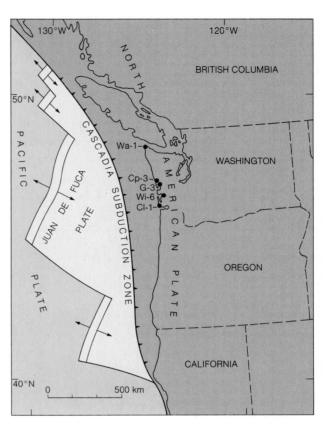

FIGURE B14.1 The Cascadia subduction zone separates the down-going Juan de Fuca Plate from the overriding North American Plate. Evidence of sudden, catastrophic sinking of near-coastal lowlands has been discovered along the Washington coast at the sites marked (*Source*: After Atwater, 1987.)

6. The lithosphere is rigid and is approximately 100 km thick; it overlies a plastic zone within which seismic waves have low velocities. The low-velocity zone coincides with the asthenosphere.

7. The base of the crust is a pronounced seismic discontinuity called the M-discontinuity. Thickness of the crust ranges from 20 to 60 km in continental regions, but is only about 10 km beneath the oceans.

8. The Earth is not a perfect sphere. It bulges at the equator and is flattened at the poles.

9. The outer portions of the Earth are in approximate isostatic balance; in other words, like huge icebergs floating in water, the lithosphere "floats" on the asthenosphere.

IMPORTANT WORDS AND TERMS TO REMEMBER

Benioff zone (p. 332)

body waves (p. 320)

Bouguer correction (p. 335)

compressional wave (p. 320)

core (p. 331)

earthquake focus (p. 319)

earthquake magnitude (p. 322)

epicenter (p. 319)

fault (p. 317)
free-air correction (p. 335)

gravimeter (p. 334)
gravity anomaly (p. 335)

inertia (p. 318)
inertial seismograph (p. 318)
inner core (p. 332)
isostasy (p. 337)

kimberlite pipe (p. 329)

low-velocity layer (p. 329)

M-discontinuity (p. 327)
Modified Mercalli Scale (p. 324)
moho (p. 327)
Mohorovičić discontinuity (p. 327)

P waves (p. 321)
polymorphic transition (p. 331)

reflection (p. 325)
refraction (p. 325)
Richter magnitude scale (p. 323)

S waves (p. 321)
seismic belt (p. 332)
seismic discontinuity (p. 331)
seismic sea wave (p. 324)
seismic tomography (p. 331)
seismic waves (p. 320)
seismograph (p. 318)
seismology (p. 318)
shear wave (p. 321)
strain seismograph (p. 318)
surface waves (p. 320)

tsunami (p. 324)

QUESTIONS FOR REVIEW

1. What is the cause of most earthquakes?
2. Explain how an inertial seismograph works.
3. What is the relation between an earthquake focus and the corresponding epicenter?
4. How can an epicenter be located from seismic records?
5. What are the two different kinds of seismic body waves, and how do they differ?
6. What are seismic surface waves and how do they differ from body waves?
7. The Richter magnitude scale is used by seismologists to estimate the energy released during an earthquake. How is the estimate made?
8. Name four different ways in which earthquake damage occurs.
9. What are reflection and refraction and how do they affect the passage of *P* waves?
10. How is the base of the crust defined?
11. Under what circumstances is it possible to obtain samples from rocks in the mantle?
12. What is the low-velocity layer and why is it an important part of the explanation of plate tectonics?
13. What is the presumed reason for the pronounced seismic discontinuity at a depth of 400 km below the surface?
14. What are seismic belts and how are they related to tectonic plates?
15. How do gravity anomalies arise?
16. How is the Earth's surface topography related to isostasy?
17. Describe some evidence which proves that isostasy is operating in the Earth.

CHAPTER 15

Deformation of Rock

*A plunging anticline of Proterozoic-aged strata, 90 km southwest of Alice Springs,
Australia.*

HOW IS ROCK DEFORMED?

The Earth's internal and external activities are in balance, or nearly so. The rock cycle brings new rock to the surface about as fast as old rock is removed by erosion. The balance means that solid rocks must be capable of movement. Other than by the rise of magma, how else could new rock be brought to the surface from deep in the crust? To answer that question it is necessary to consider how solid rock can be deformed, and how that deformation controls movement of rock in the crust and mantle.

In order to discuss deformation, such as bending, flow, and fracture in rocks, it is helpful to review some of the elementary properties of solids. Knowledge of rock deformation comes largely from laboratory experiments in which cylinders or cubes of rock are squeezed and twisted under controlled conditions. The terms used to describe deformation in the laboratory are convenient ones to employ in a discussion of deformation in the crust.

Stress and Strain

When a solid is subjected to a squeezing, stretching, or twisting force, the term *stress* is used rather than *pressure* for the deforming force. Stress and pressure are measured in the same units, force per unit area. The unit of pressure, as described in front matter, is the pascal (Pa). Why use the word *stress* instead of *pressure*? As discussed in chapter 5, where the term *stress* was first introduced, stress has the connotation of direction, whereas pressure does not. In order to discuss deformation of solids, it is necessary to identify the directions of maximum and minimum stress. Pressure, on the other hand, is commonly thought of as being equal in all directions, as in a fluid, and therefore does not deform a rock.

When a substance is stressed, it responds by changing size or shape, or both. The term used to describe change in shape or size is *strain*. If the length of a stressed rod is reduced by 10 percent, we say the longitudinal strain is 10 percent. Similarly, if the volume of a solid body is decreased by

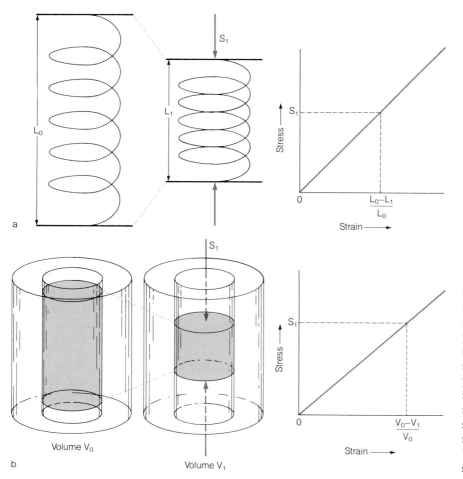

FIGURE 15.1 Hooke's law states that for elastic solids, strain is proportional to stress. (a) A spring, compressed by stress S_1, has its length reduced from L_0 to L_1. A plot of the strain ($L_0 - L_1/L_0$) against stress produces a straight line. (b) A cylinder of rock constrained by a tight metal jacket and compressed by stress S_1, has its volume reduced from V_0 to V_1. A plot of strain ($V_0 - V_1/V_0$) against stress produces a straight line.

10 percent when it is squeezed, we again say it has suffered a 10 percent volumetric strain. Stress, therefore, is a measure of the magnitude and direction of a deforming force, whereas strain is a measure of the changes in length, volume, and shape in a stressed material.

Elastic Deformation

Solids can be deformed in three basically different ways. The first is by *elastic deformation,* which is the reversible or nonpermanent deformation that occurs when an elastic solid is stretched or squeezed. The solid returns to its original shape and size when the force is removed. All solids are elastic to some degree, and rocks are no exception.

The famous British scientist Sir Robert Hooke (1635–1703) was the first to show that for elastic materials, provided the strain is not large, a plot of stress against strain yields a straight line. Hooke proved his point by using a spring (Fig. 15.1); however, his finding is equally true for rocks or for any other solid elastic body. The important point to remember is that within the elastic range, a stressed solid returns to its original size and shape when the stress is removed. This is the reason that earthquake vibrations do not generally leave any mark of their passage through rocks—the vibrations do not exceed the elastic limit of the rocks they pass through.

Deformation by Brittle Fracture

The second way the shape of a solid can be changed is by fracture. For every solid a limit exists beyond which it can no longer be deformed elastically and at which rupture will occur. A cylinder of marble, for example, when stressed parallel to its axis, will rupture at about 75 MPa (Fig. 15.2a). The marble has been deformed, and ruptured by *brittle fracture.* Fractures are permanent or irreversible deformation.

Ductile Deformation

The third way solids are deformed is by *ductile deformation* which, like brittle fracture, is irreversible. A solid exhibiting ductile deformation behaves elastically under low pressure, but a point is reached where elastic properties cease and ductile flow occurs. Ductile deformation is common in metamorphic rocks.

If a cylinder of marble is jacketed and subjected to a confining pressure while it is being stressed parallel to its long axis, a very interesting result is obtained. As shown in Figure 15.2b, the stress–strain curve rises through the elastic region; then at point Z, known as the *yield point,* the curve starts to flatten out. If, at point X', the stress is removed, the marble will change its shape along the curve X'Y. A permanent strain, equal to XY, has been induced in the marble, but fracture has not oc-

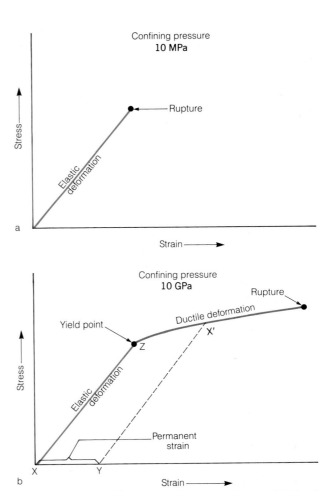

FIGURE 15.2 Typical stress–strain curves for two kinds of irreversible deformations. (a) Brittle fracture occurs after a solid has been elastically deformed, then the strength of the rock exceeded. (b) Following elastic deformation (X to Z), the yield point Z marks the onset of ductile deformation. If, at point X', the stress is removed, the solid will return to an unstressed state along path X'Y. The distance XY is a measure of the permanent, irreversible, strain produced by ductile deformation. If the stress is not released at X', but is increased and maintained, the strength of the solid is exceeded when rupture occurs.

curred. The permanent change in shape is due to ductile deformation. If instead of relieving the stress at point X' in Figure 15.2b the marble had been stressed still further, rupture would eventually have occurred. The marble would then have been changed both by ductile deformation and rupture.

Ductile Deformation versus Brittle Fracture

To evaluate deformation in rocks fully it is necessary to estimate the relative importance of brittle fracture versus ductile deformation in solids stressed beyond the limits of elastic deformation. The essential conditions controlling the relative importance of the two kinds of irreversible deformation are (1) temperature; (2) confining pressure; (3) time; and (4) composition.

Temperature

The higher the temperature, the weaker and less brittle a solid becomes. A rod of iron or glass is difficult to bend at room temperature; if we try too hard, both will break. However, both can be readily bent if they are heated to redness over a flame. At depth, rock temperature is high because of the geothermal gradient, and rock strength is reduced as a result.

Confining Pressure

The effect of confining pressure on deformation is not familiar in common experience. Confining pressure hinders the formation of fractures and so increases strength. At high confining pressures, however, it is easier for a solid to bend and flow than to break. Weakening of rock by the high temperature deep in the crust, and at the same time loss of brittleness caused by high confining pressures, are two reasons why solid rock can be bent and folded by ductile deformation.

Time

The effect of time on deformation of rock is vitally important, but as with confining pressure, it is not obvious from common experience. Stress applied to a solid is transmitted by all the constituent atoms of the solid. If the stress exceeds the strength of the bonds between atoms, either the atoms must move to another place in the crystal lattice in order to relieve the stress, or the bonds must break, which means brittle fracture occurs. Atoms in solids cannot move rapidly. Nevertheless, if the stress builds

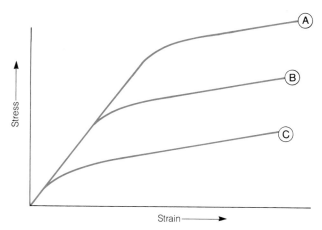

FIGURE 15.3 Typical stress–strain curves for a rock such as granite, held under a high confining pressure, but with differing temperatures and strain rates. Curve A: low temperature and high strain rate. Curve B: high temperature, high strain rate. Curve C: high temperature, low strain rate.

up slowly and gradually, and is maintained for a long period, the atoms have time to move, and the solid can slowly readjust and change shape by ductile deformation.

Strain Rate. The term used for time-dependent properties of rocks is **strain rate,** which is the rate at which a rock is forced to change its shape or volume. Strain rates are measured in terms of change of volume per unit volume per second. For example, a strain rate that is sometimes used in laboratory experiments is 10^{-6}/s, by which is meant a change in volume of one millionth of a unit volume per unit volume per second. Strain rates in the Earth are much slower than this—about 10^{-14} to 10^{-15}/s. The lower the strain rate, the greater the tendency for ductile deformation to occur.

A comparison of the influences of temperature, confining pressure, and strain rate can be seen in Figure 15.3. Low temperatures, low pressures, and high strain rates enhance brittle fracture. These are characteristic of the crust (especially the upper crust), and as a result fractures are common in upper-crustal rocks. High temperature, high pressure, and low strain rates, which are characteristic of the deeper crust and mantle, reduce the likelihood of brittle fracture and enhance the ductile properties of rock.

Composition

The composition of a rock has a pronounced effect on its strength. Composition has two aspects. First, the kinds of mineral in a rock exert a strong influence on strength properties because some minerals (such as quartz) are very strong, while others (such as calcite) are weak. Second, the presence of water reduces strength and enhances ductile properties. Two ways water reduces strength are by weakening the chemical bonds in minerals, and by forming films around mineral grains, thereby reducing the friction between grains. Thus, wet rocks have a greater tendency to be deformed in a ductile fashion than do dry rocks.

Minerals susceptible to ductile deformation are those which deform readily by creep, as ice does in a glacier (chapter 12). Minerals with pronounced ductile properties are halite, the carbonate minerals calcite and dolomite, and the sheet-structure silicate minerals such as clay, chlorite, mica, serpentine, and talc. Minerals which have minor ductile properties are those with isolated silicate tetrahedra—garnet and olivine—and the minerals polymerized in three dimensions, such as quartz and feldspar (chapter 2). Chain-structure minerals, such as amphiboles and pyroxenes, have ductile properties that are intermediate between those of mica and quartz.

Rocks that readily deform by ductile deformation are limestone, marble, shale, slate, phyllite, and schist. Rocks that tend to fail by brittle fracture rather than ductile deformation are sandstone and quartzite, granite, granodiorite, and gneiss.

Boudinage. When a sequence of interlayered rocks, some ductile and some strong but brittle, is stretched during deformation, an interesting difference in deformation style can be observed. The brittle layers fracture into elongate blocks called **boudins** (after a French word for sausage). The ductile layers flow into the fracture producing a structure called **boudinage** (Fig. 15.4).

DEFORMATION IN PROGRESS

Most deformation in the crust is too slow or too deeply buried to be observed. Large movements happen so slowly, which means at such low strain rates, they can only be measured over a few hundreds of years. Nevertheless, deformation does sometimes happen fast enough to be detected and measured. For convenience we divide large-scale deformation of the crust that can be observed into two groups: abrupt movement involving brittle fracture, in which blocks of the crust suddenly move a few centimeters or a few meters in a matter of

FIGURE 15.4 Boudinage structure in metamorphosed tuffs, Ritter Range, Sierra Nevada, California. The grade of metamorphism is greenschist facies. The broken unit is rich in epidote and was deformed by brittle fracture, whereas the enclosing units, which contain quartz, feldspar, and mica, displayed ductile deformation.

minutes or hours; and gradual movement involving ductile deformation, in which slow, steady motions occur without any abrupt jarring.

Abrupt Movement

Abrupt movement involves fracture and movement along a fracture. A fracture in a rock along which movement occurs is a **fault.** Stress builds up slowly as elastic deformation occurs in a rock; then, when the strength of the rock is exceeded, fracturing occurs. Once fracturing has started, friction inhibits continual steady slippage. Instead, stress again builds up slowly until friction between the two sides of the fault is overcome. Then abrupt slippage occurs again. If the stresses persist, the whole cycle of slow buildup, culminating in an abrupt movement, repeats itself many times. Although the extent of movement on a large fault may eventually total many kilometers, it is the sum of numerous small, sudden slips. Each sudden movement may cause an earthquake and, if the movement occurs near the Earth's surface, may disrupt and displace surface features. In doing so, the abrupt movement is readily observed and the evidence it leaves is convincing (Fig. 15.5).

Figure 15.5 is an example of horizontal movement. Abrupt vertical movements are also well documented. The largest abrupt displacement actually observed occurred in 1899 at Yakutat Bay, Alaska, during an earthquake. A stretch of the Alaskan shore (including the beach, barnacle-covered rocks, and other telltale features) was suddenly lifted as much as 15 m above sea level. This visible displacement may actually be less than the total amount, because the fault is hidden offshore and

FIGURE 15.5 An orange grove in southern California planted across the San Andreas Fault. Movement on the fault displaced the originally straight rows of trees. The direction of motion is such that trees in the background moved from left to right relative to the trees in the foreground.

the block of crust on the other side of it, entirely beneath the sea, may have moved downward, thus adding to the total displacement.

Gradual Movement

Movement along faults is usually, but not always abrupt. Measurements along the San Andreas Fault in California reveal places where steady slipping occurs, sometimes reaching a rate as high as 5 cm a year. This seems to be a case in which continuing ductile deformation may be happening at depths of 100 km or more, and brittle fracture is occurring near the surface.

Possibly no spot on the Earth is completely stationary. Measurements by U.S. Government surveyors over the past 100 years, for example, reveal great areas of the United States where the land is slowly sinking and other places where it is slowly rising (Fig. 15.6). The causes of these vast, slow movements are not all well understood, but they do prove that the solid Earth is not as rigid as it seems at first sight, and that great internal forces are continually deforming its crust.

EVIDENCE OF FORMER DEFORMATION

With such convincing evidence of present-day deformation of the Earth's crust, we might reasonably expect to find a great deal of evidence of former deformation. Studies of land and sea-bottom topography provide abundant evidence of vertical movements, and in some areas the distribution of various kinds of rock provides clear evidence that horizontal movements have occurred through distances as great as several hundred kilometers.

Not all evidence of movement and deformation observed in bedrock is as obvious as the examples cited. But once we learn to recognize it, evidence of deformation is seen to be very widespread—so much so that a special branch of geology, *structural geology,* has the study of rock deformation as its primary focus. In order to evaluate the Earth's internal activities, geologists must be able to recognize and evaluate evidence of rock deformation.

Deformation by Fracture

The brittle nature of rock is the property that leads to fracture. Rock in the crust, especially rock close to the surface, tends to be brittle. As a result, rock near the surface tends to be cut by innumerable fractures called joints (chapter 7). The fractures speed erosion, serve as channels for the circulation of groundwater, provide entryways along which magma is intruded, and in many places serve as the openings in which veins of valuable minerals are deposited.

Most fractures are small, and little or no slippage has occurred along them. They are like small cracks

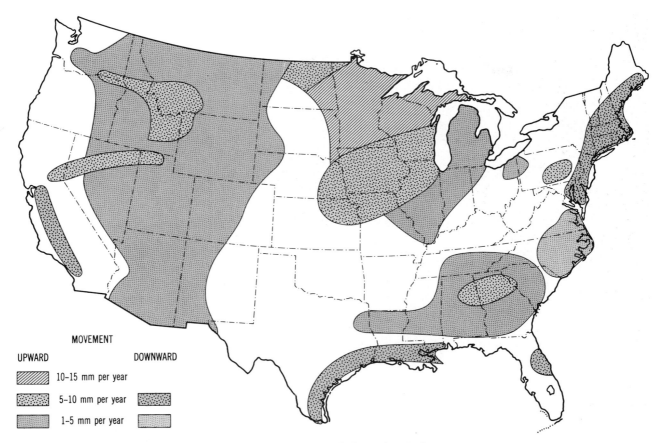

MOVEMENT

UPWARD   DOWNWARD

▨ 10–15 mm per year

▦ 5–10 mm per year

▦ 1–5 mm per year

FIGURE 15.6 Accurate measurements over a 100-year period show that in large areas of the United States the surface is slowly moving up or down. Subsidence along the coasts of California and the Gulf of Mexico is believed to have been caused in part by withdrawal of gas, oil, and water, which allows subsurface reservoirs to collapse. Uplift near the Great Lakes is a rebound effect following melting of the last ice sheet. The causes of movements in other areas are not known with certainty. Those areas in which no movement is shown are not necessarily stationary. They are simply areas in which measurements are very few. (*Source:* After Hand, 1972.)

in a pane of window glass. Along a few fractures, however, visible movement has occurred, and the fractures are then termed *faults*.

Relative Displacement

Generally, it is not possible to tell how much movement has occurred along a fault, nor which side of the fault has moved. In an ideal case, for example, if a single mineral grain or a pebble in a conglomerate has been cut through by the fault and the halves carried apart a measurable distance, the amount of movement can be determined. Yet even then it is not possible to say whether one block stood still while the other moved past it, or whether both sides shared in the movement. In classifying fault movements, therefore, geologists can determine only relative displacements; that is, one side

of a fault has moved in a given direction relative to the other side.

Hanging Wall and Footwall

Most faulting occurs along fractures that are inclined. To describe the inclination, geologists have adopted two old mining terms. From a miner's viewpoint, one wall of an inclined vein overhangs him, while the other is beneath his feet. Because veins occupy openings created by faults, we use the old miner's terms in the following way. The *hanging wall* is the surface of the block of rock above an inclined fault; the surface of the block of rock below an inclined fault is the *footwall* (Fig. 15.7). These terms, of course, do not apply to vertical faults.

a

b

c

FIGURE 15.7 Fault that displaces surface of ground. (a) Scarp made by displacement of a formerly level surface G–G' may mean that the hanging-wall block moved down from position h to h'; that the footwall block moved up from position h'; or that both blocks moved to some degree, to create the net displacement shown. (b) Scarp at the west base of Lost River Range, Idaho, formed in 1984 by abrupt movement along a fault that generated an earthquake. The whitish band of newly exposed rock marks displacement at the top of the valley fill. (c) Vertical section across the lower part of Lost River Range. Half arrows indicate relative movement of crustal blocks.

Normal Faults

Normal faults are caused by tensional forces that tend to pull the crust apart, and also by forces tending to expand the crust by pushing it upward from below. Movement on a normal fault is such that the hanging wall block moves down relative to the footwall block.

Horsts and Grabens. Commonly, two or more similarly trending normal faults enclose an upthrust or down-dropped segment of the crust. As shown in Figure 15.9, a down-dropped block is a **graben** or **rift** if it is bounded by two normal faults, or a **half-graben** if subsidence occurs along a single fault. An upthrust block is a **horst**. The central, steep-walled valley that runs down the center of the Mid-Atlantic Ridge and cuts through Iceland (Fig. 1.9) is a graben. Perhaps the world's most famous system of grabens and half-grabens is the African Rift Valley (Fig. 15.10), which runs north–south through more than 6000 km. Within parts of the Rift Valley magma has followed channelways that lead upward along the fault surfaces, creating volcanoes.

Normal faults are innumerable. Horsts and grabens are also very common, although none is as spectacular as the African Rift Valley. The north–south valley of the Rio Grande in New Mexico is a graben. The valley in which the Rhine River flows through western Europe follows a series of grabens. A spectacular example of normal faulting is found in the Basin and Range Province in Utah and Nevada. There, movement on a series of parallel and subparallel, north–south, normal faults has formed horsts and half-grabens that are now mountain ranges and sedimentary basins. The province is bounded in the east by the western edge of the Wasatch Range and continues westward to the eastern edge of the Sierra Nevada (Fig. 15.11).

Classification of Faults

Faults are grouped into classes according to (1) the inclination of the surface along which fracture has occurred, and (2) the direction of relative movement of the rock on its two sides. The common classes of faults, together with the changes in local topography they sometimes create, are listed in Figure 15.8. The standard planes of reference in classifying faults are the vertical and the horizontal. Along many faults, movement is entirely vertical or entirely horizontal, but along some faults combined vertical and horizontal movements occur.

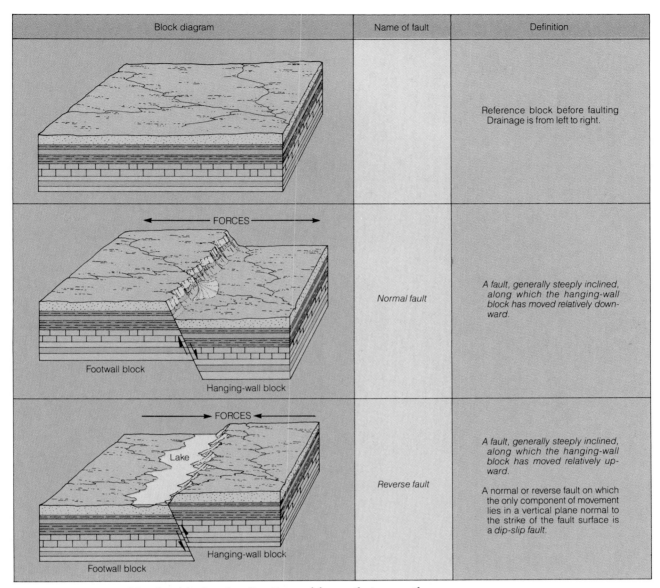

Block diagram	Name of fault	Definition
		Reference block before faulting Drainage is from left to right.
	Normal fault	A fault, generally steeply inclined, along which the hanging-wall block has moved relatively downward.
	Reverse fault	A fault, generally steeply inclined, along which the hanging-wall block has moved relatively upward. A normal or reverse fault on which the only component of movement lies in a vertical plane normal to the strike of the fault surface is a *dip-slip fault*.

FIGURE 15.8 Principal kinds of faults, the directions of forces that cause them, and some of the topographic changes they cause.

Reverse Faults

Reverse faults arise from compressive forces. Movement on a reverse fault is hanging-wall block up relative to the footwall block. Reverse fault movement pushes older rocks over younger ones, thereby shortening and thickening the crust.

Thrust Fault. A special class of reverse faults, called **thrust faults** and generally known as **thrusts,** are low-angle reverse faults with dips less than 45°. Such faults, common in great mountain chains, are noteworthy because along some of them the hanging-wall block has moved many kilometers over the footwall block. In most cases the hanging-wall block, thousands of meters thick, consists of rocks much older than those adjacent to the thrust on the footwall block (Fig. 15.12). The strata above some thrusts lie nearly parallel to those beneath, and so may appear, deceptively, to represent an unbroken sequence.

Strike–Slip Faults

One famous strike–slip fault has already been mentioned–the San Andreas Fault (Fig. 15.5). **Strike–slip faults** are those in which the principal movement is horizontal (Fig. 15.8). Fault movement is desig-

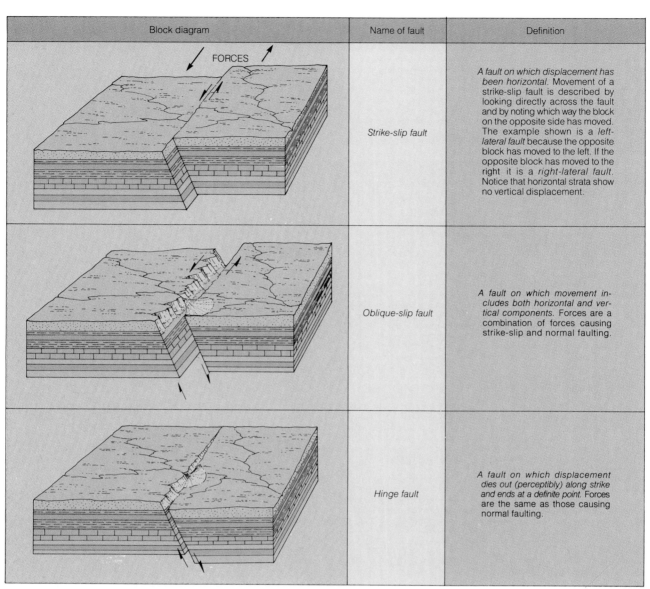

Block diagram	Name of fault	Definition
FORCES	*Strike-slip fault*	*A fault on which displacement has been horizontal.* Movement of a strike-slip fault is described by looking directly across the fault and by noting which way the block on the opposite side has moved. The example shown is a *left-lateral fault* because the opposite block has moved to the left. If the opposite block has moved to the right it is a *right-lateral fault.* Notice that horizontal strata show no vertical displacement.
	Oblique-slip fault	*A fault on which movement includes both horizontal and vertical components.* Forces are a combination of forces causing strike-slip and normal faulting.
	Hinge fault	*A fault on which displacement dies out (perceptibly) along strike and ends at a definite point.* Forces are the same as those causing normal faulting.

FIGURE 15.8 (continued).

nated as follows; to an observer standing on one fault block the movement of the other block is *left lateral* if it is to the left, or *right lateral* if it is to the right. The sense of relative motion is the same regardless of which block the observer is standing on. The San Andreas is a right-lateral strike–slip fault. Apparently, movement has been occurring along it for at least 65 million years. The total movement is not known, but some evidence suggests that it now amounts to more than 600 km.

Transform Faults.   Many of the great fracture zones that cut the oceanic ridge system are strike–slip faults. Indeed, they are so common that it has been suggested that the three major structural forms

marking sites of deformation of the Earth's crust are seafloor trenches, oceanic ridges, and strike–slip faults. These features link together to form continuous networks encircling the earth. When one feature terminates, another commences; their junction point is called a **transform**. J. T. Wilson, a Canadian scientist who first recognized the network relation, proposed that the special class of strike–slip faults that links major structural features be called **transform faults.** Close study of the strike–slip faults that offset the oceanic ridges proved Wilson's suggestion correct. As seen in Figure 15.13, movement along transform faults is a consequence of the continuous addition of new crustal material along oceanic ridges, the lateral movement of older crust

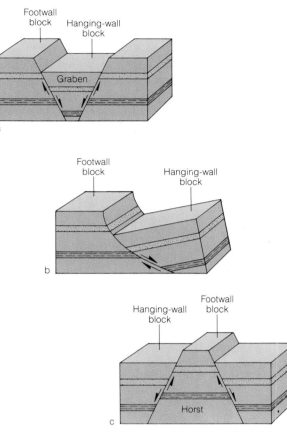

FIGURE 15.9 Horsts and grabens formed when tensional forces produce normal faults. (a) Graben. (b) Half-graben. (c) Horst.

(a)

(b)

FIGURE 15.10 The African Rift Valley. (a) Satellite image of central Kenya, reveals that the rift valley is a series of horsts and grabens. The image is about 70 km wide and 140 km long. To the east (right-hand side) is a high plateau bounded by a series of normal faults. Within the valley, normal faults run due north. Several volcanic cones are visible; magma is presumed to rise up the faults bounding the grabens. (b) The eastern wall of one of the many rifts that comprise the African Rift Valley in Tanzania. The valley floor in which Lake Manyara now lies was originally at the same height as the plateau above, but it has been lowered by movement on a normal fault. The fault surface has been modified by erosion, so the valley walls are no longer straight.

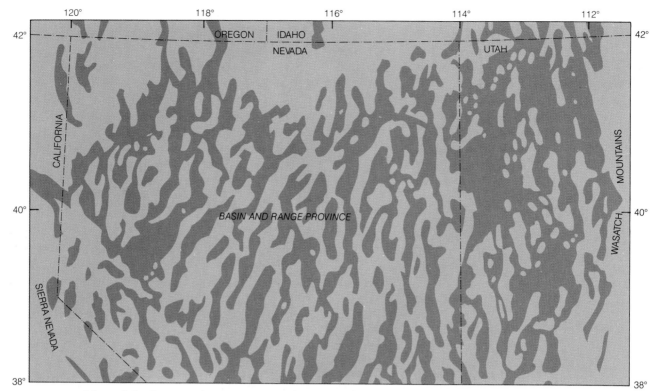

FIGURE 15.11 Map of the Basin and Range Province in Nevada and Utah. The Wasatch Mountains form the eastern border, the Sierra Nevada the western border. Green areas are mountains (horsts), brown areas basins (grabens). Boundaries of basins are normal faults. (Simplified from Geologic Map of the United States, U.S. Geological Survey, 1974.)

away from the ridge, and its consumption beneath seafloor trenches.

Evidence of Movement Along Faults

Often we find fractures in rock but cannot tell at first glance whether or not movement has occurred along them. For example, in uniform, even-grained rock such as granite, or in a pile of thin-bedded strata, no one of which is unique or distinctive, it is not possible to see the displacement of any obvious features. However, examination of the fault surface, or of rock immediately adjacent to it, sometimes reveals signs of local deformation, indicating that movement has ocucrred. Under special circumstances, even the direction of movement can be deciphered.

Slickensides.   Movement of one mass of rock past another can cause the fault surfaces to be smoothed, striated, and grooved. Striated or highly polished surfaces on hard rocks, abraded by movement along a fault, are **slickensides.** Parallel grooves and stria-

tions on such surfaces record the direction of most recent movement (Fig. 15.14a).

Fault Breccia.   Not all fault surfaces have slickensides. In many instances fault movement crushes rock adjacent to the fault into a mass of irregular pieces, forming **fault breccia** (Fig. 15.14b). Intense grinding breaks the fragments into such tiny pieces that they may not be individually visible even under a microscope. Such microbreccias are called **mylonites.** Some contain such tiny fragments they resemble chert.

Deformation by Bending

Bending may consist of broad, gentle warping that extends over hundreds of kilometers, or it might be close, tight flexing of microscopic size, or anything in between. Regardless of the volume of rock involved or the degree of warping, the bending of rocks is referred to as folding. Before discussing folds and folding, it is necessary to become familiar with the terms used to describe them.

a

FIGURE 15.12 Keystone Thrust, west of Las Vegas, Nevada. (a) Air view northward defines the fault by a color contrast in the strata. Light-colored Jurassic sandstone, forming a cliff nearly 600 m high (right), lies below the fault; dark-colored Paleozoic limestone and dolostones (left) lie above it. (b) Section drawn across area shown in the photograph.

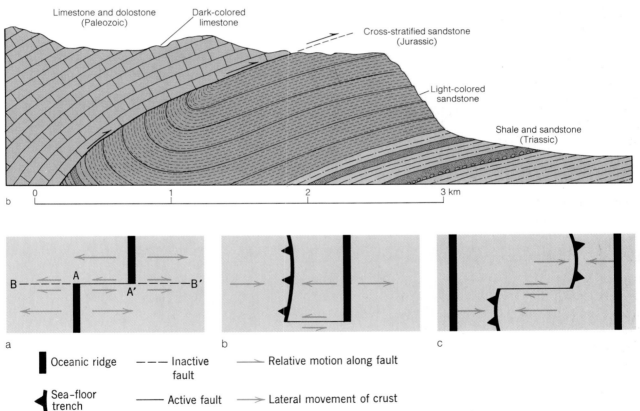

FIGURE 15.13 Transform faults. (a) An oceanic ridge offset by a transform fault. Crust on both sides of the two ridge segments moves laterally away from the ridge. Between segments of the ridge, along A–A', movement on the two sides of the fault is in opposite directions. Beyond the ridge, however, along segments A–B and A'–B', movement on both sides of the projected fault is in the same direction. A transform fault does not, therefore, cause the ridge segments to move continuously apart. (b) A transform fault joins an oceanic ridge with a seafloor trench. New crust formed at a spreading edge (the oceanic ridge) moves laterally away from the ridge and plunges back into the mantle at a place marked by the seafloor trench. (c) Transform fault joining two seafloor trenches.

(a)   (b)

FIGURE 15.14 The effects of faulting. (a) Slickensides on a fault surface, Dixie Valley, Nevada. The hanging-wall rocks have been removed by erosion, thus exposing the normal fault on which the foot wall (left) was raised relative to the hanging wall, (right). (b) Fault breccia. Angular gneiss fragments (dark) broken by faulting set in a matrix of rock flour and calcite. Titus Canyon, Death Valley.

FIGURE 15.15 A monocline in southern Utah that interrupts the generally flat-lying sedimentary strata of the wide Colorado Plateau. Along the line of flexure the strata are nearly vertical (right-hand side of photo). On both sides of the monocline the exposed layers are nearly horizontal. View looking south.

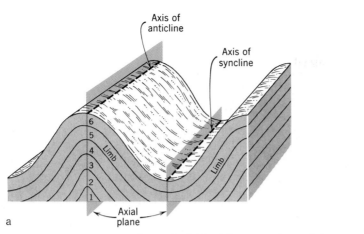
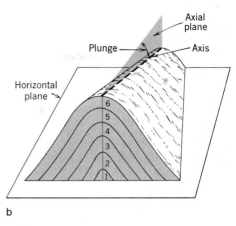

FIGURE 15.16 Features of simple folds. Upper surface of the youngest layer (6) slopes toward the axis of the syncline but away from the axis of the anticline. (a) Fold axis horizontal. (b) Fold axis plunging.

Types of Folds

The simplest fold is a *monocline*, a one-limbed flexure, on both sides of which the strata either are horizontal or dip uniformly at low angles (Fig. 15.15). An easy way to visualize a monocline is to lay a book on a table. Then drape a handkerchief over one side of the book and out onto the table. So draped, the handkerchief forms a monocline.

Most folds are more complicated than monoclines (Fig. 15.16). An upfold in the form of an arch is an *anticline.* A downfold with a troughlike form is a *syncline.* Anticlines and synclines are usually paired.

Geometry of Folds

As shown in Figure 15.16, the sides of a fold are the *limbs,* and the median line between the limbs, along the crest of an anticline or the trough of a syncline, is the *axis* of the fold. A fold with an inclined axis is said to be a *plunging fold,* and the angle between a fold axis and the horizontal is the *plunge* of a fold. An imaginary plane that divides a fold as symmetrically as possible, and that passes through the axis, is the *axial plane.*

Many folds, such as those in Figure 15.16, are nearly symmetrical. Others, however, are not symmetrical; strong deformation may create complex shapes. The common forms of folds are shown in Figure 15.17. If only fragmentary exposures of bedrock are available, it is apparent that difficulties might arise in deciding whether a given fold is overturned or not. It is necessary to know whether a layer is right-side up or upside down in order to decide which limb of a fold it is in. This is not always possible, but in some cases sedimentary structures, such as mud cracks and graded layers, do record this (Fig. 6.3). In other examples only careful, thorough mapping of all bedrock exposures can provide the answer.

Folds are often so large that examination of a single exposure does not make an observer aware he or she is seeing folded rock. Nevertheless, when all the exposures of a particular rock body are plotted and a geologic map is prepared, folds can be recognized from the distribution of the various rock types. Differences of erodibility in adjacent strata can also lead to distinctive topographic forms by which the presence of folds can be recognized (Fig. 15.18).

REGIONAL STRUCTURES

Cratons

On the scale of a continent, two distinctly different kinds of structural units can be distinguished. The first unit is a kind of core or nucleus of very ancient rock and is called a *craton.* The term is applied to a portion of the Earth's crust which has attained tectonic stability. Rocks within cratons may be deformed, but the deformation is invariably ancient (Fig. 15.19).

Orogens

Draped around cratons are the second kind of crustal building unit, *orogens* or *orogenic belts*, which

Name	Description	
Symmetrical	Both limbs dip equally away from the axial plane.	
Asymmetrical	One limb of the fold dips more steeply than the other.	
Overturned	Strata in one limb have been tilted beyond the vertical. Both limbs dip in the same direction, though not necessarily at the same angle.	
Recumbent	Axial planes are horizontal. Strata on the lower limb of anticline and upper limb of syncline are upside down.	
Isoclinal	Both limbs are essentially parallel, regardless whether the fold is upright, overturned, or recumbent.	

a

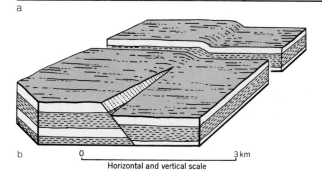

b

0   3 km
Horizontal and vertical scale

FIGURE 15.17 (a) Five kinds of folds. (b) Normal fault (front block) passes laterally into monocline (rear block).

are elongate regions of the crust that have been intensely folded and faulted during mountain-building processes. Orogens differ in age, history, size, and origin; however, all were once mountainous terrains and they are younger than the cratons they surround. Only the youngest orogenic belts are mountainous today; ancient orogenic belts, now deeply eroded, reveal their history through the kinds of rock they contain and the kind of deformation in the belt.

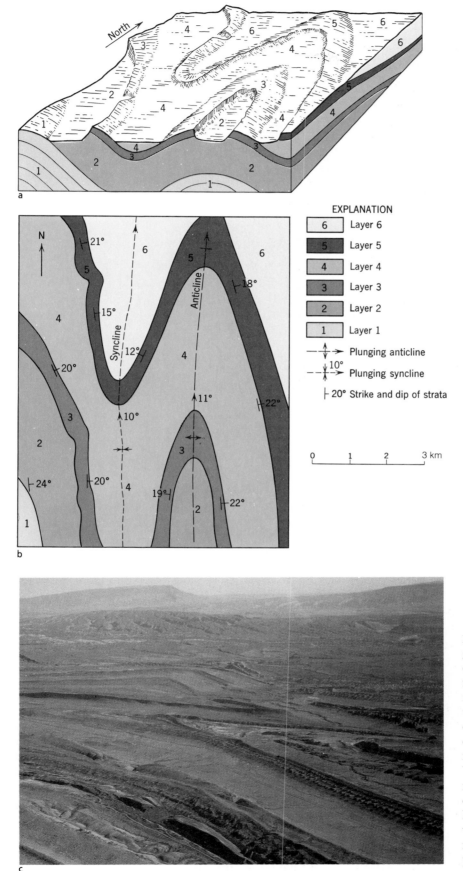

EXPLANATION

6	Layer 6
5	Layer 5
4	Layer 4
3	Layer 3
2	Layer 2
1	Layer 1

Plunging anticline

Plunging syncline

20° Strike and dip of strata

0   1   2   3 km

FIGURE 15.18 Distinctive topographic forms and patterns in the distribution of various kinds of rock reveal the presence of plunging folds. (a) Block diagram showing topographic effects. (b) Geologic map of area shown in part a. (c) Air view of the plunging anticline at the northern end of Sheep Mountain, Wyoming. The view is toward the northeast. Resistant sandstone layers make jagged, low ridges; shale layers erode readily and form the valleys. The curve of the sandstone ridges points in the direction of the plunge.

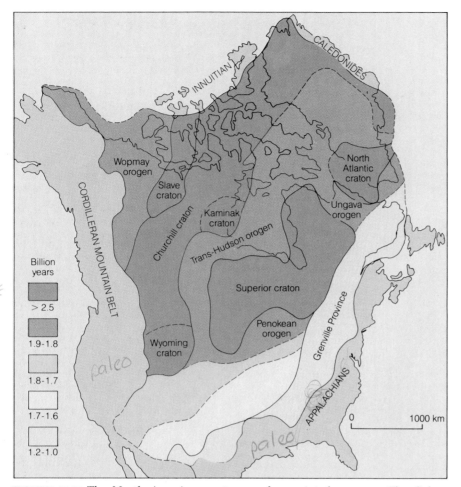

FIGURE 15.19 The North American cratons and associated orogens. The Cale-donide, Appalachian, Cordilleran, and Innuitian orogens are each younger than 600 million years. (*Source:* After Kerr, 1985.)

Continental Shields

Where the assemblage of cratons and ancient or-ogens are exposed at the surface the term *conti-nental shield* is used. That portion of a continental shield that is covered by a thin layer of little-de-formed sediments is called a *stable platform.* By careful mapping, and through radiometric dating, geologists have recognized several cratons and or-ogens in the continental shield and stable platform of North America (Fig. 15.19). Within the cratons, all rocks are older than 2.5 billion years. The oro-gens that separate the cratons are also geologically old—1.6 billion years or more. Such rocks can be observed in many places in eastern Canada, but within the United States the cratonic rocks only crop out in a small region around Lake Superior. Nevertheless, by drilling through the cover of sed-

imentary rocks on the stable platform it has been found that the cratons and orogens that separate them lie below much of the central United States and part of western Canada. The four youngest orogens, the Innuitian, Caledonide, Appalachian, and Cordilleran orogens, are draped around the assemblage of small cratons and older orogens that make up the core of the continent.

The small cratons were probably minicontinents during the Archean Eon. By about 1.6 billion years ago, the smaller fragments had become welded to-gether to form the assemblage of cratons and an-cient orogens we see today. As the larger cratonic fragments moved and collided, orogenic belts were formed along their margins. The existence of an-cient collision belts is the best evidence available to support the idea that plate tectonics operated at least as far back in time as 2 billion years ago.

YOUNG OROGENS

Today's mountain ranges are the orogens that formed during the last few hundred million years. Within any orogenic belt it is generally possible to observe that several different kinds of mountain units are present. An elongate series of mountains belonging to a single geologic unit is called a *mountain range*. Excellent examples are the Sierra Nevada in eastern California and the Front Range in Colorado. A group of ranges similar in general form, structure, and alignment, and presumably owing their origin to the same general causes, constitute a *mountain system*. The Rocky Mountain system is a great assemblage of ranges, all formed within a few tens of million years of each other, that extend northward beginning near the Mexican border, through the United States, and north to western Canada. The term *mountain chain* is used somewhat more loosely to designate an elongate unit consisting of numerous ranges or systems, regardless of similarity in form or equivalence in age. An example is the gigantic mountain chain that runs along the western edge of the Americas, from the tip of South America to northwestern Alaska, and that includes all the systems and ranges in between. This broad belt of ranges is called the American Cordillera.

Mountain Ranges

Mountain ranges display such a great variety of rocks and structures that no two are identical. If we concentrate on the details, we are in danger of seeing only the foliage and missing the forest. The most helpful way to organize thoughts about mountain ranges is to identify the single, most characteristic feature and use it for classification. On this basis it is possible to identify three principal kinds of ranges:

1. fold-and-thrust mountain ranges;
2. volcanic mountain ranges;
3. fault-block mountain ranges.

Fold-and-Thrust Mountain Ranges

Fold-and-thrust mountain ranges are orogens. They are spectacular, complex structures. Occurring in great arc-shaped systems a few hundred kilometers wide, they commonly reach several thousand kilometers in length. The words *fold* and *thrust* indicate their most characteristic features. Strata have been compressed, faulted, folded, and crumpled,

commonly in an exceedingly complex manner. Metamorphism and igneous activity are always present. Examples are widespread: the Appalachians, the Alps, the Urals, the Himalaya, and the Carpathians are all fold-and-thrust mountain systems. Indeed, the Alps, the Carpathians, and the Himalaya belong to a single gigantic system of fold-and-thrust mountain ranges formed during the Mesozoic and Cenozoic eras.

Geosynclines. All fold-and-thrust mountain systems share another feature related to their folded strata. They develop from thick piles of sedimentary strata, commonly 15,000 m or more in thickness. The strata are predominantly marine; this conclusion can be drawn from the presence of marine fossils in the sediments. In the Alps, the marine strata are mostly of deep-water origin. In the Appalachians, much of the sediment apparently accumulated in shallow water. Regardless of water depth, the kinds and thicknesses of sediments lead to two important conclusions. First, the sites of sediment deposition are predominantly oceanic. Second, the great thickness of sediment, which commonly exceeds the greatest depths of the ocean, indicates that the catchment area must have been sinking while it was being filled. The problem of such a huge catchment site for sediment had been recognized by the middle of the nineteenth century. The American geologist J. D. Dana coined the term *geosyncline*, by which he meant a great trough that has received thick deposits of sediment during its slow subsidence through long geologic periods. Dana's perception of a trough has not proven to be correct, but the name geosyncline has remained in use.

Miogeocline and Eugeocline. Modern oceanographic studies solved the geosyncline puzzle. The site of sediment deposition is not a trough. Rather, it is the margin of a continent where oceanic and continental crust are joined. It is here that sediment derived by weathering of the adjacent continent accumulates.

The shallow-water marine sediment that accumulates on the continental shelf is underlain by continental crust. Such a wedge of sediment is called a *miogeocline.* Deep-water sediment and associated volcanic rock that accumulates on, and at the foot of, the continental slope (chapter 1), is underlain in part by oceanic crust, in part by continental crust. The wedge of deep-water sediment is called a *eugeocline.* Eugeoclinal sediments tend to contain a significant amount of material derived

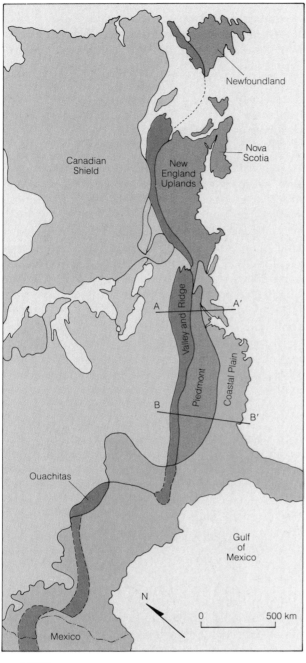

Folded and thrust sedimentary
strata

Metamorphic and
igneous rocks

Mesozoic and Cenozoic
rocks of the Coastal Plain

FIGURE 15.20 The Appalachian Mountain System runs from Newfoundland to the Mexican border. Part of the eastern and southern margins of the system is covered by younger sediments of the coastal plain. Figures 15.21 and 15.22 are cross sections drawn approximately along the lines A–A′ and B–B′, respectively.

from pyroclastic and other volcanic rocks. This is particularly true when a continental margin is adjacent to a subduction zone.

The Appalachians.   Strata in a geosyncline can be separated into miogeoclinal and eugeoclinal sediments. They are an essential phase in the formation of fold-and-thrust mountains. Before discussing the final step in the process of forming fold-and-thrust mountains (chapter 16), it is helpful to describe briefly the geology of some typical fold-and-thrust mountain systems and compare them with that of other kinds of mountain systems.

First, consider the Appalachians, a Paleozoic-age fold-and-thrust mountain system 2500 km in length that borders the east and southeast coasts of North America (Fig. 15.20) and that continues offshore, as eroded remnants, beneath the sediment of the modern continental shelf. The sediment in the miogeocline contains mud cracks, ripple marks, fossils of shallow-water organisms, and in places freshwater materials such as coal. Evidence for deposition of sediment on the continental shelf and nearshore area is strong. The sediment is underlain by a basement of metamorphic and igneous rock, and becomes markedly thicker away from the former western shore (that is, the sediment thickens from west to east).

Most, but not all, of the sediment in the old miogeocline has now been deformed. Today, if we approach the central Appalachians from western New York and western Pennsylvania, we see first the former sediment occurring as essentially flat-lying, undisturbed strata. Continuing eastward, we notice the same strata thicken and become gently folded and thrust-faulted. In eastern Pennsylvania, in the region known as the Valley and Ridge Province, the strata have been bent into broad anticlines and synclines. The province gets its name because valleys have been developed by eroding the weakest strata, composed of limestone, dolostone, and claystone (Fig. 15.21).

In the mountains further south, in Tennessee and the Carolinas, a somewhat different style of deformation is apparent. Here, thrust faults predominate (Fig. 15.22). Huge, thin slices of sedimentary strata were pushed westward, each successive slice riding upward and over earlier slices. The surface or layer along which movement occurred is known as a **detachment surface,** and the slice that moved is commonly referred to by the French name, **décollement.** A distinctive feature of a décollement is that the style of deformation above the detachment surface is usually different from

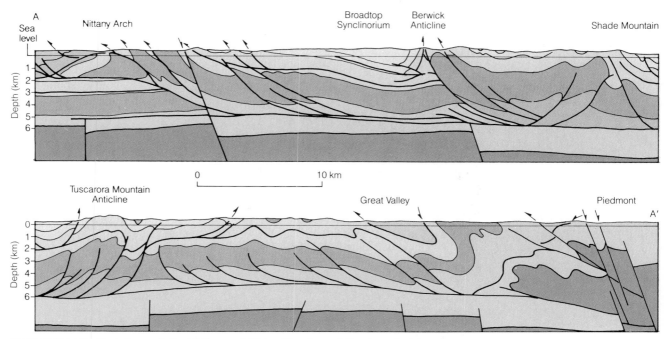

FIGURE 15.21 Section through the Valley and Ridge Province of the Appalachians in Pennsylvania, along the line A–A′ in Figure 15.20. The structure includes both folding and thrusting. The prominent stratum is a limestone of middle Cambrian to early Ordovician age. (*Source:* After Woodward, 1985.)

that below; that is, the weaker sedimentary strata above the detachment surface have been fractured, moved along thrust faults, and stacked like a series of thin cards, while the older basement rocks below tend to have resisted faulting and large-scale translation and to have been deformed by ductile deformation.

Proceeding still farther east, toward the region from which the thrust slices came, we find the core of the Appalachians. Here the ancient basement rocks can be examined. The deep-water sediments that were deposited in the eugeocline also can be seen. The strata are increasingly metamorphosed and deformation becomes increasingly intense. Folds

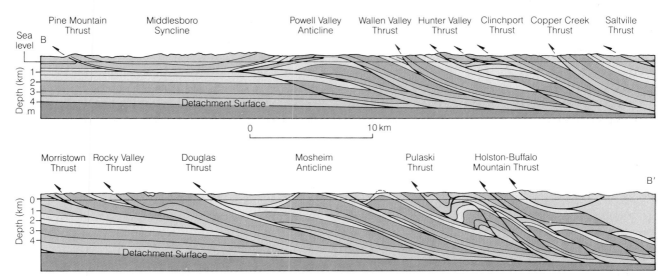

FIGURE 15.22 Section through the Valley and Ridge Province of the Appalachians in Tennessee and North Carolina, along the line B–B′ in Figure 15.20. Compare with Figure 15.21. Development of décollement by thrust faulting is predominant in the southern Appalachians. (*Source:* After Woodward, 1985.)

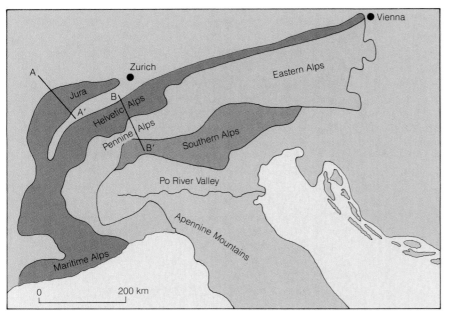

FIGURE 15.23 Map of the major units of the Alps in Switzerland and Austria. The mountains were formed as a result of compressive forces operating in a southeast to northwest direction. Figure 15.24a and b are along the lines A–A′ and B–B′, respectively.

become isoclinal and then overturned (Fig. 15.17), and faulting is prevalent. In places, fragments of the old basement can be seen to have been thrust up over younger sedimentary strata. Finally, we reach a region where intense metamorphism has occurred and where granite batholiths have been emplaced.

The Alps. We naturally ask how well the Appalachian picture can be applied to other fold-and-thrust mountains. The answer is that similar features are found in all of them. The Alps and associated mountain ranges in southern Europe (Fig. 15.23) were formed later than the Appalachians, during the Mesozoic and Cenozoic eras. Nevertheless, the two systems have many features in common. For instance, the Jura Mountains, which mark the northwestern edge of the Alps, have the same folded form and origin as the Valley and Ridge Province (Fig. 15.24). Also, the Jura Mountains were formed from shallow-water miogeoclinal sediments. In the high Alps, which correspond to the now deeply eroded Appalachians that can be seen in Connecticut, Vermont, Virginia, and Maryland, thrusting appears to have developed on a much grander scale than in the Appalachians (Fig. 15.24). The high Alps are composed of deeper-water marine, eugeoclinal strata.

The Canadian Rockies. The Canadian Rocky Mountains, a magnificent mountain system much less eroded than the Appalachians, can also be compared. A section through the Canadian Rockies

at about the latitude of Calgary, Alberta (Fig. 15.25), reveals all the features described for the Appalachians. A central zone has been intensely metamorphosed. In it, parts of the older basement rocks have been thrust upward, and on the margins folding and thrust faulting are evident. How this happens in a fold-and-thrust mountain range is discussed more fully in chapter 16. The thrust sheets in the Canadian Rockies moved eastward, away from the core zone. It is apparent that each sedimentary unit becomes thinner as it is followed from west to east, indicating that the eastern portion was the miogeocline, while the core zone coincides with the thickest part of the old geosyncline, and thus was the eugeocline.

Volcanic Mountain Ranges

Some of the world's most beautiful and scenic mountains are volcanoes: Mount Fuji, Mount Etna, Mount Rainier, Mount Mayon, and Mount Kilimanjaro are examples. Volcanic mountains differ in a fundamental way from fold-and-thrust mountains in that they are formed by deposition of new pyroclastic and volcanic rock and not by deformation of preexisting crust. Although volcanic mountain ranges are found on land, they are far more abundant on the seafloor. In some chains of seafloor mountains, such as the chain of volcanoes that forms the Hawaiian Islands, the higher peaks protrude above sea level. In other chains (for example, in the oceanic ridges) the entire chain is submerged.

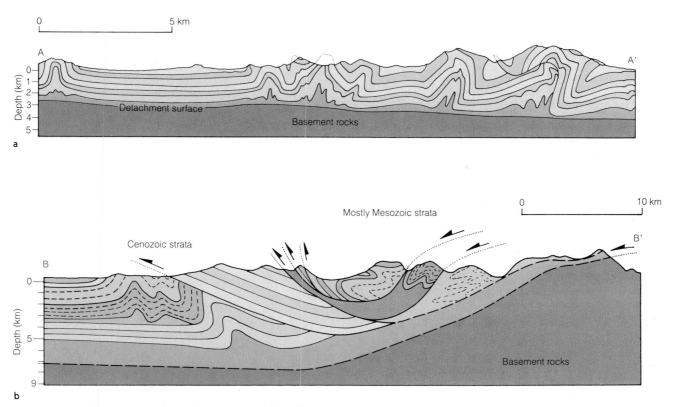

FIGURE 15.24 Two sections through portions of the Alps. (a) Section through the Jura Mountains along line A–A' in Figure 15.23. The weak sedimentary rocks were deformed by slippage along the detachment surface. (b) Section in central Switzerland along the line B–B' in Figure 15.23. Strata have moved northward along great thrust faults which later were themselves folded. A major thrust fault separates overlying strata from basement rocks. (*Source:* After Heim, 1922.)

Oceanic Island Arcs. The locations, shapes, and extents of volcanic mountain chains are all controlled by plate tectonics. Those controls will be discussed further in chapter 16. One special class of volcanic mountain chain, an *oceanic island arc,* is a great arcuate belt of andesitic and basaltic vol-

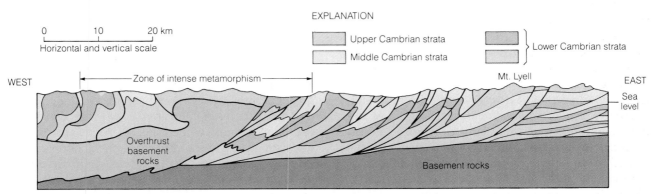

FIGURE 15.25 Section through the Canadian Rocky Mountains at about the latitude of Calgary, Alberta. The zone of intense metamorphism coincides with the region of maximum uplift and maximum deformation. Farther east, where strata become progressively thinner, the pile has been greatly thickened by movement along thrust faults. The sense of movement is such that each fault block has moved toward the east, riding over the block beside it. (*Source:* After Price and Mountjoy, 1970.)

canic islands formed over subducting oceanic crust. Some island arcs are 2000 km or more in length. The Aleutian Islands are a conspicuous island arc; another arc runs from Kamchatka through the Kurile Islands and down through Japan; yet another consists of the islands of Sumatra, Java, Sumba, and Timor.

Continental Volcanic Arcs.   Where a subduction zone occurs beneath an edge of a continent, the volcanoes that form are located on continental crust. Instead of an oceanic island arc, therefore, the result is a chain of stratovolcanoes on land—a *continental volcanic arc.* An example of a volcanic mountain chain formed in this manner is the Andes, a system of andesitic volcanic mountain ranges that border the western margin of South America. Another example is the Cascade Range in the northwestern United States (Fig. 15.26). This is a range of young, predominantly andesitic volcanoes, extending from Lassen Peak, California (at the south end) to Mount Baker, Washington (more than 900 km farther north) and also includes several volcanic peaks in southern British Columbia. The largest volcanoes (all active during the past few million years, and some such as Mount St. Helens, still active today) were erupted onto a platform of older, folded, and deeply eroded rocks. The building of the mountains is controlled by wet partial melting of subducted oceanic crust (chapter 3). The piece of oceanic crust that is being subducted, and is responsible for forming the Cascade Range, is the Juan de Fuca Plate.

Fault-Block Mountain Ranges

In many parts of the world isolated mountain ranges stand abruptly above surrounding plains. Study reveals that these ranges are frequently separated from the intervening lowland areas by normal faults of great displacement. The ranges seem to be giant pieces of crust punched upward from below. These are *fault-block mountains*. Rock within the mountains commonly contains evidence that either former fold-and-thrust mountains, or former volcanic mountains once occupied the same sites, but that erosion had worn them down before the fault blocks formed. An example of a mountain range of this kind is the Sierra Nevada of California. It was originally a continental volcanic arc that formed above a subduction zone. By 20 million years ago the original range was eroded down to a surface of low relief that was stable for millions of years. Then, in the Cenozoic Era, the region again became active and the modern range was lifted up as a series of

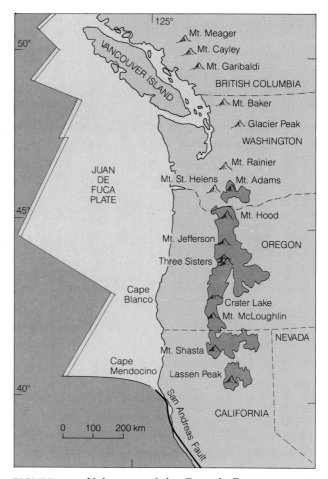

FIGURE 15.26 Volcanoes of the Cascade Range, a continental volcanic arc. Each volcano has been active during the last 2 million years. (*Source:* After Tabor and Crowder, 1969.)

great fault blocks. How and why are incompletely understood. One suggestion is that the eroded remnants of the ancient mountain range still had a root of low-density rocks beneath them, so that an upward-pressing force existed because of isostasy (chapter 14). If the isostatic force were not great enough to overcome the friction that opposed movement on faults, nothing would happen. However, if movement of the lithosphere caused a slight warping or twisting of the crust, the friction might be overcome and the fault blocks could then rise to form the mountains.

Basin and Range Province.   The Basin and Range Province is one of the most extensive fault-block mountain systems in the world (Fig. 15.11). It contains a spectacular development of fault-bounded mountain ranges (horsts), separated by sediment filled valleys (grabens), in Idaho, Oregon, Nevada,

FIGURE 15.27 Basin and Range topography near the western margin of the province. On the far left is the Panamint Range in California. On the far right is the Amargosa Range on the California–Nevada border. Death Valley is the light-colored area in the center. This is a false-color satellite image.

Utah, Arizona, New Mexico, California and northern Mexico (Fig. 15.27).

The geology of the Basin and Range Province is quite complex and has led to long-term controversies over the origin of the ranges. More than one process has apparently been involved because both thrust faults and normal faults are present, but overall, it appears that the Basin and Range has undergone extension through tensional, or pull-apart, forces. Just how much extension has oc-curred remains in question, but some authorities suggest the width of the province has increased by 15 percent or more. The cause of the extension is also a matter of open debate. Many authorities point to the volcanism that accompanied the faulting as evidence that rising magma may have heated and distended the crust, thereby forming the kind of structures shown in the simplified diagram in Figure 15.28.

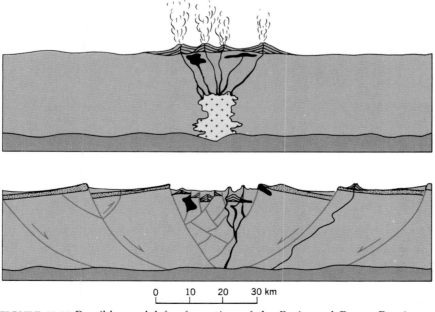

FIGURE 15.28 Possible model for formation of the Basin and Range Province. Warping of the crust and extrusion of large quantities of lava and volcanic ash were followed by collapse along a series of steeply inclined normal faults that become flatter at depth. (*Source:* After Mackin, 1968.)

The Sierra Nevada marks the western boundary of the Basin and Range Province. The geological history of the Sierra Nevada is probably closely linked to that of the Basin and Range Province. One suggestion how this may have come about is shown in Figure 15.29. The province is underlain by Paleozoic sedimentary strata deposited on Precambrian rocks. Following a period of folding, during the late Paleozoic, the region was deeply eroded during Mesozoic and Cenozoic times when it supplied sediment to form strata that later developed into the Coast Ranges of California and the Rocky Mountains. Volcanism commenced in the region more than 60 million years ago, but starting about 25 million years ago extensive volcanic activity commenced and the region broke up into a series of blocks 30 to 40 km in width and as much as 150 km long, bounded by steeply inclined faults.

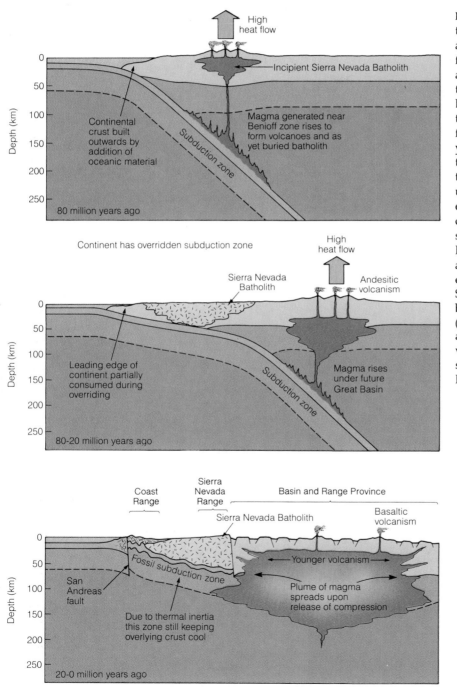

FIGURE 15.29 One interpretation of the way the Sierra Nevada and the adjacent Basin and Range Province formed. Prior to 80 million years ago an active chain of volcanoes lay along the western edge of the continent. Beneath the volcanic mountain range the Sierra Nevada Batholith was formed. Between 80 and 30 million years ago, North America overrode the subduction zone and the portion of the subducted crust undergoing partial melting moved easterly. The volcanic mountain chain was eroded and volcanism started in the Basin and Range Province. Between 30 million years and the present, North America overrode part of the mid-ocean ridge. Subduction ceased and the plate boundary became a transform fault (the San Andreas Fault). The Basin and Range Province is disrupted by volcanism and block faulting as shown in Figure 15.27. (*Source:* After Henyey and Lee, 1976.)

Essay

MEASURING HIDDEN FAULTS

Many active faults are hidden from view but nevertheless reveal their presence through earthquakes. Some are deep in the crust and do not reach the surface because they are covered by ductile layers of younger rocks, others are in the oceanic crust and are covered by seawater. For studies such as the prediction of earthquakes, and determination of how a given block of crust is stressed, it is necessary to know whether such hidden faults have motions that are normal, reverse, or strike–slip.

The information needed to solve the problem has two parts. First, it is necessary to know the orientation of the fault plane. Second, the relative directions of motion of the fault blocks must be determined.

The orientation of an active fault can be determined by plotting the positions of the foci of the earthquakes it generates. It is more difficult to determine the directions of movement of the two fault blocks. But it can be done, and the information needed to do so is contained in the arrival records of seismic body waves.

Consider the *P*-wave record. If the first arrival is a compressive pulse, the release of elastic energy, and the fault motion, must be *toward* the seismograph (Fig. B15.1a). If it is an expansion, the fault motion must be away from the seismograph. In Figure B15.1b the effect of an earthquake caused by movement on a strike–slip fault is shown. It is apparent that the first *P*-wave motion observed depends on the location of the seismograph. The fault movement can be determined by plotting first motions from several seismographs.

The radiation pattern of body waves is in three dimensions. It is not possible to distinguish up–down movement from back–forth movement using *P* waves alone. A 90° ambiguity is inherent in the *P*-wave radiation pattern. Because two possibilities exist—up–down versus back–forth—two independent measurements are needed. But *S*-wave oscillations also carry the signature of the direction of the first motion. They provide the needed independent data and can thus be used to resolve the ambiguity.

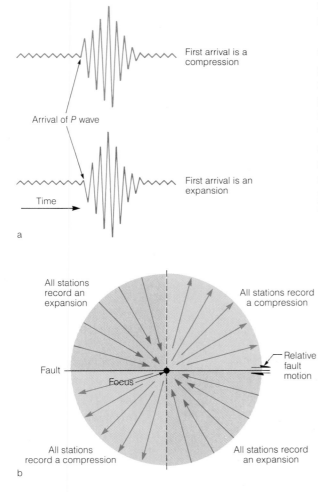

First arrival is a compression

Arrival of *P* wave

First arrival is an expansion

Time

a

All stations record an expansion

All stations record a compression

Fault

Focus

Relative fault motion

All stations record a compression

All stations record an expansion

b

FIGURE B15.1 Initial motion of seismic body waves used to determine direction of movement on a fault. (a) Initial motion of a *P* wave detected by a seismograph is either a push away from the focus (that is, arrival of a compression), or a pull toward the focus (arrival of an expansion). (b) Plotting the first motions detected at a number of seismograph stations allows the direction of movement on a fault to be uniquely determined. The example shown is for right-lateral motion on a strike–slip fault.

SUMMARY

1. Solids can be deformed in three different ways: by elastic deformation (no permanent change), by ductile deformation (folding), and by brittle deformation (fracturing).

2. High confining pressure and high temperatures enhance ductile properties. Low temperatures and low confining pressure enhance elastic properties and failure by fracture when the elastic strength is exceeded.

3. The rate at which a solid is deformed (strained) also controls style of deformation. High strain rates lead to fractures; low strain rates cause folding.

4. Fractures along which slippage occurs are called faults. Normal faults are caused by tensional forces, reverse and thrust faults by compressional forces. Strike–slip faults are vertical fractures which have horizontal motion.

5. Two major structural units can be discerned in the continental crust. Cratons are ancient portions of the crust that are tectonically stable. Surrounding the cratons are orogenic belts of highly deformed rock, marking the site of mountain ranges.

6. Mountain ranges can be divided into three kinds based on their principal geological feature: fold-and-thrust mountain ranges, volcanic mountain ranges, and fault-block mountain ranges.

7. Fold-and-thrust mountain ranges are formed along subduction edges of plates by the compression and thickening of sediments accumulated along continental margins.

8. Volcanic mountain ranges are formed by magmatic activity along spreading centers and above subduction zones.

9. Fault-block mountain ranges form by isostatic reactivation of ancient, deeply eroded fold-and-thrust mountain ranges and by tensional stretching of the crust.

IMPORTANT WORDS AND TERMS TO REMEMBER

anticline (p. 357)
axial plane (of a fold) (p. 357)
axis (of a fold) (p. 357)

boudin (p. 347)
boudinage (p. 347)
brittle fracture (p. 345)

confining pressure (p. 346)
continental shield (p. 360)
continental volcanic arc (p. 366)
craton (p. 357)

décollement (p. 362)
detachment surface (p. 362)
ductile deformation (p. 345)

elastic deformation (p. 345)
eugeocline (p. 361)

fault (p. 347)
fault-block mountain range (p. 366)
fault breccia (p. 354)

fold-and-thrust mountain range (p. 361)
footwall (p. 349)

geosyncline (p. 361)
graben (p. 350)

half-graben (p. 350)
hanging wall (p. 349)
horst (p. 350)

isoclinal fold (p. 358)

limb (of a fold) (p. 357)

miogeocline (p. 361)
monocline (p. 357)
mountain chain (p. 361)
mountain range (p. 361)
mountain system (p. 361)
mylonite (p. 354)

normal fault (p. 351)

oceanic island arc (p. 365)
orogen (p. 357)
orogenic belt (p. 357)
overturned fold (p. 358)

plunge (p. 357)
plunging fold (p. 357)

recumbent fold 358
*relative displacement
(on a fault)* (p. 349)
reverse fault (p. 351)
rift (p. 350)

slickensides (p. 354)
stable platform (p. 360)
strain (p. 344)
strain rate (p. 346)
stress (p. 344)
strike–slip fault (p. 351)
structural geology (p. 348)
symmetrical fold (p. 358)

QUESTIONS FOR REVIEW

1. What is the relationship between stress and strain in an elastic solid?

2. Describe two ways by which a solid can be irreversibly deformed when the elastic limit is exceeded.

3. What properties determine whether a solid fails by brittle fracture or is deformed by ductile deformation?

4. Name three rock types that readily deform by ductile deformation.

5. Name three rock types that tend to fail by brittle fracture.

6. Name three prominent topographic features on the Earth that are comprised of a system of grabens and horsts.

7. Describe the way a transform fault works. Why are they called transform faults?

8. Draw a sketch of an anticline and mark the axis, axial plane, and limbs.

9. What are cratons and how do they differ from orogens?

10. What is a miogeocline and how is it related to a eugeocline?

11. Name three fold-and-thrust mountain ranges.

12. How do fault-block mountain ranges originate?

13. What is the tectonic environment of an oceanic island arc? Name two examples of modern oceanic island arcs.

14. What is the origin of the Cascade Range; the Sierra Nevada; the Appalachians?

15. What evidence indicates that plate tectonics has been operating for at least the last 2 billion years of Earth history?

C H A P T E R 16

Dynamics of the Crust

The Alpine Fault in Awatere Valley, New Zealand. This active strike-slip fault is slicing the South Island of New Zealand in two pieces.

A NEW PARADIGM

The scientific revolution that was initiated by the hypothesis of plate tectonics started in the late 1960s. The revolution shows no signs of abating. Every corner of geology has to be looked at through new eyes, every deduction reexamined, every conclusion rethought, and every question asked again and again. Throughout this book we have tried to integrate the abundant fruits of the revolution; they appear in almost every chapter. Plate tectonics is no longer just a hypothesis; scientists working for NASA have now measured the velocities of plate motions using satellites and lasers. Geological deductions were proved correct—plates really are moving with measured velocities between 1.5 and 7 cm/yr.

As with all successful concepts in science, plate tectonics has provided simplified explanations for many seemingly complex problems. It has provided testable answers for long-standing problems (such as the origin of ocean basins), and it has shown that seemingly unrelated features (such as midocean ridges and grabens on continents) are actually closely related.

SEARCH FOR A SOLUTION

One cannot but wonder why continents have their peculiar shapes, why ocean basins are where they are, why mountain ranges, earthquake belts, volcanoes, and many other major features occur where they do. Such wondering prompted many scientists to think that there might be a single, underlying cause for the whole array of the Earth's major features.

During the nineteenth century people favored the idea that the Earth was originally hot—a molten mass—and that it had been gradually cooling, contracting, and that the crust had been gradually compressed. They pointed to fold-mountain ranges and seismic belts as the places where most of the contraction now occurs and has occurred in the past. Contraction did explain some features, but it did not help with questions about the shapes and the distribution of continents. Nor did it help explain the great rift valleys and other features where the crust was clearly in a state of tension rather than compression.

When it was realized that the Earth's interior is kept hot by the decay of natural radioactive isotopes, some scientists suggested that the Earth might actually be heating up. A much smaller Earth, they suggested, could once have been covered largely by continental crust. Heating would cause the Earth to expand and the continental crust would then crack and break into fragments. As expansion continued, the cracks would grow into ocean basins, and through the cracks basaltic magma would rise up from the mantle to build new oceanic crust. Although the theory of an expanding Earth did not easily account for fold-mountain ranges, it did offer a plausible explanation for the approximately parallel coastlines of adjacent continents, such as Africa and South America. Nevertheless, the expansion theory has other fundamental problems. No evidence exists (such as an increase in the rate of production of magma) to suggest that the Earth is heating. To get around the flaws in both the expansion theory and the contraction theory, the effects of other forces on the crust were examined. The centrifugal force caused by the Earth's rotation and the gravitational pull of the Moon were both suggested as forces that could possibly deform continents, just as the Earth's rotation and the Moon's gravity influence the motions of ocean currents. Calculations showed, however, that both forces are far too weak to cause large-scale deformation, so the ideas were abandoned. By the middle of the twentieth century all reasonable suggestions concerning the shapes and positions of continents seems to have been exhausted. The time was ripe for a totally new approach.

Wegener and Continental Drift

One key suggestion was made early in the twentieth century, soon after the contraction theory had collapsed. Alfred Wegener, a German meteorologist, suggested in 1912 that continents drift slowly across the surface of the Earth, sometimes breaking into pieces and sometimes colliding with each other. Lateral motion could produce both compressive and tensional forces at the same time. The front edge of the moving continent would be in a state of compression, the interior and tail ends would be in a state of tension.

Pangaea

Wegener's theory of **continental drift** originated when he attempted to explain the striking parallelism of the edges of the shorelines on the two sides of the Atlantic Ocean, especially the shores of Africa and South America. Other bits of favorable evidence were quickly found. These bits of evidence supported the hypothesis that the world's landmasses had once been joined together in a sin-

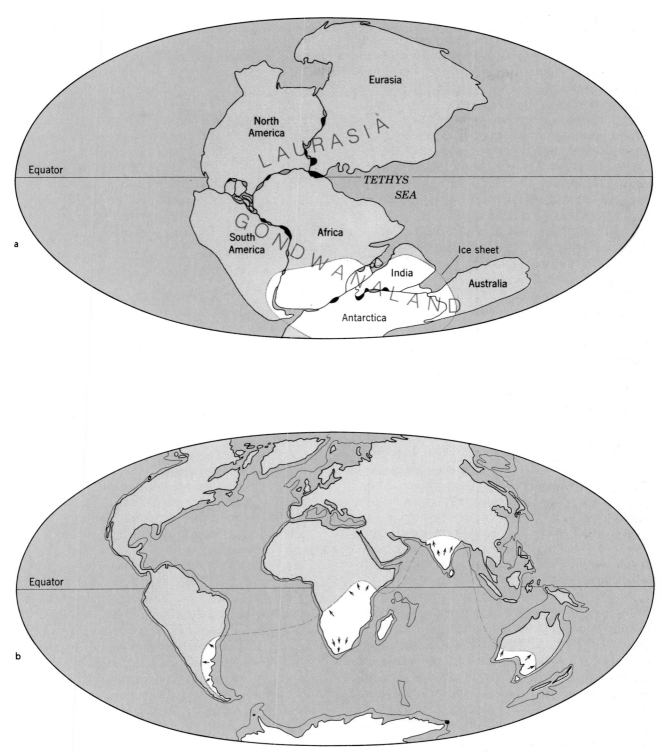

FIGURE 16.1 The continents attained their present shapes when Pangaea broke apart 200 million years ago. (a) Shape of Pangaea, determined by fitting together pieces of continental crust along a contour line 2000 m below sea level. It is the line along which continental crust meets oceanic crust. In a few places some overlap (black) occurs: in others, small gaps (red) are found. These are places where existing maps are poor or where later events have modified the shapes of continental margins. The white area is the region affected by continental glaciation 300 million years ago. (b) Present continents and the 2000-m contour below sea level. The white area is where evidence of the old ice sheets exists. Arrows show directions of movement of the former ice. The dashed line joining the glaciated regions indicates how large the ice sheet would have to be if the continents were in their present positions at the time of glaciation.

gle great supercontinent which was dubbed *Pangaea* (pronounced Pan-jee-ah, meaning "all lands") (Fig. 16.1). The northern half of Pangaea is called *Laurasia*, the southern half is *Gondwanaland*. According to the hypothesis, Pangaea was somehow disrupted during the Triassic, and its fragments (the continents of today) slowly drifted to their present positions. Proponents of the theory likened the process to the breaking up of a sheet of ice that floats in a pond. The broken pieces, they argued, should all fit back together again, like pieces of a giant jigsaw puzzle. Figure 16.1a shows that a jigsaw reconstruction indeed works well.

One impressive line of evidence presented by Wegener is that a continental ice sheet covered parts of South America, southern Africa, India, and southern Australia 300 million years ago (Fig. 16.1b). The ice sheet resembled the one that covers Antarctica today; evidence of its existence is so well preserved that thousands of glacial striations (chapter 12) reveal the directions in which the ice flowed. However, if 300 million years ago continents were in the positions they occupy today, the ice sheet would have had to cover all the southern oceans, and in places would even have had to cross the equator! A glacier of such huge size could only mean that the world climate was exceedingly cold. Yet if the climate had been cold, why had no evidence of glaciation at that time been found in the Northern Hemisphere? The dilemma would be explained neatly by continental drift. Three hundred million years ago the regions covered by ice lay in high, cold latitudes. Indeed those regions were adjacent to the South Pole (Fig. 16.1a). At that time, therefore, the Earth's climates need not have been greatly different from those of today.

Despite the impressive evidence that favored the drifting of continents, many scientists remained unconvinced, largely because the fluidlike properties of the asthenosphere had not been discovered. Indeed, the concept of a lithosphere and an asthenosphere having different physical properties without parallel changes in composition was hardly suspected. Wegener struggled with the problem of how solid, brittle rocks could move through or over other solid, brittle rocks. The fact that continental rocks differed from oceanic rocks was known, and of course the fact that the continental land surface stood high above the seafloor was obvious. Putting these facts together, Wegener suggested that, despite problems of strength, continental crust must somehow slide over oceanic crust. Opponents, many of whom were geophysicists, quickly argued that because of the great frictional resistance to such sliding, rigid continental crust simply could not

slide over rigid oceanic crust without both crusts disintegrating. The process is like trying to slide two sheets of coarse sandpaper past each other.

Wegener died in 1930, and although debate continued, its pace slowed down. True, the geological evidence of glaciers suggested that continents might have moved; however, geophysicists, studying the Earth's physical properties, became convinced that movement could not have occurred, and elaborate explanations were erected to explain the geological evidence. The situation became a stalemate. More and more scientists discarded the hypothesis that continents had drifted because an acceptable mechanism could not be found to explain the movement.

Apparent Polar Wandering

A turning point in the debate came during the 1950s. From the mid-1950s to the mid-1960s geophysicists made a number of remarkable discoveries. The first of the key discoveries arose through studies of paleomagnetism. As previously discussed, when certain igneous and sedimentary rocks form, they become weakly magnetized (chapter 6), and preserve a "fossil" record of the Earth's magnetic field at the *time* and *place* of formation. Three essential bits of information are contained in the fossil magnetic record. The first is the polarity—whether the magnetic field was normal or reversed at the time of formation. The second piece of information is the direction of the magnetic pole at the time the rock formed. Just as a free-swinging magnet today will point toward the magnetic poles, so does paleomagnetism record the direction of the magnetic poles from the point of rock formation. The third piece of information, and the one that provides the needed information to say how far away the magnetic pole lay, is the magnetic inclination, which is the angle with the horizontal assumed by a freely swinging bar magnet (Fig. 16.2). Note in Figure 16.2 that the magnetic inclination varies regularly with latitude, from zero at the magnetic equator to 90° at the magnetic pole. The paleomagnetic inclination is therefore a record of the place between the pole and the equator (that is, the magnetic latitude) where the rock was formed. Once the magnetic latitude of a rock and the direction of the magnetic poles are known, the position of the magnetic pole at the time the rock is formed is determined.

Geophysicists who were studying paleomagnetic pole positions during the 1950s found evidence suggesting that the poles wandered all over the globe. They referred to the strange plots of paleo-pole positions as paths of *apparent polar wandering.* The geophysicists were puzzled by evidence

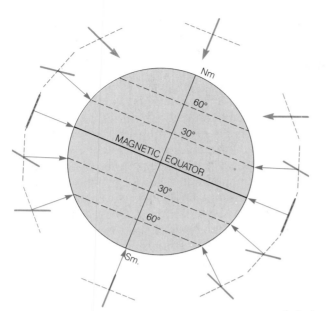

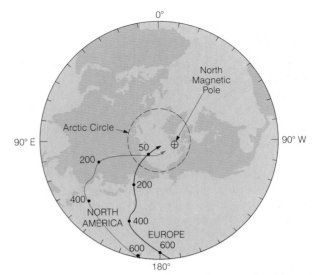

FIGURE 16.2 Change of magnetic inclination with latitude. The solid red line is the inclination taken up by a free-swinging magnet. The dashed blue line indicates a horizontal surface at each point. Averaging a number of estimates of magnetic paleolatitudes will give a good estimate for the true paleolatitude of a rock unit.

FIGURE 16.3 Curves tracing the apparent path followed by the north magnetic pole through the past 600 million years. Numbers are millions of years before the present. The curve determined from paleomagnetic measurements in North America (red curve) differs from that determined by measurements made in Europe (black curve). Wide-ranging movement of the pole is unlikely; therefore, it is concluded that it is not the pole but the continents that have moved. (*Source:* After Northrop and Meyerhoff, 1963.)

for polar wandering. The Earth's magnetic field, they knew, was caused by motions in the liquid outer core, and the motions in turn were caused by the Earth's rotation.

Although the magnetic poles might wobble a little, they should always remain close to the poles of rotation. Determination of magnetic latitude should therefore be a good approximation to the geographic latitude. When it was discovered that the path of apparent polar wandering measured in North America differed from that in Europe (Fig. 16.3), geophysicists were even more puzzled. Somewhat reluctantly, they concluded that because it is unlikely that the magnetic poles have moved, it is more likely that the continents and the magnetized rocks had moved instead. In this way the hypothesis of continental drift was revived, but a mechanism to explain how the movement occurred was still lacking.

Seafloor Spreading

Help came from an unexpected quarter. All the early debate about continental drift, and even the data about apparent polar wandering, had centered on evidence drawn from the continental crust.

In 1962 Harry Hess of Princeton University hypothesized that the topography of the seafloor could be explained if the seafloor moves sideways, away

from the oceanic ridges. His hypothesis came to be called the theory of *seafloor spreading* and was soon proved correct. Once again it was geophysicists who used paleomagnetism to provide the proof.

Hess postulated that magma rose from the interior of the Earth and formed new oceanic crust along the midocean ridge. He could not explain what made the crust move away from the ridge, but he nevertheless proposed that it did and that as consequence the oceanic crust became older the further it moved. Two tests for the Hess theory were soon proposed. One test was suggested by J. Tuzo Wilson, a Canadian geophysicist. Wilson argued that long-lived magma sources seemed to be present deep in the mantle and as a consequence lines of volcanic islands should form as the seafloor moves over the magma source. In such a line of volcanic islands there should be a steady progression in the age of volcanism on the islands. As discussed in chapter 17 the Hawaiian chain of islands provides a striking confirmation of Wilson's suggestion. A second and more powerful test of the Hess theory was proposed by three geophysicists: Frederick Vine (who was a student at the time), Drummond Matthews (Vine's mentor), and Lawrence Morley (a Canadian scientist who made an independent discovery). The Vine–

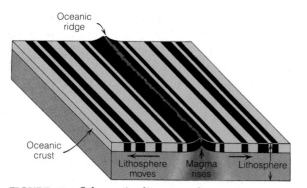

FIGURE 16.4 Schematic diagram of oceanic crust. Lava extruded along an oceanic ridge forms new oceanic crust. As lava cools, it becomes magnetized with the polarity of the Earth's field. Successive strips of oceanic crust have alternate normal polarity (black) and reversed polarity (yellowish-green).

Matthews–Morley suggestion concerned thermoremanent magnetism of the oceanic crust.

When lava is extruded at the oceanic ridge the rock it forms becomes magnetized and acquires the magnetic polarity that existed at the time it cooled through the Curie point. New lava continually makes new oceanic crust, and the crust is continually moving away from the oceanic ridge. The oceanic crust should therefore contain a continuous record of the Earth's magnetic polarity. Magnetic data gathered as a result of antisubmarine defense re-

search proved this point. The crust is, in effect, two very slowly moving, symmetrical magnetic tape recorders, one each side of the midocean ridge, in which successive strips of oceanic crust are magnetized with normal and reversed polarity (Fig. 16.4). Seafloor magnetism can be measured with instruments carried in ships or airplanes; an example of the results is given in Figure 16.5. It was a simple matter to match the sort of pattern observed in Figure 16.5b with the record of magnetic polarity, such as that shown in Figure 6.16. The distinctive magnetic striping allowed the age of any place on the seafloor to be determined. Because the ages of magnetic polarity reversals had been so carefully determined, magnetic striping also provided a means to estimate the speed with which the seafloor moved. In places it was found to be remarkably fast, reaching values as high as 10 cm/yr.

PLATE TECTONICS: A POWERFUL THEORY

Proof that the seafloor moves was the spur needed for an all-embracing theory covering both continental and oceanic crust to emerge. It was quickly forthcoming. Although a lot of the evidence to support the theory came from geophysicists, all branches of geology combined to provide the needed evi-

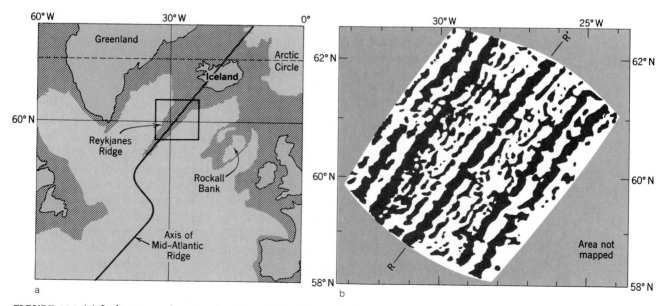

FIGURE 16.5 (a) Index map showing location of Reykjanes Ridge, a portion of the Mid-Atlantic Ridge southwest of Iceland. (b) Map of the magnetic striping of rock on the sea floor. R–R′ is the center line of Reykjanes Ridge. Strips of rock with normal polarization (black) alternate with reversely polarized rock (white). (*Source:* After Heirtzler et al., 1966.)

dence. The essential points in formulating a theory of plate tectonics were, first, that the low-velocity zone (soon identified with the asthenosphere) is exceedingly weak and has viscous fluidlike properties. The second point was that the rigid lithosphere is strong enough to form coherent slabs which can slide sideways over the weak, underlying asthenosphere. These two points answered the objection that the geophysicists had directed at Wegener—movement must occur without massive resistance from friction. The lithosphere is much thicker than the crust, however, so one consequence of the theory was that as the lithosphere moved, the crust was simply rafted along as a passenger. Continents move, to be sure, but they only do so as portions of larger plates, not as discrete entities.

The third essential point answered by the theory of plate tectonics concerned the destruction of old oceanic crust. If, as the theory of seafloor spreading required, new oceanic crust is continually created along the midocean ridges, either the Earth must be expanding and the oceans must be getting larger, or an equal amount of old crust must necessarily be destroyed in order to maintain an Earth of constant size. The clue was provided by the previously unexplained Benioff zones (chapter 14). These

slanting zones of deep earthquake foci are the places where old, cold lithosphere is sinking back into the asthenosphere. In this way, cool and still brittle lithosphere can sink to great depths. In a simplified form, the basic elements of plate tectonics are shown in Figure 16.6.

Structure of a Plate

The surface of the Earth is covered by six large and many small plates of lithosphere, each about 100 km thick, sliding over the fluidlike asthenosphere (Fig. 1.10). The plates are rigid, or nearly so, moving as single coherent units; that is, the plates do not crumple and fold like wet paper, but act more like semirigid sheets of plywood floating on water. The plates may flex slightly, causing gentle up- or down-warping of the crust, but the only places where intense deformation occurs is at the edges along which plates impinge on each other. Such plate margins are *active zones;* plate interiors are *stable regions.*

Plates have three kinds of margins: *divergent margins* along which two plates move apart from each other; *convergent margins* along which two plates move toward each other; and *transform fault margins* along which two plates simply slide past

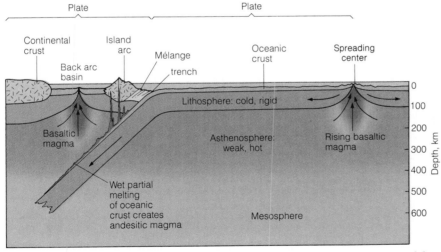

FIGURE 16.6 Cross section through a tectonic plate. Lithosphere is capped by oceanic crust formed by basaltic magma rising from the asthenosphere. Moving laterally, the lithosphere accumulates a thin layer of marine sediment and eventually starts sinking into the asthenosphere. At the point of sinking, an oceanic trench is formed and sediment deposited in the trench, plus sediment from the moving plate, is compressed and deformed to create a mélange. The sinking oceanic crust eventually reaches the temperature where wet partial melting commences and forms andesitic magma that rises to form an island arc of stratovolcanoes on an adjacent plate. Behind the island arc, tensional forces lead to the development of a back-arc basin.

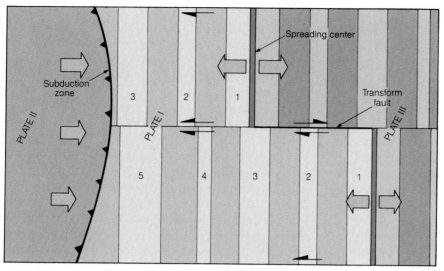

FIGURE 16.7 Simplified structure of a plate. The spreading center is offset by the transform fault that divides the plate. Bars are magnetic time lines of oceanic crust. Broad arrows show the direction of plate motion.

TABLE 16.1 *Kinds of Plate Margins and Characteristic Features*

Crust on Each Plate	Feature	Kind of Margin		
		Divergent	Convergent	Transform Fault
Oceanic–Oceanic	Topography	Oceanic ridge with central rift valley	Seafloor trench	Ridges and valleys created by oceanic crust
	Earthquake	All foci less than 100 km deep	Foci from 0 to 700 km deep	Foci as deep as 100 km
	Volcanism	Basaltic pillow lavas	Andesitic volcanoes in an arc of islands parallel to trench	Volcanism rare; basaltic along "leaky" faults
	Example	Mid-Atlantic Ridge	Tonga-Kermadec Trench; Aleutian Trench	Kane Fracture
Oceanic–Continental	Topography	—	Seafloor trench	—
	Earthquake	—	Foci from 0 to 700 km deep	—
	Volcanism	—	Andesitic volcanoes in mountain range parallel to trench	—
	Example	(No examples)	West coast of South America	(No examples)
Continental–Continental	Topography	Rift valley	Young mountain range with folded crust	Fault zone that offsets surface features
	Earthquake	All foci less than 100 km deep	Foci as deep as 300 km over a broad region	Foci as deep as 100 km throughout a broad region
	Volcanism	Basaltic and rhyolitic volcanoes	No volcanism. Intense metamorphism and intrusion of granitic plutons	No volcanism
	Example	African Rift Valley	Himalaya, Alps	San Andreas Fault

each other (Fig. 16.7). Each margin creates distinctive topography in its vicinity and is associated with a distinctive kind of earthquake activity and volcanism. The features are summarized in Table 16.1 and briefly discussed in the following.

Divergent Margins

As defined in chapter 1, a ***divergent margin*** or ***spreading center*** marks the new growing edges of adjacent plates. It is a line along which two plates move apart from each other, and along which new lithosphere is created. Spreading centers are places where crust is being stretched by tensional forces. The kinds of faults associated with tensional forces are normal faults.

Earthquakes along spreading centers tend to have low magnitudes and shallow foci. This is so because the ductile asthenosphere comes close to the surface beneath a spreading edge. The kind of volcanic activity along a spreading center is almost always basaltic.

Magnetic Records and Plate Velocities

The magnetic polarity record implanted in oceanic crust at a spreading center provides useful clues. Working outward from an active center, the crust becomes progressively older. The first magnetic reversal recorded away from the crest of a midocean ridge is that which occurred 730,000 years ago (Fig. 6.16). Subsequent reversals are located in succession away from the ridge. The oldest reversals so far found in oceanic crust date back to the middle Jurassic, about 165 million years ago. When the

positions and ages of the magnetic time lines have been located (as shown in Figure 16.8), it is apparent that plate velocities can be calculated.

Relative versus Absolute Velocities. From the symmetrical spacing of magnetic time lines on both sides of a midocean ridge it appears that both plates move away from a spreading center at equal rates. Appearances can be deceiving, however. The same pattern of magnetic time lines in Figure 16.8 would be observed if the African Plate were stationary and both the Mid-Atlantic Ridge and the North American Plate were moving westward. Later in this chapter evidence will be presented to substantiate the suggestion that midocean ridges do indeed move. All that can be deduced from magnetic time lines is the ***relative velocity*** of one plate to that of another. An answer to the question of ***absolute velocities*** requires more information.

Variations Among Plate Velocities. The relative motions across some midocean ridges are much greater than they are across others (Fig. 16.9). The reasons for high relative velocities are not known with certainty, but they appear to be related to the amount of continental crust sitting on a plate. Plates that do not carry a large load of continental crust tend to have high relative velocities. This is the case for the Pacific and Nazca plates. Plates with large loads of continental crust, such as the African, North American, and Eurasian plates, have low relative velocities.

The second reason that plate velocities vary from place to place has to do with the geometry of motion on a sphere. One might think, intuitively, that

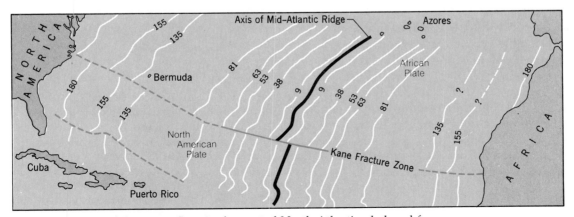

FIGURE 16.8 Age of the ocean floor in the central North Atlantic, deduced from magnetic striping. Numbers give ages in millions of years before the present. The Kane Fracture Zone, observed near the center of the oceanic ridge, continues across the Atlantic and causes consistent offsetting of the age contours. (*Source:* After Pitman and Talwani, 1972.)

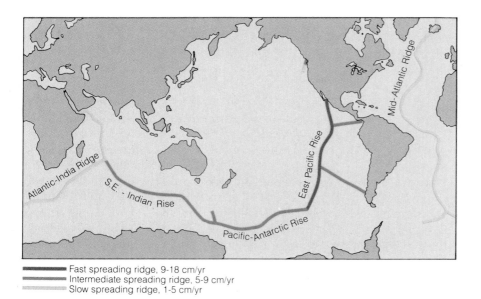

Fast spreading ridge, 9-18 cm/yr
Intermediate spreading ridge, 5-9 cm/yr
Slow spreading ridge, 1-5 cm/yr

FIGURE 16.9 Spreading rates of principal midoceanic ridges. Fast spreading rates mean plates move away from each other between 9 and 18 cm a year. Intermediate rates are 5 to 9 cm a year; slow rates are 1 to 5 cm a year.

all points on a plate move with the same velocity, but that is incorrect. Our intuitions would only be correct if plates of lithosphere were flat and moved over a flat asthenosphere (like plywood floating on water); then, all points on the plate *would* move with the same velocity. However, plates of lithosphere are pieces of a shell on a spherical Earth, so they are curved, not flat. In the geometry of a sphere, any movement on the surface is a rotation about an axis of the sphere. A consequence of rotation and, therefore, of a curved plate moving over the surface of a sphere is that different parts of a plate move with different velocities.

To picture how points on a plate move with different velocities, imagine a plate so large that it forms a hemispherical cap covering half the Earth (Fig. 16.10). The cap moves independently of the Earth's rotation and rotates instead about an axis of its own, colloquially called a *spreading axis.* In the figure, point P, where the spreading axis reaches the surface, is a *spreading pole.*

No plate is large enough at present to cover half the Earth, nor does any plate rotate around a spreading pole in the center of the plate. But the principle is the same for a small plate as it is for a hemispherical cap. Consider Plate A in Figure 16.10. The motion of Plate A is from east to west around the spreading axis. Point A″, close to the spreading pole, must move more slowly than point A′, more distant from the pole.

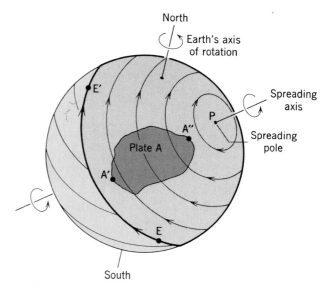

FIGURE 16.10 Movement of a curved plate on a sphere. The movement of each plate of lithosphere on the Earth's surface can be described as a rotation about a spreading axis. Point P has no velocity of movement because it is the fixed point around which the hemispherical cap rotates. Point E, at the edge of the cap, has a high velocity because it must move completely around the Earth, along path E–E′, during a single revolution of the cap. Any point on the cap between points P and E has an intermediate velocity that is slower if the point is closer to P, faster if it is closer to E. (*Source:* Adapted from Wyllie, 1976.)

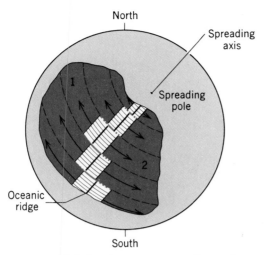

North

Spreading axis

Spreading pole

1

2

Oceanic ridge

South

FIGURE 16.11 Relation between spreading axis, oceanic ridge, and transform faults in two adjacent plates. Plates 1 and 2 have a common spreading center offset by transform faults (red). Each segment of the oceanic ridge lies on a line of longitude that passes through the spreading pole. Each transform fault lies on a line of latitude with respect to the spreading pole. The width of new oceanic crust (striped shading) increases away from the spreading pole.

The motion of each of the Earth's plates can be described in terms of rotation around a spreading axis, and the velocity of each point on the plate depends on its distance from the spreading pole. One consequence of differing velocities of motion is this: The width of new oceanic crust that borders a divergent margin between two plates increases with distance from the spreading pole (Fig. 16.11). A further consequence is that each segment of an oceanic ridge lies on a line that passes through the spreading pole. Such a line is analogous to a line of longitude—as if the spreading pole were the Earth's pole of rotation. Each transform fault that offsets the oceanic ridge lies on a line analogous to a line of latitude around the spreading pole. The relation between transform faults and spreading poles can be used to determine the position of the spreading pole of each plate. The same property can be used to determine the positions of former spreading poles from the positions of old transform faults and, therefore, to determine whether a plate has, at some time, changed its direction of motion.

Topography of the Seafloor

The topography of the seafloor is controlled by the growth and movement of plates. Two prominent features in particular are related to spreading cen-

ters. The first feature is the midocean ridge. The shape of the ridge is strongly influenced by the rate of spreading.

Fast spreading rates, 9 to 20 cm/yr, mean that new oceanic crust is created very rapidly. This in turn means that magma must rise rapidly and continually from below and that large magma chambers must lie at shallow depths below the center of the ridge. As a result, a fast-spreading center like the East Pacific Rise is thermally inflated and stands high above the seafloor. By contrast, a slow-spreading center like the Mid-Atlantic Ridge is cooler and less inflated. The overall ridge still stands high above the deep ocean floor, but the central graben is wider and more pronounced.

The second prominent feature is the ocean floor itself. A large fraction of the heat that escapes from the Earth's interior does so along spreading centers. As a result, not only the midocean ridges, but also the adjacent seafloor, are high points because the lithosphere beneath them is thermally expanded. As lithosphere moves away from a midocean ridge, it cools and contracts. As contraction occurs, the depth of the seafloor increases. A constant depth is reached after about 100 million years, by which time oceanic lithosphere has cooled and reached thermal equilibrium (Fig. 16.12). To a first approximation, therefore, the depth of the ocean floor below sea level provides an estimate of the age of the oceanic crust.

Convergent Margins

Subduction zones were defined in chapter 1 as the edges along which plates of lithosphere turn down into the mantle. They are **convergent margins** where two plates move toward each other.

Island Arcs

The Earth's surface area is not increasing. This means that the production of new lithosphere at spreading centers must be balanced by the destruction of old lithosphere at subduction zones. Such destruction occurs when one plate sinks downward beneath the other at an angle of 20 to 60° to the horizontal (Fig. 16.6). As the plate descends it is heated up and eventually reaches a temperature at which wet partial melting commences. This process forms andesitic magma. Rising to the surface, the magma forms a chain of stratovolcanoes. As discussed in chapter 15, if the stratovolcanoes form on oceanic crust, the chain is called an **oceanic island arc**, if it is built on continental crust it is called a **continental**

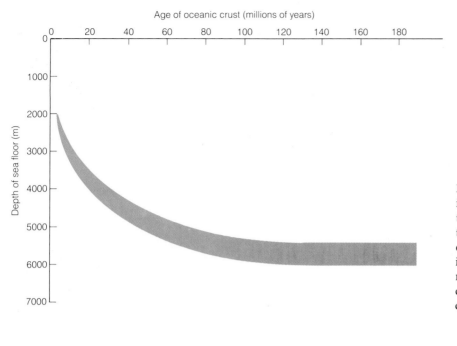

FIGURE 16.12 Depth of the seafloor in the world oceans as a function of the age of the oceanic crust. Near the spreading center, young lithosphere is thermally expanded. As it moves away from the crest of the midocean ridge, the lithosphere cools and contracts. The ocean becomes deeper as a result.

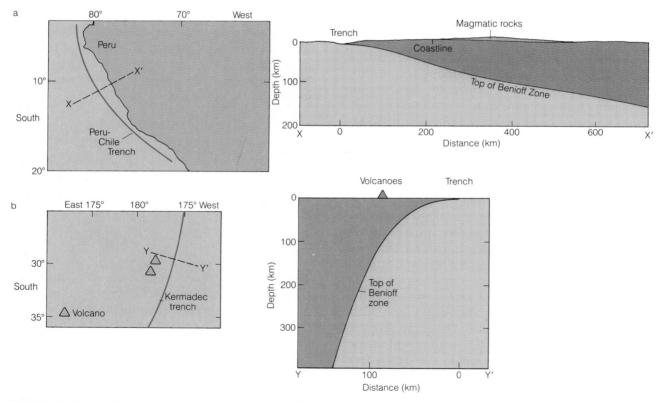

FIGURE 16.13 The steepness of the angle at which lithosphere sinks at a subduction zone controls the curvature of the trench and of the magmatic arc. (a) Beneath Peru the top of the Benioff zone dips at a shallow angle, and the adjacent Peru–Chile Trench has a pronounced curvature. (b) The Benioff zone associated with the Kermadec Trench in the southwest Pacific has a steep dip and the trench has only a slight curvature.

volcanic arc. Regardless of location, the origin is the same in both cases. It is an arc-shaped region of magmatic activity called a *magmatic arc* parallel to the seafloor trench and separated from it by a distance of 150 to 300 km, the distance depending on the angle of dip of the descending plate (Fig. 16.13). Island arc, continental volcanic arc, and magmatic arc are names given to different features arising from the same process—magmatic activity produced by wet partial melting of subducting oceanic crust.

The Japanese islands and the Aleutians are modern-day island arcs. There are many other island arcs around the edge of the Pacific Ocean. Examine Figure 3.12 and it becomes apparent that the Andesite Line coincides with island arcs. Note, too, that although each arc is part of a circle, some arcs are parts of a circle with a large radius and some are highly curved and are parts of a circle with a smaller radius. The radius of curvature is an indication of the angle at which lithosphere is plunging back into the mantle. If the angle of plunge were perpendicular to the Earth's surface, an island arc would be straight. If the angle of plunge is almost flat, the island arc has a pronounced curvature (Fig. 16.13).

Careful determination of the age of oceanic crust being subducted shows that old, cold, and therefore dense crust forms island arcs that have a large radius of curvature. Young oceanic crust that has still not reached thermal equilibrium, and that is less dense than older, colder crust, forms arcs with short radii of curvature. This observation is a very informative one, because it suggests that oceanic lithosphere might be sinking under its own weight through the hot, weak asthenosphere. This means, further, that old, cold lithosphere, when capped by oceanic crust, must be more dense than the hot, plastic asthenosphere. The older and colder the lithosphere, the faster the rate of sinking, and the steeper the angle of the Benioff zone. As lithosphere sinks, it must start to heat up. Earthquakes can occur in the down-going slab as long as it is cool enough to be brittle. Even with a sinking rate of 8 cm/yr, calculations show that lithosphere loses its brittle properties by the time it reaches a depth of 670 km (Fig. 16.14). This is probably the reason that earthquake foci are no deeper than 670 km.

Mélange

Many features occur as a result of deformation along convergent margins. A distinctive feature of some margins is the development of a *mélange*, a chaotic

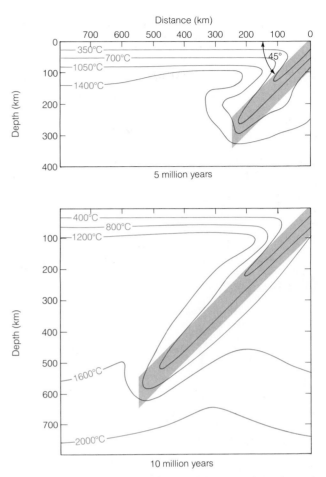

FIGURE 16.14 Computer-aided calculations of the fate of a descending slab of cool lithosphere. A plate 100 km thick, descending at an angle of 45° and a rate of 8 cm/yr, will cool the surrounding mantle but will slowly become heated as it sinks. Contours depict the temperature. Between 600 and 700 km the temperature of the tip of the descending slab reaches the temperature of the adjacent mantle and earthquakes cease. (*Source:* After Hsui and Toksöz, 1979.)

mixture of broken and jumbled rock (Fig. 16.15). Once a subduction zone forms and a seafloor trench is created, sediment accumulates in the trench. A sinking plate drags sedimentary rock downward beneath the overriding plate. Sedimentary rock has a low density. As a result it is buoyant and cannot be dragged down very far. Caught between the overriding and sinking plates, the sediment becomes shattered, crushed, sheared, and thrust-faulted to form a mélange (Fig. 16.16). As the mélange thickens, it becomes metamorphosed. The cold sedimentary rocks are dragged down so rapidly that they remain cooler than adjacent rock at the same depth. The kind of metamorphism that is common in many mélange zones, therefore, is

FIGURE 16.15 The Lichi Formation, near Taitung, Taiwan is a chaotic mélange of mudstone, sandstone, and shale produced by thrusting associated with subduction of the Philippine Plate beneath Taiwan.

that shown in Case C of Figure 5.15—a high-pressure, low-temperature metamorphism distinguished by blue schists and eclogites. The blue color comes from a bluish amphibole called glaucophane.

Outer-Arc Ridges and Outer-Arc Basins

Between the trench and the magmatic arc, both of which are prominent topographic features, two less prominent features are present along many con-

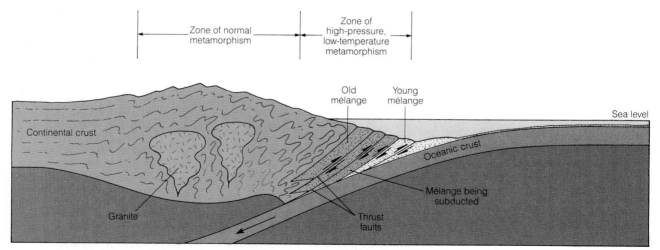

FIGURE 16.16 Mélange is formed when young sediment in a trench is smashed by moving lithosphere and dragged downward in slices bounded by thrust faults. As successive slices are dragged down, older mélange, closer to the continent, is pushed back up. The process is like lifting a deck of cards by adding new cards at the base of the deck.

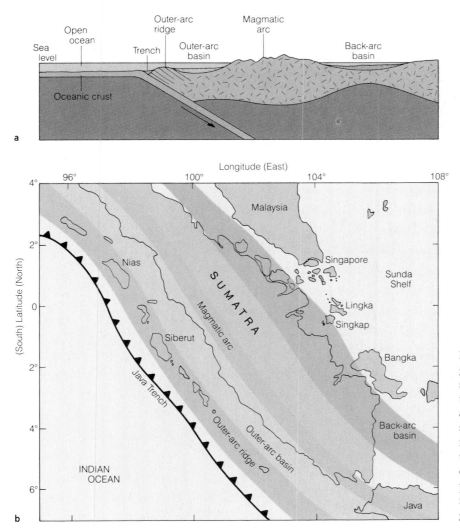

FIGURE 16.17 Features of a convergent plate margin. (a) Idealized cross section showing the most distinctive features. The outer-arc and back-arc basins both tend to be filled with sediment derived from the magmatic arc. The outer-arc ridge tends to be underlain by mélange. (b) Map of portion of Sumatra showing the positions of the major topographic features in a present-day convergent plate boundary.

vergent margins. The two features occur together and are named, respectively, the outer-arc ridge and outer-arc basin (Fig. 16.17). An *outer-arc ridge* (also called a five-arc ridge) is commonly underlain by mélange and is caused by a local thickening of the crust due to thrust faulting at the edge of the overriding plate. The *outer-arc* (or *fore-arc*) *basin* is a low-lying region between the outer-arc ridge and the magmatic arc.

Back-Arc Basin

When the sinking rate of a subducting plate is faster than the forward motion of the overriding plate, part of the overriding plate can be subjected to tensional stress. The leading edge of the overriding plate must remain in contact with the subduction edge or else a huge void would open. What happens is that the overriding plate grows slowly larger at a rate equal to the difference in velocities between the two plates. Most commonly, this process

is manifested by a thinning of the crust and an opening of an arc-shaped basin behind the magmatic arc (Fig. 16.17). Basaltic magma may rise into a so-called *back-arc basin* and a small region of new oceanic crust may even form. Because of the proximity of stratovolcanoes above the magmatic arc, an abundant source of sediment is generally available; hence both back-arc and outer-arc basins tend to be filled with a mixture of volcanic rocks and clastic sediments.

Transform Fault Margins

The faults at the margins of plates are *transform faults* (Fig. 15.13). They are huge, vertical, strike–slip faults cutting down into the lithosphere. They can form when either a new divergent or a convergent margin fractures the lithosphere (Fig. 16.18). Neither compressional nor tensional forces are associated with the faults; they are simply margins along which two plates slide past each other. The sliding

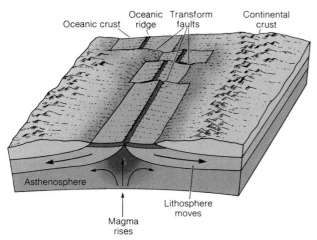

FIGURE 16.18 Transform faults form when an oceanic ridge first forms. Shapes of continental margins reflect faults now found along the oceanic ridge.

margins smash and abrade each other like two giant strips of sandpaper, so the faults are marked by zones of intensely shattered rock. Where the faults cut oceanic crust, they make elongate zones of narrow ridges and valleys on the seafloor. When transform faults cut continental crust, they too influence the topography; however, the features are less pronounced than on the seafloor. Transform faults on the land tend to be marked by parallel or nearly parallel faults in a zone that can be as much as 100 km wide.

The sliding movement of transform faults causes a great many shallow-focus earthquakes, some of them of high magnitude. Most transform faults do not have any volcanic activity associated with them. Occasionally, however, a small amount of plate separation does occur and a "leaky" transform results in a small amount of volcanism.

The best-known transform fault in North America is the San Andreas Fault in California. The many earthquakes that disturb California are caused by movement along it. Some of those earthquakes, including the one that devastated San Francisco in 1906, have been particularly destructive, and it is probable that future quakes will be just as devastating. As long as the plates continue to move, activity must occur along the San Andreas Fault, and residents of California can expect more earthquakes.

The San Andreas Fault is the largest of several transform faults that offset the segment of midocean ridge called the East Pacific Rise. Figure 16.19 shows how the transform faults and segments of the East Pacific Rise separate the North American Plate from the Pacific Plate. Figure 16.19

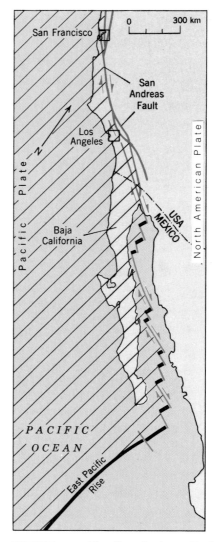

FIGURE 16.19 The San Andreas Fault is one of several transform faults that offset an oceanic ridge (East Pacific Rise). The faults and ridge segments separate the Pacific Plate (left) from the North American Plate (right). The two plates are sliding past each other, causing frequent earthquakes in California.

also shows that the San Andreas is only one of several faults that break the continental crust. The others are subsidiary to the San Andreas, however, and are part of a fault zone that is roughly parallel to the main fault. Movement along the San Andreas Fault arises from movement between the North American and Pacific Plates. The peninsula of Baja California and the portion of the state of California that lies west of the San Andreas Fault are on the Pacific Plate. That plate is moving northwest, relative to the North American Plate, at a rate of several centimeters per year. In about 10 million years Los Angeles will have moved far enough north so as to be opposite San Francisco. In about 60 million years, at the present rate of movement, the segment of continental crust on which Los Angeles lies will have become separated completely from the main mass of continental crust that comprises North America.

PLATE TECTONICS AND CONTINENTAL CRUST

Seafloor spreading and plate tectonics were proved correct using evidence from the oceanic crust. This result should hardly be surprising. Spreading centers are found beneath the sea. The velocity, direction of motion, and age of a plate are most convincingly established by the thermoremanent magnetism of the oceanic crust. New crust is made beneath the sea and the system is balanced by the subduction of cool, dense oceanic crust at convergent margins. Plate tectonics would probably operate even if there were no continental crust at all. In a sense, continental crust is simply a passenger rafted on large plates of lithosphere. But it is a passenger that is buffeted, stretched, fractured, and altered by the ride. Someone once characterized continental crust as the product of bump-and-grind tectonics. Each bump between two fragments forms an orogenic belt, each grind a strike–slip fault, each stretch a thinning of continental crust and a rift valley. Scars left on the continental crust by bump-and-grind tectonics are evidence of former plate motions. This is fortunate because the most ancient crust known to exist in the ocean dates only from the mid-Jurassic Period. Indeed, the only direct evidence concerning geological events more ancient than the mid-Jurassic comes from the continental crust.

Continental Margins

Fragmentation, drift, and the welding together of continental crust are inevitable consequences of plate tectonics. Evidence of fragmentation and welding is most strikingly preserved along the compressed or stretched margins of the fragments of continental crust. It is helpful to review briefly the features associated with the five principal kinds of margins that bound continental crust. They are

1. *Passive continental margins,* of which the Atlantic Ocean margins of the Americas, Africa, and Europe are examples.

2. *Continental convergent margins,* of which the Andean coast of South America is an example.

3. *Continental collision margins,* for which the Alpine–Himalayan mountain chain provides an example.

4. *Transform fault margins,* which are exemplified by the San Andreas Fault in California and the Alpine Fault that slices through the South Island of New Zealand.

5. *Accreted terrane margins,* consisting of island arc and other small masses of crust added to an existing continental margin. An example is the northwest margin of North America from northern California to Alaska.

Passive Continental Margins

A new ocean basin forms by the rifting of continental crust following the sequence illustrated in Figure 16.20. This process can be seen in the Red Sea, which is a young ocean with an active spreading ridge running down its axis (Figs. 16.21). Indeed, the Red Sea is such a young ocean that the ridge is still partially covered by sediment.

The sequence of sediments deposited in the Red Sea rift starts with the deposition of clastic nonmarine sediments. These nonmarine sediments are followed by evaporites and then marine shales. The sequence is distinctive and apparently arises in the following manner. Basaltic magma, associated with formation of the new spreading edge, heats and expands the lithosphere so that a plateau forms with an elevation of as much as 2.5 km above sea level. When tensional forces split the crust and form a rift, there is a pronounced topographic relief between the plateau and the floor of the rift. The earliest rifting of the Red Sea must have been very much like the African Rift Valley today. Before the rift floor sank low enough for seawater to enter, clastic, nonmarine sediments such as conglomerates and sandstones were shed from the steep valley walls and accumulated in the rift. Associated with these sediments are basaltic lavas, dikes, and sills, formed by magma rising up the normal faults. As the rift widened, a point was reached where seawater entered. The early flow was apparently restricted and the water was shallow, resembling a shallow lake more than an ocean. The rate of evaporation would have been high and as a result strata of evaporite salts were laid down on top of the clastic, nonmarine sediments. Finally, as rifting continued and the depth of the seawater increased, normal clastic marine sediments were deposited.

MODERN EXAMPLE

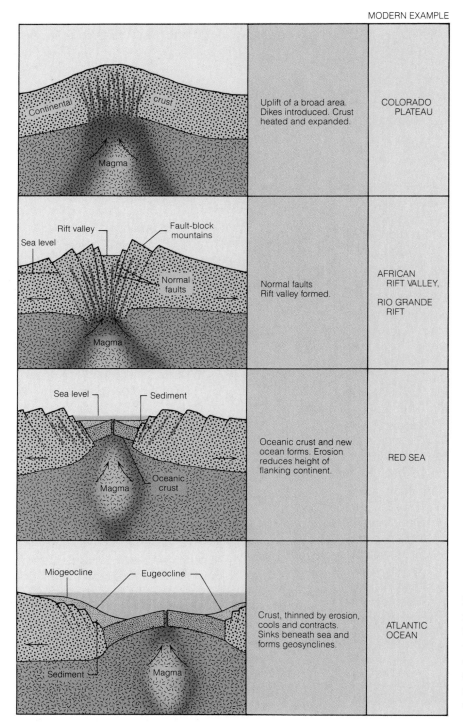

	Uplift of a broad area. Dikes introduced. Crust heated and expanded.	COLORADO PLATEAU
Normal faults Rift valley formed.	AFRICAN RIFT VALLEY, RIO GRANDE RIFT
Oceanic crust and new ocean forms. Erosion reduces height of flanking continent.	RED SEA
Crust, thinned by erosion, cools and contracts. Sinks beneath sea and forms geosynclines.	ATLANTIC OCEAN

FIGURE 16.20 The rifting of continental crust to form a new ocean basin bounded by passive continental margins. The process of rifting can cease at any stage. It is not necessarily correct to conclude, therefore, that the African Rift Valley will open to form a new ocean.

This is the stage the Red Sea is in today. Eventually, if further rifting exposes new oceanic crust, the Red Sea will evolve into a younger version of the Atlantic Ocean. Sedimentation will form a shallow-water miogeocline and a deep-water eugeocline (chapter 15) along both sides of the ocean.

Triple Junctions and Rifts.   Notice in Figure 16.21 that the Gulf of Aden, the Red Sea, and the north-

ern end of the African Rift Valley meet at angles of 120°. The meeting point is a plate triple junction formed by three spreading edges. Two of the edges, the Gulf of Aden and the Red Sea, are active and still spreading. The third, the African Rift Valley, is apparently no longer spreading and probably will not evolve into an ocean. What will remain on the African continent is a long, narrow sequence of grabens filled primarily with nonmarine sedi-

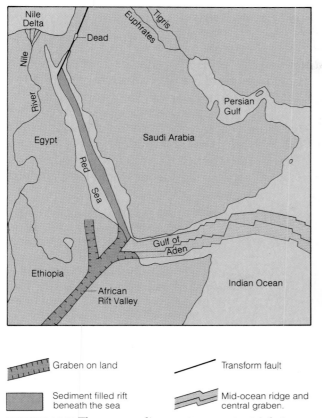

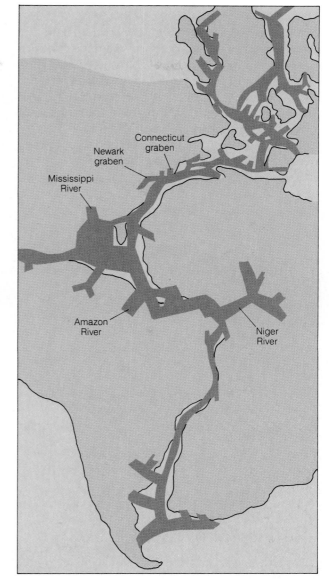

Graben on land

Transform fault

Sediment filled rift beneath the sea

Mid-ocean ridge and central graben.

FIGURE 16.21 Three spreading centers meet a triple junction. Two, the Gulf of Aden and the Red Sea, are actively spreading, whereas the third, the African Rift, appears to be a failed rift that will not develop into an open ocean. The spreading edge down the center of the Red Sea is covered by sediment.

FIGURE 16.22 Map of a closed Atlantic Ocean showing the rifts formed when Pangaea was split by a spreading center. The rifts on today's continents are now filled with sediment. Some of them serve as the channelways for large rivers. (*Source:* After Burke, 1980.)

ment. The formation of three-armed rifts with one of the arms not developing into an ocean is apparently a characteristic feature of continental rifting caused by new spreading edges. This can be seen from Figure 16.22 which shows the reassembled positions of the continents flanking the Atlantic Ocean prior to breakup. Note that some of the world's largest rivers flow down valleys formed by undeveloped rifts associated with the opening of the Atlantic Ocean. The pattern of rifts is a distinctive feature arising from plate tectonics.

Continental Convergent Margins

Subduction of oceanic crust beneath continental crust produces deformation of a continental margin (together with a distinctive style of metamorphism and deformation of sediments deposited in the trench, and characteristic magmatic activity). Sediment subjected to deformation in such a setting forms a mélange. The tectonic setting in which sediments are subjected to high-pressure, low-tem-

perature metamorphism is in mélange produced at a subduction zone. Adjacent to and parallel with the belt of mélange the thickened edge of continental crust is also metamorphosed, but in that case it is normal, regional metamorphism (chapter 5). One distinctive feature of a continental convergent margin, therefore, is a pair of parallel metamorphic belts.

A second characteristic feature of many continental convergent margins is an arc of stratovolcanoes built on top of the continental crust. Modern examples can be seen in the chains of volcanoes in the Andes and the Cascade Range (Fig. 15.26). Where the volcanoes have been eroded and the deeper parts of the underlying magmatic arc are exposed, granitic batholiths can be observed. They are remnants of the magma chambers that once fed stratovolcanoes far above. The strings of huge, elongate batholiths that run from southern California to northern British Columbia (Fig. 3.32) provide striking examples of deeply eroded magmatic arcs formed along a fossil subduction margin (Fig. 15.29).

Continental Collision Margins

When continental crust is part of a plate of lithosphere that is being subducted beneath the margin of a second piece of continental crust, the two continental fragments will eventually collide. The collision sweeps up the sediment accumulated along the leading edges of both continents and forms a fold-and-thrust mountain system. The suture zone between the two masses of deformed sediment is commonly marked by the presence of serpentinites

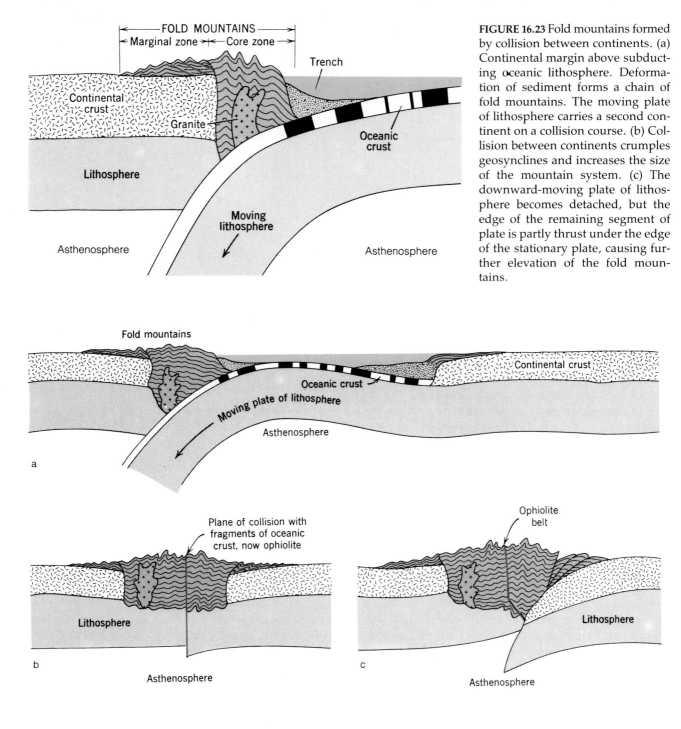

FIGURE 16.23 Fold mountains formed by collision between continents. (a) Continental margin above subducting oceanic lithosphere. Deformation of sediment forms a chain of fold mountains. The moving plate of lithosphere carries a second continent on a collision course. (b) Collision between continents crumples geosynclines and increases the size of the mountain system. (c) The downward-moving plate of lithosphere becomes detached, but the edge of the remaining segment of plate is partly thrust under the edge of the stationary plate, causing further elevation of the fold mountains.

formed by alteration of sheared and deformed fragments of ophiolites caught up in the collision (Fig. 16.23). Fold-and-thrust mountain systems may also have stratovolcanoes, batholiths, and paired metamorphic belts associated with them, because a collision margin must be a continental convergent margin prior to collision.

One distinctive feature of a fold-and-thrust mountain system formed by collision is that the new mountain system lies in the interior of a major landmass. A modern example is the great Himalayan mountain chain formed by the collision of India with Asia. Another can be seen in the Alps, which were formed by the collision of Africa and Europe starting in early Mesozoic time. Older examples are provided by the Ural mountains in the U.S.S.R. and by the Appalachians (Fig. 16.24), each of which was formed by Paleozoic collisions.

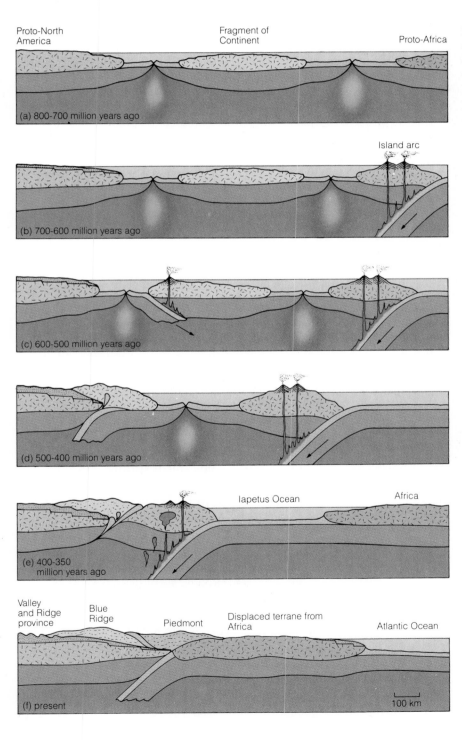

FIGURE 16.24 A suggested sequence of events explaining the evolution of the southern Appalachians in terms of a plate tectonic model. A sequence of subduction zones, collisions, and thrusting produced the present-day structure. Iapetus is the posthumous name of the ocean that disappeared about 350 million years ago, when Africa collided with North America. Iapetus was one of the minor Greek gods and father of Atlas and Prometheus.

Proto-North America · Fragment of Continent · Proto-Africa

(a) 800-700 million years ago

Island arc

(b) 700-600 million years ago

(c) 600-500 million years ago

(d) 500-400 million years ago

Iapetus Ocean · Africa

(e) 400-350 million years ago

Valley and Ridge province · Blue Ridge · Piedmont · Displaced terrane from Africa · Atlantic Ocean

(f) present

100 km

Several differences exist between orogenic belts formed along continental convergent margins and those formed by collision. The paired metamorphic belts are asymmetric with respect to the subduction zone. A collision zone, on the other hand, is roughly symmetrical because deformed continental crust is present on both sides of the collision zone (Fig. 16.23).

It seems reasonable to conclude that all of the now deeply eroded orogenic belts around the world were once fold-and-thrust mountain chains. A further conclusion is that such orogenic belts were formed either along continental convergent or continental collision edges and that they mark former plate margins.

Transform Fault Margins

Giant strike–slip faults provide the most direct and convincing evidence that large lateral motions have occurred in the past. It is rarely possible to prove that ancient strike–slip faults were actually transform faults that connected spreading centers or subduction zones but the inference is strong that they were. This inference arises from the manner in which the presently active large, strike–slip faults originated. As seen in Figure 16.25, the San Andreas Fault apparently formed when the westward-moving North American continent overrode portions of the East Pacific Rise. The San Andreas Fault is the transform fault that connects the two remaining segments of the old spreading center.

Accreted Terrane Margins

Plate motions can raft small fragments of crust tremendous distances. Eventually, any fragment that is not consumed by subduction will be added (accreted) to a larger continental mass. Some of the small fragments are island arcs formed by subduction of oceanic crust beneath oceanic crust. Other fragments form when they are sliced off the margin of a large continent, much as the San Andreas Fault is slicing a fragment off North America today. Other combinations of volcanism, rifting, faulting, and subduction can also form small fragments of crust that are too buoyant to be subducted. In the western Pacific Ocean there are many such small fragments of continental crust; examples include the island of Taiwan, the Philippine islands, and the many islands of Indonesia. Each fragment, called a ***terrane***, is a geological entity characterized by a distinctive stratigraphic sequence and structural history. The ultimate fate of all terranes is to be accreted to a larger continental mass. Accreted terranes, then, modify a preexisting subduction, collision, or transform fault margin by the addition of rafted-in, exotic blocks of crust. An accreted terrane is always fault bounded and differs markedly in its geological features from adjacent terranes.

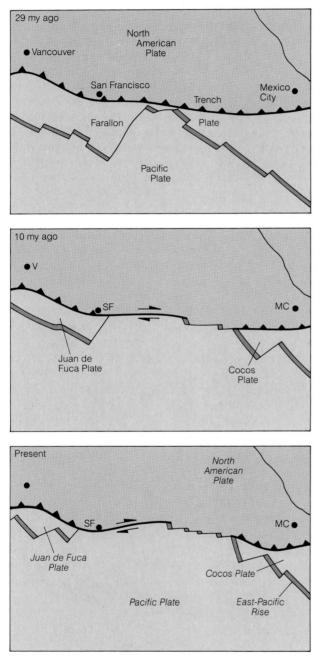

FIGURE 16.25 Diagrams showing the origin of the San Andreas Fault. Twenty-nine million years ago the edge of North America overrode a portion of the Farallon Plate, creating two smaller plates in the process, the Cocos Plate and the Juan de Fuca Plate. The San Andreas Fault is the transform fault that connects the remaining pieces of the severed spreading ridge. (*Source:* From Stewart, 1978.)

The western Pacific is a place where the accretion process is apparently in progress today (Fig. 16.26). The western margin of North America is an example of a young, accreted margin. Using a combination of lithologic and paleontologic studies, structural analysis, and paleomagnetism, many terranes have now been identified. They have all been accreted since the early Mesozoic, about 200 million years ago (Fig. 16.27). One of the terranes is an ancient seamount, another is an older limestone platform formed on some other continental

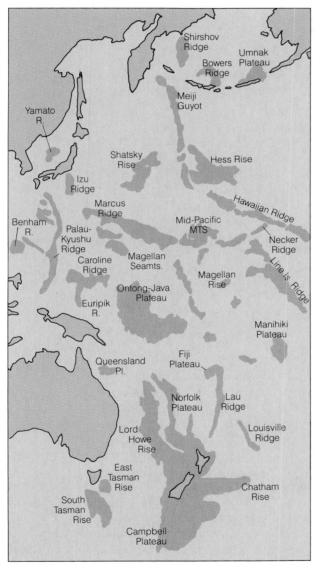

FIGURE 16.26 Regions of high seafloor topography and islands in the western Pacific Ocean. As the Pacific Plate moves westerly, and the Australian–Indian Plate continues to move north, it is probable that most of these oceanic islands and plateaus will be accreted to the moving continents, increasing the sizes of the continents in the process and causing orogenic deformations. (*Source:* After Ben-Avraham et al., 1981.)

margin. Still others are fragments of island arcs and even fragments of old metamorphic rocks. Paleomagnetic studies suggest that some terranes have moved 5000 km or more, and that once accreted, the process of movement did not necessarily cease. Later motion along transform faults caused still further reorganizations.

The recognition of accreted terrane margins is a relatively new discovery. In a sense, it is a second stage of complexity in the plate tectonic revolution. A great deal still remains to be discovered concerning terranes. How to recognize a terrane, how to work out where it came from, and how it was moved, are all challenges facing geologists. Only recently has the existence of accreted terranes of Paleozoic age been recognized along the eastern margin of North America. Although it is a new and exciting concept, accreted terrane margins of continents seems to be yet another distinctive piece of evidence that can be used to prove the existence of ancient plate motions.

HOT SPOTS AND ABSOLUTE MOTIONS

It was pointed out earlier in this chapter that plate motions determined from magnetic time lines are only relative motions. In order to determine absolute motions, an external frame of reference is necessary. A familiar example of absolute versus relative motion occurs when one automobile overtakes another. If observers in the two automobiles could only see each other, and could not see the ground or any fixed objects outside their cars, they could only judge the *difference* in velocity between the two cars. One car could be traveling at 50 km/h, the overtaking car at 55 km/h, but all that the observers could determine is that the *relative velocity* difference is 5 km/h. On the other hand, if the observers could measure velocity with respect to a stationary or fixed reference such as the ground surface, they could determine that the *absolute velocities* were 50 and 55 km/h, respectively.

We would be constrained to determine only relative plate velocities if a fixed reference framework did not exist. Fortunately, a reasonable framework does exist. During the last century, the American geologist James Dwight Dana observed that the age of volcanoes in the Hawaiian island chain increased from southeast to northwest (Fig. 17.8). As mentioned earlier in this chapter, J. Tuzo Wilson suggested that the age of the volcanic islands recorded the movement of the seafloor. Wilson postulated

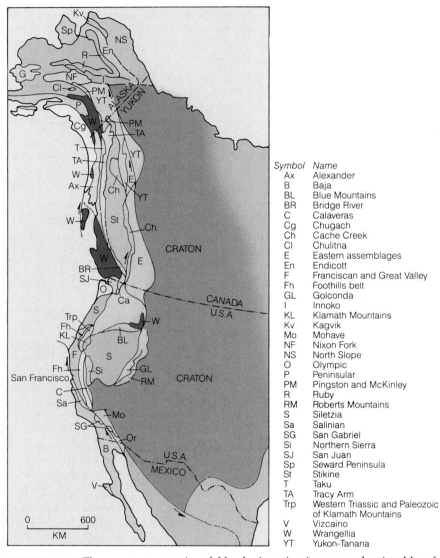

Symbol	Name
Ax	Alexander
B	Baja
BL	Blue Mountains
BR	Bridge River
C	Calaveras
Cg	Chugach
Ch	Cache Creek
Cl	Chulitna
E	Eastern assemblages
En	Endicott
F	Franciscan and Great Valley
Fh	Foothills belt
GL	Golconda
I	Innoko
KL	Klamath Mountains
Kv	Kagvik
Mo	Mohave
NF	Nixon Fork
NS	North Slope
O	Olympic
P	Peninsular
PM	Pingston and McKinley
R	Ruby
RM	Roberts Mountains
S	Siletzia
Sa	Salinian
SG	San Gabriel
Si	Northern Sierra
SJ	San Juan
Sp	Seward Peninsula
St	Stikine
T	Taku
TA	Tracy Arm
Trp	Western Triassic and Paleozoic of Klamath Mountains
V	Vizcaino
W	Wrangellia
YT	Yukon-Tanana

FIGURE 16.27 The western margin of North America is a complex jumble of terranes accreted during the Mesozoic and Cenozoic eras. Each terrane is fault-bounded and is a distinct geological entity. Some terranes, such as Wrangellia (W), were fragmented during the accretion process and now occur in several different fragments. (*Source:* Adapted from Beck et al., 1980.)

that a deep, long-lived magma source lies somewhere far down in the mantle. Because the lithosphere moves, a volcano can only remain in contact with the magma source for about a million years.

Wilson made his suggestion as a way of testing seafloor spreading. However, it was not long before he realized that if hot spots do exist deep in the mantle, and do have long-continued lives, they might provide a series of fixed points against which absolute plate motions can be measured. If lithosphere moves over a fixed hot spot a chain of volcanoes should result. If lithosphere is stationary, a long-lived volcano should be found above each hot

spot. More than a hundred hot spots have now been identified (Fig. 16.28). Using them for reference, it has been found that the African Plate must be very nearly stationary because volcanoes there seem to be very long lived. Because the African Plate is almost completely surrounded by spreading edges, and because the relative velocities along the encircling ridges are closely matched, it must be concluded that the Mid-Atlantic Ridge is moving westward and that the oceanic ridge that runs up the center of the Indian Ocean is moving to the east. If the absolute motion of the African Plate is zero or nearly so, the Mid-Atlantic Ridge in the

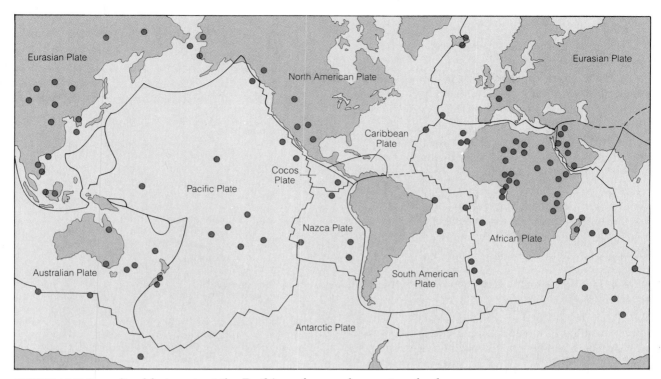

FIGURE 16.28 Long-lived hot spots at the Earth's surface, each a center of volcanism, are believed to lie above deep-seated sources of magma in the mantle. Because the magma sources lie far below the lithosphere, and do not move laterally, hot spots can be used to determine the absolute motions of plates.

southern Atlantic Ocean must be moving westward at the rate of about 2 cm/yr, and the absolute velocity of the South American Plate must be 4 cm/yr.

The Australian–Indian Plate is moving almost directly northward. All other plates, with the exception of the stationary African Plate, are moving in approximately eastward or westward directions. Several plates do not have subduction margins and must therefore be increasing in size. Most of the modern subduction zones are to be found around the Pacific Ocean along the edge of the Pacific Plate, so much of the oceanic lithosphere that is now being destroyed is in the Pacific. It follows then, that the Indian Ocean, the Atlantic Ocean, and most other oceans must be growing larger, while the Pacific Ocean must be steadily getting smaller.

CAUSES OF PLATE TECTONICS

Just as Alfred Wegener could not explain what made continents drift, we are still unable to say *exactly why* plates of lithosphere move. Until the driving force is explained, plate tectonics must remain a kinematic description of what occurs without knowing why it happens. The situation is analogous to knowing the details of shape, color, size, and speed of an automobile but not knowing what makes it run. But meanwhile we can hypothesize about the causes of the motion and test the hypotheses by making detailed calculations based on the laws of nature.

The lithosphere and asthenosphere are inevitably bound together. If the asthenosphere moves, it will make the lithosphere move, just as movement of sticky molasses will move a piece of wood floating on its surface. So too will movement of the lithosphere cause movement in the asthenosphere below. Such is our state of uncertainty that we cannot yet separate the relative importance of the two effects. However, on one point we can be quite certain: moving lithosphere has kinetic energy. That energy must come from somewhere. The source of the energy is apparently the Earth's internal heat, and the way much of the heat energy seems to reach the surface is by convection in the mantle. What has not yet been discovered is the precise way convection and plate motions are linked.

Convection in the Mantle

The mantle is solid rock; however, it is hot and apparently weak enough so that under slow strain rates even small stresses will make it flow like a very sticky viscous liquid. Like a liquid, too, the mantle must be subject to convection currents when a local source of heat causes a mass of rock to become heated to a higher temperature than surrounding rock. The heated mass expands, becomes less dense, and rises very slowly at rates as low as 1 cm/yr. To compensate for the rising mass, cooler, more dense material must flow downward. The laws of nature indicate that the rate at which heat reaches the Earth's surface can only be accounted for if convection in the mantle brings heat from the deep interior.

Several kinds of convection cells within the mantle have been suggested. The first kind is a cell in which all movement is confined to the asthenosphere and the lithosphere. The mantle below about 670 km would have, in this model, very little motion but would serve as a giant stove to heat the asthenosphere. Each plate of lithosphere would be the top of a giant convection cell. This means that there would have to be several convection cells; their sizes and shapes are indicated by the sizes and shapes of plates. The masses of hot rock are postulated to rise vertically beneath oceanic ridges, then turn sideways, flowing horizontally as plates. The farther the rock moves away from the ridge, the cooler it gets. Eventually, the lithosphere becomes so cool and so dense that it sinks back into the asthenosphere. The place where sinking occurs is beneath a seafloor trench (Fig. 16.29).

Many problems surround the suggestion that convection cells are confined to a few hundred kilometers in the mantle. One problem concerns the plates themselves. As previously discussed, ocean ridges move; some plates are expanding, others are shrinking. It is very difficult to understand how convection cells could move and change size.

A second suggestion, one that avoids the problem of heat sources in the asthenosphere, is that convection cells involve the entire mantle, and that the heat brought up comes from the outer core (Fig. 16.29b). Just how localized transfer of heat takes place between core and mantle is not known. The flowing motions and sideways movement of lithosphere would be the same in a whole mantle convection pattern as in an upper-mantle-limited convection. The main differences are the sizes of the convection cells in the two cases and, in the latter case, the notion that convection can involve not only the weak asthenosphere, but also the stronger

mesosphere below. These larger convection cells are even more difficult to reconcile with moving oceanic ridges and plates of varying size.

A third possibility, and to many people a more likely one, is that some sort of stacked convection system exists. If the two convection systems shown

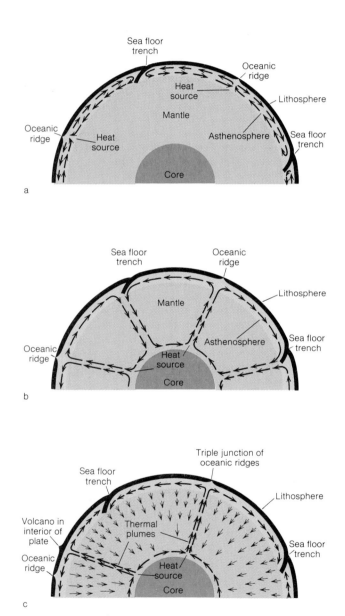

FIGURE 16.29 Three suggested mechanisms by which convection and flow in the asthenosphere might move plates of lithosphere. (a) Convection is confined to the asthenosphere. (b) Convection involves the entire mantle. (c) Thermal plumes rise from the mantle–core boundary and cause local hot spots at the Earth's surface.

in Figures 16.29a and b were combined, large, deep convection cells would supply heat to drive smaller and shallower cells. The boundary between the two cell systems need not be the base of the asthenosphere. Some scientists have argued that it is more likely to be at a depth of 670 km where there is a pronounced seismic discontinuity.

Even stacked convection cells leave many unanswered problems. One concerns the existence of long-lived, deep-seated sources of magma indicated by the hot spots. Some scientists suggest that if local hot regions do occur on the core–mantle boundary, they will be small, roughly circular spots. Instead of producing a large convection cell, they maintain, a small hot spot will cause a long cylinder of hot rock, a few hundred kilometers in diameter, to rise. They refer to the vertically rising cylinder as a *thermal plume* and suggest that plumes are the sites of the hot spots that cause long-continued volcanism (such as the Hawaiian Islands). All upward motion could be accounted for by no more than 20 thermal plumes, although more probably exist. When a plume reaches the base of the lithosphere, it creates a local hot spot from which a little magma rises. However, most of the convecting plume would spread laterally and flow horizontally in all directions beneath the lithosphere. By this suggestion, the asthenosphere becomes, in essence, the top of a whole series of plume-driven convection cells. Return flow to balance the concentrated upward flow in the plumes would not necessarily involve well-defined, down-flowing plumes, but could be accomplished by slow downward movement of the entire mantle. Movement of plates of lithosphere by thermal plumes is more difficult to visualize because flow of the asthenosphere should be equal in all directions away from the hot spot. This means that plate motion must somehow involve the lithosphere itself.

The preceding discussion about convection is speculative. Evidence from seismic tomography and heat flow indicates that convection of some sort does occur beneath the lithosphere. Even so, it is difficult to see how plate motions can be due entirely to convection. For this reason, most scientists agree with the hypothesis that the motion of the lithosphere is due to a combination of processes, and that convection is only one of the processes. One important thing that convection must do is keep the asthenosphere hot and weak by bringing up heat from the deep mantle and core. In this sense at least, convection is essential for plate tectonics.

Movement of the Lithosphere

Three different forces might play a role in making the lithosphere move. The first is a push away from a spreading center. Rising magma at a spreading center creates new lithosphere, and in the process it pushes the plates sideways (Fig. 16.30). Once the process is started, it would tend to keep itself going. The problem is that pushing involves compression, whereas the structure of the crust along a midocean ridge indicates a state of tension.

A second way by which lithosphere could be made to move is by dragging rather than pushing. Proponents of the dragging idea point out that a descending tongue of old, cold lithosphere must be more dense than the hot asthenosphere surrounding it. Because rock is a poor conductor of heat, they urge, the temperature at the center of a descending slab can be as much as 1000°C cooler than the mantle at depths of 400 to 500 km. The dense slab of lithosphere must then sink under its own weight and exert a pull on the entire plate.

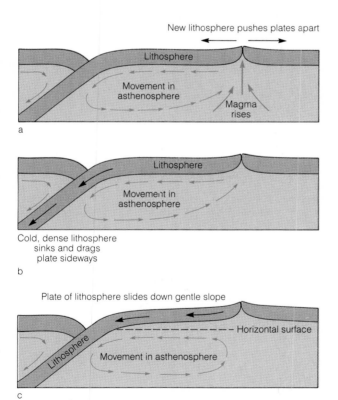

FIGURE 16.30 Three suggested mechanisms by which lithosphere might move over the asthenosphere. (a) Magma, rising at a spreading edge, exerts enough pressure to push the plates of lithosphere apart. (b) A tongue of cold, dense lithosphere sinks into the mantle and pulls the rest of the plate behind it. (c) A plate of lithosphere slides down a gently inclined surface of asthenosphere.

This is somewhat like a heavy weight that hangs over the side of a bed and is tied to the edge of a sheet. The weight falls and pulls the sheet across the bed. To compensate for the descending lithosphere, rock in the asthenosphere must flow slowly back to the spreading edge (Fig. 16.30).

However, both the pushing and the dragging mechanisms have problems. Plates of lithosphere are brittle, and they are much too weak to transmit large-scale pushing and pulling forces without major deformation occurring. Deformation is not present.

The third possible mechanism for movement of a plate of lithosphere is for it to slide downhill away from the spreading center. The lithosphere grows cooler and thicker away from a spreading center.

As a consequence, the boundary between the lithosphere and the asthenosphere must slope away from the spreading center. If the slope is as little as 1 part in 3000, the weight of the lithosphere could cause it to slide down the slope at a rate of several centimeters per year (Fig. 16.30).

At present there is no way to choose between the three lithosphere mechanisms. Calculations suggest that each operates to some extent, so that the entire process is possibly more complicated than we now imagine. The prevailing idea at present is that the sinking of old, cold lithosphere starts the process and then the other processes combine to keep it going. Only future research will resolve the question.

Essay

THE PLATE MOSAIC AND PRESENT-DAY PLATE MOTIONS

To a first approximation, plates of lithosphere behave as rigid bodies. This means that plates do not stretch and shrink, like rubber sheets. The distance between New York and Chicago, both on the North American Plate, remains fixed, even though the plate may flex and warp up and down. Of course, the distances between places on adjacent plates do change due to plate motions. Using velocities calculated from magnetic time lines, Figure B16.1 shows the inferred relative motions of plates today. The motions recorded by magnetic time lines can only be inferred to be today's motions unless actual measurements show the plates really are in motion. The space age has made it possible to get that proof. Using laser beams bounced off satellites we can measure the distance between two spots on the Earth with an accuracy of about 1 cm. As seen in Figure B16.1, velocities measured by laser ranging agree very closely with velocities calculated from magnetic time lines. The agreement proves the plates move steadily, not by starts and stops.

Magnetic time lines can be used both for calculating relative plate velocities and also for reconstructing the world map in times past. Magnetic time lines are symmetrical with respect to a spreading ridge and parallel to the ridge that created them (Figs. 16.4 and 16.8). The reason for this is straightforward. Each magnetic time line marks the edge of an earlier spreading center. Thus, two magnetic time lines have the same age, but lying on opposite sides of an oceanic ridge, can be brought together to show the configuration of plates as they were at an earlier time. By such means the

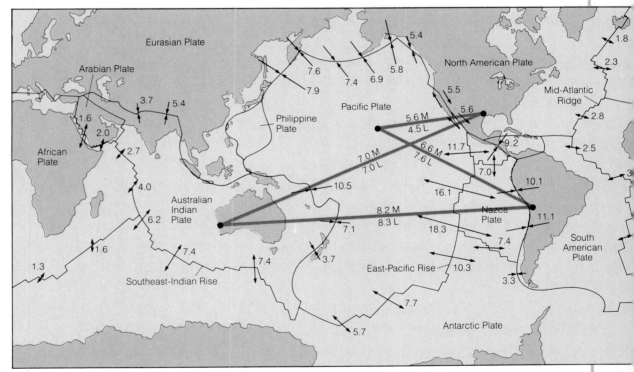

FIGURE B16.1 Present-day plate velocities in centimeters per year, determined in two ways. Numbers along the midocean ridges are mean velocities indicated by magnetic reversals. A velocity of 16.1, as shown for the East Pacific Rise, means that the distance between any point on the Nazca Plate and any point on the Pacific Plate increases, on the average, by 16.1 cm each year in the direction of the arrow. The long red lines connect stations used to determine plate motions by means of satellite laser ranging (L) techniques. The measured velocities between stations are very close to the average velocities estimated from magnetic reversals (M). (*Source:* Adapted from a NASA report. Geodynamic Branch, May 1986.)

opening of ocean basins—and movements of continents as a result of plate motions—can be reconstructed. Figure B16.2 shows a recon-struction of the opening of the southern Atlantic Ocean determined in this fashion.

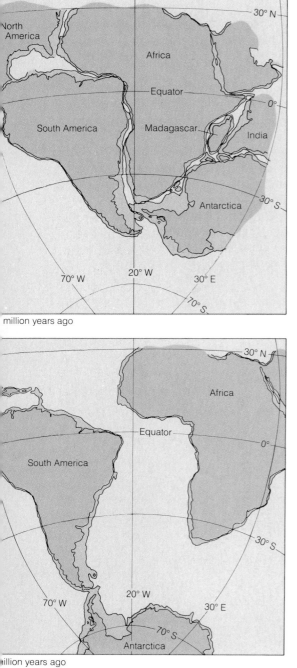

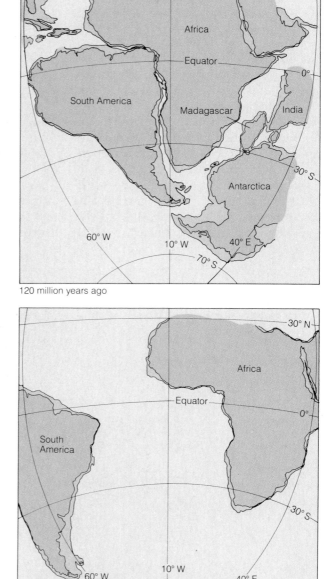

FIGURE B16.2 Magnetic data were used to plot the opening of the southern part of the Atlantic Ocean as South America and Africa drifted apart. About 200 million years ago, the present continents were joined together to form Pangaea. When the pieces are fitted back together along a line 2000 m below sea level, very few overlaps (shaded), or gaps (dark) remain. Notice how the continents move relative to the equator and the way Antarctica slowly drifts south. (*Source:* After Owen, 1983.)

SUMMARY

1. Abundant evidence proves that continents have not remained fixed on the Earth's surface, but have moved repeatedly from place to place.

2. The lithosphere is broken into six large and many smaller plates, each about 100 km thick, and each slowly moving over the top of the weak asthenosphere beneath it.

3. Each plate is bounded by three different kinds of margins. Divergent margins (= spreading centers) are those where new lithosphere forms. Plates move away from them. Convergent margins (= subduction zones) are lines along which plates compress each other and where lithosphere capped by oceanic crust is subducted back into the mantle. Transform fault margins are lines where two plates slide past each other.

4. Movement of a plate can be described in terms of rotation across the surface of a sphere. Each plate rotates around a spreading axis. The spreading axis does not necessarily coincide with the Earth's axis of rotation.

5. Because plate movement is a rotation, the velocity of movement varies from place to place on the plate.

6. Each segment of oceanic ridge that marks a divergent margin of a plate lies on a line of longitude passing through the spreading pole.

Each transform fault margin of a plate lies on the line of latitude of the spreading pole.

7. The mechanism that drives a moving plate is not known, but apparently it results from a combination of convection in the mantle plus forces that act on a plate of lithosphere.

8. There are five kinds of continental margins: passive, convergent, collision, transform fault, and accretion margins.

9. Passive margins develop by rifting of the continental crust. The Red Sea is an example of a young rift.

10. A characteristic sequence of sediments forms along a passive margin, starting with clastic nonmarine sediments, followed by marine evaporites, and then marine clastic sediments.

11. Continental convergent margins are the locale of paired metamorphic belts, chains of stratovolcanoes (= magmatic arc) and linear belts of granitic batholiths.

12. Collision margins are the locations of the fold-and-thrust mountain systems.

13. Accreted terrane margins arise from the addition of blocks of crust brought in by subduction and transform fault motions.

IMPORTANT WORDS AND TERMS TO REMEMBER

absolute velocity (of a plate) (p. 381)
accreted terrane (p. 394)
active zones (p. 379)
apparent polar wandering (p. 376)

back-arc basin (p. 387)

continental collision margin (p. 392)
continental convergent margin (p. 391)
continental drift (p. 374)
convection (p. 398)
convergent margin
   (= subduction zone) (p. 379)

divergent margin
   (= spreading center) (p. 379)

fore-arc basin (outer-arc basin) (p. 387)
fore-arc ridge (outer-arc ridge) (p. 387)

Gondwanaland (p. 376)

hot spot (p. 397)

Laurasia (p. 376)

magmatic arc (p. 385)
magnetic equator (p. 377)
magnetic inclination (p. 377)
magnetic striping (p. 378)
mélange (p. 385)

outer-arc basin (p. 387)

outer-arc ridge 387

Pangaea 374
passive continental margin (p. 389)

relative velocity (of a plate) (p. 381)

seafloor spreading (p. 377)
spreading axis (p. 382)

spreading center (p. 381)
spreading pole (p. 382)
stable regions (p. 379)

terrane (p. 394)
thermal plume (p. 398)
transform fault margin (p. 379)
triple junction (p. 390)

QUESTIONS FOR REVIEW

1. Long ago it was suggested that the location and shape of mountain ranges and other major topographic features were the result of contraction as the Earth cooled. Why is that explanation incorrect?

2. Who was Alfred Wegener and what revolutionary idea did he suggest?

3. Explain how the apparent wandering of magnetic poles through geologic history can be used to help prove continental drift.

4. What are the main features of seafloor spreading?

5. What are the three kinds of margins that bound tectonic plates?

6. How are the velocities of tectonic plates determined?

7. Do magnetic time lines provide relative or absolute plate velocities?

8. What is a spreading pole and how does plate velocity depend on the position of a plate relative to the spreading pole?

9. Why does the radius of curvature differ from one island arc to another?

10. What is a mélange and what kind of metamorphism is associated with mélanges?

11. Draw a cross section through the lithosphere at a convergent plate margin and mark the positions of the outer-arc basin, the back-arc basin, and the magmatic arc.

12. Name the five kinds of continental margins and describe how they form.

13. Describe the sequence of events that lead to the opening of a new ocean basin flanked by two passive continental margins.

14. How does an accreted terrane margin form? Name a continental margin that was modified by terrane accretion.

15. How are the absolute velocities and directions of motions of tectonic plates determined?

16. What is a thermal plume? How might it be connected with plate tectonics?

17. Name three forces that act on lithosphere that might cause it to slide over the asthenosphere.

The Evolution of Landscapes

Streams plunge 500 m down a vertical cliff at the head of a valley on the island of Hawaii as they slowly erode into the flanks of an ancient volcano.

DYNAMICS OF LANDSCAPE EVOLUTION

One of the most fascinating and unique attributes of our planet is its amazing variety of natural landscapes. Who could fail to be impressed by a view of the majestic snow-covered Himalaya rising abruptly from the plains of India, or the lofty Andes of South America with their array of active volcanic cones. Equally impressive are the vast subtropical deserts of north Africa, Australia, and the Arabian Peninsula, the dense, flat jungle terrain of South America's Amazon basin, or the undulating glaciated landscape of eastern Canada and the north-central United States with its myriad lakes and streams.

Even a casual look at landscapes is likely to raise questions in our mind. Why, for example, can we see clear evidence of the sculptural effects of running water in areas now so dry that they have no flowing streams? Why do the largest streams of south-central Asia flow *across* the mighty Himalaya? Why do some islands in the Pacific Ocean rise to heights of 4 km above sea level while many others barely reach the height of an average man?

These and other questions lead to more fundamental ones: How can such a diversity of landscapes be explained? Are landscapes eternal, as many of our forebears once thought, or are they transient, undergoing evolutionary change with the passage of time? What can landscapes tell us about the nature and history of the Earth's mobile lithosphere, and the fluctuations of global climate through geologic time?

Factors Controlling Landform Development

Process

We have already seen that distinctive landforms result from the activity of various surface processes. A sand dune has a form that is different from that of a moraine. We can distinguish, as well, between an alluvial fan and a delta on the basis of their form. In each case the active process and depositional environment lead to a unique end product, or landform. Process, then, is one factor that helps dictate the character of landforms.

Climate

Climate, in turn, helps determine which processes are active in any area. Climate also controls vegetation cover, which influences the effectiveness of erosive processes (Fig. 10.16). Distinctive land-scape regions can be identified that are dominated by landforms resulting from one or several surface processes. However, because climates have fluctuated through time, the active surface processes have also changed repeatedly, both in time and space. The latest large-scale change occurred at the end of the last glacial age, with the result that some existing landscapes primarily reflect former conditions rather than those of the present.

Lithology

Within any climatic zone, a given surface process may interact with surface materials differently, depending on their lithology. Rock types that are less erodible than others commonly produce more prominent landforms than those more susceptible to erosion. Any one rock type, however, may behave differently under different climatic conditions. For example, limestone may underlie valleys in moist climatic zones where dissolution is effective, but in dry desert areas the same rocks may form bold cliffs.

Structure

Structure also plays an important role. With a little experience, a person trained in geology can easily identify structural features at the Earth's surface visible from an airplane. Due to differential erosion, certain folded or faulted beds stand high or control the drainage pattern in such a way that they impart a grain or pattern to the landscape, thereby disclosing the underlying structure (Fig. 17.1). A well-jointed or fractured rock is likely to be more susceptible to weathering, mass-wasting, and erosion than massive rock of the same composition, and typically will form more subdued terrain.

Relief

Relief of the land (its range of altitude) is another primary control on landscape development, which in turn is determined by the tectonic environment. In tectonically active regions where rates of uplift are high, mountains reach great altitudes, slopes are steep, and erosion rates and sediment yields are high. Such landscapes tend to be extremely dynamic. Measured rates of **denudation**, the sum of the weathering, mass-wasting, and erosional processes that result in the progressive lowering of the Earth's surface, generally are high. By contrast, in areas far removed from active tectonism, where relief is low, erosional processes tend to operate at

FIGURE 17.1 Ancient sedimentary strata of the McDonnell Ranges in central Australia have been broadly folded and differentially eroded into a series of aligned ridges.

TABLE 17.1 *Spatial Classes of Landscape Elements*

Order	Class	Examples	Typical Width
1	Major Earth features	Continents, ocean basins, lithospheric plates	>1000 km
2.	Major landform provinces	Mountain ranges, plateaus, sedimentary basins	100–1000 km
3	Major structural or constructional geologic features	Volcanoes, structural domes, fault-block mountains	10–100 km
4	Large-scale erosional and depositional units	Large valleys, ice-sheet moraines, small deltas	1–10 km
5	Medium-scale erosional and depositional units	Floodplains, cirques, drumlins, cinder cones, alluvial fans	100–1000 m
6	Small-scale erosional and depositional units	Sand dunes, eskers, taluses, beaches	10–100 m
7	Minor terrain features	Solifluction lobes, small gullies, channels of braided streams, rockfall boulders	1–10 m
8	Microterrain features	Small slumps, small-scale patterned ground	10–100 cm
9	Minor roughness features	Glacially grooved bedrock, sand ripples	1–10 cm
10	Microroughness features	Glacial striations, differentially weathered minerals in a rock	<1 cm

slower rates, and changes take place more gradually. However, even in nontectonic areas, relatively rapid changes of sea level or regional isostatic movements resulting from changing ice and water loads may initiate and control significant changes in the landscape over broad areas.

Time

Finally, the concept of landscape evolution necessarily involves the element of time. Although some landscape features can develop rapidly, even catastrophically, others quite obviously develop only over long geologic intervals. We know this, or at least we infer this, from measurements of surface processes now operating and by dating deposits that place limits on the ages of specific landforms or land surfaces.

Landscape Equilibrium

Change is implicit in the concept of landscape evolution. Presumably landforms or landscapes will experience change if a change occurs in any of the controlling variables. Change may be started by a tectonic event that causes a landmass to be uplifted, or by a drop of sea level that causes streams to

assume new gradients. It can be initiated by a shift in climate that modifies the relative effectiveness of different surface processes. A change may also result as a stream, eroding downward through weak rock, suddenly encounters massive hard rock beneath.

Over short intervals of time, rates of change may vary due to natural fluctuations of the magnitude and intensity of surface processes. Over longer intervals, the rate of change may increase due to more rapid tectonic uplift, or experience a gradual decrease as a land surface is progressively worn down and approaches the level of the sea.

Does a landscape, then, ever achieve a state of perfect equilibrium in which no change takes place? The answer apparently is no, for we have abundant evidence that the Earth's surface is now, and very likely always has been, a dynamic surface, constantly experiencing changes in response to the natural motions of the lithosphere, hydrosphere, and atmosphere.

Scales of Space and Time

Landscapes can be examined at different spatial scales ranging from very large to very small (Table 17.1). The largest obvious features of the Earth are

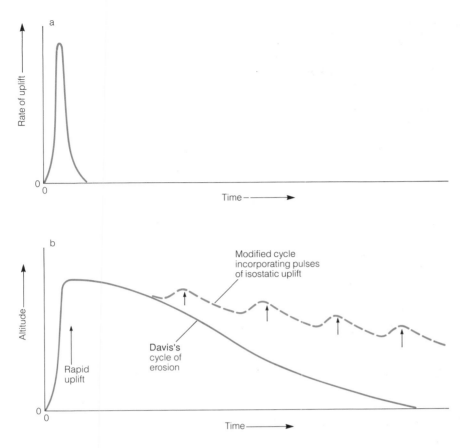

FIGURE 17.2 W. H. Davis's cycle of erosion. (a) Davis's cycle of erosion involves a pulse of rapid uplift during which uplift rate increases sharply, then declines. (b) As uplift rate increases, the land increases rapidly in altitude. As uplift rate drops, the uplifted land is gradually reduced by erosion. If isostatic adjustments to the removal of sediment are incorporated, a longer time is required to erode the land surface to low altitude.

continents and ocean basins, but we can also recognize major structural elements such as lithospheric plates, the boundaries of which do not always coincide with continental landmasses. Within the continents we can further identify major topographic features, most of which are related to broad-scale structure. These include mountain ranges, sedimentary basins, plateaus, and major rift systems. At smaller scales we can recognize further subdivisions that reflect specific structural, eruptive, lithologic, erosional, or depositional units. At the smallest scales of subdivision, single processes may be operative and changes may take place very rapidly. At the largest scales, however, a complex interplay of many different processes and factors occurs and major landscape features change very slowly. Thus, a landslide deposit or a lava flow are formed almost instantaneously in the geologic sense, whereas a mountain range will evolve only over millions of years.

In the case of small-scale features, geologists can observe and measure the active process, and see the resulting landform evolve. On much larger spatial scales, direct observation becomes impossible because of the extremely long spans of time over which major landscape features develop. Accordingly, we are left with several options. On the one hand we can observe processes operating under natural conditions and extrapolate their measured rates back through time. However, this is an uncertain approach at best, for we know that surface environments, and the magnitude and effectiveness of various surface processes, have differed greatly in the past. A second option is to substitute space for time; that is, we can examine landscapes in different stages of evolution and interpolate rates of processes between stages. Finally, we can construct theoretical models to explain the origin of the large-scale landforms and regional landscapes that are observed today.

Cycle of Erosion

The most influential theory of landscape evolution was proposed by an American geographer, W. M. Davis, in the late nineteenth century. Davis called his model the *cycle of erosion*, implying that it had a beginning and an end. A cycle was initiated by rapid uplift of a landmass, with little accompanying erosion, so that the initial relief was large. Erosion then progressively sculpted the land and reduced its altitude until it was worn down close to sea level (Fig. 17.2).

Davis deduced that a landscape passed through a series of stages. During the earliest stage, streams cut down vigorously into the uplifted landmass and produced sharp, V-shaped valleys, thereby increasing the local relief. Gradually, the original gentle upland surface is consumed as the drainage system expands and valleys become deeper and wider. During the next stage the land achieves its maximum local relief. Streams that have reached a graded condition begin to meander in their valleys, and valley slopes are gradually worn down by mass-wasting and erosion. In the final stage, the landscape consists of broad valleys containing wide floodplains, stream divides are low and rounded, and the landscape is worn down ever closer to sea level.

Davis's theory attracted wide attention and formed the basis for most interpretations of landscape evolution during the following decades. Elaborations of the theory subsequently were made to account for landscapes strongly influenced by lithologic, structural, and climatic controls. Davis also visualized interruptions of erosion cycles due to climatic fluctuations or to renewed uplift. Others later pointed out that isostatic response of the crust to the progressive transfer of sediment from the land to the ocean should lead to periodic uplifts, requiring that a cycle be longer than Davis envisaged (Fig. 17.2).

Landscape Evolution and Plate Tectonics

The major landscape features of the Earth have developed over long intervals of time as the lithosphere has evolved and continents have been continually rearranged. The lateral motions and resulting collisions of lithospheric plates, leading to the generation of orogenic belts, have provided much of the driving force for landscape change over hundreds of millions of years.

Crustal Convergence

Davis visualized a normal erosion cycle as beginning with a brief, sharp pulse of uplift, followed by a long interval of crustal quiet as the land was gradually worn down toward sea level (Fig. 17.2a). However, when a new belt of crustal convergence develops, uplift is likely to begin slowly and increase gradually to some maximum average value that is dependent on absolute rates of plate motion and on compensating isostatic adjustments (Fig. 17.3a). Such rates could continue as long as the relative plate motion is maintained, perhaps for

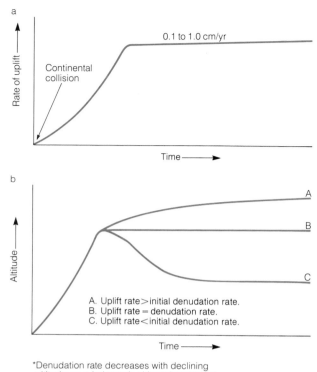

*Denudation rate decreases with declining altitude, so a balance is ultimately achieved between uplift and denudation.

FIGURE 17.3 Model of landscape evolution adjacent to a convergent plate boundary. (a) Rate of uplift increases following continental collision and is maintained at a level that is determined by the rate of convergence. (b) Altitude of the land surface increases following collision and then, depending on the relationship of uplift rate to denudation rate, (1) continues to rise slowly; (2) maintains a relatively constant altitude; or (3) declines in altitude to some steady-state condition.

many millions or tens of millions of years. Under such conditions, a landmass would be likely to increase gradually in altitude as the rate of uplift increases. Once a maximum uplift rate is reached, the land would slowly continue to increase in altitude if the uplift rate exceeded the denudation rate. Alternatively, it might maintain a certain average altitude if uplift and denudation rates were balanced, or it might decline in altitude if the denudation rate exceeded the uplift rate. Over time, the rising landmass might become deeply dissected by streams, resulting in extreme relief, as in the case of the Himalaya in Nepal (see below).

Spreading Rates and Uplift Rates

The model presented above is not completely realistic, for spreading rates and average uplift rates vary through time. Studies have shown, for example, that since late in the Cretaceous Period, the

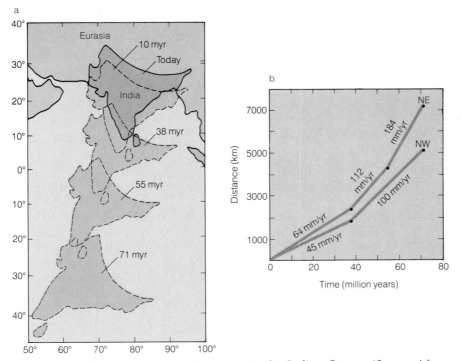

FIGURE 17.4 Variations in spreading rates in the Indian Ocean. (*Source:* After Molnar, Chen, and Tapponnier, 1981.) (a) Position of India with respect to Eurasia at different times since the Late Cretaceous Period, based on magnetic anomalies on the ocean floor. The two landmasses are thought to have collided between about 50 and 40 million years ago. (b) Distance of the northeast and northwest tips of India from their present positions as a function of time. About 40 to 50 million years ago, before collision, the rate of movement was more than 100 mm/yr, whereas afterward it decreased to about half as much.

rate of seafloor spreading in the Indian Ocean has undergone significant change (Fig. 17.4). Prior to about 40 million years ago, India moved progressively toward the Eurasian continent at rates averaging 100 to 200 mm/yr. As the two landmasses collided, the average rate of convergence slowed to only about 45 to 64 mm/yr (Fig. 17.4b).

Measured uplift rates in orogenic belts are also variable. The highest rates range between 1 and about 10 mm/yr, averaged over intervals of several thousand to several million years. However, it appears likely that average values have changed through time as rates of seafloor spreading and plate convergence varied. Each such change is likely to lead to a compensating adjustment in landscapes as they begin to evolve toward a new condition of equilibrium.

Models of Landscape Evolution

The example of landscape development shown in Fig. 17.3 is but one of many that could be postulated. By varying the rates of uplift, denudation, and isostatic adjustment to account for interactions of lithospheric plates on various time scales, geologists can develop different models to represent landscape evolution under a range of tectonic conditions.

Landscapes of Low Relief

The ultimate reduction of a landmass to low altitude is likely to occur only if changes in plate motion lead to diminishing orogeny and tectonic uplift ceases. Denudational processes can then gradually lower the relief. Examples of such landscapes can be found in the world's shield areas where the roots of ancient mountain systems have been exposed as the crust has thinned through the action of long-continued erosion and compensating isostatic adjustment.

Widespread erosional landscapes having low relief and low altitude are not commonplace. This must either mean that the Earth's crust has been very active in the recent geologic past or that such landscapes take an extremely long time to develop. Estimates have been made of the time it would take to erode a landmass to or near sea level by extrapolating current denudation rates into the past. Such estimates must take into account two important factors. Studies have shown that rates of denudation are strongly related to altitude and relief, implying that as the land is lowered the rate of denudation will decline. Furthermore, the eroding land will rise isostatically as the crust adjusts to transfer of sediment from the land to the sea. Both factors tend to increase the time it takes for the final reduction of a landmass to low altitude. Assuming that no tectonic uplift is taking place, estimates of the time it would take to reduce a landmass about 1500 m high to near sea level range from approximately 15 to 110 million years.

Although the Earth's crust has been extremely active during late Cenozoic time, could there have been earlier intervals of relative crustal quiet when low-relief surfaces did develop? Many ancient land surfaces are preserved in the geologic record as unconformities. Some can be traced over thousands of square kilometers and can be shown to possess only slight relief (Fig. 6.1). Associated with such surfaces are weathering profiles that imply long intervals of continuous weathering at relatively low altitude. Such buried paleolandscapes offer evidence of times when broad areas were eroded to low relief and may provide important clues about the history of lithospheric plates in the remote past.

Landscape Evolution and Climatic Cycles

Climate plays a fundamental role in determining the types and intensities of surface geologic processes at work on a landscape. Significant changes of climate can leave a record of changing conditions in landforms and sediments. On the time scale of glacial and interglacial ages, cyclic changes of surface environmental conditions related to astronomical variations of the Earth's orbit (chapter 12) have left their imprint on landscapes over much of the planet.

The growth and decay of continental ice sheets has repeatedly modified the land over which they moved. Streams have alternately deposited and then eroded sediment along their courses. Falling and rising sea level linked to changes in global ice volume have strongly influenced the evolution of coastal regions. Eolian sedimentation related to glacier expansion has alternated with interglacial periods of

landscape stability and soil formation. Changing temperature and precipitation patterns have led to differences in vegetation cover and to corresponding changes in the type, magnitude, and frequency of mass-wasting processes acting on the land surface. Thus, the landscapes we see about us often have had a complex history in which both plate tectonics and changing climate have played an important role.

LANDSCAPES OF PLANET EARTH

Every landscape has a story to tell. Each is the product of a past history of geologic events and climatic change which together have shaped the present surface. The following examples provide but a brief glimpse of the varieties of landscapes on our planet and the geologic forces that have generated them.

The Roof of the World

The lofty, glacier-clad Himalaya, rising steeply from the plains of India, includes within its array of peaks the highest summits in the world (Figs. 17.5 and 17.6). To the north of the range lies the vast upland of the Qinghai-Xizang (or Tibetan) Plateau (Fig. 17.6a) which, with average altitudes of 4000 to 5000 m, is the highest extensive land area on Earth (Fig. 17.6b). Both the Himalaya and the adjacent plateau are geologically young features, still in the process of formation. Together they provide an important example of what happens when converging lithospheric plates cause continents to collide and thereby produce a major mountain system. In this part of central Asia we can see young orogenic landscapes that are literally changing before our eyes as denudational processes attack the rapidly rising crustal rocks.

Drainage Patterns

Some streams that drain the high mountains and adjacent plateau are subsequent streams that follow belts of erodible rock or prominent structures to which they have become adjusted. Others are consequent streams and flow down the regional slope of the land. However, segments of four of the largest streams cut sharply across the trend of major structures as they flow from the southern plateau through deep valleys in the high mountains and out onto the Indian plains (Fig. 17.6a).

The northwest-flowing Indus River drains the northern slope of the western Himalaya, as well as

FIGURE 17.5 Oblique view of the Himalaya and the Qinghai-Xizang Plateau rising above the plains of India, as seen from an orbiting spacecraft.

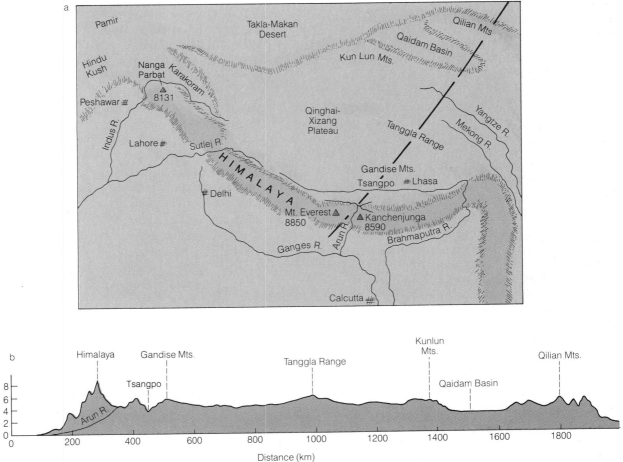

FIGURE 17.6 Himalayan drainage systems. (a) Major streams draining the Himalaya and southern Qinghai-Xizang Plateau. The Indus, Sutlej, Arun, and Tsangpo-Brahmaputra rivers originate on the plateau and flow across the structural axis of the Himalaya in deep valleys. (b) Topographic profile showing the high altitude of the plateau and the course of the Arun River as it crosses the axis of the Himalaya.

the lofty Karakoram and other nearby ranges beyond the western extremity of the plateau; however, it then changes course to pass directly across the mountains in a deep canyon beside the high massif of Nanga Parbat (8131 m). The Sutlej River originates north of the Himalaya, and after paralleling the crest for nearly 300 km turns abruptly southwest to cross the range axis between peaks that reach altitudes of more than 6000 m. The Arun River also begins north of the range crest, but crosses it through a deep canyon that lies 6500 m below the nearby summits of Chomolungma (Mt. Everest) (8850 m) and Kanchenjunga (8590 m). The north slope of the central and eastern Himalaya, as well as the southern plateau, are drained by the Tsangpo, which flows eastward more than 1000 km before it turns sharply southward across the range and then travels west along its southern base as the Brahmaputra River.

Drainage Evolution

Two hypotheses have been proposed to explain the cross-structural courses of these rivers. The first assumes the drainage to have originated as a series of consequent streams flowing down the north and south flanks of the mountain range as it began to rise. This initial pattern was modified because the most powerful south-flowing streams, due to greater discharge and steeper slopes on that flank of the range, were able to cut vigorously headward past the range crest and capture drainage on the northern side. The alternative hypothesis proposes an antecedent origin for these streams, the courses of which are believed to predate the Himalaya. When major uplift began, they were able to deepen their channels at a rate that matched or exceeded the rate of uplift and so could maintain their courses.

Several lines of evidence favor the hypothesis of antecedent origin. For example, as the upper Arun River turns south toward the Himalaya it flows across a body of hard gneiss in a remarkable deep gorge. Yet adjacent to it is a belt of weak, erodible schist. Had the river acquired its course by progressive headward erosion, it most reasonably would have cut back into the more erodible rock instead of through the hard gneiss, which could have been avoided. On the other hand, if the mountains rose across its ancestral course, the river would have become intrenched in the hard rocks as its channel was deepened by vigorous downward erosion.

Evidence of Major Uplift

Chinese geologists have suggested that the Himalaya did not exist as a major topographic feature before the middle of the Miocene Epoch (about 20 million years ago), and that the plateau remained at altitudes of less than about 2000 m throughout most of the Pliocene Epoch (circa 5.3 to 1.6 million years ago). Pliocene plant fossils have been found on the plateau at altitudes of 4000 to 6000 m in what are now alpine and cold desert environments. They include many subtropical forms, modern examples of which live at altitudes of less than 2000 m. Remains of a primitive horse adapted to subtropical forests further point to a mild, humid climate in the plateau region at that time. These fossils are widespread throughout Asia, suggesting that the horses could migrate freely over the whole continent at a time when no major mountain barriers existed.

Ancient karst topography has also been discovered high on the plateau in the frigid periglacial zone. It is similar to karst now developing at low altitudes in the warm monsoonal zone of southern

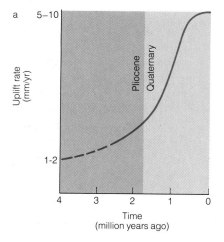

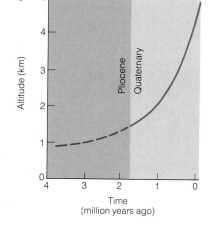

FIGURE 17.7 Changes in uplift rate and altitude of central Asia since the Pliocene Epoch. (a) Estimated uplift rate of Qinghai-Xizang Plateau during the last 4 million years, showing increasing rate during the Quaternary Period. (b) Increase in altitude of the plateau during the Quaternary, as inferred from geologic evidence.

China and is regarded, therefore, as further evidence of substantial uplift since its formation during the Pliocene.

The rate of uplift apparently increased at the beginning of the Quaternary Period when forests on the plateau were replaced by a treeless cool-climate vegetation and when thick alluvial deposits began to accumulate as the rising mountains were eroded.

Rates of Uplift

If the varied evidence has been correctly interpreted, the plateau and the bordering Himalaya must have experienced uplift of at least 3000 to 4000 m during the last 2 million years or less, with especially rapid uplift occurring during the last few hundred thousand years (Fig. 17.7a). Relatively low altitudes during the Pliocene should correlate with low rates of uplift (Fig. 17.7b), whereas high uplift rates during the Quaternary elevated both the Himalaya and the interior plateau to great altitude. Prior to large-scale Quaternary uplift, major rivers may have flowed south, draining the low-lying plateau. As uplift accelerated, only the largest streams were able to maintain their courses across the rapidly rising mountain system.

A Dynamic Landscape

Are the Himalaya and the plateau still growing today? Estimates of current uplift rates in the high mountains range from 2 to more than 5 mm/yr. Maximum denudation rates for drainage basins that lie entirely or partly within the Himalaya are between 0.5 and 2.5 mm/yr. Even if these figures are not very accurate, they nevertheless suggest that in some sectors the mountain system is increasing in altitude more rapidly than erosional forces can tear it down. It also means that the Himalaya and the great plateau to the north will remain areas of dynamic change as the landscapes continue to evolve.

Volcanoes in the Sea

Although we cannot directly observe the development of major second- and third-order landscape features (Table 17.1) because of the immense time involved, it would be instructive to examine a simple series of landforms that represent successive stages of landscape development. On the continents such landform series are not easy to find or interpret because of lithologic and structural complexities. However, nearly ideal examples do exist in midocean volcanic island chains. The Hawaiian Islands and the associated Emperor Seamounts, which together extend 6000 km across the Pacific Ocean Basin, are among the most instructive (Fig. 17.8).

Degree of Erosion

The American geologist James Dwight Dana first pointed out, in the mid-nineteenth century, that the Hawaiian Islands appear to increase in age up the chain from Hawaii to Kauai (Fig. 17.8). Dana based his judgment on the degree to which the islands have been dissected by erosion. The island of Hawaii, which includes several active volcanoes among the five that comprise it, is relatively unaffected by erosion except along part of its windward seacoast where some of the oldest lavas are found. The volcanoes of Maui and Molokai have been strongly eroded by streams and by wave action, while significant parts of the summits of Oahu, Kauai, and Niihau have been eroded away.

Ages of the Islands

Dana's hypothesis was later proved correct when lavas from the islands were dated by the K/Ar method and their magnetic polarity was measured. The oldest rocks exposed on Hawaii are less than 730,000 years old, the age of the last major magnetic reversal. The best available ages for the volcanoes of Maui, Lanai, and Kahoolawe are 0.8 to 1.9 million years, of Molokai 1.8 to 1.9 million years, of Oahu 2.6 to 3.7 million years, and of Niihau and Kauai 4.9 to 5.1 million years (Fig. 17.8). Geologic mapping has shown that the volcanoes have had rather similar histories, with broad, relatively simple shields of basalt being constructed as the islands emerged from the sea. By examining the eroded landscapes on successively older volcanoes, one can obtain a reasonable understanding of how the most ancient Hawaiian landscapes probably evolved.

The emergent islands of the Hawaiian group represent only the youngest and highest part of the chain. To the northwest lie a series of small islets and rocks that are remnants of former larger volcanic islands. Farther up the chain are found small atolls, including those of the Midway group, which consist of thick coral reefs that cap volcanic rocks lying at depths of 150 m or more. Still older volcanoes form the Emperor Seamounts, which trend northward toward the west end of the Aleutian Islands. The *seamounts* are isolated, submerged volcanic islands that rise many kilometers above the ocean floor. Fossils in sediment dredged from the top of a seamount at the northern end of the chain indicate an age of at least 70 million years. These now-eroded and submerged volcanoes must once have resembled the islands of Hawaii. Because fossils found in reef rocks atop the seamounts are of types that live only in shallow water, the reefs and the underlying volcanos must have experienced progressive submergence over millions of years.

Origin of the Island Chain

The Hawaiian–Emperor chain is believed to have formed as the Pacific Plate moved slowly north and then northwest across a midocean hot spot above which frequent and voluminous eruptions built a succession of volcanoes. Once formed, each volcanic island is carried slowly away from the hot spot toward cooler crust and into deeper water (Fig. 17.9). Ultimately, each now-submerged volcano reaches the Aleutian Trench and is accreted to the adjacent island arc.

Initial construction of a volcanic edifice proceeds rapidly. Based on the calculated average eruption rate ($0.05 \text{ km}^3/\text{yr}$) and volume ($42,500 \text{ km}^3$) of the island of Hawaii, it is estimated that less than a million years was required for its construction (Fig. 17.10). This is consistent with the known age of the oldest rocks on the island. During this constructional phase the rate of upbuilding far exceeds the rate of denudation, and the landscape assumes the form of coalescing shield volcanoes. As the vol-

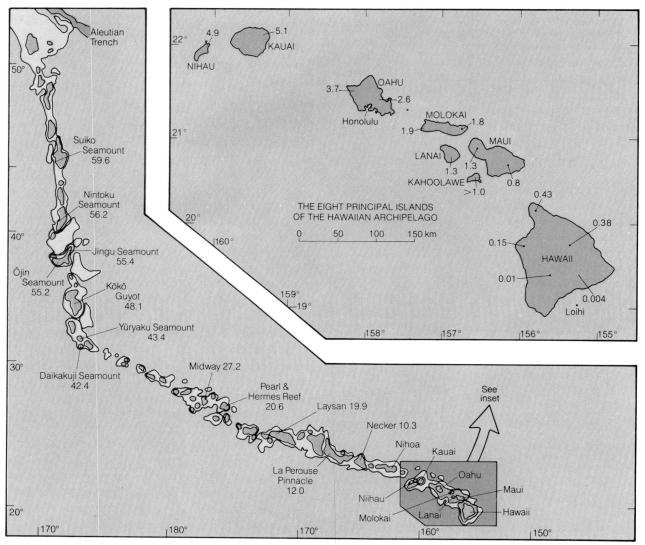

FIGURE 17.8 Hawaiian–Emperor chain, showing oldest reliable ages (in millions of years) for basaltic rocks of the volcanic shields. (*Source:* After Clague and Dalrymple, 1987.)

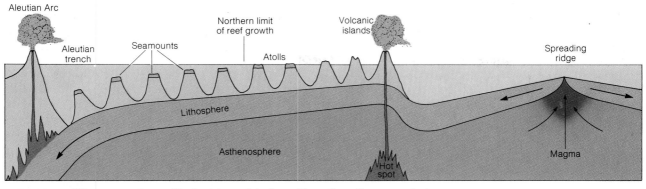

FIGURE 17.9 Diagrammatic profile (not to scale) along Hawaiian–Emperor chain illustrating how emergent volcanic islands are transformed into atolls and then seamounts as the Pacific Plate moves slowly away from the mid-Pacific hot spot toward the Aleutian Trench.

canic pile accumulates, the localized added weight on the seafloor causes isostatic subsidence of the ocean crust. This helps to limit the maximum altitude to which a volcanic island can rise. As the basaltic phase ends, and subsequent explosive eruptions of more silica-rich lavas decrease in frequency, erosion begins to take its toll.

Stream Dissection and Coastal Erosion

The moist trade winds are intercepted by the high Hawaiian volcanoes, which results in abundant precipitation on their windward slopes. Deep canyons are cut by streams whose discharge is enhanced by groundwater issuing from aquifers that

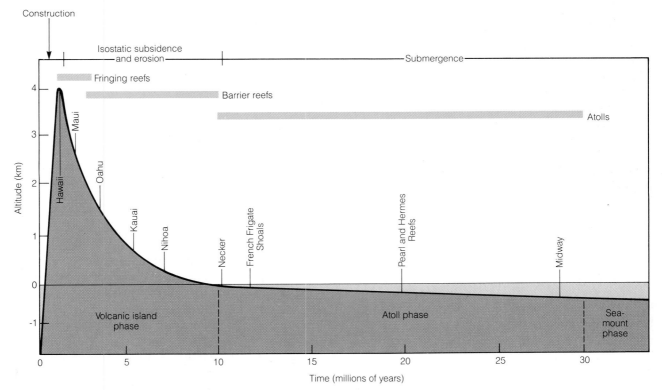

FIGURE 17.10 Change in altitude of a volcanic island as it is built up from the seafloor and then, through isostatic subsidence and erosion, is gradually reduced to sea level. Continued submergence occurs as the island moves away from the hot spot on the slowly cooling and subsiding lithosphere and becomes first an atoll and then a seamount. Mid-Pacific islands, from Hawaii to Midway, provide examples of different stages in the evolutionary sequence.

are confined within the porous sequence of thin lava flows by numerous near-vertical dikes. Simultaneously, waves attack the exposed coasts, producing steep cliffs that are prone to landsliding. In this manner an island becomes deeply dissected and its margins are worn back by wave attack. As volcanic activity diminishes and finally ceases, denudational processes and continuing submergence bring the landmass ever closer to sea level. In the case of the Hawaiian chain, this entire process takes about 10 million years (Fig. 17.10).

Coral Reefs and Seamounts

During the emergent phase of a tropical island's history, fringing reefs develop like those seen on many of the Hawaiian islands (Fig. 13.20a). These are gradually transformed into barrier reefs as the land slowly subsides. The Hawaiian Islands lack well-developed barrier reefs and enclosed lagoons, but good examples are found to the southwest in the Society Islands (Fig. 13.20b). As the rocky remains of the volcano subside beneath sea level, the bordering reef becomes an atoll dotted with small sandy islets (Fig. 13.20c). In the Hawaiian chain, this phase takes another 15 to 20 million years before slow northwestward movement of the Pacific Plate carries an atoll out of the tropical zone of coral growth and into ever-deeper water.

At this point the subaerial island landscape ceases to exist, except as a relict beneath the ocean where it may continue to experience slow modification with passing time. It has taken some 25 to 30 million years since initial outpouring of lava on the seafloor for the island to be transformed into a seamount, and it will take more than twice this long for the volcano and its coralline cap to complete its northward ride and be carried into the deep-ocean trench.

Glacial Landscapes of Northeastern North America

An airline passenger traveling the polar route from London to Vancouver or Seattle, who can arrange for a window seat, will be treated to one of the most scenic and geologically spectacular flights in the world. After passing Iceland astride the Mid-Atlantic Ridge, seas littered with icebergs, and the rugged highlands and enveloping ice sheet of Greenland, the plane takes a course across a landscape that emerged from beneath a vast continental glacier less than 10,000 years ago. First comes Baffin Island with its deep fjords (Fig. 12.14) and ice

caps that contain the last remnants of a thick ice sheet that covered northeastern North America during the last glacial age. On the west coast of the island, and on many other smaller islands north of Hudson Bay, can be seen multitudes of raised beaches that record isostatic uplift of the land as the ice sheet thinned and retreated (Fig. 14.24). West of Hudson Bay the plane passes over large areas of bare, scoured bedrock and elongate ridges and giant grooves that mark the passage of the glacier many millennia ago (Fig. 12.15). Ahead in the distance, the sun reflects off immense lakes, their basins eroded by successive continental ice sheets along the contact between hard metamorphic rocks of the Canadian Shield and more erodible sedimentary strata that overlie them. Before reaching the Rocky Mountains, the airplane passes over rolling morainal topography, dotted with numerous kettle lakes, that formed near the margin of the retreating ice sheets.

Landscape Zones

The different landscapes along the plane's route are identifiable along other transects across the continent. On a map, they form concentric zones characterized by distinctive landforms (Fig 17.11). Along the northeastern edge of the glaciated region, in the easternmost Canadian arctic islands, the landscape is characterized by large, glacially eroded troughs separated by uplands bearing few signs of glacial action. Moving southwest, there next comes an extensive interior zone in which most landscape features are attributable to ice scour; the belt of maximum erosion sweeps across southeastern and central Canada near the outside of this zone. Beyond lies a broad depositional zone that runs from New England across the upper midwestern United States and northern Great Plains into the Canadian Prairies. Within it end moraines and dead-ice deposits largely mask the underlying bedrock.

A model that attempts to explain this concentric landscape zonation relates each zone to conditions at the base of the former ice sheet (Fig. 17.12). In the far north, where little or no glacial erosion is evident, the ice sheet is presumed to have been very cold and frozen to its bed. Selective linear erosion is associated mainly with uplands where ice on high plateaus was cold and frozen at its base, but thicker ice in adjacent troughs was melting at its base and was able to erode. In the interior of the glacier, the ice was so thick that its great weight led to pressures high enough to produce basal melting. This warm basal ice was able to flow across

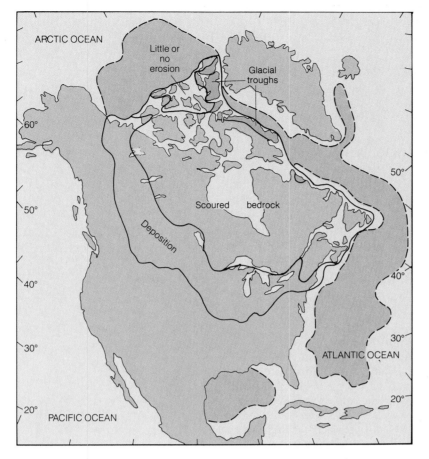

FIGURE 17.11 Map showing limit of last great Pleistocene ice sheet in eastern North America and concentric zones of deposition and erosion inside it. Outlined zones offshore represent principal regions where glacially eroded fine-grained sediments accumulated on the seafloor. (*Source:* After Bell and Laine, 1985; Sugden, 1978.)

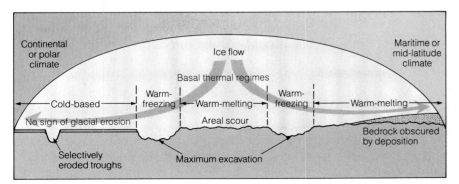

FIGURE 17.12 Diagrammatic profile through a continental ice sheet showing path of ice flow (bold arrows), conditions at the base of the glacier, and generalized erosional effects. Note the contrast in conditions along a margin having polar or continental climate with those at a margin marked by more temperate or maritime climate. (*Source:* After Chorley et al., 1984.)

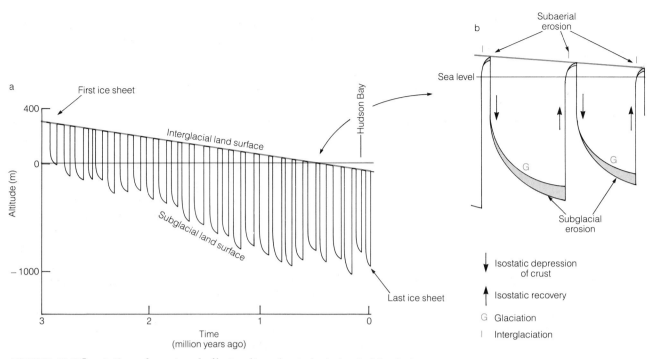

FIGURE 17.13 Isostatic and erosional effects of ice-sheet glaciation in North America. (a) Schematic model showing changes in altitude of land surface in Hudson Bay region owing to isostatic depression of the crust, subglacial erosion, and isostatic recovery as ice sheets repeatedly form, reach a maximum size, and disappear. Net erosion over several million years has lowered land surface by several hundred meters. (b) Expanded view of two glacial–interglacial cycles. Subaerial erosion during interglaciations is minimal compared to erosion by ice during glaciations.

and scour its bed. The ice became colder in the outer part of this zone where the glacier was thinner. As it began to freeze, debris was plucked off the bed, leading to increased erosion. In the more temperate environments along the western and southern margin of the ice sheet, the basal ice was again largely warm and melting. In this zone transported debris was deposited to form morainal topography.

Repeated Glaciations

The glacial landforms of the various zones we see are related largely to effects of the last glaciation and are comparatively young features. The gross form of the land, however, is due to repeated glaciations during the last 2 to 3 million years. Within the outer depositional zone are superimposed layers of glacial drift separated by weathering horizons. Each layer provides evidence of ice-sheet expansion and subsequent contraction. However, large gaps occur in this depositional record, so the exact number of ice-sheet invasions is not known. Some

former ice sheets were as large as the most recent one, but others apparently were smaller. The boundaries between erosional and depositional zones are therefore likely to have shifted from one glaciation to the next.

Amount of Erosion

How much has this landscape been changed as a result of glaciation? Attempts have been made to estimate the total erosion since glaciation began, but the results vary considerably depending on what is measured. Estimates based only on the volume of sediment found within the depositional zone of the ice sheet clearly underestimate the total, for sediment was also carried beyond the glacial limit by meltwater streams and some was reworked by the wind. The bulk of the fine load reached the adjacent ocean basins where it accumulated on the seafloor (Fig. 17.11). If all the known sediment eroded from the area of the former ice sheet is summed, it totals close to 1.6 million km^3. This is the equivalent of eroding away a layer of solid rock

120 m thick from the glaciated area. Of course, the rock was not eroded equally from all the landscape. Instead, denudation was probably greatest in the interior parts of the glaciated zone where glacial erosion is likely to have been most effective and the ice was active for the longest time. It is possible that the landscape has been eroded down at least several hundred meters since the first ice sheet formed (Fig. 17.13).

Cyclic Landscape Development

The present landscape of this large region of North America has developed, therefore, very late in geologic history. Although the average rate of denudation over the last 3 million years has been quite low, the actual rate has been quite variable, because long intervals marked by active glacial erosion and deposition alternated with shorter intervals when less-effective nonglacial agencies were operating

(Fig. 17.13). If the last glacial cycle is representative of earlier ones, then during each a large portion of central Canada may have been continuously covered by glacier ice for many tens of thousands of years. The outer parts of the glaciated region were covered for shorter intervals when the ice sheets expanded to their greatest size. During interglacial ages, which may have averaged no more than about 10,000 years long, the land was largely free of ice. At such times, nonglacial processes did little to modify the landscape which remained, as it is today, largely a relict of the previous glaciation.

Landscape development in this region of fluctuating ice cover thus has been marked by pulses of erosion and deposition alternating with brief intervals of relative landscape stability. This vast landscape is one in which locally significant changes in surface altitude have been related to the waxing and waning of continental ice sheets that resulted in isostatic depression and subsequent rise of the underlying crust.

Essay

WHEN AUSTRALIA WAS WET

The landscape of Western Australia is dominated by a vast Precambrian shield composed of some of the most ancient rocks on the Earth. Rocks older than 2.5 billion years crop out over an area of about a million km² where the surface altitude generally lies between 100 and 500 m.

Landforms are strongly controlled by rock type, structure, and relative erodibility. Granites tend to form low, slightly rounded outcrops. Intervening greenstones display little relief except where thin beds of iron-rich chert, which are more resistant to weathering, form low, sharp ridges. Quartzites and layered iron formations are very resistant to erosion and form bold outcrops.

Most of the Western Australia Shield receives only about 200 to 300 mm of precipitation a year. Because of such arid conditions, the region is characterized by sparse vegetation and poorly developed stream systems. Rainfall soon evaporates, and the little that does run off accumulates in shallow salt-encrusted playas. No large rivers flow off the shield.

Weathering as deep as hundreds of meters suggests that the shield must have been exposed at the surface for a very long time, during which its altitude remained low, and that at some time in the past it must have had a wetter climate. The land surface has now reached a state of near-equilibrium and the shield is a region with a denudation rate close to zero.

At many places on the shield it is possible to find remnants of an old lateritic surface (Fig. B17.1). Laterites are characteristically formed on landscapes having low relief, and in warm regions with high rainfall. They are forming today, for example, in such places as southeastern Asia and the Amazon Basin of tropical South America. The Western Australia laterites are not forming in today's arid climate. Their age is thought to be at least early Tertiary and possibly as old as Jurassic. Therefore the shield has been a region of low relief for more than 30 million years and possibly as long as 200 million years. Furthermore, the landscape must have experienced a significant change of climate since the laterite formed.

FIGURE B17.1 Subdued landscape of Western Australia Shield with Tertiary laterite at the surface.

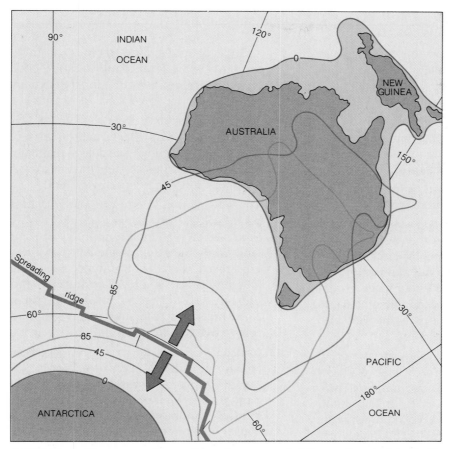

FIGURE B17.2 The progressive northward movement of the continent of Australia (including New Guinea) away from the spreading ridge to the south has carried it into more tropical latitudes. Positions of the continent, and of the Antarctic continent south of the spreading ridge, are shown at 85 and 45 million years ago and at present.

Today the shield lies between 20° and 35° S latitude. These are the latitudes where dry subtropical air causes major deserts to form. However, Australia is slowly drifting north. The spreading rate between the Australian and Antarctic plates is about 7 cm/yr. The absolute motion of the Antarctic Plate (relative to a fixed point on the Earth) is believed to be small, probably no more than 1 to 2 cm/yr. This suggests that Australia is moving north with an absolute motion of about 5 cm/yr (Fig. B17.2). The climate of the world in the early Tertiary was much warmer than it is today, so at the measured average rate of movement, 40 million years ago the Western Australia Shield would have been some 2000 km south of its present position in a temperate zone of high rainfall like that of southern Chile today. A warm early Tertiary climate and high rainfall probably explains the origin of the old laterite. Like a ship slowly sailing north, the Western Australia Shield is passing through different climatic zones. Today it lies in the dry subtropical belt. However, if the present direction and rate of motion continue for another 20 million years, it will have entered a zone of tropical climate.

SUMMARY

1. The development of distinctive landforms is influenced by process, climate, lithology, rock structure, relief, and time.

2. In tectonic settings, high uplift rates and steep slopes lead to high rates of denudation. In stable regions, denudation rates may be low, but landscape changes can be initiated by relative movements of land and sea level.

3. Landscapes evolve over long intervals of time toward conditions of equilibrium.

4. Small landforms that develop over short time periods can be observed in the process of formation. The evolution of major landscape elements over long time intervals can be inferred by extrapolating modern rates of processes, observing sequential landscapes in different stages of evolution, or by developing theories and models.

5. W. M. Davis's "cycle of erosion" attempted to explain landscape evolution as a series of stages through which an uplifted landmass was progressively reduced to a surface of low relief. Elaborations of this theory invoked the specific influences of lithology, structure, climate, and isostatic movements on landscape development.

6. Landscape evolution is now being reassessed in terms of plate tectonics theory. Different models are needed to explain the development of landscapes under a wide range of tectonic environments.

7. Reduction of landmasses by denudation to low altitude requires long periods of time. Although few modern examples can be cited, ancient unconformities of regional extent are evidence of former low-relief landscapes.

8. Major streams that drain the Qinghai-Xizang Plateau and cross the structural axis of the Himalaya apparently were able to maintain their courses by vigorous erosion and keep pace with the rapidly rising mountain system.

9. The Hawaiian–Emperor chain of volcanoes has been built as the Pacific Plate traveled north, then northwest across a midplate hot spot. Island landscapes reflect the complex effects of volcanism, isostatic subsidence, and fluctuating climate and sea level. Volcanic islands ultimately subside beneath sea level to become coral atolls and then submerged seamounts as they move toward the Aleutian Trench.

10. Concentric landscape zones in northeastern North America reflect conditions at the base of a vast ice sheet during the last glacial age. Landscape development over the last several million years has been marked by episodes of glacial erosion and deposition that alternated with intervals of relative landscape stability during interglacial times.

11. Landforms of the Western Australia Shield are strongly controlled by lithology and structure. Deep weathering and a widespread laterite of Tertiary age developed when the shield lay farther south in a zone of moister climate.

IMPORTANT WORDS AND TERMS TO REMEMBER

cycle of erosion (p. 411)   *denudation* (p. 408)   *landscape equilibrium* (p. 410)   *relief* (p. 408)

QUESTIONS FOR REVIEW

1. How does climate help determine the course of landscape evolution?

2. What factors control denudation rates?

3. Why are landscapes unlikely ever to reach a condition of perfect equilibrium?

4. Why has Davis's "cycle of erosion" received decreasing emphasis since the theory of plate tectonics was formulated?

5. Why might rates of uplift in tectonic regions change through time?

6. What is the evidence in the geologic record that some regions have been reduced to low relief and low altitude at times in the past?

7. How do climatic changes on a glacial–interglacial time scale control the nature and

magnitude of surface geologic processes involved in landscape evolution?

8. What explanation(s) can be advanced for the courses of major rivers that drain the southern Qinghai-Xizang Plateau *across* the structural axis of the Himalaya?

9. Why do the Hawaiian Islands increase systematically in age northwestward from the island of Hawaii?

10. What role does subsidence play in the evolution of the Hawaiian–Emperor Chain?

11. How can you explain the concentric zonation of landscapes in eastern North America, centered in the region of Hudson Bay?

12. How might one estimate the average thickness of rock eroded from the region of ice-sheet glaciation in North America?

13. How does plate tectonics help explain the existence of a widespread ancient laterite in the cratonic region of Western Australia?

QUESTIONS FOR REVIEW OF PART 3

1. Draw a section through a hypothetical continent; show its main features: continental shield, platform, orogenic belts, continental shelf, continental slope, and continental rise.

2. What criteria might be used to determine whether an accreted terrain came from nearby or had been moved many thousands of kilometers?

3. According to plate tectonics there are three different kinds of plate boundaries.

    a. Name the three kinds of boundaries.

    b. For each kind of boundary illustrate with a simple sketch

        i. the relative motion of the plates on either side of the boundary;

        ii. the characteristic earthquake activity;

        iii. typical land or undersea topography;

        iv. characteristic magmatic activity.

    c. Name a modern example of each kind of boundary.

    d. How would you recognize an ancient plate boundary?

4. With which kind of plate boundary is a mélange zone associated? How does it form, and what kind of metamorphism is sometimes associated with mélange zones?

5. What is the difference between absolute and relative velocities of tectonic plates? How are absolute velocities measured? Are plate velocities determined by the use of satellites and lasers absolute or relative velocities?

6. Draw a section showing the relation between a deep-sea trench, a fore-arc basin, a magmatic arc, and a back-arc basin. What kind of sediments would you expect to find in a fore-arc basin? Would sediments in fore-arc basins be different from those in back-arc basins?

7. In the context of plate tectonics, explain the following: (a) the Cascade Range; (b) the San Andreas Fault; (c) earthquakes with foci deeper than 300 km; (d) rocks of similar age and structure in eastern Canada and western Norway.

8. What conditions must prevail for the development of widespread erosion surfaces close to sea level?

9. Describe why the Hawaiian Islands–Emperor seamount chain provides a good example for analyzing how landscapes evolve over geologically long periods of time.

10. What kind of volcanism is characteristic of island arcs? How is the magma that causes arc volcanism generated? Identify at least three modern island arcs.

11. Describe in words or by sketches the sequence of events that take place when a spreading edge splits continental crust and a rift develops. Identify the way sediments change as the rift widens and becomes an ocean basin. Can you name modern examples of the steps you have outlined?

A blending of natural materials in a Moghul tomb, Agra, India. Screens are carved from individual slabs of marble 5 cm thick. Colored minerals inset into marble walls produce intricate patterns.

The Earth's Resources

CHAPTER 18

Resources of Energy and Minerals

Cube-shaped crystals of galena and reddish-colored sphalerite on calcite, located in Joplin, Missouri.

MATERIALS AND ENERGY

The industrial age is a time of enormously increased use of fuels and metals. Another term would be *the age of intensive use of energy in industry*. Stone Age people had an industry too: They chipped and flaked mineral substances (mostly pieces of flint) to make tools and weapons. However, theirs was an industry with a low input of energy because the energy was supplied by human muscle. Although they were limited by this fact and by a very narrow choice of materials with which to work, we must not underestimate them. They were as intelligent as we are, lacking only the skill that comes from accumulated experience.

This lack of skill was gradually overcome by Stone Age people and their descendants. Here are a few of their accomplishments, discovered and dated by archaeologists:

4000 B.C. Chaldeans, who lived at the head of the Persian Gulf in a province of ancient Babylonia, had become skilled workers in metals such as gold, silver, copper, lead, and tin.

3000 B.C. Eastern Mediterranean people were making glass, glazed pottery, and porcelain.

2500 B.C. Babylonians were using petroleum instead of wood for fuel.

1100 B.C. Chinese were mining coal and were drilling wells hundreds of feet deep for natural gas.

These accomplishments were arts learned by experience, and they implied the substitution of metals, glass, and other substances for stone. However, the making of a copper implement used more energy than the manufacture of a stone tool. Yet more was needed to make objects of glass. So with each advance, more energy and, therefore, more fuel were needed. Most of the energy for working iron, copper, lead, and other materials came from wood fuel and from the muscles of men and animals. But as time passed other fuels came to be used, such as coal, oil, and natural gas. The new fuels were more efficient than the old. By using them, each man could produce many more implements than he could have produced with wood fuel or with muscle. Therefore, the new fuels sparked rapid growth of the amounts of metals used.

Uses of Energy

Metals to build machines and the energy needed to power the machines are won from the Earth. Supplies of both metals and energy have been formed through geological processes. Geological processes therefore influence the daily life of each and every one of us.

Metals and energy are complementary. Machines made of metal require fuels to supply the energy to run them. As more machines are built (in order to develop transportation systems, to build houses, or to till fields and grow crops) so does the use of energy increase.

A healthy, hard-working person can produce just enough muscle energy to keep a single 100-watt light bulb burning during an 8-hour working day. It costs about 10 cents to purchase the same amount of energy from the local electrical source. Viewed strictly as machines, humans aren't worth much. By comparison, the amount of machine and electrical energy used each 8-hour working day in North America could keep 300 of those bulbs burning for every person living there.

To see where all the energy is used it is necessary to look at society as a whole and sum up all the energy that is employed to grow and transport food, make clothes, cut lumber for new homes, light streets, heat and cool office buildings, and to do myriad other things. The uses can be grouped into three categories: transportation, home and industry (meaning all manufacturing and raw material processing, plus the growing of foodstuffs), plus commercial uses. The present-day uses of energy in the United States are summarized in Figure 18.1.

How much energy do all the people of the world use? The total is enormous. The supplementary energy drawn annually from the major fuels such as coal, oil, and natural gas, and from nuclear power plants is 2.6×10^{20} J. Nobody keeps accurate accounts of all the wood and animal dung burned every day in the cooking fires of Africa and Asia, but the amount has been estimated to be so large that when they are added, the world's total energy consumption rises to about 3.0×10^{20} J. This is equivalent to the burning of 2 metric tons of coal or 10 barrels of oil for every living man, woman, and child each year!

Supplies of Energy

The chief sources of energy consumed in highly industrialized nations are few: the fossil fuels (coal, oil, natural gas), hydroelectric power, nuclear energy, wood, wind, and a very small amount of muscle energy. As recently as a century ago, wood was an important fuel in industrial societies, but now it is used mainly for space heating in some dwellings.

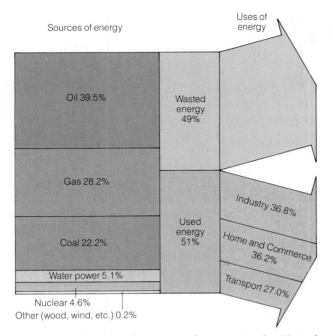

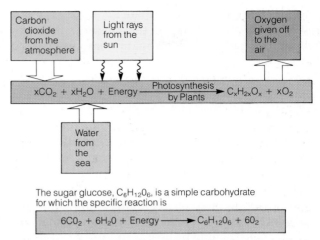

The sugar glucose, $C_6H_{12}O_6$, is a simple carbohydrate for which the specific reaction is

$$6CO_2 + 6H_2O + Energy \longrightarrow C_6H_{12}O_6 + 6O_2$$

FIGURE 18.2 Photosynthesis is the process by which plants combine carbon dioxide and water, using the Sun's energy to produce carbohydrates which have the general chemical formula $C_xH_{2x}O_x$, where x is always a whole number.

FIGURE 18.1 Uses and sources of energy in the United States based on data gathered in 1985. Wasted energy arises both from inefficiencies of use and from the fact that the laws of thermodynamics impose a limit to the efficiency of an engine, and therefore of the fraction of energy that can be usefully employed.

The breakdown of energy sources and uses in the United States is shown in Figure 18.1. In Europe, both energy supply and use are different from that shown in Figure 18.1: Coal accounts for about half the energy consumed, and industry for more than 40 percent of the use. For the world as a whole, with its hundreds of millions of agricultural workers, the breakdown is different again; wood is a more substantial source of energy, and fossil fuels, especially oil and gas, account for less.

Because the production and use of the Earth's mineral supplies requires energy to do the necessary work, we will first discuss the principal sources of energy, and then the sources of mineral supplies.

FOSSIL FUELS

The term *fossil fuel* refers to the remains of plants and animals, trapped in sediment that can be used for fuel. Fossil fuels occur in many ways, depending on the kind of sediment, the kind of organic matter trapped, and the changes that have occurred during the long geological ages since the organic matter was trapped.

Photosynthesis

Essentially all living organisms derive their energy from the Sun. The principal energy-trapping mechanism is photosynthesis. Plants use the Sun's energy to combine water and carbon dioxide to make carbohydrates and oxygen as shown in Figure 18.2.

The fuel that keeps animals alive and active is the organic compounds in plants; animals are, therefore, secondary consumers of trapped solar energy. When one animal eats another, a little bit of trapped solar energy is once again passed along. When plants or animals die and decay, oxygen from the atmosphere combines with carbon and hydrogen in the organic compounds to form H_2O and CO_2 once again. In the process the trapped energy is released, so the photosynthesis reaction is reversed.

Burial of Organic Matter

The rates at which organic matter is formed through photosynthesis, and broken down by decay, are essentially the same; if they were not essentially equal, the world would soon be covered by increasingly deep piles of organic matter. However, the growth and decay rates are not *exactly* the same. In many sediments, a little organic matter is buried before it is completely removed by decay. In this way some of the solar energy becomes stored in rocks—hence, the term *fossil fuel*. The amount of trapped organic matter is far less than 1 percent of the organic matter formed by growing plants and animals. However, from the late Proterozoic (about 600 million years ago) to the present, the total amount of trapped organic matter has grown to be very large.

Proteins, Lipids, and Carbohydrates. The kind of organic matter that is trapped in sediment plays an important role in the kind of fossil fuel that forms. In the ocean, microscopic photosynthetic phytoplankton and bacteria are the principal sources of trapped organic matter. Shales do most of the trapping. Bacteria and phytoplankton contribute mainly organic compounds called *proteins, lipids,* and *carbohydrates* and it is these compounds that are transformed (mainly by heat) to oil and natural gas.

Resins, Waxes, Lignins, and Cellulose. On land, it is large plants such as trees, bushes, and grasses that contribute most of the trapped organic matter; they contain carbohydrates but they are also rich in resins, waxes, and lignins, which tend to remain solid and form coals.

Kerogen. In many shales, burial temperatures never reach the levels at which the original organic molecules are completely broken down. Instead, what happens is that an alteration process occurs in which waxlike substances with large molecules are formed. This material, called *kerogen,* is the substance in oil shales; it can be converted to oil and gas by applying sufficient heat.

Coal

The combustible sedimentary rock we call coal is the most abundant of the fossil fuels. Most of the coal that is mined is eventually burned under boilers to make steam for electrical generators, or it is converted into coke, an essential ingredient in the smelting of iron ore and the making of steel. In addition to its use as a fuel, coal is a raw material from which nylon and many other plastics, plus a multitude of organic chemicals, can be made.

Origin of Coal

Coal occurs in strata (miners call them seams) along with other sedimentary rocks, mostly shale and sandstone. A look through a magnifying glass at a piece of coal reveals the shapes of bits of fossil wood, bark, leaves, roots, and other parts of land plants, chemically altered but still identifiable (Fig. 18.3). A definition of *coal* is a black sedimentary rock consisting chiefly of decomposed plant matter

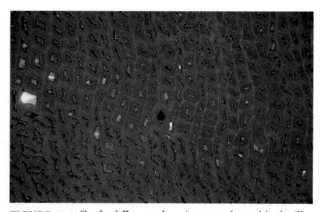

FIGURE 18.3 Coal of Pennsylvanian age from Nashville, Illinois, showing cellular structure in a fossilized fragment of wood from a plant called *Lycopod periderm.* The field of view is approximately 0.3 cm across.

and containing less than 40 percent inorganic matter.

Coal Swamps. It was recognized long ago that places where coal accumulated were ancient swamps, because (1) a complete physical and chemical gradation exists from coal to peat, which today accumulates mainly in swamps, and (2) only under swamp conditions is the conversion of plant matter to coal chemically probable. On dry land and in running water, oxygen is abundant and dead plant matter gradually rots away. However, under stagnant or nearly stagnant swamp water, oxygen is used up and not replenished. Instead, the plant matter is attacked by anaerobic bacteria which partly decompose it by splitting off some of the oxygen and hydrogen. These two elements escape, combined in various gases, and the carbon gradually becomes concentrated in the residue. Although they work to destroy the vegetal matter, the bacteria themselves are destroyed before they can finish the job, because the poisonous acid compounds they liberate from the dead plants kill them. This could not happen in a stream because the flowing water would bring in new oxygen to decompose the plants and would also dilute the poisons and permit the bacteria to complete their destructive process.

Coalification. With the destruction of bacteria the first stage of coal has been reached and the plant matter has been converted to *peat.* But as the peat is buried beneath more plant matter and beneath accumulating sand, silt, or clay, both the temperature and pressure increase. These bring about a series of continuing changes called *coalification* (Fig. 18.4). The peat is compressed, water squeezed out,

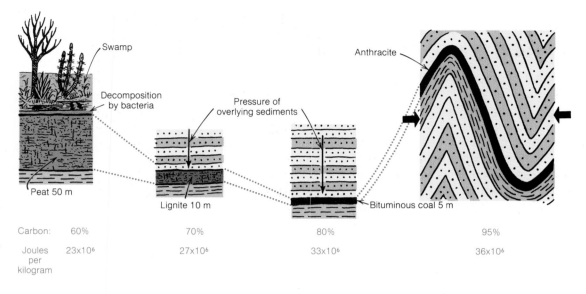

FIGURE 18.4 Plant matter is converted into coal by decomposition, pressure, and heat. By the time it has become bituminous coal, a layer of peat has decreased to one-tenth of its original thickness. During the same time the proportion of carbon has increased and the calorific value (the amount of heat energy obtained when a kilogram of coal is burned) has risen continually.

and the volatile organic compounds such as methane (CH_4) escape leaving an increased proportion of carbon. The peat is converted successively into *lignite, subbituminous coal,* and **bituminous coal.** These coals are sedimentary rocks. However, a still-later phase, *anthracite,* is a metamorphic rock.

Since anthracite occurs in folded strata that have been subjected to low-grade metamorphism, we infer that it has undergone a further loss of volatiles and carbon concentration as a result of the pressure and heat that accompany folding and metamorphism.

Because of its low content of volatiles, anthracite is hard to ignite, but once alight, it burns with almost no smoke. In contrast, lignite is rich in volatiles, burns smokily, and ignites so easily that it is dangerously subject to spontaneous ignition.

In certain regions where metamorphism has been intense, coal has been changed so thoroughly that it has been converted to graphite, in which all volatiles have been lost, leaving nothing but carbon. Graphite, therefore, will not burn in an ordinary fire.

Occurrence of Coal

A coal seam is a flat, lens-shaped body corresponding to the area of the swamp in which it originally accumulated. Most coal seams are 0.5 to 3 m thick, although some reach more than 30 m. They tend to occur in groups. In western Pennsylvania, for example, 60 beds of bituminous coal are found. This indicates that the coal must have formed in a slowly subsiding site of sedimentation.

Coal swamps seem to have formed in many sedimentary environments, of which two predominated. One consists of slowly subsiding basins in continental interiors and the swampy margins of shallow inland seas formed at times of high-sea stands. This is the home environment of the bituminous and subbituminous coal seams in Utah, Montana, Wyoming, and the Dakotas. The other consists of continental margins with wide continental shelves that were also flooded at times of high sea level. This is the environment of the bituminous coals of the Appalachian region. That same environment exists along the east coast of North America today, and it is there that one of the largest modern peat swamps is to be found. The Dismal Swamp, in Virginia and North Carolina, with an area of 5700 km^2, contains an average thickness of 2 m of peat and is a modern analogue to swamps in which material was deposited that became coal.

Cyclothems. In many coal basins, swamps were formed repeatedly as the sea level rose and fell, or as the local area was uplifted and submerged. What is observed today is a repetitive sequence of sedimentary rocks—sandstone, shale, coal, sandstone, shale, coal, and so on—reflecting cyclic deposition. Each unit is a repeated sequence of peat and clastic sediments called a *cyclothem.* As a result of cyclic deposition, most of the world's coal basins contain many coal seams that overlie one another.

Coal-Forming Periods

Although peat can form under even subarctic conditions, it is clear that the luxuriant plant growth needed to form thick and extensive coal seams develops most readily under a tropical or semitropical climate. Unless Dismal Swamp lasts for an exceptionally long time, even that dense growth is probably insufficient to produce, ultimately, a coal seam as thick as some of the seams in Pennsylvania. This means that either the global climate was very much warmer, or the swamps in which most of the world's coal seams formed were within 30° of the equator, when the plant matter accumulated. Probably both effects were involved.

Peat formation has been widespread and more or less continuous from the time land plants first appeared about 450 million years ago during the Silurian Period. The size of peat swamps has varied greatly, however, and so, as a consequence, has the amount of coal formed. By far, the greatest period of coal swamp formation occurred during the Carboniferous and Permian, when Pangaea existed. The great coal beds of Europe and the eastern United States formed at this time, when the plants of coal swamps were giant ferns and the so-called scale trees (gymnosperms). The second great period of coal deposition peaked during the Cretaceous, but commenced in the early Jurassic and continued until the mid-Tertiary. The plants of the coal swamps during this period were flowering plants (angiosperms), much like flowering plants today.

Petroleum: Oil and Natural Gas

Oil

Rock oil is one of the earliest products our ancestors learned to use. Natural oil seeps occur in the valleys of the Tigris and Euphrates Rivers in Iraq and the people living in those regions have used the oil since before recorded history. Some of the ways Babylonians, Assyrians, and other ancient people used oil were (1) to make bitumen glue to hold arrowheads on spears; (2) to set tiles; (3) to make a mortar for cementing building bricks together; (4) to waterproof boats; and (5) to embalm bodies. In later times, people in the Arab world and China used oil for many additional purposes. However, the major use of oil really started about 1847 when a merchant in Pittsburgh, Pennsylvania, started bottling and selling rock oil from natural seeps to be used as a lubricant. Five years later, in 1852, a Canadian chemist discovered that distillation of both rock oil and coal yielded kerosene, a liquid that could be used in lamps. This discovery spelled doom for candles and whale-oil lamps. Wells were soon being dug by hand near Oil Springs, Ontario, in order to increase the yield of oil. In Romania, using the same hand-digging process, oil production in 1857 was 2000 barrels a year.* In 1859, the first oil well was drilled in Titusville, Pennsylvania. On August 27, 1859, at a depth of 21.2 m oil-bearing strata were encountered and up to 35 barrels of oil a day were pumped out. Oil was soon discovered in West Virginia (1860), Colorado (1862), Texas (1866), California (1875), and many other places.

Natural Gas. The earliest known use of natural gas was more than 3000 years ago in China, where gas seeping out of the ground was collected and transmitted through bamboo pipes to be used to evaporate salt water in order to recover salt. Modern uses of gas started in the early seventeenth century in Europe, where gas made from wood and coal was used for illumination. Commercial gas companies were founded as early as 1812 in London, and 1816 in Baltimore. The stage was set for the exploitation of an accidental discovery at Fredonia, New York, in 1821. A water well drilled in that year produced not only water, but bubbles of a mysterious gas. The gas was accidentally ignited and produced such a spectacular flame that a new well was drilled on the same site, and wooden pipes were installed to carry the gas to a nearby hotel, where 66 gas lights were installed. By 1872, natural gas was being piped as far as 40 km from its source.

Petroleum. Oil and gas are the two chief kinds of petroleum. *Petroleum* is defined as gaseous, liquid, and semisolid substances, occurring naturally and consisting chiefly of chemical compounds of carbon and hydrogen. Oil and gas occur together and are searched for in the same way. Therefore, discussion of oil pools, oil exploration, and the origin of oil should be understood to mean not only oil but gas as well.

---

*A barrel is equal to 42 U.S. gal and is the volume generally used when commercial production of oil is discussed.

Occurrence of Oil

Oil Pool. Oil possesses two important properties that affect its occurrence: It is fluid, and it is lighter than water. Oil is produced from pools (an ***oil pool*** is an underground accumulation of oil and gas in a reservoir limited by geologic barriers). The word *pool* sometimes gives a wrong impression because an "oil pool" is not a lake of oil. It is a body of rock in which oil occupies the pore spaces.

Oil Field. A group of pools of similar type, or a single pool in an isolated position, constitutes an ***oil field.*** The pools in a field can be side by side or one on top of the other.

Accumulation. For oil or gas to accumulate in a pool, five essential requirements must be met: (1) A ***reservoir*** must be present to hold the oil, and this rock must be permeable so that the oil can percolate into it. (2) The reservoir rock must be overlain by a layer of impermeable ***roof rock,*** such as shale, to prevent upward escape of the oil, which is floating on groundwater. (3) The reservoir rock and roof rock must form a ***trap*** that holds the oil and prevents it from moving any farther under the pressure of the water beneath it (Fig. 18.5).

Although the presence of a reservoir, roof, and trap are essential, they do not guarantee a pool. In many places where they occur together, drilling has shown that no pool exists, generally because of lack of a source from which oil could enter the trap. So, to the foregoing requirements two others must be added. (4) A ***source rock*** must provide oil; and (5) the deformation that forms the trap must occur *before all the oil has escaped* from the reservoir rock.

Origin of Oil

Petroleum is a product of the decomposition of organic matter of both plant and animal origin. This statement is based principally on two observations: (1) oil possesses optical properties known only in hydrocarbon compounds derived from organic matter; and (2) oil contains nitrogen and certain compounds (porphyrins) that scientists believe can only originate in living matter.

Oil is nearly always found in marine sedimentary strata. Indeed, in places on the seafloor, particularly on the continental shelves and at the bases of the continental slopes, sampling has shown that fine-grained sediment contains up to 8 percent organic matter. Thus, geologists conclude that oil

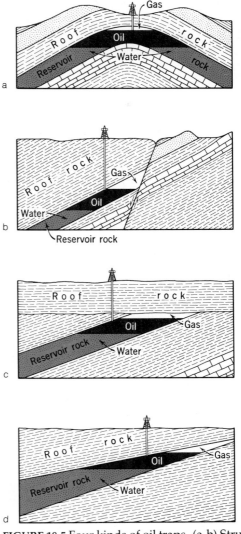

FIGURE 18.5 Four kinds of oil traps. (a,b) Structural traps; (c,d) stratigraphic traps. In part c an unconformity marks the top of the reservoir; in part d a porous stratum thins out and is overlain by an impermeable roof rock. Gas (white) overlies oil (black), which floats on groundwater (blue), saturates reservoir rock, and is held down by roof of claystone. Oil fills only the pore spaces in the rock.

originated as organic matter deposited with marine sediment.

A long and complex chain of chemical reactions is apparently involved in the conversion of the original organic constituents to crude petroleum. In addition, chemical changes may occur in oil and gas even after they have migrated into their reservoirs. This explains why chemical differences exist between the oil in one pool and that in another.

Migration. The migration of oil needs more explanation. The sediment in which oil substance is

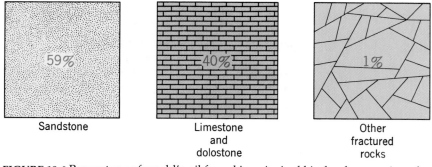

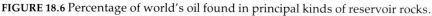

FIGURE 18.6 Percentage of world's oil found in principal kinds of reservoir rocks.

accumulating today is rich in clay minerals, whereas most of the strata that constitute oil pools are sandstones (consisting of quartz grains), limestones and dolostones (consisting of carbonate minerals), and much-fractured rock of other kinds (Fig. 18.6). It seems obvious, therefore, that oil forms in one kind of material and at some later time migrates to another.

Oil migration is analogous to the movement of groundwater. When oil is squeezed out of the clay-rich sediment in which it originated and enters a body of sandstone or limestone somewhere above, it can migrate more easily than before. Sandstone is more permeable than any clay-rich rock. Also, the force of molecular attraction between oil and quartz or carbonate minerals is less strong than that between water and quartz or water and carbonate minerals. Hence, because oil and water do not mix, water remains fastened to the quartz or carbonate grains while oil occupies the central parts of the larger openings. Because it is lighter than water, the oil tends to glide upward past the carbonate- and quartz-held water. In this way it becomes segregated from the water; when it encounters a trap, it can form a pool.

Most of the oil that forms in sediments does not find a suitable trap and eventually makes its way, along with artesian water, to the surface. It is estimated that no more than 0.1 percent of all the organic matter originally buried in a sediment is eventually trapped in an oil pool. Most of it escapes to the surface. It is not surprising, therefore, that the highest ratio of oil pools to volume of sediment is found in rock no older than 2.5 million years, and that nearly 60 percent of all the oil so far discovered has been found in strata of Cenozoic age (Fig. 18.7). This does not mean that older rocks produced less oil. It simply means that oil in older rocks has had a longer time in which to escape.

Distribution

Petroleum, like coal, is widespread but distributed unevenly. The reasons for the uneven distribution are not as obvious as they are with coal. Suitable source sediments for oil are very widespread and seem as likely to form in subarctic waters as in tropical regions. The critical controls seem to be a supply of heat to effect the conversion of solid organic matter to liquid and gaseous forms, and the formation of a suitable trap before the petroleum has leaked away.

Conversion of solid organic matter to oil and gas happens within a specific range of depth and temperature defined by the geothermal gradients shown

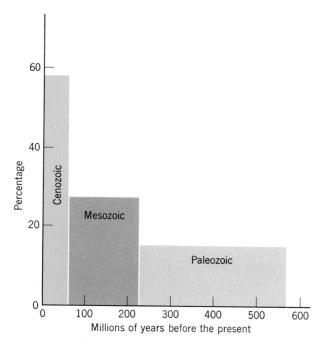

FIGURE 18.7 Percentage of world's total oil production from strata of different ages.

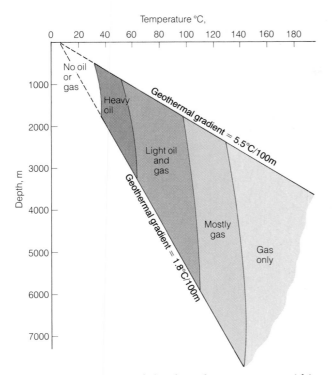

FIGURE 18.8 Regions of depth and temperature within which the generation and trapping of oil and gas occur.

types and structures are similar. Using this approach, and considering all the sedimentary basins of the world (Fig. 18.9), experts estimate that somewhere between 1500 and 3000 billion barrels of oil have already been, or eventually will be, discovered and produced. So far, over 500 billion barrels of oil have been produced.

Heavy Oils and Tars

Oil that is exceedingly viscous and thick will not flow and cannot be pumped. Colloquially called **tar** or **asphalt,** heavy, viscous oil acts as a cementing agent between mineral grains. The tar can be recovered from the tar sands only if the sand is heated to make the tar flow. The resulting tar must then be processed to recover the valuable gasoline fraction. The largest known occurrence of tar sands is in Alberta, Canada, where the *Athabasca Tar Sand* covers an area of 5000 km^2 and reaches a thickness of 60 m (Fig. 18.10). Similar deposits almost as large are known in Venezuela and in the U.S.S.R.

Oil Shale

Another source of petroleum consists of solid organic matter (kerogen) enclosed in fine-grained sedimentary rock. If the organic matter is heated, the solid breaks down and liquid and gaseous hydrocarbons, similar to those in oil, can be distilled out. All sedimentary rock contains some organic matter, but to be considered an energy resource the organic matter must yield more energy than that required for the processes of mining and distillation. The only kind of sedimentary rock that contains sufficient solid organic matter to be given any attention is shale, and only those shales that yield 40 or more liters of distillate per ton can be considered, because the energy needed to mine and process a ton of shale is equivalent to that created by burning 40 liters of oil.

The world's largest deposit of rich oil shale is in the United States. During the Eocene Epoch many large, shallow lakes existed in basins in Colorado, Wyoming, and Utah; in three of them was deposited a series of rich organic sediments that are now the Green River Oil Shales (Fig. 18.10). The richest shales were deposited in the lake in Colorado. These shales are capable of producing as much as 240 liters of oil per ton. Scientists of the U.S. Geological Survey estimate that oil-shale resources capable of producing 50 liters or more oil a ton in the Green River Oil Shales alone total about 2000 billion barrels of oil.

in Figure 18.8. If a thermal gradient is too low— that is, less than 1.8°C/100 m—conversion does not occur. If the gradient is above 5.5°C/100 m, conversion to gas starts at such shallow depths that very little trapping occurs. Once oil and gas have been formed, they will only accumulate in pools if suitable traps are present. Most oil and gas pools are found beneath anticlines; the timing of the folding event is therefore a critical part of the trapping process. If folding occurs after petroleum has formed and migrated, pools cannot form. The great oil pools in the Middle East arose through the fortunate coincidence of a high thermal gradient and the development of anticlinal traps during the collision of Europe and Asia with Africa.

How much oil is there in the world? This is an extremely controversial question. Approximately 600 billion barrels of oil have been discovered, but a great deal remains to be found by drilling. Unlike coal, for which the volume of strata in a basin of sediment can be accurately estimated, the volume of undiscovered oil can only be guessed at. The way guesses are made is to use the accumulated experience of a century of drilling. Knowing how much oil has been found in an intensively drilled area, such as eastern Texas, experts make estimates of probable discoveries in other regions where rock

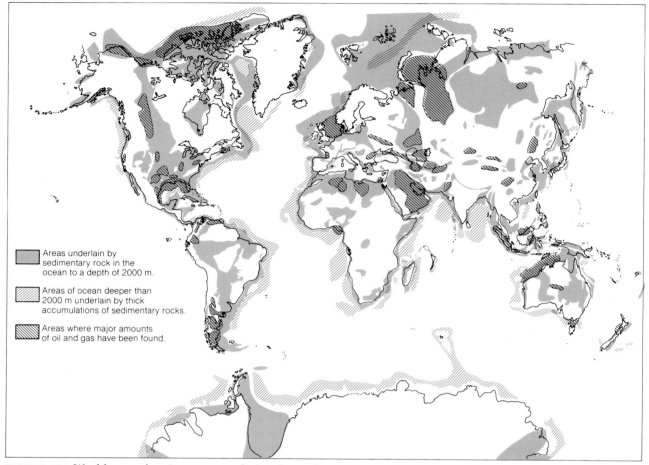

FIGURE 18.9 World map showing areas underlain by sedimentary rock and regions where large accumulations of oil and gas have been located. Where the ocean is deeper than 2000 m, sedimentary rock has yet to be tested for its oil and gas potential.

Legend:
- Areas underlain by sedimentary rock in the ocean to a depth of 2000 m.
- Areas of ocean deeper than 2000 m underlain by thick accumulations of sedimentary rocks.
- Areas where major amounts of oil and gas have been found.

Rich resources of oil shale in other parts of the world have not been adequately explored, but there is a huge deposit in Brazil, in the Irati Shale. Another very large deposit is known in Australia, and others have been reported in such widely dispersed places as South Africa and China.

How Much Fossil Fuel?

Are supplies of fossil fuels adequate to meet future demands? If we use a barrel of oil as our unit of measurement, we can compare quantities of all fossil fuels directly. Thus, approximately 0.22 ton of coal produces the same amount of heat energy as one barrel of oil. The world's recoverable coal reserves of $13,800 \times 10^9$ tons are equivalent to about 63,000 billion barrels of oil.

Considering the approximate world-use rate of barrels of oil (30 billion barrels a year), and comparing the estimated recoverable amounts of fossil fuels (Table 18.1), it is apparent that only coal seems to have the capacity to meet long-continued demands.

OTHER SOURCES OF ENERGY

Three sources of energy other than fossil fuels have already been developed to some extent: the Earth's plant life (so-called biomass energy), hydroelectric energy, and nuclear energy. Several others—such as the Sun's heat, winds, tides, and the Earth's internal heat—have been tested and developed on a limited basis. None has yet been developed on a large scale in the United States, but the day may not be far in the future when each will become locally important.

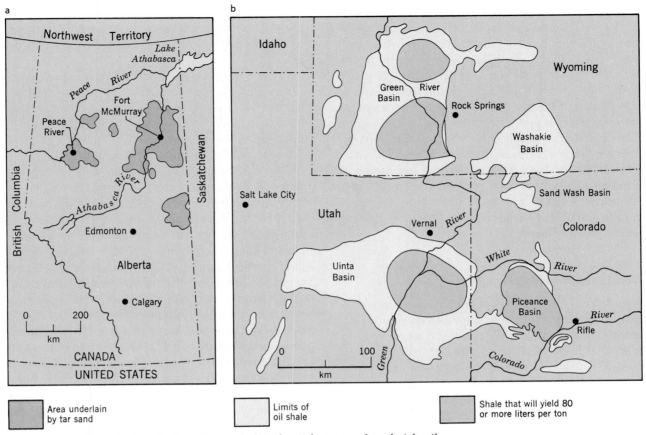

FIGURE 18.10 Areas in North America underlain by rich tar sand and rich oil shale. (a) Athabasca Tar Sand in Alberta; (b) Green River Oil Shale, originally deposited as sediment rich in organic matter. The sediment accumulated in freshwater lakes that formed in shallow basins in Colorado, Wyoming, and Utah. Compaction and cementation of the sediment formed the shale seen today.

Biomass Energy

Scientists working for the United Nations estimate that wood and animal dung used for cooking and heating fires now amounts to energy production of 4×10^{19} J annually. This is approximately 14 percent of the world's total energy use. The great-est use of wood as a fuel occurs in developing countries where the cost of fossil fuel is very high in relation to income.

Measurements made on living plant matter indicate that new plant growth on land equals 1.5×10^{11} metric tons of dry plant matter each year. If

TABLE 18.1 *Amounts of Fossil Fuels Possibly Recoverable Worldwide. (Unit of comparison is a barrel of oil)*

Fossil Fuel	Total Amount in Ground (billions of barrels)	Amount Possibly Recoverable (billions of barrels)
Coal	About 100,000	62,730[a]
Oil and gas (flowing)	1500 to 3000	1500 to 3000
Trapped oil in pumped-out pools	1500 to 3000	0 to ?
Viscous oil (tar sands)	3000 to 6000	500 to ?
Oil shale	Total unknown. Much greater than coal.	1000 to ?

[a] 0.22 ton of coal = 1 barrel of oil.

all of this were burned, or used in some other way as a *biomass* energy source, it would produce almost nine times more energy than the world uses each year. Obviously, it is not possible to do this because forests would have to be destroyed, plants could not be eaten, and agricultural soils would be devastated. The danger in using biomass energy on a large scale is that food supplies could suffer. Nevertheless, controlled harvesting of fuel plants could probably increase the fraction of the biomass now used for fuel without serious disruption to food supplies. In several parts of the world, such as Brazil, China, and the United States, experiments are already underway to develop this obvious energy source.

Hydroelectric Power

Hydroelectric power is recovered from the potential energy of water in streams slope as they flow down to the sea. Water ("hydro") power is an expression of solar power because it is the Sun's heat energy that drives the water cycle. That cycle is continuous; so energy obtained from flowing water is also continuous (unlike coal and oil, it cannot be used up). However, to convert the power of flowing water into electricity efficiently, it is necessary to dam streams. Because a stream carries a suspended load of sediment, the reservoir behind the dam eventually will become filled up, often within 50 to 200 years. Lake Nasser, the reservoir held behind the great Aswan Dam on the river Nile, will be almost half filled with silt by the year 2025. Thus, although water power is continuous, the reservoirs needed for the conversion of water power to electricity have limited lifetimes.

Water power has been used in small ways for thousands of years, but only in the twentieth century has widespread use been made for generating electricity. All the water flowing in the streams of the world has a total amount of recoverable energy that has been estimated as 9.2×10^{19} J/yr. This is an amount of energy equivalent to burning 15 billion barrels of oil per year.

Nuclear Energy

Nuclear energy is the heat energy produced during controlled transformation of suitable radioactive isotopes (a process called *fission.*) Three of the same radioactive atoms that keep the Earth hot by spontaneous radioactive decay—^{238}U, ^{235}U, and ^{232}Th—can be mined and used in this way. Fission is accomplished by bombarding the radioactive atoms

with neutrons, thus accelerating the rate of disintegration and the release of heat energy. The device in which this operation is carried out is called a *pile.*

When ^{235}U fissions, it not only releases heat and forms new elements but also ejects some neutrons from its nucleus. These neutrons can then be used to induce more ^{235}U atoms to fission, and a continuous chain reaction occurs. The function of a pile is to control the flux of neutrons so that the rate of fission can be controlled. When a chain reaction proceeds without control, an atomic explosion occurs. Controlled fission, therefore, is the method used by nuclear power plants, and a tremendous amount of energy can be obtained in the process. During fissions one gram of ^{235}U produces as much heat as the burning of 13.7 barrels of oil. Unfortunately, however, ^{235}U is the only natural radioactive isotope that will maintain a chain reaction, and is the least abundant of the three radioactive isotopes. Only one atom of each 138.8 atoms of uranium in nature is ^{235}U. The remaining atoms are ^{238}U, which will not sustain a chain reaction. All the present nuclear power plants use ^{235}U. However, if ^{238}U is placed in a pile with ^{235}U that is undergoing a chain reaction, some of the neutrons will bombard the ^{238}U and convert it to a new isotope, plutonium-239 (^{239}Pu). This new isotope can, under suitable conditions, sustain a chain reaction of its own. The pile in which the conversion of ^{238}U takes place is called a *breeder reactor.* The same kind of device can be used to convert ^{232}Th into a new isotope, ^{233}U, that also will sustain a chain reaction.

Already there are nearly 200 nuclear plants operating around the world. They utilize the heat energy from fission to produce steam, which in turn drives turbines and generates electricity. Approximately 7.6 percent of the world's electrical power is derived from nuclear power plants. In France, more than half of all the electrical power comes from nuclear plants; the fraction is rising sharply in some other European countries and Japan too. The reason for the increase is obvious. Japan and most European countries do not have adequate supplies of fossil fuels in order to be self-sufficient.

Many problems are associated with nuclear energy. The isotopes used in power plants are the same isotopes used in atomic weapons, so a security problem exists. The possibility of a power plant failing in some unexpected way creates a safety problem. The Chernobyl disaster in 1986 in the U.S.S.R. is an example of such an event. Finally, the problem of safe burial of dangerous radioactive waste matter must be faced. Some of the waste matter will retain dangerous levels of radioactivity for thousands of years.

Geothermal Power

Geothermal power, as the Earth's internal heat flux is called, has been used for more than 50 years in Italy and Iceland and more recently in other parts of the world, including the United States. The steam produced in hot-spring areas can be used to power generators in the same way as steam produced in coal- and oil-fired boilers. To capture steam in hot-spring areas, we need to drill into the hot underground reservoirs that feed the springs, bring the steam or hot water to the surface in pipes, and feed it into a power plant (Fig. 18.11). The places where sources of heat are close enough to the surface to produce steam, and therefore where geothermal energy can be developed, are mostly in areas of current or recent volcanic activity, where magma or hot intrusive igneous rock is close to the surface and can serve as a source of heat. In the United States these areas include California, Nevada, Montana, Wyoming, New Mexico, Utah, Alaska, and Hawaii. In other countries, in addition to Italy and Iceland, large resources of geothermal energy exist in Japan, U.S.S.R., New Zealand, several countries in Central America, Ethiopia, and Kenya. Not surprisingly, most of the world's geothermal steam reservoirs are around the margins of plates, because plate margins are where most of the recent volcanic activity has occurred.

A depth of 3 km seems to be a rough lower limit for big geothermal steam and hot-water pools. It is estimated that the world's geothermal reservoirs can yield about 8×10^{19} J—equivalent to burning 13 billion barrels of oil. This estimate incorporates the observation that in New Zealand and Italy only about 1 percent of the energy in a geothermal reservoir is recoverable. If the recovery efficiency were to rise, the estimate of recoverable geothermal resources would also rise.

Interesting geothermal experiments are now being conducted in New Mexico. In the Jemez Mountains on the edge of an extinct (but still hot) volcano, scientists have drilled deep into the hot rock. They then shattered the hot rock with explosives to create an artificial reservoir, and pumped water through to produce steam. The first test was only partly successful. A major difficulty was that water did not flow uniformly through the hot rock but instead followed narrow flow paths, and the rocks that lined them soon cooled down. Further tests

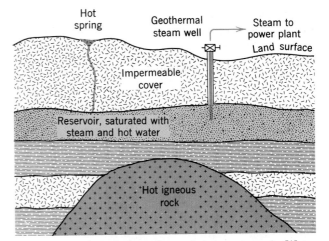

FIGURE 18.11 A typical geothermal steam reservoir. Water in a permeable aquifer, such as sandstone, is heated by magma or hot igneous rock. As steam and hot water are withdrawn through the well, cold water flows into the reservoir through the aquifer.

are now planned, not only in the Jemez Mountains, but in France, England, and other countries also.

USES OF MATERIALS

We now turn to the mineral substances that provide *materials* from which engines and a myriad of other necessary things can be made. The number and diversity of such substances is so great that to make a simple classification covering all of them is almost impossible. Nearly every rock and mineral can be used for something. A society such as ours, that possesses an energy-intensive industry, not only requires a diverse group of metals for machines, but also demands a host of nonmetallic mineral products (such as shale and limestone for making cement, gypsum for making plaster, salt for making chemical compounds, and calcium phosphate [apatite] for making fertilizer). To supply all needs, the people of the world now mine several hundred kinds of mineral products. The amounts used are truly enormous, and are still increasing. In the United States the quantity of iron and steel used each year is equivalent to 500 kg for every man, woman, and child. The use of sand and gravel amounts to 3000 kg per person; of crushed stone, 3900 kg; of aluminum, 20 kg; and of copper, 8 kg. When the mineral substances mined for energy (petroleum and coal) are added to all the rest, we get a total of 15,000 kg, or 15 metric tons of mineral substances that are mined and used *each*

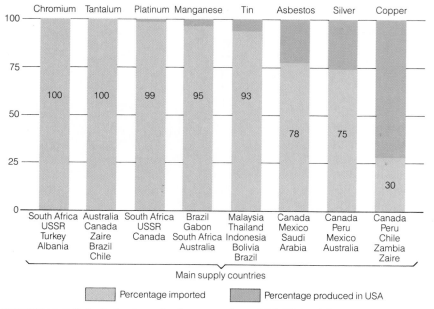

FIGURE 18.12 Selected mineral substances for which United States consumption exceeds production. The difference must be supplied by imports. Data are plotted for 1980, but the percentage changes little from year to year. (*Source:* Data from U.S Bureau of Mines, 1980).

year for every person living in the United States. Because living standards vary around the world, the quantity of material mined annually for every person in the world is less, 3750 kg.

Supplies of Minerals

Many industrialized nations possess a strong mineral base; that is, they are rich in many kinds of ***mineral deposits*** (any volume of rock containing an enrichment of one or more minerals) which they are exploiting vigorously. Yet no nation is entirely self-sufficient in mineral supplies, and so each must rely to some extent on other nations to fulfill its needs (Fig. 18.12).

Mineral resources have three distinctive aspects that differ from plant and animal resources such as those produced through agriculture, grazing, forestry, and fisheries. First, occurrences of usable minerals are limited in abundance and distinctly localized at places within the Earth's crust. This is the main reason why no nation is self-sufficient where mineral supplies are concerned. Because usable minerals are localized, they must be searched out, and the search ranges over the entire globe. The special branch of geology concerned with discovering new supplies of usable minerals is called, appropriately, ***exploration geology.***

Second, the quantity of a given material available in any one country is rarely known with accuracy,

and the likelihood that new deposits will be discovered is difficult to assess. As a result, production over a period of years can be difficult to predict. Thus, a country that today can supply its need for a given mineral substance may face a future in which it will become an importing nation. Britain is a good example of this. A little more than a century ago, Britain was a great mining nation, producing and exporting such materials as tin, copper, tungsten, lead, and iron. Today, most of the known deposits have been worked out and Britain, once self-sufficient in most minerals, is almost entirely an importing country.

Third, unlike plants and animals that are cropped yearly or seasonally, deposits of minerals are depleted by mining, which eventually exhausts them. Therefore, minerals have only a "one-crop" availability per occurrence; this disadvantage can be offset only by finding new occurrences or by making use of scrap—that is, by reusing the same material repeatedly.

These peculiarities of the mineral industry place a premium on the skills of the geologists, prospectors, and engineers who play the essential roles in finding and mining of mineral substances used by society. The task of finding and mining is accomplished through the application of the basic principles that have been set forth in this book, and by the use of additional specialized knowledge concerning the origin and distribution of mineral deposits. Much ingenuity has been expended in bringing the production of minerals to its present state. Because known deposits are being rapidly exploited, while demands for minerals continue to grow, we can be sure that even more ingenuity will be needed in the future.

Ore. Minerals for industry are sought in deposits from which the desired substances can be recovered least expensively. The more concentrated the preferred minerals, the more valuable the deposit. In some deposits the desired minerals are so highly concentrated that even very rare substances such as gold and platinum can be seen with the naked eye. For every desired mineral substance a level of concentration exists below which the deposit cannot be worked economically. To distinguish between profitable and unprofitable mineral deposits, the word *ore* is used, meaning an aggregate of minerals from which one or more minerals can be extracted profitably. It is not always possible to say exactly how much of a given mineral must be present in order to constitute an ore. For example, two deposits may contain the same iron mineral and be the same size, but one is ore and the other is not. The reasons are many; for example, the deposit could be too deeply buried, or so remote that the costs of mining and transport are so high that the final product, metallic iron, is not competitive with iron from other deposits. Furthermore, as both costs and market prices fluctuate, a particular aggregate of minerals may be an ore at one time but not at another.

Gangue. Along with ore minerals from which the desired substances are extracted, there are other minerals, collectively termed *gangue* (pronounced gang). These are the nonvaluable minerals of an ore. Familiar minerals that commonly occur as gangue are quartz, feldspar, mica, calcite, and dolomite.

The ore problem has always been twofold: (1) to find the ores (which altogether underlie an infinitesimally small proportion of the Earth's land area); and (2) to mine the ore and get rid of the gangue as cheaply as possible. Getting rid of gangue and the mining itself are both technical problems; engineers have been so successful in solving them that some deposits now considered ore are only one-sixth as rich as were the lowest-grade ores 100 years ago (Fig. 18.13). Notice, however, that the curve in Figure 18.13 has risen up from its lowest value, reached during the 1970s. The reason for this is that overproduction of copper around the

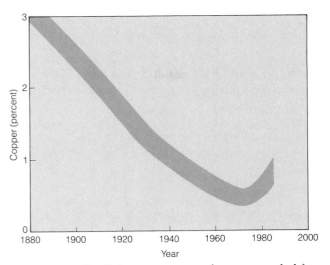

FIGURE 18.13 Declining percentage of copper needed for a mineral deposit to be ore. The large reduction occurred because large-volume mining with efficient machines led to a steady reduction of mining costs. The sharp upturn, which started in the mid-1970s and continues into the 1980s, is due to increased fuel costs of the 1970s and the worldwide economic recession of the 1980s.

world, combined with an economic recession, produced a glut of newly mined copper. This, in turn, drove the price of copper down and led to the closing of many mines and, in particular, those mines with ore containing the lowest percentage of copper.

ORIGIN OF MINERAL DEPOSITS

Ores are mineral deposits because each of them contains a local enrichment of one or more minerals or mineraloids. The reverse is not true, however. Not all mineral deposits are ores. *Ore* is an economic term, while mineral deposit is a geological term. How, where, and why a mineral deposit forms is the result of one or more geological processes. Whether or not a given mineral deposit is an ore is determined by how much we human beings are prepared to pay for its content. Fascinating though the economics of ores and mining are, they cannot be explored in this volume. Instead, discussion is limited to the origin of mineral deposits without necessary regard to questions of economics.

In order for a deposit to form, some process or combination of processes must bring about a localized enrichment of one or more minerals. A convenient way to classify mineral deposits is through the principal concentrating process. Minerals become concentrated in five ways:

1. Concentration by hot, aqueous solutions flowing through fractures and pore spaces in crustal rock to form *hydrothermal mineral deposits.*

2. Concentration by magmatic processes within a body of igneous rock to form *magmatic mineral deposits.*

3. Concentration by precipitation from lake water or seawater to form *sedimentary mineral deposits.*

4. Concentration by flowing surface water in streams or along the shore to form *placer* or *detrital mineral deposits.*

5. Concentration by weathering processes to form *residual mineral deposits.*

Hydrothermal Mineral Deposits

Many of the most famous mines in the world contain ores that were formed when their essential minerals were deposited from hot-water solutions, commonly called *hydrothermal solutions.* Probably more mineral deposits have been formed by deposition from hydrothermal solutions than by any other mechanism. However, despite the importance of such deposits, the origins of the solutions are often difficult to decipher. Deposition occurs deep underground where it cannot be seen; by the time a deposit is finally uncovered by erosion, the solution that formed it is no longer present. Nevertheless, clues have been found so that many details of the process of deposition are now understood.

Composition of the Solutions

The principal ingredient of hydrothermal solutions is water. The water is never pure and always contains dissolved within it salts such as sodium chloride, potassium chloride, calcium sulfate, and calcium chloride. The amounts vary, but most solutions range from about the saltiness of seawater (3.5 percent dissolved solids by weight) to about ten times the saltiness of seawater. A hydrothermal solution is therefore a brine, and brines, unlike pure water, are capable of dissolving minute amounts of seemingly insoluble minerals such as gold, chalcopyrite, galena, and sphalerite.

Origins of the Solutions

Brines have many sources. One way they form is by the cooling and crystallization of magma formed by wet partial melting (chapter 3). Most of the water that causes the wet partial melting is released when such a magma solidifies. Instead of being pure water, however, it carries in solution the most soluble constituents in the magma, such as NaCl and elements that do not readily enter quartz, feldspar, and other common minerals by atomic substitution. The solution is a brine, and examples of the elements carried in solution are gold, silver, copper, lead, zinc, mercury, and molybdenum.

High temperatures increase the effectiveness of brines to form hydrothermal mineral deposits. It is not surprising, therefore, that many mineral deposits are associated with hot volcanic rocks that were invaded by deep circulating groundwater. Nor is it surprising that a great many mineral deposits are found in the upper portions of volcanic piles where they were deposited when upward-moving hydrothermal solutions became cool and precipitated the ore minerals.

Volcanogenic Massive Sulfide Deposits. Submarine volcanic eruptions are common along midocean ridges and above subduction zones. One very important class of mineral deposit, called *volcanogenic massive sulfide deposits*, forms as a result. The hydrothermal solutions in this case are believed to start out as seawater rather than as water in magma, because basalt is a "dry" magma. All the cracks and openings in volcanic rocks in the oceanic crust are saturated by seawater. As magma rises through the oceanic crust toward a site of submarine eruption, the crust is heated, and its contained seawater evolves into a hydrothermal solution.

The ore–mineral constituents in massive sulfide deposits apparently come from the igneous rocks in the oceanic crust. Heated seawater reacts with the rocks it is in contact with, causing small changes in mineral composition. For example, feldspars are changed to clays and epidote, and pyroxenes are changed to chlorites. As the minerals are transformed, trace metals such as copper and zinc, present by atomic substitution, are released and become concentrated in the hot seawater.

Hydrothermal solutions formed beneath the sea can become so hot that they rise rapidly through fractures and form jetlike eruptions of hot, hydrothermal solutions into cold seawater (Fig. 18.14). Many such eruptions have been observed along the East Pacific Rise. When a jetting hydrothermal solution cools, it deposits minerals such as pyrite, chalcopyrite, sphalerite, and galena in a massive blanket around the erupting vent (Fig. 18.15). Vol-

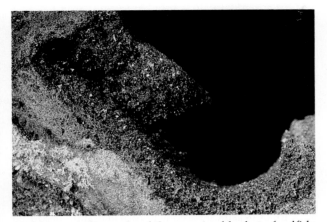

FIGURE 18.15 A sample of the massive blanket of sulfide minerals that form around the seafloor vents at 21° N. This specimen is part of a system of chimneys. Lining the chimney is pyrite, and scattered through the pyrite are grains of chalcopyrite and sphalerite. Surrounding the massive pyrite is a mixture of gypsum, pyrite, and chalcopyrite. The specimen is 28 cm across.

FIGURE 18.14 A so-called "black smoker" photographed at a depth of 2500 m below sea level on the East Pacific Rise, at 21° N latitude. Seawater descends along fractures in rocks of the oceanic crust, becomes heated by magma chambers beneath the midocean ridge, and then rises convectively again to the seafloor. The "smoker" seen here has a temperature of 320°C. The rising hot water is actually clear; the black color is due to fine particles of iron sulfide and other minerals precipitated from solution as the plume is cooled through contact with cold seawater.

canogenic massive sulfide deposits have been found in rocks as old as 3 billion years and can be observed to be forming today; they are possibly the most common of all kinds of hydrothermal mineral deposits.

Hydrothermal solutions having similar compositions can apparently form in many different ways. Which ore constituents are carried in solution depends on the kinds of rocks involved in the formation of the solution. For example, copper and zinc are present in pyroxenes by atomic substitution, so that the pyroxene-rich rocks of the oceanic crust yield solutions charged with copper and zinc. Most massive sulfide deposits are copper- and zinc-rich as a result.

The important questions concerning hydrothermal solutions are not *where* the water came from

or how it became a brine, but rather *where did the ore constituents come from* and *what made the solutions precipitate their soluble mineral load?*

Causes of Precipitation

When a deposit-forming solution moves slowly upward, as with groundwater percolating through a confined aquifer, the solution cools very slowly. If dissolved minerals were precipitated from such a slow-moving solution, they would be spread out over great distances and would not be sufficiently concentrated to form a useful mineral deposit. But when a solution flows rapidly, as in an open fracture or a mass of shattered rock or any other place where flow is less restricted, cooling can be sudden and happen over short distances. Rapid precipitation and a concentrated mineral deposit are the result. Other effects such as boiling, a rapid decrease in pressure, composition changes of the solution caused by reaction with adjacent rock, and dilution by mixing with groundwater can also cause rapid precipitation and form concentrated deposits. When valuable minerals are present, an ore is the result.

Examples of Precipitation

Many veins containing valuable minerals are found in regions of volcanic activity, particularly where the volcanism is rhyolitic or andesitic. This is so

FIGURE 18.16 Dark-colored wall-rock alteration produced in a calcareous siltstone by hydrothermal solutions. When the hot solutions flowed through joint-controlled openings, they introduced iron and other chemical elements, allowing an assemblage of amphiboles, pyroxenes, and garnets to grow.

because volcanism heats solutions very near the surface, making them effective ore-formers. The famous gold deposit at Cripple Creek, Colorado, was formed in a small caldera and the huge tin deposits in Bolivia are all localized in and around stratovolcanoes. In each case, the magma chamber that fed the volcano also served as the source of the hydrothermal solutions that rose up and formed the mineralized veins in the overlying rocks. Formation of hydrothermal solutions by volcanism also occurs, as we have seen, when volcanism takes place beneath the sea. The gold deposits at Kirkland Lake in Ontario and Kalgoorlie in Western Australia were formed apparently during subma-

rine volcanic activity. So too were the famous copper deposits of Cyprus, and those in New Brunswick, Canada, and many other places.

When a body of granitic magma cools, it is a source of heat just as the magma chamber beneath a volcano is—and it is also a source of hydrothermal solutions that are released by crystallization. Such solutions move outward from a cooling intrusive body. They will flow through any fracture or channel, altering the surrounding rock in the process and commonly depositing valuable minerals (Fig. 18.16). Many famous ore bodies are associated with shallow intrusive rocks. The tin deposits of Cornwall, England, and the copper deposits at Butte, Montana; Bingham, Utah; and Bisbee, Arizona, are examples.

Metallogenic Provinces. Many of the ore bodies formed by hydrothermal solutions are alike in that they are related in one way or another to igneous activity. Igneous activity tends to be concentrated along both spreading center and subduction zone boundaries of plates. It is not surprising, therefore, that certain kinds of ore deposits are also concentrated near present or past plate boundaries (Fig. 18.17).

The relation between plate boundaries and ore deposits provides an explanation for the existence of **metallogenic provinces**. These are limited regions of the crust within which mineral deposits occur in unusually large numbers. A striking example is the metallogenic province that runs along the western side of the Americas. Within the province is the world's greatest concentration of large

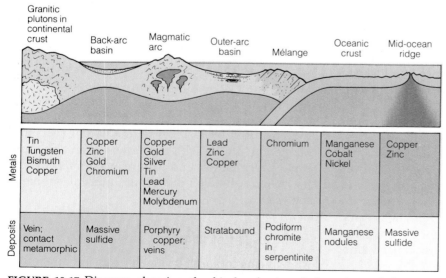

FIGURE 18.17 Diagram showing the kinds of mineral deposits and the most important metals concentrated in relation to tectonic plates.

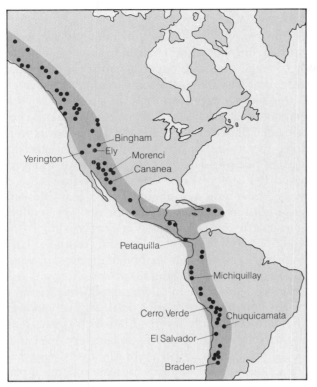

FIGURE 18.18 Metallogenic province along the western edge of the Americas. The province parallels the edge of the North and South American plates and is due to the igneous activity caused by subduction. Many of the world's largest copper deposits occur in the province. Some important deposits are named.

copper deposits (Fig. 18.18). These deposits are associated with porphyritic igneous rock (chapter 3) and are called **porphyry coppers**. The igneous bodies (and, therefore, the deposits themselves) formed as a consequence of subduction because they are in, or adjacent to, old stratovolcanoes.

Magmatic Mineral Deposits

The processes of partial melting and fractional crystallization in magmas (chapter 3) are, by their very definitions, ways of separating some substances from others. These special circumstances lead sometimes to the creation of large and potentially valuable mineral deposits.

Pegmatites

When a magma undergoes differentiation by fractional crystallization, the residual melt becomes progressively enriched in chemical elements that are not removed in the early-crystallizing minerals. Separation and crystallization of the remaining melt produces an igneous rock that contains the concentrated elements. An important example of this kind of concentration process is provided by pegmatites, an especially coarse-grained kind of igneous rock. These unusual rocks form by extreme differentiation of deep-seated granitic magma. Commonly, they contain significant enrichments of rare elements such as beryllium, tantalum, niobium, uranium, and lithium.

Crystal Settling

One form of magmatic differentiation occurs when dense minerals accumulate on the floor of a magma chamber. The process is called crystal settling, and in some cases the segregated minerals make desirable ores. Most of the world's chromium ores were formed in this manner by accumulation of the mineral chromite ($FeCr_2O_4$) (Fig. 18.19). The largest known chromite deposits are in South Africa, Zimbabwe, and the U.S.S.R. Similarly, vast deposits

FIGURE 18.19 Layers of chromite (black) and anorthite (white) formed by magmatic segregation during the crystallization of the Bushveld Igneous complex. The location of this unusually fine outcrop is the Dwars River, South Africa.

of ilmenite ($FeTiO_3$), a source of titanium, were formed by magmatic differentiation. Large deposits occur in the Adirondack Mountains.

A form of concentration similar to crystal settling occurs when, for reasons not clearly understood, certain magmas separate into two immiscible magmas. One, a sulfide liquid that is rich in copper and nickel, sinks to the floor of the magma chamber because it is denser. After cooling and crystallization the resulting igneous rock has a copper or nickel ore at the base. The world's greatest known concentration of nickel ore, at Sudbury, Ontario, is believed to have formed in this fashion. Other great nickel deposits in Canada, Australia, and Zimbabwe formed in the same manner.

Sedimentary Mineral Deposits

The term *sedimentary mineral deposit* is applied to any local concentration of minerals formed through processes of sedimentation. Any process of sedimentation can form localized concentrations of minerals, but it has become common practice to restrict use of the term *sedimentary* to those mineral deposits formed through precipitation of substances carried in solution.

Evaporite Deposits

The most important way in which sedimentary mineral deposits form is by evaporation of lake water or seawater. The layers of salts that precipitate as a consequence of evaporation are called *evaporite deposits.*

Examples of salts that precipitate from lake waters of suitable composition are sodium carbonate (Na_2CO_3), sodium sulfate (Na_2SO_4), and borax ($Na_2B_4O_7 \cdot 10H_2O$). Such salts are used in many manufacturing processes such as production of paper, soap, certain detergents, antiseptics, and chemical for tanning and dyeing.

Much more common and important than nonmarine evaporites are the marine evaporites formed by evaporation of seawater. The most important salts that precipitate from seawater are gypsum ($CaSO_4 \cdot 2H_2O$), halite ($NaCl$), and carnallite ($KCl \cdot MgCl_2 \cdot 6H_2O$). Low-grade metamorphism of evaporite deposits causes another important mineral, sylvite (KCl), to form. Marine evaporite deposits are widespread; in North America, for example, strata of marine evaporites underlie as much as 30 percent of the entire land area (Fig. 18.20). Much of the salt that is used, the gypsum used for plaster, and potassium used in plant fertilizers are recovered from marine evaporites.

Iron Deposits

Sedimentary deposits of iron minerals are widespread, but the amount of iron in average seawater is so small that such deposits cannot have formed from seawater that is the same as today's seawater.

All sedimentary iron deposits are tiny by comparison with the class of deposits characterized by the Lake Superior iron deposits. The deposits were long the mainstay of the United States steel industry but are declining in importance today as imported ores displace them. The deposits are of early Proterozoic age, about 2 billion years old, and are found in sedimentary basins on every craton, particularly in Labrador, Venezuela, Brazil, the U.S.S.R., India, South Africa, and Australia. Lake Superior-type deposits are one of the most unusual kinds of chemical sediment known. They are sediments of wholly chemical (or possibly biochemical) origin and they are free of detritus, even though they are commonly interbedded with clastic sedimentary strata. The deposits contain very fine bands of recrystallized chert (Fig. 18.21). Every aspect of the Lake Superior deposits indicates chemical precipitation. Because the deposits are so large, it is inferred that the iron must have been transported in surface water, but the cause of precipitation remains problematic. The inference forces a further conclusion. Seawater 2 billion years ago must have differed from today's seawater. The most important difference probably involved the oxygen content, and this in turn means the atmosphere must have been different from today's atmosphere. It must have contained less oxygen, so the surface waters were less oxygenated than those of today and could transport large amounts of dissolved ferrous iron.

Manganese Deposits

Precipitation from seawater is also the origin of most of the world's manganese deposits. The two largest deposits occur in the U.S.S.R. They are of early Cenozoic age. All the ore minerals are oxides.

Phosphorus Deposits

Sedimentary deposits of phosphorus minerals form through the precipitation of apatite [$Ca_5(PO_4)_3(OH,F)$] from seawater. The surface waters of the ocean are depleted in phosphorus because fish and other animals that live in the water extract phosphorus to make bone, scales, and other parts of their bodies. When the animals die and sink to the seafloor, their bodies slowly decay and release the contained phosphorus to the deep ocean water.

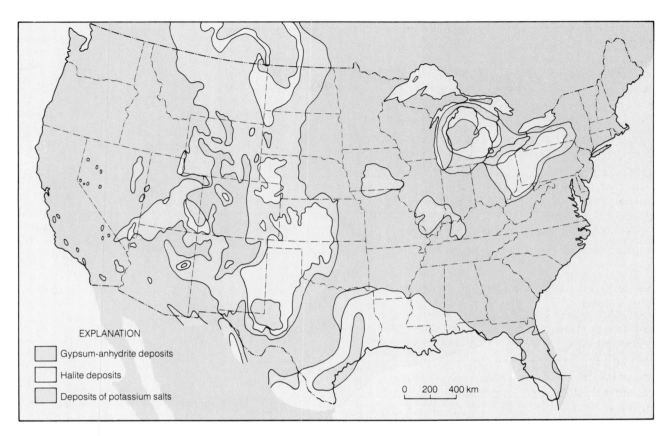

FIGURE 18.20 Portions of the United States known to be underlain by marine evaporite deposits. The areas underlain by gypsum and anhydrite do not contain halite. Areas underlain by halite are also underlain by gypsum and anhydrite. Areas underlain by potassium salts are also underlain by halite and by gypsum and anhydrite. (*Source:* After U. S. Geological Survey, 1973.)

FIGURE 18.21 Siliceous iron-rich sediments of the Brockman Iron Formation, Hamersley Range, Western Australia. Typical of the Lake Superior-type iron formations, the white layers are largely chert. The darker bluish- and reddish-colored layers consist largely of iron-rich silicate, oxide, and carbonate minerals.

If such phosphorus-rich waters are brought to the surface by rapid upwelling, precipitation of apatite can occur. Phosphorus-rich sediments are forming today off the west coasts of Africa and South America, but the process has been much more common at times in the past—particularly when shallow epicontinental seas were prevalent, as was the case, for example, along the margins of the old Tethys Sea (Fig. 16.1a) during the Mesozoic and Cenozoic eras.

Placer Deposits

The famous California goldrush of 1849 followed the discovery that the sand and gravel in the bed of a small stream contained bits of gold. Similar gold-bearing gravels are found in many parts of the world. The gravels themselves are sometimes rich enough to be ores, but even when they are too lean the gold is a clue that a source must lie upstream. Indeed, many mining districts have been

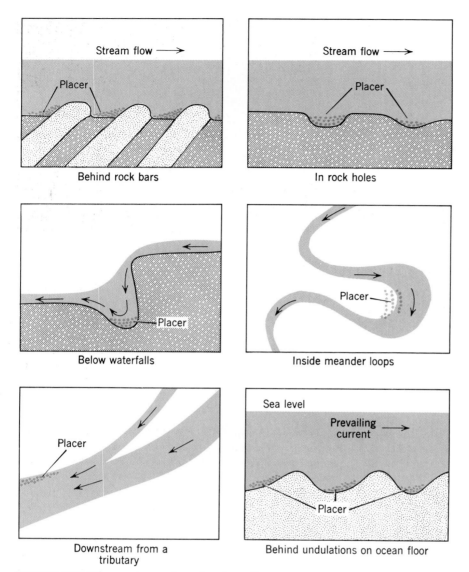

FIGURE 18.22 Placers occur where barriers allow stream water to carry away the suspended load of lightweight particles while trapping denser particles carried in the bed load. Placers can form wherever water moves, but are most commonly associated with streams.

discovered by following trails of gold and other minerals upstream to their sources in veins in bedrock. Because pure gold is heavy (density, 19 g/cm³), it is deposited from the bed load of a stream very quickly, while quartz, with a density of only 2.65 g/cm³, is washed away. As most silicate minerals are light by comparison with gold, grains of gold become mechanically concentrated in places where the velocity of stream flow is least (Fig. 18.22). A deposit of heavy minerals concentrated mechanically is a **placer**. Besides gold, other heavy, durable metallic minerals form placers. These include minerals that occur as pure metals, such as platinum and copper, as well as tinstone (cassiterite, SnO_2) and nonmetallic minerals such as diamond and ruby

and sapphire (both gem forms of corundum). Even if a vein contains a low percentage of gold or cassiterite, the placer it yields may be quite rich. In order for minerals to become concentrated in placers they must not only be dense, they must be resistant to chemical weathering and not readily susceptible to cleaving as the mineral grains are tumbled in the stream.

Every phase of the conversion of gold in a fissure vein into placer gold has been traced. Chemical weathering of the exposed vein releases the gold, which then moves slowly downslope by mass-wasting. In some places mass-wasting alone concentrates gold or cassiterite sufficiently to justify mining these metals.

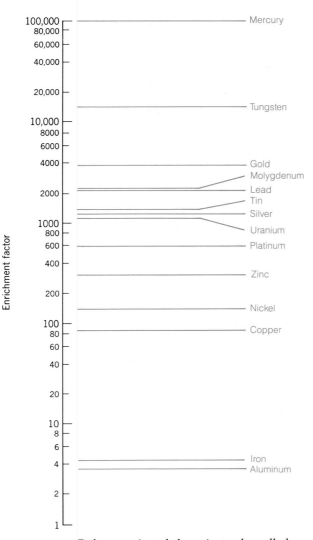

FIGURE 18.25 Before a mineral deposit can be called an ore, the percentage of valuable metal in the deposit must be greatly enriched above its average percentage in the Earth's crust. The enrichment is greatest for metals that are least abundant in the crust, such as gold and mercury. As mining and mineral processing become more efficient and less expensive, it is possible to work leaner ore and enrichment factors decline. Note that the scale is a magnitude (logarithmic) scale, in which the major divisions increase by multiples of ten.

atomic substitution (chapter 2). Atoms of the scarce metals (such as nickel, cobalt, and copper) can readily substitute for more common atoms (such as magnesium and calcium). In order for a mineral deposit to form, therefore, some gathering and concentrating agent such as a hot brine must react with the rock-forming minerals and leach the scarce metals from them. The brine must then transport the metals in solution and deposit them as separate

minerals in a localized place. With such a complicated chain of events, it is hardly surprising that deposits of geochemically scarce metals are rarer and very much smaller than deposits of the abundant metals.

Most minerals that are concentrated to form deposits of the scarce metals are sulfides; a few, such as the most important minerals of tin and tungsten, are oxides. In the case of gold, platinum, palladium, and a few less common elements, the metal itself is the most important mineral. Most scarce metal deposits form as hydrothermal and magmatic mineral deposits. In the case of the precious metals gold and platinum, placer concentration is also important.

Nonmetallic Substances

The great diversity of nonmetallic substances makes it impossible to devise an unambiguous scheme of classification. Some substances have several different uses.

Chemical Materials

Most of the materials used by the chemical industry are organic substances (coal, oil, and gas). They are the raw materials used to make petrochemical products such as plastics, drugs, pesticides, synthetic fibers, and countless other products. Many inorganic substances are used also. The most important are sodium chloride, sodium carbonate, sodium sulfate, and minerals such as borax $(Na_2B_4O_7 \cdot 10H_2O)$ that contain boron. Each of these is recovered from evaporites. Fortunately, the supplies of all of the chemical substances are very large.

Fertilizer Materials

The three essential fertilizer elements are (1) nitrogen, which is recovered by chemical means from the atmosphere; (2) potassium, which is recovered as the soluble mineral sylvite (KCl) from marine evaporite deposits; and (3) phosphorus, recovered as apatite $[Ca_5(PO_4)_3(OH,F)]$.

Calcium carbonate and sulfur must also be added to soils to keep a balance between acidity and alkalinity favorable for maximum plant growth. Calcium carbonate is produced from the abundant limestone strata around the world, but sulfur is more restricted, being recovered from fumaroles (chapter 3), certain gypsum deposits in marine evaporites, the H_2S recovered from many natural gas wells, and certain high-sulfur coals.

Building Materials

Besides cut stone, crushed stone, and sand and gravel, of which the world has enormous resources, the main building materials are cement (manufactured from shale and limestone), gypsum (used for plaster), and clay (used for tile and brick). Asbestos (a variety of serpentine used for wallboards, siding, and insulation) was, until a few years ago, used extensively in the United States and Canada. Because asbestos creates health hazards, its use in North America is declining rapidly. With the exception of asbestos, which is formed principally as an alteration product of peridotites by hot groundwater, supplies of building materials are enormous.

Ceramic and Abrasive Materials

We sometimes forget how extensively ceramic materials (including glass) are used in our everyday lives. Some ceramics have properties that make them desirable substitutes for certain metals, so it is reasonable to anticipate that the uses of ceramics will grow. Supplies of the essential raw materials for ceramics—certain kinds of clays, feldspar, and quartz—are all abundant and widespread.

Abrasives also play a widespread and vital part in our lives. Modern industry employs machines that work accurately and efficiently at high speeds. Such machines require precision grinding and exact shaping and polishing of the machine parts; for these purposes a wide variety of abrasives is needed. The abrasives must be hard enough to cut metals and tough enough so they will not rapidly fracture during the grinding process. The principal abrasive minerals are: quartz and garnet, both of which are common rock-forming minerals; corundum, which is rare as a mineral but easily synthesized, and diamond. It is not commonly appreciated that about 80 percent of all the diamonds produced are used as abrasives, and only 20 percent are cut as gems.

FUTURE SUPPLIES OF MINERALS

Mineral deposits are exploited in the least expensive way possible. For some deposits this means underground mines in which miners drill and blast away at narrow veins (Fig. 18.26); the quantities of material dug from the ground in this fashion are truly enormous. For other deposits, exploitation means huge open pits from which the ore is removed by truck or train (Fig. 18.27). Yet for still

FIGURE 18.26 Two miners digging out a vein of silver ore, Potosi, Bolivia. The width of the vein corresponds to the width of the opening the miners are standing in.

FIGURE 18.27 Mining copper from an open pit. An aerial view of the Sierrita Pit, near Tucson, Arizona. Mining proceeds by moving each bench back, thus making the overall pit wider, and allowing a new bench to be started at the bottom of each pit. Each bench is about 40 m high. The size can be judged from the large trucks visible at the bottom of the pit.

others, such as some salt bodies, exploitation may mean solution in hot water pumped down a drill hole and recovery of salt by evaporation from the resulting brine. All kinds of mining are becoming increasingly efficient. Once a deposit is discovered, it can be worked quickly and efficiently. But how can enough deposits be discovered? Minerals for industry, like fossil fuels, are being used at ever-increasing rates; and mineral deposits, as previously stressed, are a one-crop-only resource.

At present, new mineral deposits are being found about as fast as old ones are being exhausted. But the new finds are mostly in places such as Chile, Australia, and Siberia, where intensive prospecting has begun rather recently. Antarctica is the only continent where intensive prospecting has yet to begin.

Modern prospectors are aided by sophisticated geophysical and geochemical devices that can sense mineral deposits buried by as much as 150 m of sediments. They can only do so under ideal conditions, however, and so far they have been unsuccessful where the cover is more than 150 m thick. Considering that about half of the Earth's surface is covered by sediment that is more than 150 m thick, it must be concluded that a great many mineral deposits remain to be found. First, however, it must be discovered how to find them; then it must be asked whether the cost of finding and mining such buried deposits will be worthwhile. The answer to that question is one for the future.

Even so, the number of mineral deposits in the crust is fixed. Every time one is mined out, one less is available for the future. The rate at which mineral resources are used is growing and the number of deposits left to be found is shrinking. Perhaps, therefore, when all parts of the Earth's surface have been prospected intensively, people will have to look to kinds of deposits not presently exploited. For geochemically scarce metals (such as mercury, silver, gold, and molybdenum) that time may come soon. For these scarcer materials, careful conservation of supplies will some day have to be practiced, substitutes will have to be found for some of their applications, and efficient recycling will become essential. For materials that are abundant in the crust (such as iron, aluminum, common salt, and sulfur), supplies will probably always be sufficient to meet demand.

Probably the future of mineral supplies will be similar to the future of energy supplies. A sufficiency will be found, but present patterns of use will change. Some materials will become more important, others less important, and yet other substances, not used today, will assume major importance. But one prediction can safely be made: mining operations will get bigger, more materials will be used, and pollution will probably be intensified. Environments will come under even heavier pressure than that which endangers them today. In short, human beings will continue to be an increasingly important factor in geological processes.

Essay

GOLD MINING

It will never be known who among our ancestors was the first to pick up a grain of gold and fashion it into decorative jewelry. What is known, however, is that throughout all of recorded history, and far back into prehistoric times of the very earliest civilizations, gold was sought and prized.

Gold is a *noble metal,* which means that it is resistant to chemical reactions and thus does not corrode or dissolve. Indeed, gold has so little tendency to corrode that most of the gold that has ever been mined is still in use. When a gold object is no longer needed, it is simply melted down and the gold is reused. Gold that was mined by the Romans 2000 years ago is quite possibly present in a piece of new jewelry that might be purchased today.

It is estimated that all gold produced throughout history is about 130,000 metric tons. The annual world production now is about 1200 metric tons. By contrast, the *annual* world production of copper is in excess of 8 million metric tons!

The density of gold is 19.3 g/cm^3, so 130,000 metric tons of gold is equivalent to a cube of gold that is only 18.5 meters on an edge. It seems incredible that so small a volume of metal should be the result of the labors of the millions of people who have sought and mined gold through the ages. Perhaps it is just scarcity that gives gold its allure and that drives miners to seek more gold.

Through the middle years of the 1980s, the price of gold fluctuated between about $14 and $16 a gram. This led to a boom in gold prospecting and to the discovery of a large number of new ore deposits in the United States, Canada, Australia, the Pacific islands, and elsewhere. Despite all of the new discoveries, one country, South Africa, continues to dominate the world's gold production. In 1987 South Africa supplied about 42 percent of all the gold produced in the world, and it is estimated that 35 percent of all the gold ever mined has come from South Africa.

The South African gold deposits are giant placers and they have many unusual features. The ore is a series of gold-bearing conglomerates (Fig. B18.1), laid down 2.7 billion years ago in the shallow marginal waters of a marine basin. Associated with the gold are detrital pyrite and detrital uranium minerals. So far as size and richness are concerned, nothing like the deposits in the Witwatersrand Basin have been discovered anywhere else. Nor has the source of all the placer gold been discovered, so it is not possible to say why so much of the world's minable gold should be concentrated in this one sedimentary basin.

Mining in the Witwatersrand Basin had reached a depth of 3600 m (11,800 feet) by 1987. This is the deepest mining in the world and there are plans to continue mining to depths as great as 4500 m. Despite such ambitious plans, the heyday of gold mining in South Africa has probably passed. The deposits are running out of ore. South Africa's contribution to the world's gold production will slowly decrease in the years ahead.

FIGURE B18.1 Detrital gold is recovered from ancient placer deposits of the Witwatersrand, South Africa. The gold is found at the base of conglomerate layers interbedded with finer-ground sandstone, here seen in weathered outcrop at the site where gold was first discovered in 1886.

SUMMARY

1. Ancient industry, based largely on wood and muscle, has given way to industry that uses energy intensively and is based mainly on fossil fuels.

2. Coal originated as plant matter in ancient swamps, and is both abundant and widely distributed.

3. Oil and gas probably originated as organic matter sedimented on seafloors and decomposed chemically. Later these fluids moved through reservoir rocks and were caught in geologic traps to form pools.

4. Heavy, nonflowing oil (tar) can be recovered by mining techniques. The amount of tar in the world probably equals the amount of flowing oil.

5. When heated, part of the solid organic matter found in shale will convert to oil and gas. Oil from shales is the world's largest resource of fossil fuel. Unfortunately, most shale contains so little solid organic matter that more oil must be burned to heat the shale than is produced by the conversion process.

6. Nuclear energy is derived from atomic nuclei of radioactive isotopes, chiefly uranium. The nuclear energy available from naturally occurring radioactive elements is the single largest energy resource now available.

7. Other sources of energy currently used to some extent are geothermal heat, the tides, energy from flowing streams, winds, and the Sun's heat.

8. The mineral industry is based mainly on local concentrations of useful minerals in mineral deposits.

9. When a mineral deposit can be worked profitably it is called an ore.

10. Mineral deposits form when minerals become concentrated in one of five different ways: (1) by concentration through hydrothermal solutions to form hydrothermal mineral deposits; (2) by concentration through magmatic cooling and crystallization processes to form magmatic mineral deposits; (3) by concentration from lake water or seawater to form sedimentary mineral deposits; (4) by concentration in flowing water to form placers; and (5) by concentration through weathering processes to form residual deposits.

11. Deposits of geochemically abundant metals, which make up 0.1 percent or more of the crust, form in several ways. The amounts available for exploitation are enormous.

12. Deposits of geochemically scarce metals, present in the crust in amounts less than 0.1 percent, form mainly as hydrothermal and magmatic deposits. Amounts of scarce metals available for exploitation are limited and geographically restricted.

13. Gold, platinum, tinstone, diamonds, and other minerals are commonly found mechanically concentrated in placers.

14. Nonmetallic substances are used mainly in the chemical industry for fertilizers, for building materials, and for ceramics and abrasives.

IMPORTANT WORDS AND TERMS TO REMEMBER

anthracite (p. 435)
asphalt (p. 439)

bauxite (p. 453)
biomass energy (p. 442)
bituminous coal (p. 435)
breeder reactor (p. 442)

carbohydrates (p. 434)
coal (p. 434)
coalification (p. 434)

coal swamp (p. 435)
crystal settling (p. 449)
cyclothem (p. 436)

evaporite deposit (p. 450)
exploration geology (p. 444)

fission (p. 442)
fossil fuel (p. 433)

gangue (p. 445)

geochemically abundant metals (p. 454)
geochemically scarce metals (p. 454)
geothermal power (p. 443)

heavy oil (p. 439)
hydroelectric power (p. 442)
hydrothermal mineral deposit (p. 446)
hydrothermal solution (p. 446)

kerogen (p. 434)

lignite (p. 435)
lipids (p. 434)

magmatic mineral deposit (p. 449)
metallogenic province (p. 448)
migration of oil (p. 437)
mineral deposit (p. 444)

natural gas (p. 436)
noble metal (p. 458)
nuggets (p. 452)

oil (p. 436)
oil field (p. 437)

oil pool (p. 437)
oil shale (p. 439)
ore (p. 445)

peat (p. 434)
pegmatite (p. 449)
petroleum (p. 436)
pile (p. 442)
placer (p. 452)
porphyry copper deposit (p. 449)
proteins (p. 434)

reservoir (p. 437)

residual concentration (p. 453)
residual mineral deposit (p. 453)
roof rock (p. 437)

sedimentary mineral deposit (p. 450)
source rock (p. 437)
subbituminous coal (p. 435)

tar (p. 439)
tar sand (p. 439)
trap (for oil) (p. 437)

volcanogenic massive sulfide deposit (p. 446)

QUESTIONS FOR REVIEW

1. What is the meaning of the term *fossil fuel*? Name four different kinds of fossil fuel.

2. Several key steps are needed in order for coalification to proceed. What are they?

3. During what two periods in the Earth's history was most coal formed? Can you offer explanations for why this happened, and why the coal is now found where it is?

4. Five steps are needed in order for an oil pool to form. What are they?

5. What kind of rocks serve as source rocks for petroleum? In what kinds of rocks does petroleum tend to be trapped? Why?

6. Can you offer an explanation for the fact that oil drillers find more petroleum per unit volume of rock in Cenozoic rocks than in Paleozoic rocks of the same kind?

7. Oil shales are rich in organic matter. Can you explain why such shales have not served as source rocks for petroleum?

8. What are two possible limitations to the development of hydroelectric power?

9. What are mineral deposits? Describe five different ways by which a mineral deposit can form.

10. What factors determine whether a mineral deposit is ore or not?

11. How do hydrothermal solutions form and how do they form mineral deposits?

12. Briefly describe the formation of three different kinds of sedimentary mineral deposit.

13. What factors control the concentration of minerals in placer deposits?

14. Compare and contrast mineral deposits of the geochemically abundant and geochemically scarce metals.

A time exposure of stars viewed over Mt. Everest. Star trails are curved due to the Earth's rotation.

Beyond Planet Earth

The Planets: A Review

The rings of Saturn. Slight variations in composition of the rings are recorded in orange and ultraviolet wavelengths from a distance of 8.9 million km.

THE SOLAR SYSTEM

Interest in visible stars dates back to prehistory. Our ancestors noticed that a few stars seem to wander in completely different paths from the annual progression of most of their fellows across the sky. The Greeks mapped the paths of these strange objects and called them *planetai*, or wanderers. The Romans named the objects after their gods: Saturn, Jupiter, Mars, Venus, and Mercury. We now know that the curious wandering paths of the **planets** result from orbital motions around the Sun and that the other objects in the heavens are other suns, so far distant that they seem to occupy fixed places in the sky. Three other planets—Uranus, Neptune, and Pluto—were unknown to our ancestors because they are visible only through telescopes. Finally, we now know that other planets besides the Earth have moons of their own. A few vital statistics of the planets are given in Figure 19.1.

Planetology

The first person to view the sky through a telescope was Galileo. The year was 1609 and his homemade device was crude by comparison with a modern child's toy telescope. Galileo was astonished to see mountains on the Moon and large flat areas that looked to him like seas. He observed several moons circling Jupiter and he saw disk-shaped rings around Saturn. He also saw that Venus, like the Moon, had phases—full, half, quarter—and that huge, dark spots appeared every now and then on the surface of the Sun. The discoveries electrified Galileo's contemporaries, just as visits to the Moon by astronauts and images of distant planets sent back by unmanned spaceships have electrified the present generation.

In 1957, when the first artificial satellite was placed in orbit around the Earth, a brand-new scientific specialty emerged. **Planetology**, devoted to a comparative study of the Earth with the Moon and other planets, has taught us much about the Earth's earliest days—that time, before 3.8 billion years ago, for which most of the record seems to have been erased by the operation of the rock cycle. In the present chapter we present a brief account of the enormous successes of planetology. The discoveries of the past 30 years have motivated scientists to seek a unifying theory of origin for all suns and their planets. If the scientists are successful, it may one day be possible to say which of the billions of other suns, visible in the heavens, have planets like the Earth, and perhaps even to say which, if any, of those planets may have biospheres of their own.

Planets and Moons

The solar system consists of the Sun, nine planets, 50 known moons, a vast number of asteroids, millions of comets, and innumerable small fragments of rock called meteorites. All of the objects in the solar system move through space in smooth, regular orbits, held in place by gravitational attraction. The planets, asteroids, and comets circle the Sun while the moons circle the planets.

The distances between the planets are immense. Vast distances, like vast stretches of time, are difficult to comprehend. To put the solar system into perspective, think of the Sun as a basketball. The nearest planet, Mercury, would be a speck of dust about 12 m away. Earth would be a grain of sand about 1 mm in diameter at a distance of 30 m, Saturn a pebble the size of a grape nearly 300 m away, and Pluto, the most distant planet, would be another grain of sand 1200 m away. The comets, which are mixtures of rock, ice, and dust, lie farther still, beyond the orbit of Pluto, and they would form a cloud of minuscule dust specks about 1600 m away!

Terrestrial Planets. The planets can be separated into two groups based on their densities and compositions. The innermost planets, Mercury, Venus, Earth, and Mars, are small, rocky, and dense. Each has a density of 3 g/cm³ or more. They are similar in composition and are called the **terrestrial planets** because they are similar to **terra** (Latin for the Earth).

Jovian Planets. The planets more distant from the Sun are much larger (with the exception of Pluto), yet much less dense than the terrestrial planets. The masses of Jupiter and Saturn, for example, are 317 and 95 times the mass of the Earth, but their densities are only 1.3 and 0.7 g/cm³, respectively. The **jovian planets**, Jupiter, Saturn, Uranus, Neptune, and Pluto, take their name from *Jove*, an alternate designation for the Roman god Jupiter. They all probably have rocky cores that resemble terrestrial planets, but, with the exception of Pluto, most of their planetary masses are contained in thick atmospheres of hydrogen, helium, and other gases. It is the thick atmospheres that keep the densities of the giant planets low. In the case of Pluto, the low density probably results from a thick outer layer of ice.

Rocky matter exists in the outer regions of the solar system because some of the moons that circle

a

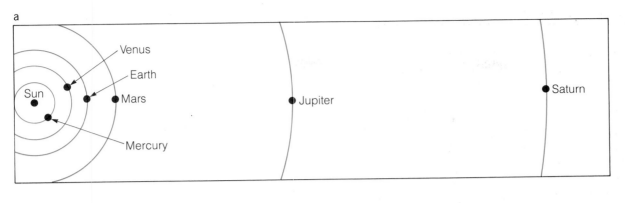

b

c

	Mercury	**Venus**	**Earth**	**Mars**	**Jupiter**	**Saturn**	**Uranus**	**Neptune**	**Pluto**
Diameter (km)	4880	12,104	12,756	6787	142,800	120,000	51,800	49,500	6000?
Mass (Earth = 1)	0.06	0.81	1	0.11	317.9	95.2	14.6	17.2	0.003
Density (water = 1)	5.4	5.2	5.5	3.9	1.3	0.7	1.2	1.7	1.1
Number of moons	0	0	1	2	15	15	15	2	0
Length of day (in Earth hours)	1416	5832	24	24.6	9.8	10.2	11	16	153
Period of one revolution around Sun (in Earth years)	0.24	0.62	1.00	1.88	11.99	29.5	84.0	165	248
Average distance from sun (millions of kilometers)	58	108	150	228	778	1427	2870	4497	5900

FIGURE 19.1 Orbits and properties of the planets. (a) The orbits of the planets around the Sun are all very close to the same plane. (b) Arranged in order of their relative positions, outward from the Sun, the planets are shown in their correct relative sizes. The Sun, 1.6 million km in diameter, is 13 times larger than Jupiter, the largest planet. (c) Numerical data concerning the orbits and properties of the planets.

the giant, gassy planets have high densities like that of the terrestrial planets. Two of Jupiter's moons, Io and Europa, have densities between 3.2 and 3.5 g/cm^3, and are close to the density of the Earth's Moon. Space exploration has revealed that two of Jupiter's low-density moons, Callisto and Ganymede, have thick blankets of ice surrounding small, rocky cores. As discussed later in this chapter, the progression outward from the Sun, from dense, rocky planets to less dense, giant gassy planets with ice-shrouded moons, provides an important clue to the way the solar system formed.

Meteorites and Asteroids

Meteorites. **Meteorites** are small stony or metallic objects from interplanetary space that impact a planetary surface. They can only be observed in motion when they flash into view as they plunge through the Earth's atmosphere. They, too, follow orbits around the Sun.

The orbits of the planets around the Sun are elliptical. The planets all revolve around the Sun in the same direction and the orbits lie in nearly the same plane as the orbit of the Earth. The orbit of Pluto is an exception because it is tilted at 17° to the *ecliptic*, which is the name given to the plane of the Earth's orbit. Most of the moons revolve around the planets in the same direction as the planets revolve around the Sun.

Titius–Bode Rule. The solar system occupies a region in space that is at least 12 billion km in diameter. Measuring outward from the Sun, the **Titius–Bode rule** states that the distance to each planet is approximately twice as far as the next inner one. Thus, the distance from the Sun to Mercury is 58 million km, while the distance to Venus, the next planet, is 107 million km.

Asteroids. The Titius–Bode rule seems to break down between Mars and Jupiter. Mars is 226 million km from the Sun, Jupiter is 775 million km. This rule suggests that a planet should be found about 450 million km from the Sun. No planet is present. What is found instead are at least 100,000 asteroids, small, irregularly shaped rocky bodies that have orbits lying between the orbits of Mars and Jupiter. The asteroids are either fragments of a planet that once existed and was somehow broken up, or they are rocky fragments that failed to gather into a planetary mass.

The asteroids and meteorites are dense, rocky objects. They are believed to have formed in the inner regions of the solar system, in the same way and at the same time as the terrestrial planets.

Cratered Surfaces

The most important processes that shape the surface of the Earth can be divided into three groups: tectonic processes, magmatic processes, and the surficial processes of weathering, mass-wasting, and erosion. To varying degrees each group of processes plays, or has played, a role in shaping the surfaces of all the rocky planets and moons in the solar system. However, in a planetary context, a fourth process must be added—the process of impact cratering.

Impact Cratering

The process by which a planetary surface is deformed as a result of a transfer of energy from a bolide to the planetary surface is known as **impact cratering**. A **bolide** is defined as an impacting body; it can be a meteorite, an asteroid, or a comet.

The velocities of meteorites have been measured between 4 and 40 km/s. If a large bolide had such a velocity, the amount of energy released on impact would be enormous. It has been calculated, for example, that a bolide 30 m in diameter and traveling at a speed of 15 km/s would, on impact, release as much energy as the explosion of 4 million tons of TNT. The resulting impact crater would be the size of Meteor Crater in Arizona—1.2 km across and 200 m deep (Fig. 19.2). Cratering is a very rapid geological process; the Meteor Crater event is estimated to have lasted about 1 min.

No large, natural impact crater has been observed as it formed. What is known about the process of cratering comes largely from laboratory experiments. The sequence of events is illustrated in Figure 19.3. As a high-speed bolide impacts and penetrates the surface it causes a jet of debris to be ejected at high velocity away from the point of impact. At the same time, the impact compresses the underlying rocks and sends intense shock waves outward. The pressures produced by the shock waves from a large bolide are so great that the strength of the rock is exceeded and a large volume of crushed and brecciated material results. In very large impact events, local melting and even vaporization may occur. Once the compressive shock waves have passed, a rapid expansion or decompression occurs. Expansion causes more material to be ejected from the impact crater and

FIGURE 19.2 Meteor Crater, near Flagstaff, Arizona. The crater was created by the impact of a bolide about 20,000 years ago. It is 1.2 km in diameter and 200 m deep. Note the raised rim of the crater wall and the blanket of ejecta thrown out of the crater.

produces a blanket of ejecta that surrounds the crater and thins away from the rim.

Note, in Figure 19.3, that the stratigraphy of layers in the ejecta blanket adjacent to the crater rim has been overturned. In the very largest impact structures, the central crater is circled by one or more raised rings of deformed rock. The outer rings are presumed to form as a result of the initial compression (Fig. 19.4). Following the immediate impact event, a number of postimpact events tend to modify the crater. Crater walls may slump (Fig. 19.5), isostatic rebound may produce changes in the floor and rim of the crater, erosion may fill the crater with debris, and in some instances, magma may rise along fractures produced by the impact, and lava may fill the crater.

Approximately 200 impact craters have been identified on the Earth. However, impact events must have been much more common than this small number implies. The reason so few craters have been found is that weathering and the rock cycle continually erase the evidence. On most planetary

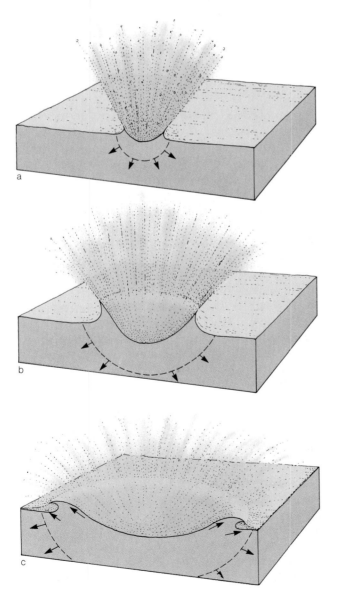

FIGURE 19.3 The shapes of impact craters are the same regardless of the angle of entry of the impacting bolide. The shape develops in three stages. (a) The initial bolide contact ejects a high-velocity jet of near-surface material. (b) The passage of shock waves through the bedrock produces high pressures and the compression of strata. In places the rock strength is exceeded and fracturing and brecciation result. Decompression throws broken rock out of the crater. (c) Strata along the rim of the crater are folded back and overturned by decompression. The ejected debris forms a circular blanket around the crater. (*Source:* After Gault et al., 1968.)

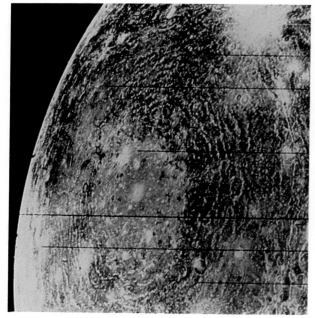

FIGURE 19.4 Multiple concentric rings around Valhalla, an impact crater on Callisto, a moon of Jupiter. The crater (central bright spot) is 600 km in diameter. The rings were formed as a result of thrust faults caused by the compression of the giant impact. The surface of Callisto is ice. Rings are also observed around giant craters on rocky planetary bodies, but they are neither so pronounced nor so numerous as the rings around Valhalla.

FIGURE 19.5 A large impact crater on the Moon. The crater is more than 200 km in diameter; the oblique photo was taken from a manned spacecraft. The highlands in the distance are part of the ejecta blanket. The stepped terraces in the middle distance were formed as a result of postcratering collapse of the crater rim. The hills in the foreground lie at the center of the crater and were formed by rebound of the crater floor during decompression.

surfaces a very different situation prevails. On most of the terrestrial planets, and on many of the rocky moons, atmospheres are absent, and tectonic activity has either ceased or is so slow that a rock cycle probably does not operate. As a result, the most striking features to be seen on most of the solid surfaces of the solar system are impact craters, some of which are more than 4 billion years old. The largest craters are 1000 km or more in diameter. They range down to the size of a pin's head. Such tiny craters are produced by dust-sized bolides.

THE TERRESTRIAL PLANETS

Besides impact cratering, volcanism has been widespread on each of the terrestrial planets and the Moon. On Mercury, the Moon, Mars, and Venus, it is clear that volcanism is predominantly basaltic. As discussed later in the chapter, only on Venus is it possible that other kinds of volcanism, similar to those found on the Earth, may have occurred. However, Venus is cloud-covered and difficult to study, so the question remains open. What space

exploration has made clear, and what has been confirmed by analysis of lunar and martian rocks, is that basaltic volcanism is a very common process in the solar system. This means, presumably, that each planet has gone through (or is still going through) a stage in its developmental history when internal heating caused partial melting. The product of this partial melting is basaltic magma. The products of partial melting are apparently similar on all of the rocky bodies in the solar system; this is suggestive evidence that the parent bodies all have similar compositions. This means, in turn, that the differences observed today between the rocky planets must reflect factors such as size and distance from the Sun in addition to composition.

Mercury

Mercury, the innermost planet, is so close to the Sun that it can only be seen just before sunrise or just after sunset. Telescopic viewing is difficult under such circumstances and as a result, almost everything that is known about Mercury comes from observations made during fly-by missions of unmanned spacecraft. Mercury has a diameter of 4880 km and is just a little larger than the Moon. It rotates slowly about its axis, and has a density of 5.4 g/cm^3.

To account for the high density it is necessary to conclude that Mercury has a metallic core about

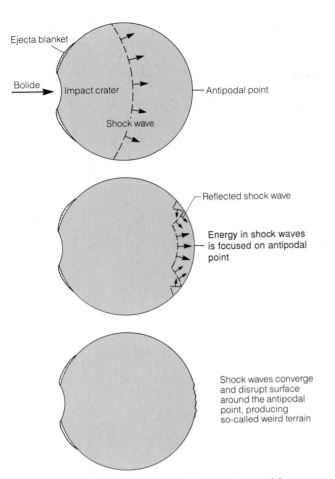

FIGURE 19.6 Origin of the "weird" terrain on Mercury. The terrain is exactly opposite the giant impact crater, Caloris. Compressive waves caused by the impact traveled through the body of the planet, and surface waves traveled around the edge. Both sets of waves were focused at the antipodal point, and they severely disrupted the surface. (Source: After Schultz and Gault, 1975.)

3600 km in diameter, and that this core accounts for 80 percent of the planet's mass. Mercury's core alone is the size of the Moon. Images of the surface, sent back by spacecraft, show that Mercury is heavily pockmarked by ancient impact craters, that it lacks an atmosphere, and that there is no evidence of moving plates of lithosphere. The largest impact basins are filled with basaltic lava flows; this means that Mercury had a period of magmatic activity. However, the lava plains are not crumpled and deformed, so the magmatic activity was not followed by tectonic activity.

The largest impact structure on Mercury is the Caloris Basin, 1300 km in diameter. On the far side of Mercury, exactly opposite Caloris, the surface is jumbled into a "weird," hilly terrain. Scientists who have studied Mercury believe that the weird terrain was produced by compressive shock waves from the bolide that formed Caloris. The waves passed completely through the planet and disrupted the far side (Fig. 19.6).

The most extraordinary feature about Mercury is the presence of a magnetic field about one one-hundreth as strong as that of the Earth. The field is dipolar and the magnetic axis coincides with the axis of rotation. The origin of the magnetic field is a puzzle. If Mercury's core is molten, the magnetic field could be caused by the fluid motions of the planet's rotation. There are two objections to this idea. The first is that Mercury rotates so slowly— once every 59 days—it is difficult to see how fluid motions strong enough to produce such a magnetic field could exist in the core. The second concerns tectonism: If Mercury has a huge core of molten iron, the interior must be very hot. Why then, is Mercury not tectonically active? The puzzle of the magnetic field remains unanswered.

Venus

Venus is the planet most like the Earth in size and mass. However, the similarities are fewer than the differences. Venus is enveloped in a cloudy atmosphere of carbon dioxide that is a hundred times more dense than the Earth's atmosphere. The clouds prevent direct observation of the venusian surface. Information comes from several spacecraft that have landed on the surface and radioed back information, and from radar measurements of the surface topography.

Landing spacecraft have reported that the surface temperature of Venus is astonishingly hot, about 500°C. At this temperature metals such as lead, zinc, and tin are in a molten state. The explanation of the high temperature is evident. First, Venus is closer to the Sun than the Earth. However, more important is the fact that the carbon dioxide in Venus's atmosphere acts like the glass of a greenhouse: It lets the Sun's rays through to heat the surface, but serves as a barrier that prevents heat from leaving. Russian spacecraft have operated in this intense heat long enough to send back a few local images of the surface of Venus. The images show masses of broken rock fragments, each about 20 cm across, covering the surface. Several of the Russian spacecraft carried out quick chemical analyses before they were overcome by the high temperatures. In each case, the analyses suggest that the rocks analyzed were basaltic in composition. A

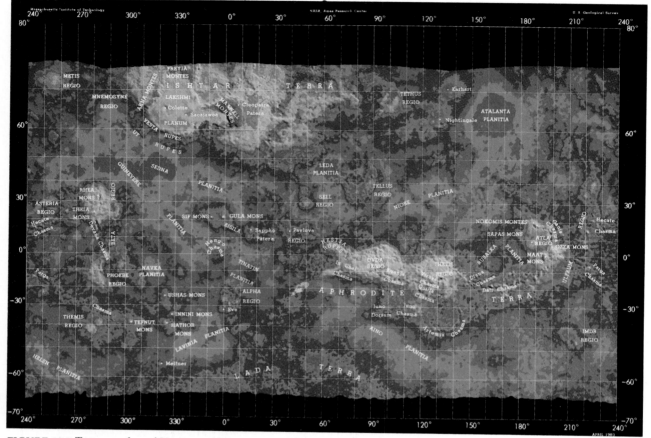

FIGURE 19.7 Topography of Venus as determined by radar. Blue areas are lowlands; green, yellow, and red, in that order, represent increasingly high topography. If continental crust exists on Venus, it is represented by the yellow and red highlands.

few analyses indicated that some rocks on Venus have relatively high potassium contents. This observation suggests that Earthlike magmatic differentiation, probably following Bowen's reaction series, may have been operating.

Sensitive radar measurements of Venus's shape show that large circular structures (presumably, impact craters) are present, and that much of the planet is nearly smooth (Fig. 19.7), with a total relief from the bottom of the deepest chasm to the top of the highest peak of 13 km. By comparison, the total relief on the Earth is 20 km. Although it is still too early to be sure, the distribution of lowlands and high plateaus in Figure 19.7 suggests to some experts that plate tectonics is active on Venus. If that conclusion is correct, the red, yellow, and green areas in Figure 19.7 represent continental crust, while the blue areas are basaltic plains equivalent to oceanic crust.

Visiting spacecraft discovered that Venus lacks a magnetic field. Because the density of Venus is

very close to the Earth's density, Venus must have an iron core and (presumably, like the Earth's) at least part of it is molten. The reason for the lack of a magnetic field is probably to be found in the slow rate of rotation of Venus about its axis—once every 243 days. The rotation is apparently too slow to cause fluid motion in the core.

One can speculate that the history of Venus might be like the early part of the Earth's history. Magmas were created, convection probably occurred in the mantle, and a period of moving lithospheric plates ensued. Some scientists have suggested that Venus is actually more advanced in its history than the Earth, that erosion has proceeded further, and that future study of Venus will provide clues to the Earth's future. The observation that large impact craters are present does not support this idea. All such suggestions are at best speculation, and we must await future spacecraft missions and more reliable observations before it is possible to draw any definite conclusions about the Earth's twin.

The Moon

The Earth's Moon is unique in the solar system. Because its diameter is 3476 km, only a little less than that of Mercury, the Moon is often described as a small terrestrial planet, and the Earth–Moon pair is described as a double planet. The largest moons of Jupiter and Saturn, although bigger than our Moon, are tiny in comparison with the sizes of their giant neighbors.

Because the Moon is a small, dense, rocky planet, it probably formed in the inner regions of the solar system, just as the other terrestrial planets did. If ideas about the way planets form are even partly correct, the Moon's structure and composition should be similar to those of the other terrestrial planets. Information about the Moon, therefore, should help in an understanding of the Earth and the other terrestrial planets.

Structure

Each time astronauts visited the Moon, the measurements they made yielded clues about its structure. The most informative measurements are seismic waves. Compared to earthquakes, moonquakes are weak and infrequent. Moonquakes large enough to be detected by instruments placed on the Moon by astronauts number fewer than 400 per year; on the Earth the same instruments would record nearly a million quakes per year. Despite the low frequency of moonquakes and the weakness of the seismic waves they generate, the quakes reveal a good deal about the lunar structure. Some of the most important points are: (1) on the side of the Moon that faces the Earth (at least on those parts that have been visited by astronauts) there is a crust about 65 km thick; (2) the Moon is layered (Fig. 19.8); (3) covering the surface is a layer of regolith that ranges from a few meters to a few tens of meters thick; (4) below the regolith is a layer (about 2 km thick) of shattered and broken rock produced by the continual rain of large and small bolides; (5) below the broken-rock zone is about 23 km of basalt, then 40 km of a feldspar-rich rock; and (6) at a depth of 65 km the velocities of seismic waves increase rapidly, indicating that the lunar crust overlies mantle.

The scarcity of moonquakes (and their weakness) immediately suggests that processes such as volcanism and plate tectonics, the causes of most earthquakes, are not happening on the Moon. Some moonquakes are caused by meteorites hitting the Moon. However, most quakes occur in groups, and at the times when its elliptical orbit brings the Moon closest to the Earth—the very moment when gravitational forces between the Earth and the Moon are strongest. This suggests that most moonquakes result from the gravitational pull that the Earth exerts on the Moon. The pull causes slight movement along cracks, each one causing a tiny moonquake. Foci of moonquakes have been recorded as deep as 1000 km; this observation, too, is informative. It means that unlike the Earth, the Moon is rigid enough at 1000 km for elastic deformation to store energy and for brittle failure to occur. This in turn means that the lunar lithosphere must be at least 1000 km thick and that the Moon's asthenosphere, if it exists at all, lies very deep within the Moon's body. The thick, rigid lithosphere makes it most unlikely that there is any present-day tectonic activity on the Moon.

The Moon's density is 3.3 g/cm^3, whereas that of the Earth is 5.5 g/cm^3. Assuming that the Moon's core, like the Earth's, is mainly iron, it can be calculated that its radius can be no larger than about 700 km. If the core were larger, the Moon's overall density would have to be higher.

The Lunar Surface

The Moon's surface can be divided into two general categories: (1) *highlands*, mountainous areas that appear to us as light-colored patches; and (2) *maria* (the "seas" seen by Galileo), smooth lowland areas that appear to us as dark-colored regions (Fig. 19.9). The highlands are regions of intense cratering. The maria are covered by basaltic lava flows (Fig. 19.10). In places, the lavas fill and cover the impact craters. The highlands must therefore be older than the maria. Note, in Figure 19.9, that the maria are circular or nearly so. Detailed study shows that they are actually the sites of giant impacts that must predate the basaltic flows that now fill them.

In the highlands, mountains soar tens of thousands of meters above the maria, and in places stand even higher than the Earth's mountains. Because lunar mountains are big, one way to obtain clues about the Moon's interior regions is to learn whether isostasy operates as it does on the Earth, and whether the great lunar mountains have roots. The test is best made by measuring variations in the Moon's gravitational pull. Such measurements were made while spacecraft orbited the Moon; they indicate that the mountains do have roots and that the highlands are isostatically balanced. We can infer, then, that when the lunar mountains formed, at least one of the Moon's outer layers must have

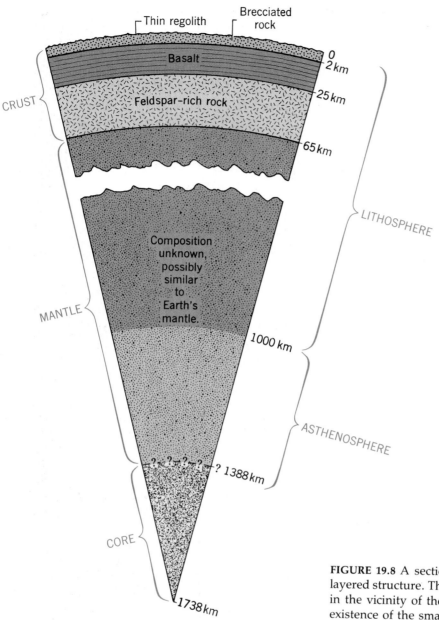

FIGURE 19.8 A section through the Moon showing the layered structure. The crust is known with certainty only in the vicinity of the astronauts' landing sites, and the existence of the small core is still uncertain.

been sufficiently fluidlike and plastic to enable the highlands to float.

Rock and Regolith

Astronauts returned to the Earth with a treasure trove of samples of rock and regolith. From these have come many clues about the Moon's composition and history.

Igneous Rock. The most interesting samples are the igneous rocks; in terms of age and composition, three different kinds are found. The first and oldest consists of feldspar-rich rocks such as anorthosite,

a variety of igneous rock formed by extreme magmatic differentiation, and consisting largely of calcium-rich plagioclase. Radiometric dates of these oldest rocks, which come from the highlands, indicate they could have been formed as long ago as 4.5 billion years, only 100 million years after the formation of the Moon. The second kind of igneous rock is basalt that contains high concentrations of potassium and phosphorus; this too is 4 billion or more years old. The potassium-rich basalts likewise come from highland areas and they are lava flows that seem to represent the last igneous activity in those areas. The third kind of igneous rock is also basalt, but it is rich in iron and titanium rather than

FIGURE 19.9 Face of the moon as seen from Earth. Light-colored areas pitted with craters are lunar highlands. Dark-colored, lowland areas are maria (plural of mare), formed when basaltic lava flowed out to fill the craters made by exceptionally large impacts. Copernicus, Kepler, and Tycho are three prominent, young meteorite craters that formed after the mare basins were filled. The impacts that made them splashed bright-colored rays of rocky debris over ancient basalt. The sites of the six Apollo lunar landing missions are indicated. (*Source:* Mosaic LEM-1, 3rd edition, 1966, made from telescopic photographs by U.S Air Force.)

potassium. Titanium- and iron-rich lavas have been found only in the maria (Fig. 19.11a) and are dated radiometrically at 3.2 to 3.8 billion years. We infer that such material underlies each mare to a depth of about 25 km. Mare basalt, then, seems to have formed several hundred million years after the highlands had formed. Even though magmas do not seem to occur on the Moon today, the mare basalts prove that magmas similar to those on the Earth have existed on the Moon for a long time.

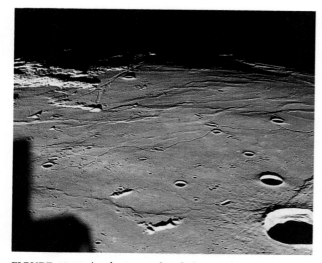

a

FIGURE 19.10 A photograph of the surface of a mare. Taken by an astronaut during the *Apollo 11* mission, the photograph shows portions of Mare Tranquillitatis. The irregular ridges, looking like long sand dunes, are ancient basaltic lava flows. The impact crater on the lower right-hand side is called Maskelyne.

b

Regolith.   The lunar regolith is a mixture of gray pulverized rock fragments and small particles of dust, many of which are glassy. Its composition is essentially that of the lunar igneous rocks. Regolith covers all parts of the lunar surface like a gray shroud (Fig. 19.11b), as though giant hammers had crushed the surface rock. Indeed the surface has been hammered, but by the bolides that continually strike its surface. The impacts are unhindered by the moderating influences of an atmosphere.

FIGURE 19.11 Two kinds of lunar samples brought back by astronauts. (a) Basalt, containing numerous vesicles formed when gases escaped during cooling and crystallization of lava. Collected during the *Apollo 12* mission, this sample is typical of the kinds of basalt found in the maria. (b) Regolith, a mixture of many rock and mineral types, together with glassy fragments produced by the bombardment of the lunar surface by meteorites. Only coarser grains from the regolith, from 1 to 2 mm across, are shown in the picture.

History of the Moon

Accretion.   The history of the Moon begins about 4.6 billion years ago, by which time the Moon had formed as a solid body. The available evidence does not prove exactly how the Moon was formed; but it does indicate that the final stages of formation involved the impacting of innumerable bolides, large and small, because ancient impact craters are still present. Probably the entire growth of the Moon occurred by **accretion**, the process by which solid bodies gather together to form a planet.

Magma Ocean.   By 4.6 billion years ago the Moon had accreted to about its present size. However, as it accreted and grew larger, the strength of its gravitational attraction increased, so that the speeds at which accreting bolides reached the lunar surface grew ever greater. Eventually, speeds became so great that each impact generated a large amount of heat. Near the end of the accretion process, so much heat was generated that an outer layer, 150 to 200 km thick, of the Moon's body was apparently melted. Thus, the Moon had a solid interior with a molten outer shell, in effect, a **magma ocean**. The period of rapid accretion soon ended, and the magma ocean began to cool and crystallize. The first mineral to crystallize was plagioclase feldspar; and because it was lighter than the parent magma, it floated. Soon a thick crust, rich in feldspar crystals, floated in the remaining liquid which eventually became the upper part of the lunar mantle. Bolides fall on the lunar surface today; so it can be inferred that bolides must also have been falling while the crust was forming. The largest impacts must have broken the crust, letting liquid from below ooze out to form the ancient potassium-rich lava flows. Samples of those ancient products of magmatic differentiation were found by the astronauts in the lunar highlands.

Mare Basins. The magma ocean would have crystallized within about 400 million years. By about 4 billion years ago, therefore, the highland crust had formed, and the major activity on the Moon had become the incessant rain of bolides that pitted and pocked the surface. Some bolides, larger than others, made exceptionally large impact scars. These huge circular cavities, eventually became the *mare basins* (Fig. 19.9).

Mare Basalts. While the Moon's surface was being bombarded by bolides, its interior regions were slowly heating up. If we use the composition of meteorites that fall on the Earth as a way to judge the amount of radioactivity present in the materials from which the Moon and the terrestrial planets formed, the amount of radioactivity in the Moon was small, but nevertheless it was sufficient to cause partial melting in the upper mantle, beginning about 3.8 billion years ago. So formed, the magma worked its way up to the surface along fractures caused by impacts of the largest meteorites. When it reached the surface, the magma filled the impact basins and formed the basalt flows now seen in the maria. By 3 billion years ago, the extrusion of mare basalt flows had ceased.

Except for the continuing rain of bolides, the Moon has remained a tectonically and magmatically dead planet.

Lessons for the Earth

Did the Earth ever look like the Moon? It probably did. The Moon supports the idea that the terrestrial planets formed by accretion of solid bodies. The Earth is larger than the Moon, so its gravitational pull is stronger. Presumably, therefore, an even thicker magma ocean covered the Earth at the end of the accretion process. Perhaps the earliest crust began to form through cooling of that magma. However, all traces of the Earth's primitive crust have been lost. The reason is not hard to find.

Because radioactive heating of the Earth has continued to create magma, probably the earliest crust has been remelted and reabsorbed.

Mars

Mars, with a diameter of 6787 km, has only one-tenth of the Earth's mass. Despite its size, Mars is earthlike in many ways. It rotates once every 24.6 h, so the length of Martian day is nearly the same as the length of an Earth day. Also, Mars has an atmosphere, although it is only one one-hundredth as dense as the Earth's and consists largely of carbon dioxide. Mars has polar "ice" caps (Fig. 19.12), consisting mostly of frozen carbon dioxide ("dry ice"), plus a small amount of water ice. Like the Earth, Mars has seasons, and the diameters of the ice caps alternately grow and shrink with the coming of winter and summer.

Mars lacks a dipolar magnetic field. Like the Moon it has a density somewhat less than that of the Earth. However, when allowance is made for the fact that Mars is smaller than the Earth, and that internal pressures are less, the difference in density is small. This means that the composition of the Earth and Mars must be similar and that Mars has a core. Presumably that core is solid. If it were molten, Mars would probably have a magnetic field.

Unfortunately, no other information is available about the layering of Mars. One of the spacecraft that landed on Mars in 1976 carried a seismometer with it, but no marsquakes occurred during the months the seismometer was active, so seismic evidence is unavailable.

Surface Features

The topography of Mars is extraordinary, and many features have yet to be adequately explained. The most obvious surface features are shown in Figure 19.13.

Southern Hemisphere. Approximately half the martian surface, the southern hemisphere, is densely cratered and resembles the surfaces of the Moon and Mercury. The largest impact crater discovered in the solar system, Hellas, with a diameter of nearly 2000 km, is on Mars. Scientists who have studied the images of the Martian surface sent back by unmanned spacecraft conclude that the southern hemisphere is covered by ancient crust similar to the crust of the lunar highlands, and presumably of the same age. The dense population of craters records a rain of bolides prior to 4 billion years ago.

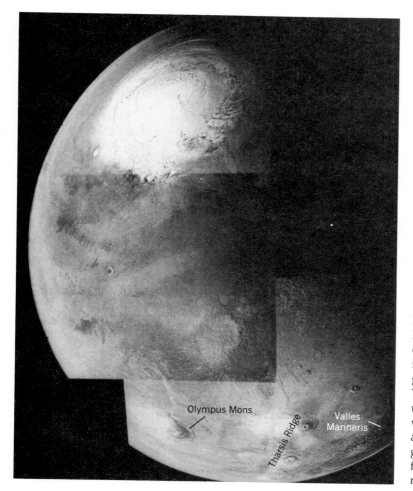

Olympus Mons

Tharsis Ridge

Valles
Marineris

FIGURE 19.12 The northern hemisphere of Mars in the late spring seen from a distance of 13,700 km. The figure is a mosaic of images from *Mariner 9*. The northern polar cap is greatly shrunken from its winter maximum. The origin of the spiral-shaped structure is uncertain but is probably related to spiral wind patterns. Near the base of the image are several shield volcanoes, including the giant Olympus Mons. The great canyonlike feature at lower right is Valles Marineris, named for the *Mariner* spacecraft.

Northern Hemisphere. The northern hemisphere presents a very different picture. Craters are sparse and large areas are relatively smooth. The obvious conclusion to draw is that some process (or processes) has produced a much younger surface. The probable answer to the puzzle is provided by the widespread evidence of volcanism in the northern hemisphere.

Olympus Mons and the Shield Volcanoes. At least 20 huge shield volcanoes and many smaller cones have been discovered. The giant among them is Olympus Mons whose basal diameter is 600 km, approximately the distance from Boston to Washington, D.C. Olympus Mons stands 27 km above the surrounding plains and is capped by a complex caldera that is 80 km across (Fig. 19.14). Mauna Loa, the largest volcano on the Earth, is also a shield volcano, but it is only 225 km across and 9 km high above the adjacent seafloor. It is estimated that the amount of volcanic rock in Olympus Mons exceeds all of the volcanic rock in the entire Hawaiian chain of volcanoes.

The presence of a huge volcanic edifice such as Olympus Mons implies several things. First, in order for such a huge volcano to form, long-lived sources of magma must be present in the Martian interior. Second, the magma source must remain connected to the volcanic vent for a very long time. This in turn means that the Martian lithosphere must be stationary. In short, plate tectonics is apparently not operating on Mars. A third implication is that the Martian lithosphere must be thick and strong. If it were not, it would be bowed down by the weight of Olympus Mons. Isostasy is apparently not operating, or if it is operating, the isostatic response is very slow.

Are volcanoes active on Mars today? This intriguing question cannot be answered exactly, but it is possible that some volcanism persists. The evidence comes from the volcanic cones. If they were old features, they would be pitted by impact craters. Craters are rare on the volcanic slopes, leading some experts to conclude that the youngest flows on Olympus Mons are probably less than 100 million years old.

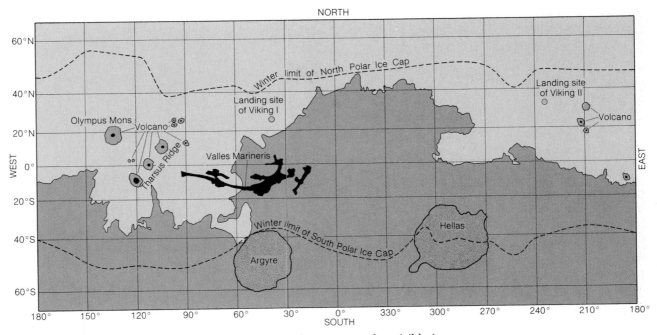

NORTH

SOUTH

FIGURE 19.13 Mars on a Mercator projection. Some features are also visible in Figure 19.12. The southern hemisphere is occupied by densely cratered terrain (red) believed to be ancient crust. The northern hemisphere is smoother and less cratered; the crust there is believed to be younger. It was on the edge of the smooth, northern hemisphere in a region called Chryse Planitia, that *Viking 1* landed on July 20, 1976. *Viking 2*, which also landed on the northern hemisphere, came down a few months later on Utopia Planitia. Several large shield volcanoes, each crowned by a caldera, occur in the younger crust, and two very large impact structures, Argyre and Hellas, occur in the ancient crust. The feature labeled Valles Marineris is the largest canyon in a region of vast canyons and gorges. The winter limit of the northern polar cap is much farther south than the edge of the cap seen in later spring in Figure 19.12. (*Source:* Adapted from a map, based on *Mariner 9* photographs, prepared by U.S. Geological Survey.)

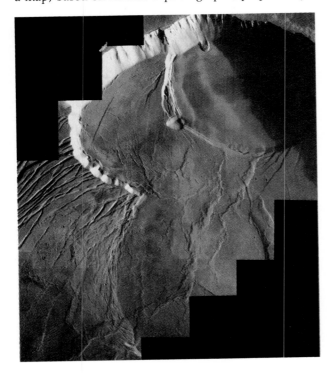

FIGURE 19.14 Detail of the summit caldera of Olympus Mons. Several phases of caldera collapse have occurred. The irregular ridges in the center foreground are basaltic flows.

Valles Marineris. Adjacent to Olympus Mons is Valles Marineris, a system of canyons that dwarf the Grand Canyon in Arizona. If the same feature were present on the Earth it would stretch from San Francisco to New York. The great canyons are a series of giant grabens, but what formed them is not known. Possibly they formed when great crustal up-warping occurred, or when the crust subsided into openings left empty when magma was extruded to build the huge volcanoes.

Erosion

Wind Erosion. As remarkable as the giant volcanoes and canyons are, even more remarkable features can be seen in the images returned by orbiting

FIGURE 19.15 An image of the Martian surface from *Viking 2*. The red color indicates that the surface rocks have undergone oxidation as a result of weathering.

spacecraft, and by the two Viking landers. *Viking 1* landed in a region of the northern hemisphere called Chryse Planitia, and *Viking 2* landed in a somewhat similar region called Utopia Planitia (Fig. 19.13). In both cases the images sent back showed a reddish-brown surface covered by loose stones and windblown sand (Fig. 19.15). Both landers made chemical analyses of the regolith on which they landed, and in both cases the results indicated clays and a sulfate mineral (probably gypsum). Weathering has obviously influenced the Martian surface and wind continually blows the weathering products around. Indeed, when the spacecraft *Mariner 9* was nearing Mars in 1971, a sandstorm of spectacular proportions was raging which continued unabated for several months (Essay, Chapter 11). It seems likely that on Mars, wind-driven sand is now the main agent of erosion.

Water Erosion. Still more spectacular than the Martian sand is evidence that water or some other liquid has influenced the surface of Mars (Fig. 19.16). Valleys look much like those cut by intermittent desert streams on the Earth. They meander, they branch, and they have braided patterns and other features characteristic of valleys made by running water. Some of the features look as though they were caused by gigantic floods. Yet Mars now lacks rainfall, streams, lakes, and seas. The sparse H_2O that does occur on Mars is in the atmosphere, or exists near the poles condensed as ice. The Martian surface is too cold for liquid water to exist.

Some have suggested that ice might be present beneath the surface dust as permafrost, and that in former times the Martian climate warmed up, causing the ice to melt, and creating torrential floods. Others suggest that Mars simply does not have sufficient ice in the polar caps or in permafrost to melt and cut great stream channels. Some other explanation must be found, they insist. The origin of the martian channels remains unsolved.

Geological History

The early history of Mars must have been much like that of the Earth and the Moon. The terrain of the southern hemisphere is equivalent to the lunar highlands. Then, as in both the Earth and the Moon, radioactive heating inside Mars created magma by partial melting. Because Olympus Mons and its mates are shield volcanoes, we infer that the magma they extruded was of low viscosity and, therefore, was basalt. This means that magmatic differentiation has occurred, but we cannot yet say when the volcanism started. To be able to do that we must have radiometric dates.

Planetary differentiation and volcanism would have released volatiles such as H_2O and CO_2. Because Mars is compositionally like the Earth, it can be calculated that the amount of H_2O released would have been sufficient to cover a smooth, uniform Mars with water to a depth of 50 to 100 m. It is possible, therefore, that very early in Martian history—a time corresponding to the Archean on the Earth—rains may have fallen, lakes and streams may have existed, and water erosion may have

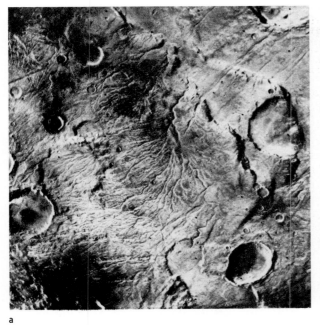

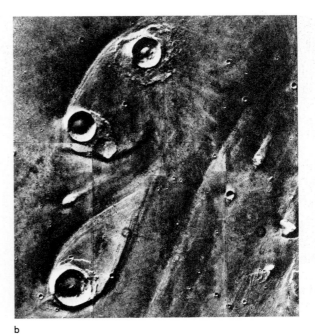

a   b

FIGURE 19.16 Part of the surface of Mars, showing features that have been interpreted as ancient stream channels. (a) Channels that branch in dendritic patterns. The channels are most prominently developed in the old, cratered terrain of the southern hemisphere. The width of the image is about 320 km. (b) Part of a channel system in Chryse Planitia. The teardrop-shaped plateaus formed behind resistant plateaus when floodwaters moved from lower left to upper right. Largest crater is about 10 km in diameter.

occurred. For some reason the Martian atmosphere became thinner. Probably this happened because weathering and erosion locked up H_2O and CO_2 in clay and carbonate minerals. It was a one-way process: H_2O and CO_2 were not released to the atmosphere again because a rock cycle did not operate and therefore did not cause recycling. The temperature would have dropped as the atmosphere thinned. Liquid water eventually became unstable on the surface of Mars. However, H_2O could still exist as ice, and this is presumably the form it is in today, possibly existing as a cement between grains in the regolith. In short, Mars became a planet covered by permafrost. The torrential floods that caused the striking erosion features (Fig. 19.16) may have occurred when regions of permafrost were subjected to sudden melting, perhaps by near intrusion of magma or sudden changes in climate.

Comparison of the Terrestrial Planets

Despite the differences between the terrestrial planets, they share many features in common. The differences apparently arise through the interaction of several factors. First, the planet size controls not only the atmosphere, but also the thermal properties. Small planets cool rapidly and magmatic activity soon ceases. Large planets cool more slowly and remain magmatically active. Second, the distance of a planet from the Sun determines whether or not H_2O can exist as a liquid. The third factor is the presence or absence of life. The Earth's atmosphere is the way it is because living plants and animals play essential roles in the geochemical cycles that control the composition. If life had developed on Venus, that planet would probably have an atmosphere like the Earth's. However, life apparently did not develop, so all of the CO_2 is still in the atmosphere. On the Earth, plants and animals have been the means whereby carbon dioxide has been removed from the atmosphere and the carbon is locked up in rocks as fossil organic matter and as calcium carbonate.

THE JOVIAN PLANETS AND THEIR MOONS

Jupiter

The giant, gassy planets tell us little about the evolution of the Earth, but they provide the best-preserved samples of the gases from which all planets are believed to have formed. Thus, they reveal much about how the solar system may have formed. Jupiter is the largest and best studied of the giant planets. It has an atmosphere composed of hydrogen, helium, ammonia, and methane, plus other trace constituents. It is presumed that a rocky core exists inside the dense atmosphere.

Jupiter has about twice the mass of all the other planets combined. Had it been slightly larger, it would have reached an internal temperature high enough for nuclear burning to start, and as a result it would have been a sun. Jupiter is unusual in many ways. One unusual feature is that it gives off twice as much energy as it receives from the Sun. The reason for this seems to be that Jupiter is still undergoing gravitational contraction which gives off heat energy.

The Jovian Moons

One of the most interesting things about Jupiter is its moons, four of which are as large as, or larger, than the Earth's Moon. The moons closest to Jupiter, Io and Europa, have densities of 3.5 and 3.2 g/cm^3, respectively, indicating that they are rocky bodies.

Io. The moon closest to Jupiter, Io, is extraordinary. It is a highly colored body with shades of yellow and orange predominating, suggesting that it is covered by sulfur and sulfurous compounds (Fig. 19.17). Impact craters are absent and the reason is not hard to find—Io is volcanically active. Not only is Io volcanically active, it is by far the most volcanically active body in the solar system. Impacts by bolides certainly must occur, but the craters are quickly covered up by volcanic debris.

Io's volcanism seems to be of two kinds. The first is the familiar basaltic volcanism found so widely though the solar system. Lava plains and shield volcanoes are the result. One of the shield volcanoes, Ra Patera, is almost as large as Olympus Mons on Mars. Fresh lava flows can be seen on its slopes. The second kind of volcanism seems to involve sulfur and sulfur dioxide (SO_2). Huge orange-yellow flows of what is presumed to be molten sulfur have been seen—some are as much as 700 km long. Most striking, however, are active volcanic plumes that throw sprays of sulfurous gases and entrained solid particles as high as 300 km above the surface of Io. Nine active plumes were observed by the two spacecraft *Voyager I* and *Voyager II*, as they flew by Io (Fig. 19.18). The volcanic plumes seem to be

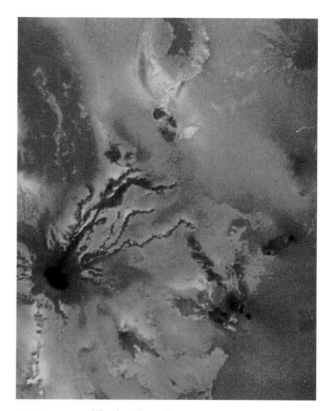

FIGURE 19.17 The bright colors on the surface of Io are caused by sulfur and sulfurous compounds given off during volcanism. The feature in the lower left of the image is a volcanic cone; numerous lava flows, basaltic in composition, radiate out from the volcano. The width of the field of view is 1000 km. The image was taken by *Voyager I* in 1979, at a distance of 128,500 km.

geyserlike in origin, but the fluid that boils and erupts is SO_2, not H_2O. It has been estimated that the plumes eject $10^{16}/g$ of fine, solid particles each year. This quantity is sufficient to bury the surface of Io with a layer of pyroclastic debris 100 m thick in a million years. No wonder no impact craters are to be seen. The process of surface renewal is much faster than it is on the Earth.

The amount of heat energy needed to drive Io's volcanoes is much greater than the heat that could be produced through radioactive decay in a stony planet. Io's volcanic heat comes from a different source—the gravitational pull exerted by the huge mass of Jupiter. As Io moves around Jupiter in an elliptical orbit it is periodically stretched more or less by the gravitational pull—more during close approach, less when far away. The bending and stretching due to the fluctuating gravitational pull generate heat, just as a copper wire becomes hot if it is bent back and forth. No other object in the

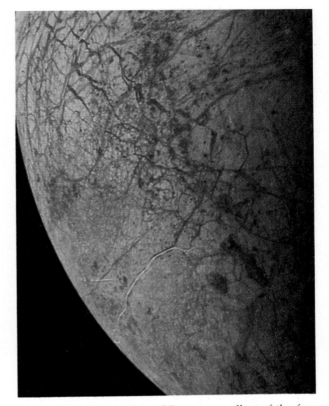

FIGURE 19.18 A volcanic eruption on Io. The volcanic plume is mostly gas, but small solid particles are also distributed by the gas. The plume rises to a height of 100 km above the surface of Io and is believed to be largely sulfur dioxide (SO_2). Several sites of active volcanism have been discovered on Io.

FIGURE 19.19 The surface of Europa, smallest of the four large moons of Jupiter, is mantled by ice to a depth of 100 km. The fractures indicate that some internal process must be renewing the surface. The dark material in the fractures apparently rises up from below. The cause of the fracturing is not known. The image was taken by *Voyager 2* in July, 1979.

solar system demonstrates the effect of tidal stresses so dramatically as Io.

Europa. Further away from Jupiter than Io are the three large moons Europa, Ganymede, and Callisto. Their densities decrease the further they are away from Jupiter. Europa has a density of 3.2 g/cm³; Ganymede, 2.0 g/cm³; and Callisto, 1.8 g/cm³. The reason for the lowered densities was discovered during the *Voyager* missions; each of the moons is sheathed with a layer of ice. The outer moons, and especially Europa, may have small metallic cores, but their densities suggest that their main masses lie in thick mantles consisting of ice and silicate minerals. Above the mantle are crusts of nearly pure ice a hundred kilometers or more thick.

Craters are rare on Europa; however, the surface is split and criss-crossed by an intricate network of fractures (Fig. 19.19). Presumably some tectonic process renews Europa's ice surface through fracture and upwelling. The upwelling may involve melting of the ice. The source of energy that results in melting is probably tidal, caused, as with Io, by the gravitational pull of Jupiter.

Ganymede. Ganymede has a much thicker sheath of ice, it is too far from Jupiter to be infuenced by tides, and it is pitted by craters. However, on Ganymede some slow-acting process is apparently renewing and reworking the ice surface, because some regions have few if any impact craters. The surface is divided into dark and light areas (Fig. 19.20). The dark areas are more heavily cratered than the light areas and, presumably, are older. Within the lighter, younger areas there are striking grooves, fractures, and grabenlike structures. It is possible that a unique kind of plate tectonics may be operating on Ganymede. The dark regions are ancient ice continents, the lighter areas are the places where ice rises convectively from the depths to create new ice crust.

Callisto. The most distant moon, Callisto, must contain the greatest proportion of ice because its

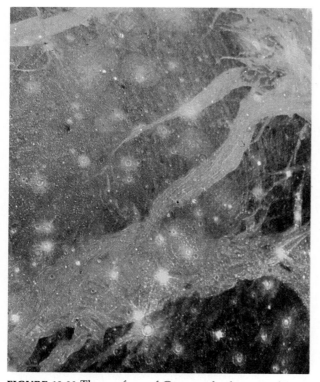

FIGURE 19.20 The surface of Ganymede, largest of Jupiter's moons, viewed from a distance of 312,000 km by *Voyager 2* in 1979. Ganymede is covered by a thick crust of ice. The dark surface is ancient ice, presumably covered by dust and impact debris. It is split into continent-sized fragments that are separated by light-colored, grooved terrains of younger ice. Ganymede is apparently tectonically active, and the grooved terrains seem to be the places where new ice rises from below, but how this happens, and what causes the grooves, is not known. The field of view is approximately 1300 km across.

FIGURE 19.21 The remarkable ring structure that surrounds Saturn. The width of the rings is 10,000 km. Note the shadow cast by the rings on the surface of Saturn.

density is the least. The surface is again icy but it must be very ancient ice as the number of impact craters is great. No evidence has yet been found to suggest that the surface of Callisto is being renewed and reworked.

Saturn

Saturn has a composition like that of Jupiter but it is not as large. Like Jupiter, Saturn radiates more energy than it receives from the Sun. The most striking feature of Saturn is its immense ring system (Fig. 19.21). The rings are a disk about 65,000 km in diameter and less than 10 km thick. They are composed of ice particles, icy snowballs, rocks, and dust ranging in diameter from a fraction of millimeter to about a meter. The rings are thought to be either the remains of a moon that got too

close to Saturn and was broken up by tidal forces, or to be material that never did aggregate to form a moon.

Saturnian Moons

Before the *Voyager* mission to Saturn in 1980, little was known of Saturn's moons. Less is known about them still than is known about the moons of Jupiter. Most of the moons are small and have low densities. In composition and structure they seem to resemble Ganymede and Callisto. They are ice-covered and cratering is extensive. Two of the moons display evidence, like that on Ganymede, suggesting that ice flow may be renewing the surfaces.

Titan. The most distinctive among the Saturnian moons is Titan, a body larger than Mercury. It is the only moon in the solar system large enough to retain a substantial atmosphere. Unfortunately, the atmosphere is an opaque, orange-colored smog that shrouds the surface from view. The composition of Titan's atmosphere is mostly nitrogen; however, ethane, acetylene, ethylene, hydrogen cyanide, and other unpleasant substances are also present.

The density of Titan is 1.9 g/cm³, which suggests that it contains about 45 percent ice and 55 percent rocky matter. Sunlight working on the atmosphere has caused ethylene and acetylene to form through photochemical reactions, and it is those compounds that produce the smog that covers Titan.

Because sunlight cannot penetrate the smog, the surface of Titan must be a cold and unfriendly place. Scientists who have studied the data sent back by *Voyager II* suggest that the surface temperature is −180°C, that Titan is covered by an ocean of liquid ethane and methane, and that continents of ice rise up from the ocean floor. Titan is stranger even than science fiction. Who could have imagined a planetary body with oceans of liquid hydrocarbons and continents of ice?

Essay

THE ORIGIN OF THE SOLAR SYSTEM

How did the solar system form? An exact answer may never be forthcoming, but the outlines of the process can be discerned from evidence obtained by astronomers, from our knowledge of the solar system today, and from the laws of physics and chemistry.

The birth throes of our Sun and its planets were the same as those of any other sun. Birth began with space that was not entirely empty. Atoms of various elements were present everywhere, even though thinly spread; they formed a tenuous, turbulent, swirling gas. When the gas thickened by a slow gathering of all the thinly spread atoms, the Sun was formed. The kinetic energy of those turbulent gas swirls eventually gave rise to the rotations of the Sun and planets. The gathering force of the gas was gravity, and as the atoms slowly moved closer together, the gas became hotter and denser. As one part of the gathering process, the Earth and the other planets formed.

More than 99 percent of all atoms in space are atoms of hydrogen and helium, the two lightest kinds of atoms. Samples of the gas are still preserved in the atmospheres of Jupiter and Saturn. Near the center of the gathering cloud of gas, the atoms became so tightly pressed and so hot that atoms of hydrogen and helium began to fuse together to form heavier elements. Fusion of light elements to form heavier ones (called nuclear burning) causes heat energy to be released. When, in the gas cloud that formed the solar system, nuclear burning of hydrogen and helium commenced, the Sun was born. The time was about 6 billion years ago. At some stage the cool outer portions of the *planetary nebula,* as the gas cloud is called, became compacted enough to allow solid objects to condense, in the same way that ice condenses from water vapor to form snow (Fig. B19.1). The solid condensates eventually became the planets. Planets nearest the Sun, where the temperatures were highest, contain only compounds that are capable of condensing at high temperatures. Those compounds consist of elements such as iron,

silicon, magnesium, and aluminum; we call them refractory elements. Planets distant from the Sun, where temperatures were lower, contain not only refractory elements but also volatile elements, such as hydrogen and sulfur, that do not condense at high temperatures but will do so at low temperatures. The farther away from the sun the condensation process occurred, the lower was the temperature and the greater the fraction of volatile elements. The most striking demonstration of this fact is the increasing amount of ice in moons, and the consequent lowering of density, the farther an object is from the Sun.

The size of a planet is also related to its distance from the Sun. A large-diameter ring of gas contains more atoms than a small-diameter ring. Planets close to the Sun formed from comparatively small amounts of gas and are, therefore, smaller than more distant planets that condensed from large rings of gas. However, near the outermost margin of the nebula, where the gas cloud was very tenuous and thin, the size of the planets becomes smaller again.

Probably all of the terrestrial planets once had a magma ocean due to the early, violent accretion process. On the Earth, all evidence of that ancient magma has been destroyed. The interiors of the earliest planets were certainly hot as a result of collision during accretion, but apparently, they were not hot enough to melt and form magma. Yet radioactive elements trapped within the planets must immediately have started to cause internal heating and to raise temperatures. As time passed, rising temperatures reached a point at which magma formed by partial melting. Opinions differ as to whether entire planets melted. Probably they did not, but partial melting certainly occurred.

Partial melting made it possible for a crust to form. As internal heating continued and as more and more magma was extruded, volcanoes gave off gases. It was these gases, mainly water vapor, carbon dioxide, methane, and possibly ammonia, that gave rise to the present atmospheres of the terrestrial planets. From the same source came the water we now find in the Earth's hydrosphere.

A great event in the Earth's history was the appearance of life. How, when, and where it happened is still problematic. It must have happened before 3.5 billion years ago because microfossils are found in rocks of that age.

Three and half a billion years ago the ocean already existed on the Earth, but the main mass of continental crust had not yet been formed. Problematic, too, is the exact manner in which life arose on the Earth but apparently did not arise on the other terrestrial planets. Perhaps future research will provide answers to even these most complex of all questions.

FIGURE B19.1 The gathering of atoms in space created a rotating cloud of dense gas that eventually became the Sun, and a surrounding disk-shaped planetary nebula. The planets formed by condensation of the planetary nebula.

SUMMARY

1. Planets close to the Sun, such as Mercury, Mars, the Earth, and Venus, are small, dense, rocky bodies. Planets farther away, such as Saturn and Jupiter, are large, low-density bodies.

2. The Moon has a layered structure. The layered structure probably formed by differentiation.

3. The Moon probably has a small core surrounded by a thick mantle, and is capped by a crust 65 km thick.

4. On the Moon, magma was formed early, but is no longer generated. The Moon is a magmatically dead planet.

5. The highlands of the Moon are remnants of ancient crust built by magmatic differentiation more than 4 billion years ago.

6. The maria (lunar lowlands) are vast basins created by the impacts of giant meteorites and later filled in by lava flows.

7. Each of the terrestrial planets went through a period of internal radioactive heating that led to generation of basaltic magma.

8. The Earth and possibly Mars and Venus are still producing magma from radioactive heating. Mars and Venus may still be tectonically active. Mars appears to be a one-plate planet, not a multiplate planet like the Earth.

9. Mars seems to be magmatically active. Olympus Mons, a shield volcano on Mars, is the largest volcano yet found in the solar system.

10. The principal eroding agent on Mars is wind-driven sand and dust. Water or some other flowing liquid cut stream channels at some time in the past.

11. Venus has about the same size and density as the Earth. It has a dense atmosphere of carbon dioxide and a surface temperature of about 500°C.

12. None of the other terrestrial planets has a climate hospitable to human life.

13. The large outer planets, Jupiter, Saturn, Neptune, Uranus, have thick atmospheres that obscure their rocky cores. Pluto, the outermost planet, is probably covered by ice.

14. The moons of Jupiter have progressively thicker outer layers of ice the further away they are from Jupiter.

15. The Sun formed when atoms in space became sufficiently compacted for nuclear burning to begin.

16. The planets formed by condensation from a disk-shaped envelope of gas (the planetary nebula) that rotated around the Sun. All planets condensed at the same time, about 4.6 billion years ago.

IMPORTANT WORDS AND TERMS TO REMEMBER

accretion (p. 476)
asteroid (p. 468)

bolide (p. 468)

crater (p. 468)
cratered surface (p. 468)

ecliptic (p. 468)
ejecta blanket (p. 469)

impact cratering (p. 468)

Jovian planets (p. 466)

lunar highlands (p. 473)

magma ocean (p. 476)
mare (p. 475)
mare basins (p. 477)
maria (p. 473)

meteorite (p. 468)

planet (p. 466)
planetary nebula (p. 486)
planetology (p. 466)

terrestrial planets (p. 466)
Titius–Bode rule (p. 468)

QUESTIONS FOR REVIEW

1. What is planetology? How is planetology related to geology?

2. Name the four terrestrial planets in order of their positions out from the Sun.

3. Besides the terrestrial planets, what other solid objects are to be found in the inner reaches of the solar system?

4. What is the most important process for shaping planetary surfaces in the solar system?

5. Briefly describe the kinds of magmatic activity that have been observed on planets and moons other than the Earth.

6. What evidence leads scientists to conclude that in the earliest days of the Moon's history, it had a molten outer layer and a solid interior?

7. What is the origin of the lunar maria?

8. How could it be possible for evidence of water erosion to exist on Mars without lakes, streams, or seas being present?

9. What is the source of the energy that makes Io, one of Jupiter's moons, the most tectonically active object in the solar system?

10. Compare the structure of Ganymede, a moon of Jupiter, with the Earth's moon. What explanation can you offer for the differences?

Units and Their Conversions

Prefixes for Multiples and Submultiples
Commonly Used Units of Measure
Length
Area

Volume
Mass
Pressure
Energy and Power
Temperature

Appendix A provides a table of conversion from older units to Standard International (SI) units.

PREFIXES FOR MULTIPLES AND SUBMULTIPLES

When very large or very small numbers have to be expressed, a standard set of prefixes is used in conjunction with the SI units. Some prefixes are probably already familiar; an example is the centimeter (which is one hundredth of a meter, or 10^{-2} m). The standard prefixes are

giga	$1,000,000,000$	$= 10^{99}$
mega	$1,000,000$	$= 10^{6}$
kilo	$1,000$	$= 10^{3}$
hecto	100	$= 10^{2}$
deka	10	$= 10$
deci	0.1	$= 10^{-1}$
centi	0.01	$= 10^{-2}$
milli	0.001	$= 10^{-3}$
micro	0.000001	$= 10^{-6}$
nano	0.000000001	$= 10^{-9}$
pico	0.000000000001	$= 10^{-12}$

One measure used commonly in geology is the nanometer (nm), a unit by which the sizes of atoms are measured; 1 nanometer is equal to 10^{-9} meter.

COMMONLY USED UNITS OF MEASURE

Length

Metric Measure

1 kilometer (km)	= 1000 meters (m)
1 meter (m)	= 100 centimeters (cm)
1 centimeter (cm)	= 10 millimeters (mm)
1 millimeter (mm)	= 1000 micrometers (μm) (formerly called microns)
1 micrometer (μm)	= 0.001 millimeter (mm)
1 angstrom (Å)	= 10^{-8} centimeters (cm)

Nonmetric Measure

1 mile (mi)	= 5280 feet (ft) = 1760 yards (yd)
1 yard (yd)	= 3 feet (ft)
1 fathom (fath)	= 6 feet (ft)

Conversions

1 kilometer (km)	= 0.6214 mile (mi)
1 meter (m)	= 1.094 yards (yd) = 3.281 feet (ft)
1 centimeter (cm)	= 0.3937 inch (in)
1 millimeter (mm)	= 0.0394 inch (in)
1 mile (mi)	= 1.609 kilometers (km)
1 yard (yd)	= 0.9144 meter (m)
1 foot (ft)	= 0.3048 meter (m)
1 inch (in)	= 2.54 centimeters (cm)
1 inch (in)	= 25.4 millimeters (mm)
1 fathom (fath)	= 1.8288 meters (m)

Area

Metric Measure

1 square kilometer (km²)	= 1,000,000 square meters (m²)
	= 100 hectares (ha)
1 square meter (m²)	= 10,000 square centimeters (cm²)
1 hectare (ha)	= 10,000 square meters (m²)

Nonmetric Measure

1 square mile (mi²)	= 640 acres (ac)
1 acre (ac)	= 4840 square yards (yd²)
1 square foot (ft²)	= 144 square inches (in²)

Conversions

1 square kilometer (km²)	= 0.386 square mile (mi²)
1 hectare (ha)	= 2.471 acres (ac)
1 square meter (m²)	= 1.196 square yards (yd²) = 10.764 square feet (ft²)
1 square centimeter (cm²)	= 0.155 square inch (in²)
1 square mile (mi²)	= 2.59 square kilometers (km²)
1 acre (ac)	= 0.4047 hectare (ha)
1 square yard (yd²)	= 0.836 square meter (m²)
1 square foot (ft²)	= 0.0929 square meter (m²)
1 square inch (in²)	= 6.4516 square centimeter (cm²)

Volume

Metric Measure

1 cubic meter (m³)	= 1,000,000 cubic centimeters (cm³)
1 liter (l)	= 1000 milliliters (ml)
	= 0.001 cubic meter (m³)
1 centiliter (cl)	= 10 milliliters (ml)
1 milliliter (ml)	= 1 cubic centimeter (cm³)

Nonmetric Measure

1 cubic yard (yd³)	= 27 cubic feet (ft³)
1 cubic foot (ft³)	= 1728 cubic inches (in³)
1 barrel (oil) (bbl)	= 42 gallons (U.S.) (gal)

Conversions

1 cubic kilometer (km³)	= 0.24 cubic miles (mi³)
1 cubic meter (m³)	= 264.2 gallons (U.S.) (gal)
	= 35.314 cubic feet (ft³)
1 liter (l)	= 1.057 quarts (U.S.) (qt)
	= 33.815 ounces (U.S. fluid) (fl. oz)
1 cubic centimeter (cm³)	= 0.0610 cubic inch (in³)
1 cubic mile (mi³)	= 4.168 cubic kilometers (km³)
1 acre-foot (ac-ft)	= 1233.46 cubic meters (m³)
1 cubic yard (yd³)	= 0.7646 cubic meter (m³)
1 cubic foot (ft³)	= 0.0283 cubic meter (m³)
1 cubic inch (in³)	= 16.39 cubic centimeters (cm³)
1 gallon (gal)	= 3.784 liters (l)

Mass

Metric Measure

1000 kilogram (kg)	= 1 metric ton (also called a tonne) (m.t)
1 kilogram (kg)	= 1000 grams (g)

Nonmetric Measure

1 short ton (sh.t)	= 2000 pounds (lb)
1 long ton (l.t)	= 2240 pounds (lb)
1 pound (avoirdupois) (lb)	= 16 ounces (avoirdupois) (oz) = 7000 grains (gr)
1 ounce (avoirdupois) (oz)	= 437.5 grains (gr)
1 pound (Troy) (Tr. lb)	= 12 ounces (Troy) (Tr. oz)
1 ounce (Troy) (Tr. oz)	= 20 pennyweight (dwt)

Conversions

1 metric ton (m.t)	= 2205 pounds (avoirdupois) (lb)
1 kilogram (kg)	= 2.205 pounds (avoirdupois) (lb)
1 gram (g)	= 0.03527 ounce (avoirdupois) (oz) = 0.03215 ounce (Troy) (Tr. oz) = 15,432 grains (gr)
1 pound (lb)	= 0.4536 kilogram (kg)
1 ounce (avoirdupois) (oz)	= 28.35 grams (g)
1 ounce (avoirdupois) (oz)	= 1.097 ounces (Troy) (Tr. oz)

Pressure

1 pascal (Pa)	= 1 newton/square meter (N/m²)
1 kilogram/square centimeter (kg/cm²)	= 0.96784 atmosphere (atm) = 14.2233 pounds/square inch (lb/in²) = 0.98067 bar
1 bar	= 0.98692 atmosphere (atm) = 10⁵ pascals (Pa) = 1.02 kilograms/square centimeter (kg/cm²)

Energy and Power

Energy

1 joule (J)	= 1 newton meter (N.m)
	= 2.390 × 10⁻¹ calorie (cal)

$$= 9.47 \times 10^{-4} \text{ British thermal unit (Btu)}$$
$$= 2.78 \times 10^{-7} \text{ kilowatt-hour (kWh)}$$

1 calorie (cal)
$$= 4.184 \text{ joule (J)}$$
$$= 3.968 \times 10^{-3} \text{ British thermal unit (Btu)}$$
$$= 1.16 \times 10^{-6} \text{ kilowatt-hour (kWh)}$$

1 British thermal unit (Btu)
$$= 1055.87 \text{ joules (J)}$$
$$= 252.19 \text{ calories (cal)}$$
$$= 2.928 \times 10^{-4} \text{ kilowatt-hour (kWh)}$$

1 kilowatt hour
$$= 3.6 \times 10^6 \text{ joules (J)}$$
$$= 8.60 \times 10^5 \text{ calories (cal)}$$
$$= 3.41 \times 10^3 \text{ British thermal units (Btu)}$$

Power (energy per unit time)

1 watt (W)
$$= 1 \text{ joule per second (J/s)}$$
$$= 3.4129 \text{ Btu/h}$$
$$= 1.341 \times 10^{-3} \text{ horsepower (hp)}$$
$$= 14.34 \text{ calories per minute (cal/min)}$$

1 horsepower (hp)
$$= 7.46 \times 10^2 \text{ watts (W)}$$

Temperature

To change from Fahrenheit (F) to Celsius (C)

$$^{\circ}\text{C} = \frac{(^{\circ}\text{F} - 32^{\circ})}{1.8}$$

To change from Celsius (C) to Fahrenheit (F)

$$^{\circ}\text{F} = (^{\circ}\text{C} \times 1.8) + 32^{\circ}$$

Tables of the
Chemical Elements and
Naturally Occurring
Isotopes

TABLE B.1 *Alphabetical List of the Elements*

Element	Symbol	Atomic Number	Crustal Abundance, Weight Percent	Element	Symbol	Atomic Number	Crustal Abundance, Weight Percent
Actinium	Ac	89	Man-made	Mercury	Hg	80	0.000002
Aluminum	Al	13	8.00	Molybdenum	Mo	42	0.00012
Americium	Am	95	Man-made	Neodymium	Nd	60	0.0044
Antimony	Sb	51	0.00002	Neon	Ne	10	Not known
Argon	Ar	18	Not known	Neptunium	Np	93	Man-made
Arsenic	As	33	0.00020	Nickel	Ni	28	0.0072
Astatine	At	85	Man-made	Niobium	Nb	41	0.0020
Barium	Ba	56	0.0380	Nitrogen	N	7	0.0020
Berkelium	Bk	97	Man-made	Nobelium	No	102	Man-made
Beryllium	Be	4	0.00020	Osmium	Os	76	0.00000002
Bismuth	Bi	83	0.0000004	Oxygen[b]	O	8	45.2
Boron	B	5	0.0007	Palladium	Pd	46	0.0000003
Bromine	Br	35	0.00040	Phosphorus	P	15	0.1010
Cadmium	Cd	48	0.000018	Platinum	Pt	78	0.0000005
Calcium	Ca	20	5.06	Plutonium	Pu	94	Man-made
Californium	Cf	98	Man-made	Polonium	Po	84	Footnote[d]
Carbon[a]	C	6	0.02	Potassium	K	19	1.68
Cerium	Ce	58	0.0083	Praseodymium	Pr	59	0.0013
Cesium	Cs	55	0.00016	Promethium	Pm	61	Man-made
Chlorine	Cl	17	0.0190	Protactinium	Pa	91	Footnote[d]
Chromium	Cr	24	0.0096	Radium	Ra	88	Footnote[d]
Cobalt	Co	27	0.0028	Radon	Rn	86	Footnote[d]
Copper	Cu	29	0.0058	Rhenium	Re	75	0.00000004
Curium	Cm	96	Man-made	Rhodium[c]	Rh	45	0.00000001
Dysprosium	Dy	66	0.00085	Rubidium	Rb	37	0.0070
Einsteinium	Es	99	Man-made	Ruthenium[c]	Ru	44	0.00000001
Erbium	Er	68	0.00036	Samarium	Sm	62	0.00077
Europium	Eu	63	0.00022	Scandium	Sc	21	0.0022
Fermium	Fm	100	Man-made	Selenium	Se	34	0.000005
Fluorine	F	9	0.0460	Silicon	Si	14	27.20
Francium	Fr	87	Man-made	Silver	Ag	47	0.000008
Gadolinium	Gd	64	0.00063	Sodium	Na	11	2.32
Gallium	Ga	31	0.0017	Strontium	Sr	38	0.0450
Germanium	Ge	32	0.00013	Sulfur	S	16	0.030
Gold	Au	79	0.0000002	Tantalum	Ta	73	0.00024
Hafnium	Hf	72	0.0004	Technetium	Tc	43	Man-made
Helium	He	2	Not known	Tellurium[c]	Te	52	0.000001
Holmium	Ho	67	0.00016	Terbium	Tb	65	0.00010
Hydrogen[b]	H	1	0.14	Thallium	Tl	81	0.000047
Indium	In	49	0.00002	Thorium	Th	90	0.00058
Iodine	I	53	0.00005	Thulium	Tm	69	0.000052
Iridium	Ir	77	0.00000002	Tin	Sn	50	0.00015
Iron	Fe	26	5.80	Titanium	Ti	22	0.86
Krypton	Kr	36	Not known	Tungsten	W	74	0.00010
Lanthanum	La	57	0.0050	Uranium	U	92	0.00016
Lawrencium	Lw	103	Man-made	Vanadium	V	23	0.0170
Lead	Pb	82	0.0010	Xenon	Xe	54	Not known
Lithium	Li	3	0.0020	Ytterbium	Yb	70	0.00034
Lutetium	Lu	71	0.000080	Yttrium	Y	39	0.0035
Magnesium	Mg	12	2.77	Zinc	Zn	30	0.0082
Manganese	Mn	25	0.100	Zirconium	Zr	40	0.0140
Mendelevium	Md	101	Man-made				

Source: After K. K. Turekian, 1969.

[a] Estimate from S. R. Taylor (1964).

[b] Analyses of crustal rocks do not usually include separate determinations for hydrogen and oxygen. Both combine in essentially constant proportions with other elements, so abundances can be calculated.

[c] Estimates are uncertain and have a very low reliability.

[d] Elements formed by decay of uranium and thorium. The daughter products are radioactive with such short half-lives that crustal accumulations are too low to be measured accurately.

TABLE B.2 *Naturally Occurring Elements Listed in Order of Atomic Numbers, Together with the Naturally Occurring Isotopes of Each Element, Listed in Order of Mass Numbers*

Atomic Number	Name	Symbol	Mass Numbers[b] of Natural Isotopes
1	Hydrogen	H	1, 2, $\boxed{3}$[c]
2	Helium	He	3, 4
3	Lithium	Li	6, 7
4	Beryllium	Be	9, $\boxed{10}$
5	Boron	B	10, 11
6	Carbon	C	12, 13, $\boxed{14}$
7	Nitrogen	N	14, 15
8	Oxygen	O	16, 17, 18
9	Fluorine	F	19
10	Neon	Ne	20, 21, 22
11	Sodium	Na	23
12	Magnesium	Mg	24, 25, 26
13	Aluminum	Al	27
14	Silicon	Si	28, 29, 30
15	Phosphorus	P	31
16	Sulfur	S	32, 33, 34, 36
17	Chlorine	Cl	35, 37
18	Argon	A	36, 38, 40
19	Potassium	K	39, $\boxed{40}$, 41
20	Calcium	Ca	40, 42, 43, 44, 46, $\boxed{48}$
21	Scandium	Sc	45
22	Titanium	Ti	46, 47, 48, 49, 50
23	Vanadium	V	$\boxed{50}$, 51
24	Chromium	Cr	50, 52, 53, 54
25	Manganese	Mn	55
26	Iron	Fe	54, 56, 57, 58
27	Cobalt	Co	59
28	Nickel	Ni	58, 60, 61, 62, 64
29	Copper	Cu	63, 65
30	Zinc	Zn	64, 66, 67, 68, 70
31	Gallium	Ga	69, 71
32	Germanium	Ge	70, 72, 73, 74, 76
33	Arsenic	As	75
34	Selenium	Se	74, 76, 77, 80, 82
35	Bromine	Br	79, 81
36	Krypton	Kr	78, 80, 82, 83, 84, 86
37	Rubidium	Rb	85, $\boxed{87}$
38	Strontium	Sr	84, 86, 87, 88
39	Yttrium	Y	89
40	Zirconium	Zr	90, 91, 92, 94, 96
41	Niobium	Nb	93
42	Molybdenum	Mo	92, 94, 95, 96, 97, 98, 100
44	Ruthenium	Ru	96, 98, 99, 100, 101, 102, 104
45	Rhodium	Rh	103
46	Palladium	Pd	102, 104, 105, 106, 108, 110
47	Silver	Ag	107, 109
48	Cadmium	Cd	106, 108, 110, 111, 112, 113, 114, 116
49	Indium	In	113, $\boxed{115}$
50	Tin	Sn	112, 114, 115, 116, 117, 118, 119, 120, 122, 124
51	Antimony	Sb	121, 123
52	Tellurium	Te	120, 122, 123, 124, 125, 126, 128, 130
53	Iodine	I	127
54	Xenon	Xe	124, 126, 128, 129, 130, 131, 132, 134, 136
55	Cesium	Cs	133
56	Barium	Ba	130, 132, 134, 135, 136, 137, 138
57	Lanthanum	La	$\boxed{138}$, 139
58	Cerium	Ce	136, 138, 140, $\boxed{142}$
59	Praseodymium	Pr	141
60	Neodymium	Nd	142, 143, $\boxed{144}$, 145, 146, 148, 150

TABLE B.2 *(Continued)*

Atomic Number[a]	Name	Symbol	Mass Numbers[b] of Natural Isotopes
62	Samarium	Sm	144, $\boxed{147}$, $\boxed{148}$, $\boxed{149}$, 150, 152, 154
63	Europium	Eu	151, 153
64	Gadolinium	Gd	$\boxed{152}$, 154, 155, 156, 157, 158, 160
65	Terbium	Tb	159
66	Dysprosium	Dy	156, 158, 160, 161, 162, 163, 164
67	Holmium	Ho	165
68	Erbium	Er	162, 166, 167, 168, 170
69	Thulium	Tm	169
70	Ytterbium	Yb	168, 170, 171, 172, 173, 174, 176
71	Lutetium	Lu	175, $\boxed{176}$
72	Hafnium	Hf	174, 176, 177, 178, 179, 180
73	Tantalum	Ta	180, 181
74	Tungsten	W	180, 182, 183, 184, 186
75	Rhenium	Re	185, $\boxed{187}$
76	Osmium	Os	184, 186, 187, 188, 189, 190, 192
77	Iridium	Ir	191, 193
78	Platinum	Pt	190, 192 , 195, 196, 198
79	Gold	Au	197
80	Mercury	Hg	196, 198, 199, 200, 201, 202, 204
81	Thallium	Tl	203, 205
82	Lead	Pb	204, 206, 207, 208
83	Bismuth	Bi	209
84	Polonium	Po	$\boxed{210}$
86	Radon	Rn	$\boxed{222}$
88	Radium	Ra	$\boxed{226}$
90	Thorium	Th	$\boxed{232}$
91	Protactinium	Pa	$\boxed{231}$
92	Uranium	U	$\boxed{234}$, $\boxed{235}$, $\boxed{238}$

[a] Atomic number = number of protons.
[b] Mass number = protons + neutrons.
[c] $\square$ indicates isotope is radioactive.

A P P E N D I X   C

Tables of the
Properties of
Common Minerals

TABLE C.1 *Properties of the Common Minerals with Metallic Luster*

Mineral	Chemical Composition	Form and Habit	Cleavage	Hardness	Specific Gravity	Other Properties	Most Distinctive Properties
Bornite	Cu_5FeS_4	Massive. Crystals very rare.	None. Uneven fracture.	3	5	Brownish bronze on fresh surface. Tarnishes purple, blue, and black. Grayish-black streak.	Color, streak.
Chalcocite	Cu_2S	Massive. Crystals very rare.	None. Conchoidal fracture.	2.5	5.7	Steel-gray to black. Dark gray streak.	Streak.
Chalcopyrite	$CuFeS_2$	Massive or granular.	None. Uneven fracture.	3.5–4	4.2	Golden yellow to brassy yellow. Dark green to black streak.	Streak. Hardness distinguishes from pyrite.
Chromite	$FeCr_2O_4$	Massive or granular.	None. Uneven fracture.	5.5	4.6	Iron black to brownish black. Dark brown streak.	Streak and lack of magnetism distinguishes from ilmenite and magnetite.
Copper	Cu	Massive, twisted leaves and wires.	None. Can be cut with a knife.	2.5–3	9	Copper color but commonly stained green.	Color, specific gravity, malleable.
Galena	PbS	Cubic crystals, coarse or fine-grained granular masses.	Perfect in three directions at right angles.	2.5	7.6	Lead-gray color. Gray to gray-black streak.	Cleavage and streak.
Gold	Au	Small irregular grains.	None. Malleable.	2.5	193	Gold color. Can be flattened without breakage.	Color, specific gravity, malleability.
Hematite	Fe_2O_3	Massive, granular, micaceous	Uneven fracture.	5–6	5	Reddish-brown, gray to black. Reddish-brown streak.	Streak, hardness.
Ilmenite	$FeTiO_3$	Massive or irregular grains.	Uneven fracture.	5.5–6	4.7	Iron-black. Brown-reddish streak differing from hematite.	Streak distinguishes hematite. Lack of magnetism distinguishes magnetite.
Limonite (*Goethite* is most common.)	A complex mixture of minerals, mainly hydrous iron oxides.	Massive, coatings, botryoidal crusts, earthy masses.	None.	1–5.5	3.5–4	Yellow, brown, black, yellowish-brown streak.	Streak.

TABLE C.1 *(Continued)*

Mineral	Chemical Composition	Form and Habit	Cleavage	Hardness / Specific Gravity		Other Properties	Most Distinctive Properties
Magnetite	Fe_3O_4	Massive, granular. Crystals have octahedral shape.	None. Uneven fracture.	5.5–6.5	5	Black. Black streak. Strongly attracted to a magnet.	Streak, magnetism
Pyrite ("Fool's gold")	FeS_2	Cubic crystals with striated faces. Massive.	None. Uneven fracture.	6–6.5	5.2	Pale brass-yellow, darker if tarnished. Greenish-black streak.	Streak. Hardness distinguishes from chalcopyrite. Not malleable, which distinguishes from gold.
Pyrolusite	MnO_2	Crystals rare. Massive, coatings on fracture surfaces.	Crystals have a perfect cleavage. Massive, breaks unevenly.	2–6.5	5	Dark gray, black or bluish black. Black streak.	Color, streak.
Pyrrhotite	FeS	Crystals rare. Massive or granular.	None. Conchoidal fracture.	4	4.6	Brownish-bronze. Black streak. Magnetic.	Color and hardness distinguish from pyrite, magnetism from chalcopyrite.
Rutile	TiO_2	Slender, prismatic crystals or granular masses.	Good in one direction. Conchoidal fracture in others.	6–6.5	4.2	Reddish-brown (common), black (rare). Brownish streak. Adamantine luster.	Luster, habit, hardness.
Sphalerite (zinc blende)	ZnS	Fine to coarse granular masses. Tetrahedron shaped crystals.	Perfect in six directions.	3.5–4	4	Yellowish-brown to black. White to yellowish-brown streak. Resinous luster.	Cleavage, hardness, luster.
Uraninite	UO_2 to U_3O_8	Massive, with botryoidal forms. Rare crystals with cubic shapes.	None. Uneven fracture.	5–6	6.5–10	Black to dark brown. Streak black to dark brown. Dull luster.	Luster and specific gravity distinguish from magnetite. Streak distinguishes from ilmenite and hematite.

TABLE C.2 *Properties of Rock-Forming Minerals with Nonmetallic Luster*

Mineral	Chemical Composition	Form and Habit	Cleavage	Hardness / Specific Gravity		Other Properties	Most Distinctive Properties
Amphiboles. (A complex family of minerals, *Hornblende* is most common.)	$X_2Y_5Si_8O_{22}(OH)_2$ where X = Ca, Na; Y = Mg, Fe, Al.	Long, six-sided crystals; also fibers and irregular grains	Two; intersecting at 56° and 124°	5–6	2.9 – 3.8	Common in metamorphic and igneous rocks. *Hornblende* is dark green to black; *actinolite*, green; *tremolite*, white.	Cleavage, habit.
Andalusite	Al_2SiO_5	Long crystals, often square in cross-section.	Weak, parallel to length of crystal.	7.5	3.2	Found in metamorphic rocks. Often flesh-colored.	Hardness, form.
Anhydrite	$CaSO_4$	Crystals are rare. Irregular grains or fibers.	Three, at right angles.	3	2.9	Alters to gypsum. Pearly luster, white or colorless.	Cleavage, hardness.
Apatite	$Ca_5(PO_4)_3(F, OH, Cl)$	Granular masses. Perfect six-sided crystals.	Poor. One direction.	5	3.2	Green, brown, blue, or white. Common in many kinds of rocks in small amounts.	Hardness, form.
Aragonite	$CaCO_3$	Massive, or slender, needle-like crystals.	Poor. Two directions.	3.5	2.9	Colorless or white. Effervesces with dilute HCl.	Effervescence with acid. Poor cleavage distinguishes from calcite.
Asbestos			See Serpentine				
Augite			See Pyroxene				
Biotite			See Mica				
Calcite	$CaCO_3$	Tapering crystals and granular masses.	Three perfect; at oblique angles to give a rhomb-shaped fragment.	3	2.7	Colorless or white. Effervesces with dilute HCL.	Cleavage, effervescence with acid.
Chlorite	$(Mg, Fe)_5(Al, Fe)_2 Si_3O_{10}(OH)_8$	Flaky masses of minute scales.	One perfect; parallel to flakes.	2 – 2.5	2.6 – 2.9	Common in metamorphic rocks. Light to dark green. Greasy luster.	Cleavage— flakes not elastic, distinguishes from mica. Color.
Dolomite	$CaMg(CO_3)_2$	Crystals with rhomb-shaped faces. Granular masses.	Perfect in three directions as in calcite.	3.5	2.8	White or gray. Does not effervesce in cold, dilute HCl unless powdered. Pearly luster.	Cleavage. Lack of effervescence with acid.

TABLE C.2 *(Continued)*

Mineral	Chemical Composition	Form and Habit	Cleavage	Hardness / Specific Gravity		Other Properties	Most Distinctive Properties
Epidote	Complex silicate of Ca, Fe and Al	Small elongate crystals. Fibrous.	One perfect, one poor.	6–7	3.4	Yellowish-green to dark green. Common in metamorphic rocks.	Habit, color. Hardness distinguishes from chlorite.
Feldspars: Potassium feldspar (*orthoclase* is a common variety)	$KAlSi_3O_8$	Prism-shaped crystals, granular masses.	Two perfect, at right angles.	6	2.6	Common mineral. Flesh-colored, pink, white, or gray.	Color, cleavage.
Feldspar	$NaAlSi_3O_8$ (albite) and $CaAl_2Si_2O_8$ (anorthite) and all compositions between.	Irregular grains, cleavable masses. Tabular crystals.	Two perfect, not quite at right angles.	6–6.5	2.6–2.7	White to dark gray. Cleavage planes may show fine parallel striations.	Cleavage. Striations on cleavage planes will distinguish from potassium feldspar.
Fluorite	CaF_2	Cubic crystals, granular masses.	Perfect in four directions.	4	3.2	Colorless, bluish green. Always an accessory mineral.	Hardness, cleavage, does not effervesce with acid.
Garnets	$X_3Y_2 (SiO_4)_3$; X = Ca, Mg, Fe, Mn; Y = Al, Fe, Ti, Cr.	Perfect crystals with 12 or 24 sides. Granular masses.	None. Uneven fracture.	6.5–7.5	3.5–4.3	Common in metamorphic rocks. Red, brown, yellowish-green, black.	Crystals, hardness, no cleavage.
Graphite	C	Scaly masses.	One, perfect. Forms slippery flakes.	1–2	2.2	Metamorphic rocks. Black with metallic to dull luster.	Cleavage, color. Marks paper.
Gypsum	$CaSO_4 \cdot 2H_2O$	Elongate or tabular crystals. Fibrous and earthy masses.	One, perfect. Flakes bend but are not elastic.	2	2.3	Vitreous to pearly luster. Colorless.	Hardness, cleavage.
Halite	NaCl	Cubic crystals.	Perfect to give cubes.	2.5	2.2	Tastes salty. Colorless, blue.	Taste, cleavage.
Hornblende			See Amphibole				
Kaolinite	$Al_2Si_2O_5(OH)_4$	Soft, earthy masses. Submicroscopic crystals.	One, perfect.	2–2.5	2.6	White, yellowish. Plastic when wet; emits clayey odor. Dull luster.	Feel, plasticity, odor.

TABLE C.2 (*Continued*)

Mineral	Chemical Composition	Form and Habit	Cleavage	Hardness / Specific Gravity	Other Properties	Most Distinctive Properties
Kyanite	Al_2SiO_5	Bladed crystals.	One perfect. One imperfect.	4.5 parallel to blade, 7 across blade / 3.6	Blue, white, gray. Common in metamorphic rocks.	Variable hardness, distinguishes from sillimanite. Color.
Mica: *Biotite*	$K(Mg, Fe)_3$-$AlSi_3O_{10}$-$(OH)_2$	Irregular masses of flakes.	One, perfect.	2.5–3 / 2.8–3.2	Common in igneous and metamorphic rocks. Black, brown, dark green.	Cleavage, color. Flakes are elastic.
Muscovite	$KAl_3Si_3O_{10}(OH)_2$	Thin flakes.	One, perfect.	2–2.5 / 2.7	Common in igneous and metamorphic rocks. Colorless, pale green or brown.	Cleavage, color. Flakes are elastic.
Olivine	$(Mg, Fe)_2SiO_4$	Small grains, granular masses.	None. Conchoidal fracture.	6.5–7 / 3.2–4.3	Igneous rocks. Olive green to yellowish-green.	Color, fracture, habit.
Orthoclase			See Feldspar			
Plagioclase			See Feldspar			
Pyroxene (A complex family of minerals. *Augite* is most common.)	$XY(SiO_3)_2$ $X = Y = Ca$, Mg. Fe	8-sided stubby crystals. Granular masses.	Two, perfect, nearly at right angles.	5–6 / 3.2–3.9	Igneous and metamorphic rocks. *Augite*, dark green to black; other varieties white to green.	Cleavage
Quartz	SiO_2	6-sided crystals, granular masses.	None. Conchoidal fracture.	7 / 2.6	Colorless, white, gray, but may have any color, depending on impurities. Vitreous to greasy luster.	Form, fracture, striations across crystal faces at right angles to long dimension.
Serpentine (Fibrous variety is *asbestos*)	$Mg_3Si_2O_5(OH)_4$	Platy or fibrous.	One, perfect.	2.5–5 / 2.2–2.6	Light to dark green. Smooth, greasy feel.	Habit, hardness.
Sillimanite	Al_2SiO_5	Long needle-like crystals, fibers.	Breaks irregularly, except in fibrous variety.	6–7 / 3.2	White, gray. Metamorphic rocks.	Hardness distinguishes from kyanite. Habit.
Talc	$Mg_3Si_4O_{10}(OH)_2$	Small scales, compact masses.	One, perfect.	1 / 2.6–2.8	Feels slippery. Pearly luster. White to greenish.	Hardness, luster, feel, cleavage.

TABLE C.2 *(Continued)*

Mineral	Chemical Composition	Form and Habit	Cleavage	Hardness	Specific Gravity	Other Properties	Most Distinctive Properties
Tourmaline	Complex silicate of B, Al, Na, Ca, Fe, Li and Mg.	Elongate crystals, commonly with triangular cross section.	None.	7–7.5	3–3.3	Black, brown, red, pink, green, blue, and yellow. An accessory mineral in many rocks.	Habit.
Wollastonite	$CaSiO_3$	Fibrous or bladed aggregates of crystals.	Two, perfect.	4.5–5	2.8–2.9	Colorless, white, yellowish. Metamorphic rocks. Soluble in HCl.	Habit. Solubility in HCl and hardness distinguish amphiboles, kyanite, sillimanite.

TABLE C.3 *Properties of Some Common Gemstones*

Mineral and Variety	Composition	Form and Habit	Cleavage	Hardness	Specific Gravity	Other Properties	Most Distinctive Properties
Beryl: *Aquamarine* (blue) *Emerald* (green) *Golden beryl* (golden-yellow)	$Be_3Al_2Si_6O_{18}$	Six-sided, elongate crystals common.	Weak.	7.5–8	2.75	Bluish green, green, yellow, white, colorless. Common in pegmatites.	Form. Distinguished from apatite by its hardness.
Corundum: *Ruby* (red) *Sapphire* (blue)	Al_2O_3	Six-sided, barrel-shaped crystals.	None, but breaks easily across its crystal.	9	4	Brown, pink, red, blue, colorless. Common in metamorphic rocks. Star sapphire is opalescent with a six-sided light spot showing.	Hardness.
Diamond	C	Octahedron-shaped crystals.	Perfect, parallel to faces of octahedron.	10	3.5	Colorless, yellow; rarely red, orange, green, blue or black.	Hardness, cleavage.
Garnet: *Almandite* (red) *Grossularite* (green, cinnamon-brown) *Andradite* (variety *demantoid* is green)						A rock-forming mineral—See Table C.2	

TABLE C.3 *(Continued)*

Mineral and Variety	Composition	Form and Habit	Cleavage	Hardness / Specific Gravity	Other Properties	Most Distinctive Properties
Opal (A mineraloid)	$SiO_2 \cdot nH_2O$	Massive, thin coating. Amorphous.	None. Conchoidal fracture.	5–6 / 2–2.2	Colorless, white, yellow, red, brown, green, gray, opalescent.	Hardness, color, form.
Quartz: (1) Coarse crystals *Amethyst* (violet) *Cairngorm* (brown) *Citrine* (yellow) *Rock crystal* (colorless) *Rose quartz* (pink) (2) Fine-grained *Agate* (banded, many colors) *Chalcedony* brown, gray) *Heliotrope* (green) *Jasper* (red)			A rock-forming mineral—See Table C.2			
Topaz	$Al_2SiO_4(OH, F)_2$	Prism-shaped crystals, granular masses.	One, perfect.	8 / 3.5	Colorless, yellow, blue, brown.	Hardness, form, color.
Tourmaline			A rock-forming mineral—See Table C.2			
Zircon	$ZrSiO_4$	Four-sided elongate crystals, square in cross-section.	None.	7.5 / 4.7	Brown, red, green, blue, black.	Habit, hardness.

Glossary

Some definitions are not included in the glossary; *units of measurement* can be found in Appendix A, *chemical elements* are listed in Appendix B, and *mineral names* are given in Appendix C.

A horizon. The uppermost horizon of a soil, generally dark colored and characterized by an accumulation of organic matter.

Aa. A rubbly fragmental lava flow, usually basaltic in composition.

Ablation. The loss of mass from a glacier.

Ablation area. A region of net loss on a glacier characterized by a surface of bare ice and old snow from which the last winter's snowcover has melted away.

Absolute velocity (of a plate). The velocity of a plate measured against a fixed, external frame of reference.

Abyssal plain. A large flat area of deep-seafloor having slopes less than about 1 m/km.

Accreted terrane. Block of crust moved laterally by strike–slip faulting or by a combination of strike–slip faulting and subduction, then accreted to a larger mass of continental crust.

Accreted terrane margins. Continental margins formed by the addition of island arcs and other small masses of crust.

Accretion. The process by which solid bodies gather together to form a planet or a continent.

Accumulation. The addition of mass to a glacier.

Accumulation area. An upper zone on a glacier, covered by remnants of the previous winter's snowfall and representing an area of net gain in mass.

Active zone. The margin of a tectonic plate where deformation is occurring.

Age. The time during which a geologic stage accumulated.

Agglomerate. A pyroclastic rock in which average particle diameters are greater than 64 mm.

Aggradation. Depositional upbuilding, as by a stream.

Alluvial fan. A fan-shaped body of alluvium typically built where a stream leaves a steep mountain valley.

Alluvium. Sediment deposited by streams in nonmarine environments.

Alpha particle (α-particle). An atomic particle expelled from an atomic nucleus during certain radioactive transformations, equivalent to an $^4_2$He nucleus stripped of its electrons.

Amorphous. A term applied to solids that lack internal atomic order.

Amphibolite. A metamorphic rock containing abundant amphibole.

Amygdule. A vesicle filled by secondary minerals deposited by groundwater.

Andesite. An aphanitic igneous rock with the composition of a diorite.

Andesite line. A line around the Pacific Ocean basin inside of which andesite is not found.

Angle of repose. The steepest angle, measured from the horizontal, at which rock debris remains stable.

Angular unconformity. An unconformity marked by angular discordance between older and younger rocks.

Anhydrous. A term applied to a substance that is H_2O-free. Opposite of hydrous.

Anion. A negative ion.

Anorthosite. A phaneritic igneous rock consisting largely of plagioclase.

Antecedent stream. A stream that has maintained its course across an area of the crust that was raised across its path by folding or faulting.

Anthracite. A metamorphic rock derived from coal by heat and pressure.

Anticline. An upfold in the form of an arch.

Aphanites. Igneous rocks in which the component grains cannot be readily distinguished with the naked eye or even with the aid of a simple hand lens.

Apparent polar wandering. The apparent motions of the magnetic poles derived from measurements of pole directions using paleomagnetism.

Aquiclude. A body of impermeable or distinctly less permeable rock adjacent to an *aquifer.*

Aquifer. A body of permeable rock or regolith saturated with water and through which groundwater moves.

Archean. The eon that follows the Hadean Eon.

Arête. A jagged, knife-edged ridge crest created where glaciers have eroded back into a ridge.

Artesian spring. Natural springs that draw their supply of water from a confined aquifer.

Artesian system. An inclined aquifer that permits water confined in it to rise to the surface in a well or along a fissure.

Artesian well. A well in which water rises above the aquifer.

Ash. Tephra in which particles have an average diameter less than 2 mm.

Ash tuff. Pyroclastic rock in which the tephra particles are less than 2 mm in diameter.

Asphalt. See *tar.*

Asteroids. Irregularly shaped rocky bodies that have orbits lying between the orbits of Mars and Jupiter.

Asthenosphere. The region of the mantle where rocks become plastic, like toffee or tar, and are easily deformed. It lies at a depth of 100 to 350 km below the surface.

Asymmetrical fold. A fold in which one limb dips more steeply than the other.

Atmosphere. The air sphere, consisting of the mixture of gases that together we call air.

Atoll. A coral reef, often roughly circular in plan, that encloses a shallow lagoon.

Atom. The smallest individual particle that retains all the properties of a given chemical element.

Atomic number. The number of protons in the nucleus of an atom.

Atomic substitution. See *ionic substitution.*

Authigenic sediment. Sediment formed in place.

Axial plane. An imaginary plane that divides a fold as symmetrically as possible, and that passes through the axis.

Axis (*of a fold*). The median line between the limbs, along the crest of an anticline or the trough of a syncline.

B horizon. A soil horizon lying below an A horizon, generally brownish or reddish in color, and commonly enriched in clay and iron oxides.

Back-arc basin (also called *inner-arc basin*). An arc-shaped basin formed behind a magmatic arc, by crustal thinning, in an overriding plate of lithosphere at a subduction zone.

Backshore. A zone extending inland from a berm to the farthest point reached by waves.

Backwash. The seaward return of water down a beach following the swash of a wave.

Bajada. A broad alluvial apron composed of coalescing adjacent fans.

Bar. An accumulation of alluvium formed in a channel where a decrease in stream velocity causes deposition.

Barchan dune. A crescent-shaped sand dune with horns pointing downwind.

Barrier island. A long island built of sand, lying offshore and parallel to the coast.

Barrier reef. A reef separated from the land by a lagoon.

Basalt. An aphanitic igneous rock with the composition of a gabbro.

Base level. The limiting level below which a stream cannot erode the land.

Batholith. A very large, intrusive igneous body of irregular shape that cuts across the layering of the rock it intrudes.

Bauxite. An aluminous laterite formed by tropical weathering. The preferred ore of aluminum.

Bay. A wide, open, curving indentation or inlet of a sea or lake into an adjacent land mass.

Bay barrier. A ridge of sand or gravel that completely blocks the mouth of a bay.

Beach. Wave-washed sediment along a coast, extending throughout the surf zone.

Beach drift. The movement of particles obliquely up the slope of a beach by the swash and directly down this slope by the backwash.

Bed. A rock-stratigraphic unit that is a subdivision of a *member*.

Bedding. The layered arrangement of rock units in a sequence of bedded rocks.

Bed load. Coarse particles that move along a stream channel.

Bedrock. The continuous mass of solid rock that makes up the crust.

Benioff zone. A narrow, well-defined zone of deep earthquake foci beneath a seafloor trench.

Berm. A nearly horizontal or landward sloping bench formed of sediment deposited by waves.

Beta particle (β-particle). An electron expelled from an atomic nucleus during certain radioactive transformations.

Biomass energy. The energy obtained through burning plant matter.

Biosphere. The totality of the Earth's organisms and, in addition, organic matter that has not yet been completely decomposed.

Bituminous coal. The highest grade of coal.

Blueschist. A metamorphic rock formed under conditions of high pressure and low temperature containing blue-colored amphiboles.

Body waves. Seismic waves that travel outward from an earthquake focus and pass through the Earth.

Bolide. An impacting body; it can be a meteorite, an asteroid, or a comet.

Bombs. Tephra with particles having average diameters greater than 64 mm.

Bond (*chemical*). The electrostatic force that holds atoms together to form compounds by sharing and transfer of electrons. See *covalent bond, ionic bond, metallic bond,* and *van der Waal's bond.*

Bottomset layer. A gently sloping, fine, thin part of each layer in a delta.

Boudin. A single broken fragment of a brittle rock enclosed in ductile material. See *boudinage.*

Boudinage. A style of deformation in layered rocks in which the brittle layers break into elongate blocks while the enclosing ductile layers stretch and flow.

Bouguer correction. A correction applied to measurements of the Earth's gravitational pull to account for the mass of material between the gravimeter and the reference surface used for free-air corrections.

Boulder train. A group of erratics spread out fanwise.

Bowen's reaction series. A schematic description of the order in which different minerals crystallize during the cooling and progressive crystallization of a magma. See *continuous* and *discontinuous reaction series.*

Braid delta. A delta composed of coarse-grained sediment built by a braided stream into a standing body of water.

Braided channel. A channel system consisting of a tangled network of two or more smaller branching and reuniting channels that are separated by islands or bars.

Breaker. An oversteepened wave that collapses in a mass of turbulent water against a shore or reef.

Breccia. A coarse-grained rock composed of indurated or cemented angular fragments of broken rock.

Breeder reactor. A nuclear reactor in which nonfissionable isotopes such as ^{238}U are converted to fissionable isotopes.

Brittle fracture. Rupture of a solid body that is stressed beyond its elastic limit.

Burial metamorphism. Metamorphism caused solely by the burial of sedimentary or pyroclastic rocks.

C horizon. A mineral horizon lying beneath the A horizon and (or) B horizon of a soil profile, often yellowish-brown in color, and consisting of weathered parent rock or sediment.

Calcareous ooze. A deep-sea pelagic sediment composed largely of calcareous skeletal remains.

Caldera. A roughly circular, steep-walled volcanic basin several kilometers or more in diameter.

Caliche. A solid, almost impervious layer of whitish calcium carbonate in a soil profile.

Calving. The progressive breaking off of icebergs from a glacier that terminates in deep water.

Calving glacier. A glacier whose terminus calves into the sea or a lake.

Capacity. The potential load a stream can carry.

Capillary attraction. The adhesive force between a liquid, such as water, and a solid.

Carbohydrates. A family of organic compounds produced by all green plants. One of the precursors of petroleum.

Carbonate shelf. A shallow marine shelf where sedimentation is dominated by carbonate-secreting organisms.

Carbonic acid. A weak acid resulting from the solution of small quantities of carbon dioxide in rain or groundwater.

Cataclastic metamorphism. Metamorphism caused solely by mechanical effects such as crushing, flattening, and elongation.

Cation. A positive ion.

Cave. A natural underground opening, generally connected to the surface and large enough for a person to enter.

Cavern. A large cave or system of interconnected cave chambers.

Cellulose. A large molecular form of carbohydrate. One of the organic compounds that is a precursor of coal.

Celsius scale. A temperature scale in which the boiling point of water is 100°, and the freezing point is 0°.

Cementation. The diagenetic process by which clastic sediments are converted to rock through deposition or precipitation of minerals in the spaces between the grains.

Cenozoic. The youngest era of the Phanerozoic Eon.

Central rift valley. A long, narrow valley at the crest of a midocean ridge.

Chemical elements. The most fundamental substances into which matter can be separated by chemical means.

Chemical recrystallization. The changes in chemical composition and growth of new minerals during metamorphism.

Chemical sediment. Sediment formed by precipitation of minerals from solution in water.

Chemical weathering. The decomposition of rocks through chemical reactions such as hydration and oxidation.

Chrons. See *magnetic chrons.*

Cirque. A bowl-shaped hollow on a mountainside open downstream, bounded upstream by a steep slope (headwall), and excavated mainly by frost wedging and by glacial abrasion and plucking.

Cirque glacier. A glacier that occupies a bowl-shaped hollow on the side of a mountain.

Clastic sediment. See *detritus.*

Claystone. A weakly indurated sedimentary rock consisting largely of clay-size particles.

Cleavage. The tendency of a mineral to break in preferred directions along bright, reflective plane surfaces.

Climate. The average weather conditions of a place or area over a period of years.

Coal. A black sedimentary rock consisting chiefly of decomposed plant matter and containing less than 40 percent inorganic matter.

Coalification. The stages by which plant matter is converted first to peat, then lignite, subbituminous coal, and bituminous coal.

Col. A gap or pass in a mountain crest where the headwalls of two cirques intersect.

Colluvium. Loose, incoherent deposits on or at the base of slopes and moving mainly by creep.

Column. A stalactite joined with a stalagmite, forming a connection between the floor and roof of a cave.

Columnar joints. Joints that split igneous rocks into long prisms or columns.

Comet. A small celestial body that circles the Sun with a highly elliptical orbit.

Compaction. A decrease in porosity and bulk of a body of sediment as additional sediment is deposited above it, or due to pressures resulting from deformation.

Competence. The size of particles a stream can transport under a given set of hydrologic conditions.

Complex ion. Two or more atoms bonded together so strongly that they act as a single ion.

Composite volcanoes. See *stratovolcanoes.*

Composition (of a mineral). The proportions of the various chemical elements in a mineral.

Compound. A combination of atoms of different elements bonded together.

Compressional waves. See *P waves.*

Conchoidal fracture. Breakage resulting in smooth, curved surfaces.

Concretion. A hard, localized body, having distinct boundaries, enclosed in sedimentary rock and consisting of a substance precipitated from solution, commonly around a nucleus.

Conduction. The means by which heat is transmitted through solids.

Cone of depression. A conical depression in the water table immediately surrounding a well.

Conglomerate. A sedimentary rock composed of clasts of rounded gravel set in a finer-grained matrix.

Consequent stream. A stream whose pattern is determined solely by the direction of slope of the land.

Contact metamorphism (also called *thermal metamorphism*). Metamorphism adjacent to an intrusive igneous rock.

Continental collision margin. A plate margin along which two continental masses collide.

Continental convergent margin. The margin of a continent that is adjacent to a subduction zone.

Continental crust. The crust that comprises the continents.

Continental drift. The slow, lateral movements of continents across the surface of the Earth.

Continental rise. A region of gently changing slope where the floor of the ocean basin meets the margin of the continent.

Continental shelf. A submerged platform of variable width that forms a fringe around a continent.

Continental shield. The portion of a craton where rocks are exposed at the surface.

Continental slope. A pronounced slope beyond the seaward margin of the continental shelf.

Continental volcanic arc. An arcuate chain of volcanoes on the continental crust.

Continuous reaction series. The continuous change of mineral composition, through solid solution, as a magma crystallizes. See *discontinuous reaction series.*

Convection. The process in liquids and gases by which hot, less dense materials rise upward, being replaced by cold, downward flowing fluids to create a convection current.

Convection current. See *convection*.

Convergent margin. The line along which two tectonic plates meet as they move toward each other. See *subduction zone*.

Coquina. A limestone composed solely or chiefly of loosely aggregated shells and shell fragments.

Cordillera. An elongate assemblage of more or less parallel mountain ranges, that may constitute the dominant mountain system of a continent.

Core. The spherical mass, largely of metallic iron, with admixtures of nickel, sulfur, silicon, and other elements, at the center of a planet.

Coriolis effect. An effect that causes any body that moves freely with respect to the rotating solid Earth to veer toward the right in the Northern Hemisphere and toward the left in the Southern Hemisphere, regardless of the initial direction of the moving body.

Correlation. Determination of equivalence, in geologic age and position, of rock units found in two or more different areas.

Covalent bond. The force that arises when two atoms share one or more electrons.

Crater. A funnel-shaped depression at the top of a volcano from which gases, fragments of rock, and lava are ejected.

Craton. A portion of the Earth's crust that has attained tectonic stability and has been little deformed for a prolonged period.

Creep. The imperceptibly slow downslope movement of regolith.

Creep of glacier ice. Slow deformation of glacier ice, with movement occurring along the internal planes of ice crystals.

Crevasse. A deep, gaping fissure in the upper surface of a glacier.

Cross section. See *geologic cross section*.

Cross strata. Strata that are inclined with respect to a thicker stratum within which they occur.

Crust. The outermost and thinnest of the Earth's compositional layers, which consists of rocky matter that is less dense than the rocks of the mantle below.

Cryosphere. The portion of the hydrosphere that is ice, snow, and frozen ground.

Crystal. A solid compound composed of ordered, three-dimensional arrays of atoms or ions chemically bonded together and displaying crystal form.

Crystal form. A geometric solid that is bounded by symmetrically arranged plane surfaces.

Crystalline. See *crystal structure*.

Crystal settling. The process by which dense minerals sink and form segregated layers of one or more minerals in a magma chamber.

Crystal structure. The geometric pattern that atoms assume in a solid. Any solid that has a crystal structure is said to be *crystalline*.

Curie point. A temperature above which all permanent magnetism is destroyed.

Cycle. A sequence of recurring events.

Cycle of erosion. A theory of landscape evolution in which a landmass is uplifted and is gradually worn down by erosion to a surface of low relief.

Cyclothem. A repeated sequence of peat and clastic sediments.

Dacite. An aphanitic igneous rock with the composition of a granodiorite.

Daughter product (= daughter). The product arising from radioactive decay. Compare *parent*.

Debris avalanche. A granular flow of regolith moving at a high velocity (≥ 10 m/s).

Debris fall. The relatively free fall or collapse of regolith from a steep cliff or slope.

Debris flow. The downslope movement of a mass of unconsolidated regolith more than one-half of which is coarser than sand.

Debris slide. The slow to rapid downslope movement of regolith across an inclined surface.

Décollement. A body of rock above the detachment surface of a thrust fault.

Decomposition (of rocks). Chemical weathering.

Deep-sea fan. Huge fan-shaped body of sediment at the base of the continental slope that spreads downward and outward to the deep seafloor.

Deflation. The picking up and removal of loose particles by wind.

Degradation. Downcutting, as by a stream.

Delta. A body of sediment deposited by a stream where it flows into standing water.

Density. The average mass per unit volume.

Denudation. The sum of the weathering, mass-wasting, and erosional processes that result in the progressive lowering of the Earth's surface.

Depositional remanent magnetism. Remanent magnetism acquired through processes of sedimentation.

Desert. Arid land, whether "deserted" or not, in which annual rainfall is less than 250 mm (10 in) or in which the evaporation rate exceeds the precipitation rate.

Desertification. The invasion of desert into nondesert areas.

Desert pavement (deflation armor). A surface layer of coarse particles concentrated chiefly by deflation.

Desert varnish. A thin, dark, shiny coating consisting mainly of manganese and iron oxides, formed on the surfaces of stones and rock outcrops in desert regions after long exposure.

Detachment surface. The surface along which a large-scale thrust fault moves.

Detrital mineral deposit. See *placer*.

Detritus (also called **clastic sediment**). The accumulated particles of broken rock and skeletal remains of dead organisms.

Diabase. A fine- to medium-grained gabbro.

Diachronous boundaries. Boundaries that vary in age in different areas.

Diagenesis. Changes that affect sediment after its initial deposition and during and after its slow transformation into sedimentary rock.

Differential weathering. Weathering that occurs at different rates or intensity as a result of variations in the composition and structure of rocks.

Dikes. Tabular sheets of intrusive igneous rock cutting across the layering of the intruded rock.

Diorite. A phaneritic igneous rock consisting mainly of plagioclase and ferromagnesian minerals. Quartz is sparse or absent.

Dip. The angle in degrees between a horizontal plane and an inclined plane, measured down from horizontal in a plane perpendicular to the strike.

Dip-slip fault. A normal or reverse fault on which the only component of movement lies in a plane normal to the strike of the fault surface.

Disaggregation. The separation of an aggregate into its component parts.

Discharge. The quantity of water that passes a given point in a stream channel per unit time.

Discharge area. Area where subsurface water is discharged to streams or to bodies of surface water.

Disconformity. Parallel layered strata separated by an irregular surface of erosion having appreciable relief.

Discontinuous reaction series. The discontinuous sequence of reactions by which early-formed minerals in a crystallizing magma react with residual liquid to form new minerals. See *continuous reaction series.*

Disintegration (of rocks). Mechanical weathering.

Dissolution. The chemical weathering process whereby minerals and rock material pass directly into solution.

Dissolved load. Matter dissolved in stream water.

Divergent margin. See *spreading center.*

Divide. The line that separates adjacent drainage basins.

Dolostone. A sedimentary rock composed chiefly of the mineral dolomite.

Drainage basin. The total area that contributes water to a stream.

Dripstone. A deposit chemically precipitated from water in an air-filled cavity.

Drumlin. A streamlined hill consisting of glacially deposited sediment and elongated parallel with the direction of ice flow.

Ductile deformation. The irreversible deformation induced in a solid that is stressed beyond its yield point, but before rupture occurs.

Dune. A mound or ridge of sand deposited by wind.

Earthflow. A granular flow of regolith with velocities ranging from 10^{-5} to 10^{-1} m/s.

Earthquake focus. The point of the first release of energy that causes an earthquake.

Earthquake magnitude. See *Richter magnitude scale.*

Earth's gravity. An inward-acting force with which the Earth tends to pull all objects toward its center.

Eccentricity (of Earth's orbit). The degree to which the shape of the Earth's orbit departs from perfect circularity.

Ecliptic. Plane of the Earth's orbit around the Sun.

Eclogite. A metamorphic rock containing garnet and jadeitic pyroxene.

Edge of consumption. The same as a *convergent margin* or *subduction zone.*

Edge of subduction. See *subduction zone.*

Ejecta blanket. Layer of broken rock surrounding an impact crater.

Elastic deformation. The reversible or nonpermanent deformation that occurs when an elastic solid is stretched and squeezed and the force is then removed.

Electrons. Negatively charged atomic particles.

Emergence. An increase in the area of land exposed above sea level resulting from uplift of the land and/or fall of sea level.

End moraine. A ridgelike accumulation of drift deposited along the margin of a glacier.

Energy. The capacity to produce activity.

Energy-level shell. The specific energy levels of orbital electrons.

Eolian. Pertaining to the wind, especially erosional and depositional processes, as well as landforms and sediments resulting from wind action.

Eon. The largest interval of geologic time.

Epicenter. That point on the Earth's surface that lies vertically above the focus of an earthquake.

Epidote amphibolite. A metamorphic rock containing both amphibole and epidote as major constituents.

Epoch. The time during which a geologic series accumulated.

Equilibrium line. A line that marks the level on a glacier where net mass loss equals net gain.

Era. A subdivision of an eon.

Erathem. A grouping of two or more systems in time-stratigraphy.

Erosion. The complex group of related processes by which rock is broken down physically and chemically and the products moved.

Erratic. A glacially deposited rock fragment whose composition differs from that of the bedrock beneath it.

Esker. A long narrow ridge, often sinuous, composed of stratified drift.

Estuary. A semienclosed body of coastal water within which seawater is diluted with fresh water.

Eugeocline. A wedge of deep-water sediment deposited at the foot of the continental slope. Commonly contains a significant amount of volcanic debris.

Evaporite. Sedimentary rock composed chiefly of minerals precipitated from a saline solution through evaporation.

Evaporite deposits. Layers of salts that precipitate as a consequence of evaporation.

Exfoliation. The spalling off of successive shells, like the "skins" of an onion, around a solid rock core.

Exploration geology. The special branch of geology concerned with discovering new supplies of usable minerals.

Exposure (also called an **outcrop**). A place where solid rock or sediment is exposed at the Earth's surface.

External processes. All the activities involved in erosion, and also in the transport and deposition of the eroded materials.

Extraterrestrial material. Material originating outside the Earth.

Extrusive igneous rock. Rock formed by the solidification of magma poured out onto the Earth's surface.

Facies. A distinctive group of characteristics, within a rock unit, that differs as a group from those elsewhere in the same unit.

Fan. See *alluvial fan.*

Fan delta. A gravel-rich delta formed where an alluvial fan builds outward into a standing body of water.

Fault. A fracture along which the opposite sides have been displaced relative to each other.

Fault breccia. Crushed and broken rock adjacent to a fault.

Fecal pellet. An organic excrement, especially of marine invertebrates, forming a component of marine sediments and some sedimentary rocks.

Ferromagnesian minerals. The common rock-forming minerals that contain iron and/or magnesium as essential constituents.

Fiord. See *fjord.*

Firn. Snow that survives a year or more of ablation and achieves a density that is transitional between snow and glacier ice.

Fissure eruption. Extrusion of volcanic rock or pyroclasts and associated gases along an extended fracture.

Fjord. A deep glacially carved valley submerged by the sea. Also spelled *fiord.*

Flash flood. A local and sudden flood of water through a stream channel, generally of relatively great volume and short duration.

Flood. A discharge great enough to cause a stream to overflow its banks.

Floodplain. The part of any stream valley that is inundated during floods.

Flow. Continuous and irreversible deformation of a geologic material resulting from the application of a high differential pressure.

Flowstone. A deposit chemically precipitated from flowing water in the open air or in an air-filled cavity.

Fluvial. Of, or pertaining to, streams or rivers, especially erosional and depositional processes of streams and the sediments and landforms resulting from them.

Foliation. A plane defined by any planar set of minerals, or bonding of minerals, found in a metamorphic rock.

Footwall. The surface of the block of rock below an inclined fault.

Fore-arc basin. See *outer-arc basin.*

Fore-arc ridge. See *outer-arc ridge.*

Foreset layer. The coarse, thick, steeply sloping part of each layer in a delta.

Foreshore. A zone extending from the level of lowest tide to the average high-tide level.

Formation. A body of rock distinctive enough on the basis of physical properties to constitute a basic unit for geologic mapping.

Fossil. The naturally preserved remains or traces of an animal or a plant.

Fossil fuel. Remains of plants and animals trapped in sediment that may be used for fuel.

Free-air correction. A correction applied to measurements of the Earth's gravitational pull in order to remove effects caused by differences in elevation.

Fringing reef. A coral reef attached to or bordering the adjacent land.

Frost heaving. The formation of ice in a confined opening within rock, thereby causing the rock to be forced apart.

Fumarole. A volcanic vent that emits only gases.

Gabbro. A phaneritic igneous rock in which olivine and pyroxene are predominant and plagioclase is the feldspar present. Quartz is absent.

Gamma rays (γ-rays). Very short wavelength electromagnetic radiation given off by an atomic nucleus during certain radioactive transformations.

Gangue. The nonvaluable minerals of an ore.

Garnet peridotite. A phaneritic igneous rock consisting largely of olivine, garnet, and pyroxene.

Geochemically abundant elements. Those chemical elements that individually comprise 0.1 percent or more, by weight, of the crust.

Geochemically scarce elements. Those chemical elements that individually comprise less than 0.1 percent by weight of the crust.

Geologic column. A composite diagram combining in chronological order the succession of known strata, fitted together on the basis of their fossils or other evidence of relative or actual age.

Geologic cross section. A diagram showing the arrangement of rocks in a vertical plane.

Geologic map. A map that shows the distribution, at the surface, of rocks of various kinds or of various ages.

Geologic time scale. A sequential arrangement of geologic time units, as currently understood.

Geology. The science of the Earth.

Geosyncline. A great trough that has received thick deposits of sediment during its slow subsidence through long geologic periods.

Geothermal gradient. The rate of increase of temperature downward in the Earth.

Geothermal power. Heat energy drawn from the Earth's internal heat.

Geyser. A hot spring equipped with a system of plumbing and heating that causes intermittent eruptions of water and steam.

Glacial drift. Sediment deposited directly by glaciers or indirectly by meltwater in streams, in lakes, and in the sea.

Glacial marine drift (also referred to as **glacial marine sediment**). Terrigenous sediment dropped onto the seafloor from floating ice shelves or from icebergs.

Glacial striations. Long subparallel scratches inscribed on a rock surface by rock debris embedded in the base of a glacier.

Glaciation. The modification of the land surface by the action of glacier ice.

Glacier. A body of ice, consisting largely of recrystallized snow, that shows evidence of downslope or outward movement due to the stress of its own weight.

Gneiss. A coarse-grained, foliated metamorphic rock, always with marked layering but with imperfect cleavage.

Gondwanaland. The southern half of Pangaea, consisting of present-day Australia, India, Madagascar, Africa, and South America.

Graben (also called a *rift*). A trenchlike structure bounded by parallel normal faults. See *half-graben*.

Graded layer. A layer in which the particles grade upward from coarse to finer.

Graded stream. A stream in which the slope has become so adjusted, under conditions of available discharge and prevailing channel characteristics, that the stream is just able to transport the sediment load available to it.

Gradient. A measure of the vertical drop over a given horizontal distance.

Granite. A phaneritic igneous rock containing quartz and feldspar, with potassium feldspar being more abundant than plagioclase.

Granitic. Any coarse-grained igneous or metamorphic rock having a texture and composition resembling that of a granite.

Granodiorite. A phaneritic igneous rock resembling a granite, in which plagioclase is more abundant than potassium feldspar.

Granular flow. A type of flow in which the weight of the flowing mass is supported by grain-to-grain contact or repeated collision between grains.

Granular texture. The interlocking arrangements of mineral grains in granitic rocks.

Granulite. A high-grade metamorphic rock consisting largely of pyroxenes.

Gravimeter (also called a *gravity meter*). A sensitive device for measuring the pull of gravity at any locality.

Gravity anomaly. Variations in the pull of gravity after correction for latitude and altitude.

Gravity meter. See *gravimeter*.

Greenschist. A low-grade metamorphic rock rich in chlorite.

Groin. A low wall, built on a beach, that crosses the shoreline at a right angle.

Groundmass. The fine-grained matrix of a porphyry.

Ground moraine. Widespread drift with a relatively smooth surface topography consisting of gently undulating knolls and shallow closed depressions.

Groundwater. All the water contained in the spaces within bedrock and regolith.

Group. In rock stratigraphy, an assemblage of formations.

Growth habit. A characteristic growth form of a mineral.

Hadean. The oldest eon.

Half-graben (also called a *rift*). A trenchlike structure formed when the hanging-wall block moves downward on a curved fault surface. See *graben*.

Half-life. The time required to reduce the number of parent atoms by one-half.

Hand specimen. A rock sample of convenient size to hold in the hand for study.

Hanging valley. A glacial valley whose mouth is at a relatively high level on the steep side of a larger glaciated valley.

Hanging wall. The surface of the block of rock above an inclined fault.

Hardness. Relative resistance of a mineral to scratching.

Hard water. Water containing an unusually high amount of calcium carbonate.

Heat. The energy a body has due to the motions of its atoms.

Heat energy. The energy of a hot body.

Heat flow. The outward flow of heat from the Earth's interior.

Hiatus. The lapse in time recorded by an unconformity.

High grade of metamorphism. Metamorphism under conditions of high temperature and high pressure.

High-grade terranes. Geologic provinces dominated by huge areas of coarse-grained, granitic rocks of both igneous and metamorphic parentage.

Hinge fault. A fault on which displacement dies out perceptibly along strike and ends at a definite point.

Highlands. See *lunar highlands*.

Horn. A sharp-pointed peak bounded by the intersecting walls of three or more cirques.

Hornfels. A hard, fine-grained rock developed during contact metamorphism of a shale.

Horst. An elevated elongate block of crust bounded by parallel normal faults.

Hot spot. A fixed point on the Earth's surface defined by long-lived volcanism.

Humus. The decomposed residue of plant and animal tissues.

Hurricane. A tropical cyclonic storm having winds that exceed 120 km/h.

Hydration. The absorption of water into a crystal structure.

Hydration rind. A thin, discolored rim on obsidian or glass resulting from water vapor diffusing into a freshly chipped surface.

Hydraulic gradient. The slope of the water table.

Hydrocarbon. Any organic compound (gaseous, liquid, or solid) consisting wholly of carbon and hydrogen.

Hydroelectric power. Energy recovered from the potential energy of rivers as they flow downward to the sea.

Hydrologic cycle. The cyclic movement of water through evaporation, wind transport, precipitation, stream flow, percolation, and related processes.

Hydrolysis. A chemical reaction in which the H^+ or OH^- ions of water replace ions of a mineral.

Hydrosphere. The "water sphere," embracing all the world's oceans, lakes, streams, water underground, and all the snow and ice, including glaciers.

Hydrothermal mineral deposit. Any local concentration of minerals formed by deposition from a hydrothermal solution.

Hydrothermal solutions. Hot brines either given off by cooling magmas, or produced by reactions between hot rock and circulating water, that concentrate ore minerals.

Hydrous. A term applied to substances that contain H_2O.

Iapetus. The name given to the ocean that disappeared when North America and Europe collided during the Paleozoic Era.

Ice cap. A dome-shaped body of ice and snow that covers a mountain highland, or lower-lying land at high latitude, and that displays generally radial outward flow.

Ice-contact stratified drift. Stratified sediment deposited in contact with supporting glacier ice.

Ice field. A broad, nearly level area of glacier ice in a mountainous region consisting of many interconnected mountain glaciers.

Ice sheet. A continent-sized mass of ice that overwhelms nearly all the land surface within its margin.

Ice shelf. Thick glacier ice that floats on the sea and commonly is located in large coastal embayments.

Igneous rock. Rock formed by the cooling and consolidation of magma.

Ignimbrite. See *welded tuff.*

Impact cratering. The process by which a planetary surface is deformed as a result of a transfer of energy from a bolide to the planetary surface.

Index fossil. A fossil that can be used to identify and date the strata in which it is found, and is useful for local correlation of rock units.

Index mineral. A mineral whose first appearance marks the outer limits of a specific zone of metamorphism.

Inert gas (also called a *noble gas*). A chemical element in which each electron shell is filled.

Inertia. The resistance a large mass has to sudden movement.

Inertial seismograph. A device for measuring earthquake waves based on inertia of a mass suspended on a sensitive spring.

Inner-arc basin. See *back-arc basin.*

Inner core. The central, solid portion of the Earth's core.

Inorganic compound. Chemical compounds that do not consist largely of the elements carbon and hydrogen.

Inselberg. Steep-sided mountain, ridge, or isolated hill rising abruptly from adjoining monotonously flat plains.

Intermediate grade of metamorphism. Metamorphism under conditions of intermediate pressures and temperatures.

Internal processes. All activities involved in movement or chemical and physical change of rocks in the Earth's interior.

Intrusive igneous rock. Any igneous rock formed by solidification of magma below the Earth's surface.

Ion. An atom that has excess positive or negative charges caused by electron transfers.

Ionic bond. The electrostatic attraction between negatively and positively charged ions.

Ionic radius. The distance from the center of the nucleus to the outermost orbiting electrons.

Ionic substitution (also called *solid solution* and *atomic substitution*). The substitution of one atom for another in a random fashion throughout a crystal structure.

Island arc (also called an *oceanic island arc*). An arcuate chain of stratovolcanoes parallel to a sea-floor trench and separated from it by a distance of 150 to 300 km.

Isoclinal fold. A fold in which both limbs are parallel.

Isograd. A line on a map connecting points of first occurrence of a given mineral in metamorphic rocks.

Isostasy. The ideal property of flotational balance among segments of the lithosphere.

Isotopes. Atoms having the same atomic number but differing numbers of neutrons.

Joint. A fracture on which movement has not occurred in a direction parallel to the plane of the fracture.

Joint set. A widespread group of parallel joints.

Jovian planets. Giant planets in the outer regions of the solar system that are characterized by great masses, low densities, and thick atmospheres consisting primarily of hydrogen and helium.

Kame. A short, steep-sided knoll of stratified drift.

Kame terrace. A terrace of ice-contact stratified drift along a valley side.

Karst topography. An assemblage of topographic forms resulting from dissolution of carbonate bedrock and consisting primarily of closely spaced sinks.

Kerogen. Insoluble, waxlike organic matter found in sedimentary rocks, especially shales.

Kettle. A basin within a body of drift created by melting out of a mass of underlying ice.

Kettle-and-kame topography. An extremely uneven terrain resulting from wastage of debris-mantled stagnant ice and underlain by ice-contact stratified drift.

Key bed. A thin and generally widespread bed with characteristics so distinctive that it can be easily recognized.

Kimberlite pipes. Narrow pipelike masses of igneous rocks, sometimes containing diamonds, that intrude the crust but originate deep in the mantle.

Kinetic energy. Energy that results from motion of an object.

Laccolith. A lenticular pluton intruded parallel to the layering of the intruded rock, above which the layers of the invaded country rock have been bent upward to form a dome.

Lacustrine. Pertaining to, produced by, or formed in a lake.

Lagoon. A bay inshore from an enclosing reef or island paralleling a coast.

Landslide. A general term covering a variety of mass-movement processes.

Lapilli. Tephra with particles having an average diameter between 2 and 64 mm.

Lapilli tuff. Pyroclastic rock in which the average diameter of tephra particles ranges between 2 and 64 mm.

Lateral moraine. An end moraine built along the side of a valley glacier.

Laterite. A hardened soil horizon characterized by extreme weathering that has led to concentration of secondary oxides of iron and aluminum.

Latitude. Part of a grid used for describing positions on the Earth's surface, consisting of parallel circles concentric to the poles. The circles are called **parallels of latitude**.

Laurasia. The northern half of Pangaea, consisting of present-day Asia, Europe, and North America.

Lava. Magma that reaches the Earth's surface through a volcanic vent.

Lava dome. See *plug dome*.

Law of faunal succession. Fossil faunas and floras succeed one another in a definite, recognizable order.

Law of original horizontality. See *original horizontality*.

Leaching. The continued removal, by water solutions, of soluble matter from bedrock or regolith.

Left-lateral fault. A strike–slip fault in which relative motion is such that to an observer looking directly at the fault, the motion of the block on the opposite side of the fault is to the left. A *right-lateral fault* has right-handed movement.

Levee. See *natural levee*.

Lignin. An amorphous organic compound present in wood.

Lignite. A low-grade coal with a calorific value between that of peat and bituminous coal.

Limbs. The sides of a fold.

Limestone. A sedimentary rock consisting chiefly of calcium carbonate, mainly in the form of the mineral calcite.

Linear dune. A long, straight, ridge-shaped dune paralleling the wind direction.

Lineation. A parallel arrangement of elongate mineral grains.

Lipids. A family of organic compounds that are one of the precursors of petroleum.

Lithification. The process that converts a sediment into a sedimentary rock.

Lithology. The systematic description of rocks in terms of mineral assemblage and texture.

Lithosphere. The outer 100 km of the solid Earth, where rocks are harder and more rigid than those in the plastic asthenosphere.

Little Ice Age. The interval of generally cool climate between the middle thirteenth and middle nineteenth centuries, during which mountain glaciers expanded worldwide.

Load. The material that is moved or carried by a natural transporting agent, such as a stream, the wind, a glacier, or waves, tides, and currents.

Local base level. Any base level, other than sea level, below which the land cannot be reduced by erosion.

Loess. Wind-deposited silt, commonly accompanied by some clay and fine sand.

Long profile. A line drawn along the surface of a stream from its source to its mouth.

Longitude. Part of a grid used for describing positions on the Earth's surface, consisting of half circles joining the poles. The half circles are called *meridians*.

Longshore current. A current, within the surf zone, that flows parallel to the coast.

Low grade of metamorphism. Metamorphism under conditions of low temperature and low pressure.

Low-grade terranes. Geologic provinces dominated by thin, highly deformed belts of lavas and sedimentary rocks that have been subjected to low grades of metamorphism.

Low-velocity layer. A region in the mantle, approximately between a depth of 100 and 350 km, where seismic wave velocities decrease.

Lunar highlands. Mountainous regions on the Moon, believed to consist of anorthosite and gabbro.

Luster. The quality and intensity of light reflected from a mineral.

Magma. Molten rock, together with any suspended crystals and dissolved gases, that forms when temperatures rise and melting occurs in the mantle or crust.

Magma ocean. Molten outer layer of the Moon formed as a result of intense rain of bolides.

Magmatic arc. An arcuate chain of magmatic activity lying above a subduction zone.

Magmatic differentiation by fractional crystallization. Compositional changes that occur in magmas by the separation of early formed minerals from residual liquids.

Magmatic differentiation by partial melting. The process of forming magmas with differing compositions by the incomplete melting of rocks.

Magmatic mineral deposit. Any local concentration of minerals formed by magmatic processes in an igneous rock.

Magnetic chrons. Periods of predominantly normal polarity (as at present), or predominantly reversed polarity.

Magnetic declination. The clockwise angle from true north assumed by a magnetic needle.

Magnetic field. Magnetic lines of force surrounding the Earth.

Magnetic inclination. The angle with the horizontal assumed by a freely swinging bar magnet.

Magnetic subchron. Short-term magnetic reversal.

Mantle. The thick shell of dense, rocky matter that surrounds the core.

Marble. A metamorphic rock derived from limestone and consisting largely of calcite.

Mare. Dark-colored lowland region of the Moon underlain by basalt.

Maria. Plural of *mare*.

Mass balance (of a glacier). A measure of the change in total mass of a glacier during a year.

Mass number. The sum of the protons and neutrons in the nucleus of an atom.

Mass-wasting. The movement of regolith downslope by gravity without the aid of a transporting medium.

M-discontinuity. See *Mohorovičić discontinuity*.

Meander. A looplike bend of a stream channel.

Meandering channel. See *meander.*

Mechanical deformation. The changes in texture of a rock due to grinding, crushing, and development of foliation during metamorphism.

Mechanical weathering. Disintegration of rocks by mechanical processes, such as frost-wedging.

Medial moraine. A linear moraine carried on and within the middle part of a glacier, generally formed by the merging of lateral moraines where two valley glaciers join.

Megascopic. Those features of rocks that can be perceived by the unaided eye, or by the eye assisted by a simple lens that magnifies up to 10 times.

Mélange. A chaotic mixture of broken and jumbled rock above a subduction zone.

Member. A rock-stratigraphic unit that is a subdivision of a formation.

Mercalli Scale. See *modified Mercalli Scale.*

Meridians. See *longitude.*

Mesosphere. The region between the base of the asthenosphere and the core/mantle boundary.

Mesozoic. The middle era of the Phanerozoic Eon.

Metallic bond. A variation of covalent bonding in which there are more electrons than are needed to satisfy bond requirements.

Metallogenic provinces. Limited regions of the crust within which mineral deposits occur in unusually large numbers.

Metamorphic aureole. A shell of metamorphic rock, produced by contact metamorphism, surrounding an igneous intrusion.

Metamorphic facies. Contrasting assemblages of minerals that reach equilibrium during metamorphism within a specific range of physical conditions belonging to the same metamorphic facies.

Metamorphic rock. Rock whose original compounds, or textures, or both, have been transformed to new compounds and new textures by reactions in the solid state as a result of high temperature, high pressure, or both.

Metamorphic zones. The regions on a map between isograds.

Metamorphism. All changes in mineral assemblage and rock texture, or both, that take place in rocks in the solid state within the Earth's crust as a result of changes in temperature and pressure.

Meteorites. Small stony or metallic objects from interplanetary space that impact a planetary surface.

Microscopic. Those features of rocks that require high magnification in order to be viewed.

Midocean ridges. Continuous rocky ridges on the ocean floor, many hundreds to a few thousand kilometers wide with a relief of more than 0.6 km.

Migmatite. A composite rock containing both igneous and metamorphic portions.

Mineral. Any naturally formed, solid, chemical substance having a definite chemical composition and a characteristic crystal structure.

Mineral assemblage. The variety and abundance of minerals present in a rock.

Mineral deposit. Any volume of rock containing an enrichment of one or more minerals.

Mineralogy. The special branch of geology that deals with the classification and properties of minerals.

Mineraloid. A mineral-like solid that lacks either a crystal structure or a definite composition, or both.

Miogeocline. A wedge of shallow-water, marine sediment accumulated on a continental shelf and continental slope.

Modified Mercalli Scale. A scale used to compare earthquakes based on the intensity of damage caused by the quake.

Moho See *Mohorovičić discontinuity.*

Mohorovičić discontinuity (also called **M-discontinuity** and **Moho**). The seismic discontinuity that marks the base of the crust.

Molecule. The smallest unit that retains all the properties of a compound.

Molecular attraction. The force that makes a thin film of water adhere to a rock surface despite the pull of gravity.

Monocline. A one-limbed flexure, on both sides of which the strata either are horizontal or dip uniformly at low angles.

Mountain chain. A large scale, elongate geologic feature consisting of numerous ranges or systems, regardless of similarity in form or equivalence in ages.

Mountain range. An elongate series of mountains forming a single geologic feature.

Mountain system. A group of ranges similar in general form, structure, and alignment, and presumably owing their origin to the same general causes.

Mud cracks. Cracks caused by shrinkage of wet mud as its surface becomes dry.

Mudflow. A flowing mass of predominantly fine-grained rock debris that generally has a high enough water content to make it highly fluid.

Mudstone. An indurated mud, similar to shale, but lacking its fine lamination and tendency to split along subparallel planes.

Mylonite. A microbreccia produced by intense grinding along a fault surface.

Natural gas. The gaseous component of petroleum. Chiefly methane.

Natural levee. A broad, low ridge of fine alluvium built along the side of a stream channel by water that spreads out of the channel during floods.

Neutron. An electrically neutral particle with a mass 1833 times greater than that of the electron.

Noble gas. See *inert gas.*

Nonconformity. Stratified rocks that unconformably overlie igneous or metamorphic rocks.

Normal fault. A fault, generally steeply inclined, along which the hanging-wall block has moved relatively downward.

Notch. See *wave-cut notch*

Nuclear energy. The heat energy produced during controlled fission or fusion of atoms.

Nucleus (of an atom). The protons and neutrons in the core of an atom.

Oblique-slip fault. A fault on which movement includes both horizontal and vertical components.

Obsidian. A natural volcanic glass.

Oceanic crust. The crust beneath the oceans.

Oceanic island arc. See *island arc.*

Oil. The liquid component of petroleum.

Oilfield. A group of oil pools, usually of similar type, or a single pool in an isolated position.

Oil pool. An underground accumulation of oil and gas in a reservoir limited by geologic barriers.

Oil shale. A shale containing kerogen that will break down to liquid and gaseous hydrocarbons on heating.

Ophiolite complex. A fragment of oceanic crust accreted during continental collision.

Ore. An aggregate of minerals from which one or more minerals can be extracted profitably.

Organic compound. Chemical compounds made from carbon and hydrogen, with or without other elements such as nitrogen and oxygen.

Original horizontality (law of). Waterlaid sediments are deposited in strata that are horizontal, or nearly horizontal, and parallel or nearly parallel to the Earth's surface.

Orogenic belts. See *orogens.*

Orogens. Elongate regions of the crust that have been intensely folded and faulted during mountain-buidling processes.

Orogeny. The process by which large regions of the crust are deformed and uplifted to form mountains.

Outcrop. See *exposure.*

Outcrop area. The area, on a geologic map, shown as occupied by a particular rock unit.

Outer core. The outer portion of the Earth's core, which is molten.

Outer-arc basin. A basin parallel to a deep-sea trench and separated from the trench by an *outer-arc ridge.* Also called *fore-arc basin.*

Outer-arc ridge. See *outer-arc basin.* Also called *fore-arc ridge.*

Outwash. Stratified drift deposited by streams of meltwater.

Outwash plain. A body of outwash that forms a broad plain.

Outwash terrace. A terrace formed by dissection of an outwash plain or valley train.

Overland flow. The movement of runoff in broad sheets or groups of small, interconnecting rills.

Overturned fold. A fold in which the strata in one limb have been tilted beyond vertical.

Oxbow lake. A crescent-shaped shallow lake occupying the abandoned channel of a meandering stream.

Oxidizing environment. A sedimentary environment in which oxygen is present and organic remains are readily converted by oxidation into carbon dioxide and water.

Pahoehoe. A smooth-surfaced lava flow usually basaltic in composition.

Paleogeography. The physical geography during past geologic times.

Paleomagnetism. Remanent magnetism in ancient rock recording the direction of the magnetic poles at some time in the past.

Paleosol. A soil that formed at the ground surface and subsequently was buried and preserved.

Paleozoic. The oldest era of the Phanerozoic Eon.

Pangaea. The name given to a supercontinent that formed by collision of all the continental crust during the late Paleozoic.

Parabolic dune. A sand dune of U-shape with the open end of the U facing upwind.

Parallel of latitude. See *latitude.*

Parallel strata. Strata whose individual layers are parallel.

Parent. An atomic nucleus undergoing radioactive decay; compare *daughter product.*

Passive continental margin. A continental margin in a plate interior.

Peat. The first stage in the conversion of plant matter to coal.

Pediment. A sloping surface, cut across bedrock and thinly or discontinuously veneered with alluvium, that slopes away from the base of a highland in an arid or semiarid environment.

Pegmatite. An exceptionally coarse-grained igneous rock, commonly granitic in composition and texture.

Pelagic sediment. Sediment consisting of material of marine organic origin.

Perched water body. A water body that occupies a basin in impermeable sediments or rocks, perched in positions higher than the main water table.

Percolation. The movement of groundwater in the saturated zone.

Peridotite. A phaneritic igneous rock consisting largely of olivine, with or without pyroxene.

Period. The time during which a geologic system accumulated.

Permeability. The capacity for transmitting fluids.

Perpendicular component of gravity. The component of gravity that acts at right angles to a slope.

Perthite. An intergrowth of plagioclase in potassium feldspar formed by the unmixing of a high-temperature solid solution.

Petrified wood. Fossilized wood, formed when wood is buried and replaced by an equal volume of mineral matter.

Petroleum. Gaseous, liquid, and semisolid substances, occurring naturally and consisting chiefly of chemical compounds of carbon and hydrogen.

Petrology. The special branch of geology that deals with the occurrence, origin, and history of rocks.

Phanerites. Igneous rocks in which the component mineral grains are distinguishable megascopically.

Phanerozoic. The eon that follows the Proterozoic Eon.

Phase transition. Atomic repacking caused by changes in pressure and temperature.

Phenocrysts. The isolated large crystals in a porphyry.

Photosynthesis. The process by which plants combine water and carbon dioxide to make carbohydrates and oxygen.

Phyllite. A well-foliated metamorphic rock in which the component platy minerals are just visible.

Piedmont glacier. A broad glacier that terminates on a piedmont slope beyond confining mountain valleys and is fed by one or more large valley glaciers.

Pile. A device in which nuclear fission can be controlled.

Pillow lava. Discontinuous, pillow-shaped masses of lava, ranging in size from a few centimeters to a meter or more in greatest dimension.

Piracy. See *stream capture.*

Placer. A deposit of heavy minerals concentrated mechanically.

Plane of the ecliptic. The plane of the Earth's orbit around the Sun.

Planetary nebula. A flattened, rotating disk of gas surrounding a proto-sun.

Planet. A large celestial body that revolves around the Sun in an elliptical orbit.

Planetology. A comparative study of the Earth with the Moon and with the other planets.

Plate tectonics. The special branch of tectonics that deals with the processes by which the lithosphere is moved laterally over the asthenosphere.

Plateau basalt. Flat lava plateau formed by fissure eruptions of basalt.

Playa. A dry lake bed in a desert basin.

Playa lake. A temporary lake that forms on the surface of a desert playa.

Plug dome (also called a *lava dome*). A volcanic dome characterized by an upheaved, consolidated conduit filling of lava.

Plunge (of a fold). The angle between a fold axis and the horizontal.

Plunging fold. A fold with an inclined axis.

Pluton. Any body of intrusive igneous rock, regardless of shape or size.

Point bar. A low arcuate ridge of sand or gravel along the inside of the bend of a meander loop.

Polar desert. A desert in the polar regions characterized by very low temperatures and precipitation.

Polar easterlies. Globe encircling belts of easterly winds in the high latitudes of both hemispheres.

Polar (cold) glacier. A glacier in which the ice is below the pressure melting point throughout, and the ice is frozen to its bed.

Polarity reversals. Changes of the Earth's magnetic field to the opposite polarity.

Polymerization. The process of linking silicate tetrahedra into large anion groups.

Polymorph. A compound that occurs in more than one crystal structures.

Pores. The innumerable tiny openings in rock that can be filled by water or other fluids.

Porosity. The proportion (in percent) of the total volume of a given body of bedrock or regolith that consists of pore spaces.

Porphyry. Any igneous rock consisting of coarse mineral grains scattered through a mixture of fine mineral grains.

Porphyry copper deposit. A class of hydrothermal mineral deposit associated with intrusions of porphyritic igneous rocks.

Potential energy. Stored energy.

Precession of the equinoxes. A progressive change in the Earth–Sun distance for a given date.

Pressure melting point. The temperature at which ice can melt at a given pressure.

Primary waves. See *P waves.*

Principle of stratigraphic superposition. See *stratigraphic superposition.*

Principle of Uniformitarianism. The same external and internal processes we recognize in action today have been operating unchanged, though at different rates, throughout most of the Earth's history.

Progradation. The outward extension of a shoreline into the sea or a lake due to sedimentation.

Prograde metamorphic effects. The metamorphic changes that occur while temperatures and pressures are rising.

Proteins. A family of organic compounds that are one of the precursors of petroleum.

Proterozoic. The eon that follows the Archean.

Proton. A positively charged particle with a mass 1832 times greater than the mass of an electron.

Pumice. A natural glassy froth made by gases escaping through a viscous magma.

P waves. Seismic body waves transmitted by alternating pulses of compression and expansion. *P* waves pass through solids, liquids, and gases.

Pyroclastic cone. A cone of pyroclastic debris surrounding a volcanic vent.

Pyroclastic rocks. Rocks comprised of fragments of igneous material ejected from a volcano, then deposited and either cemented or welded to a coherent aggregate.

Pyroclasts. Fragments extruded violently from a volcano.

Pyrometer. An optical device for measuring temperature.

Quartzite. A metamorphic rock consisting largely of quartz, and derived from a sandstone.

Radiation. Transmission of heat energy through the passage of electromagnetic waves.

Radioactivity. The process by which an unstable atomic nucleus spontaneously transforms to another nucleus.

Radiometric age. The length of time a mineral has contained its built-in radioactivity clock.

Rain shadow. A dry region on the downwind side of a mountain range where precipitation is noticeably less than on the windward side.

Recharge. The addition of water to the saturated zone of a groundwater system.

Recharge area. Area where water is added to the saturated zone.

Recrystallization. The formation of new crystalline minerals within a rock.

Recumbent fold. A fold in which the axial plane is horizontal.

Recurrence interval. The probable interval, in years, between floods of a given magnitude.

Reducing environment. An environment in which oxygen is lacking and organic matter does not decay, but instead is slowly transformed into solid carbon.

Reef. A generally ridgelike structure composed chiefly of the calcarous remains of sedentary marine organisms (e.g., corals, algae).

Regional metamorphism. Metamorphism affecting large volumes of crust, and involving both mechanical and chemical changes.

Regolith. The blanket of loose, noncemented rock particles that commonly overlies bedrock.

Regression. A retreat of the sea from the land.

Relative velocity (of a plate). The apparent velocity of one plate relative to another.

Relict sediment. A body of sediment deposited in a nearshore environment but which now lies offshore due to progressive submergence.

Relief. The range in altitude of a land surface.

Replacement. The process by which a fluid dissolves matter already present and at the same time deposits from solution an equal volume of a different substance.

Reservoir rock. A permeable body of rock in which petroleum accumulates.

Residual concentration. The natural concentration of a mineral substance by removal of a different substance with which it was associated.

Residual mineral deposit. Any local concentration of minerals formed as a result of weathering.

Resins. Hard, brittle, plant secretions. One of the precursors of coal.

Resurgent cauldron. The uplifting of the collapsed floor of a caldera to form a structural dome.

Retrograde metamorphic effects. Metamorphic changes that occur as temperature and pressure are declining.

Reverse fault. A fault, generally steeply inclined, along which the hanging-wall block has moved relatively upward.

Rhyolite. An aphanitic igneous rock with the composition of a granite.

Richter magnitude scale. A scale, based on the recorded amplitudes of seismic body waves, for comparing the amounts of energy released by earthquakes.

Rift. See *graben* and *half-graben*.

Right-lateral fault. See *left-lateral fault*.

Rind. A discolored rim of weathered rock surrounding an unweathered rock core.

Ripcurrent. A high-velocity current flowing seaward from the shore as part of the backwash from a wave.

Ripple mark. One of a series of small and fairly regular dips preserved in rock and representing a former rippled sedimentary surface.

Rock. Any naturally formed, nonliving, firm and coherent aggregate mass of mineral matter that constitutes part of a planet.

Rock cleavage (also called *slaty cleavage*). The property by which a rock breaks into platelike fragments along flat planes.

Rock cycle. The cyclic movement of rock material, in the course of which rock is created, destroyed, and altered through the operation of internal and external Earth processes.

Rock drumlin. A smooth, glacially shaped, streamlined hill having a core of rock and usually veneered with till.

Rockfall. The relatively free-falling of detached bodies of bedrock from a cliff or steep slope.

Rock flour. Fine rock particles produced by glacial crushing and grinding.

Rockslide. The sudden and rapid downslope movement of newly detached masses of bedrock across an inclined surface.

Rock-stratigraphic unit. A body of rock having a high degree of overall lithologic homogeneity. Compare *time-stratigraphic unit*.

Roof rock. A rock, such as shale, that is impermeable and caps a petroleum reservoir.

Runoff. The fraction of precipitation that flows over the land surface.

Saltation. The progressive forward movement of a sediment particle in a series of short intermittent jumps along arcing paths.

Sand sea. Vast tract of shifting sand.

Sand ripples. A series of small and rather regular ridges on the surface of a body of sand, such as a dune.

Sandstone. A medium-grained clastic sedimentary rock composed chiefly of sand-sized grains.

Saturated zone. The groundwater zone in which all openings are filled with water.

Scale (of a map). The proportion between a unit of distance on a map and the unit it represents on the Earth's surface.

Schist. A well-foliated metamorphic rock in which the component platy minerals are clearly visible.

Schistosity. The parallel arrangement of coarse grains of the sheet-structure minerals, like mica and chlorite, formed during metamorphism under conditions of differential pressure.

Sea arch. An opening through a headland, generally produced by wave erosion, that forms a bridge of rock over water.

Sea cave. A cave at the base of a seacliff produced by wave erosion.

Seafloor spreading (theory of). A theory proposed during the early 1960s in which lateral movement of the oceanic crust away from midocean ridges was postulated.

Seamount. An isolated submerged volcanic mountain standing more than 1000 m above the seafloor.

Seawall. A wall or embankment of boulders or concrete erected to prevent wave erosion along a coast.

Secondary mineral. A mineral formed later than the rock enclosing it, usually at the expense of an earlier formed primary mineral.

Secondary waves. See *S waves*.

Sediment. Regolith that has been transported by any of the external processes.

Sedimentary facies. A distinctive group of characteristics within a sedimentary unit that differs, as a group, from those elsewhere in the same unit.

Sedimentary mineral depsoit. Any local concentration of minerals formed through processes of sedimentation.

Sedimentary rock. Any rock formed by chemical precipitation or by sedimentation and cementation of mineral grains transported to a site of deposition by water, wind, or ice.

Sedimentary stratification. A layered arrangement of the particles that constitute sediment or sedimentary rock.

Seismic belts. Large tracts of the Earth's surface that are subject to frequent earthquake shocks.

Seismic sea waves (also called *tsunami*). Long wavelength ocean waves produced by sudden movement of the seafloor following an earthquake. Incorrectly called tidal waves.

Seismic stratigraphy. A cross-sectional view of crustal rocks and sediments obtained through high-resolution seismic exploration techniques.

Seismic tomography. A way of revealing inhomogeneities in the mantle by measuring slight differences in the frequencies and velocities of seismic waves.

Seismic waves. Elastic disturbances spreading outward from an earthquake focus.

Seismograph. The device used to study the shocks and vibrations caused by earthquakes.

Seismology. The study of earthquakes.

Series. In time-stratigraphy, an interval in a system.

Setting time. The moment a mineral starts accumulating a daughter product produced by radioactive decay.

Shale. A fine-grained, clastic sedimentary rock.

Shard. See *volcanic shard*.

Shear waves. See *S waves*.

Sheet erosion. The erosion performed by overland flow.

Sheeted flows. Thin sheets of lava with rapidly quenched, glassy surfaces.

Shield volcano. A volcano that emits fluid lava and builds up a broad dome-shaped edifice with a surface slope of only a few degrees.

Shore profile. A vertical section along a line perpendicular to a shore.

Silicate anion. A complex ion, $(SiO_4)^{-4}$, that is present in all common silicate minerals.

Silicate mineral. A mineral that contains the silicate anion.

Siliceous ooze. Any pelagic deep-sea sediment of which at least 30 percent consists of siliceous skeletal remains.

Sills. Tabular sheets of intrusive igneous rock that are parallel to the layering of the intruded rock.

Siltstone. A massive mudstone in which silt predominates over clay.

Sinkhole. A large solution cavity open to the sky.

Slate. A low-grade metamorphic rock with a pronounced slaty cleavage.

Slaty cleavage. See *rock cleavage*.

Slickensides. Striated or highly polished surfaces on hard rocks abraded by movement along a fault.

Sliderock. The sediment composing a talus.

Slip face. The straight, lee slope of a dune.

Slump. A type of slope failure in which a downward and outward rotational movement of rock or regolith occurs along a concave-up slip surface.

Slurry flow. A moving mass of sediment that is saturated with water trapped among the grains and transported with the flowing mass.

Snowline. The lower limit of perennial snow.

Soil. The weathered part of the regolith which can support rooted plants.

Soil profile. The succession of distinctive horizons in a soil from the surface down to the unaltered parent material beneath it.

Solid solution. See *ionic substitution*.

Solifluction. The slow downslope movement of waterlogged soil and surficial debris.

Source-rock. A sedimentary rock containing organic matter that is a source of petroleum.

Specific gravity. A number stating the ratio of the weight of a substance to the weight of an equal volume of pure water.

Spheroidal weathering. The successive loosening of concentric shells of decayed rock from a solid rock mass as a result of chemical weathering.

Spit. An elongate ridge of sand or gravel that projects from land and ends in open water.

Spreading axis. The axis of rotation of a plate of lithosphere.

Spreading center (also called a *spreading edge* and a *divergent margin*). The new, growing edge of a plate. Coincident with a midocean ridge.

Spreading pole. The point where a spreading axis reaches the Earth's surface.

Spring. A flow of groundwater emerging naturally at the ground surface.

Stable platform. That portion of a craton that is covered by a thin layer of little-deformed sediments.

Stable zone. The interior part of a tectonic plate.

Stack. An isolated rocky island or steep rock mass near a cliffy shore, detached from a headland by wave erosion.

Stage. In time-stratigraphy, an interval within a series.

Stalactite. An iciclelike form of dripstone and flowstone, hanging from cave ceilings.

Stalagmite. A blunt "icicle" of flowstone projecting upward from cave floors.

Star dune. An isolated hill of sand having a base that resembles a star in plan.

Steady state. A condition in which the rate of arrival of material or energy equals the rate of escape.

Stock. A small, irregular body of intrusive igneous rock, smaller than a batholith, that cuts across the layering of the intruded rock.

Stoping. The process by which a rising body of magma wedges off fragments of overlying rock which then sink through the magma chamber.

Strain. The measure of the changes in length, volume, and shape in a stressed material.

Strain rate. The rate at which a rock is forced to change its shape or volume.

Strain seismograph. A device for recording earthquake waves based on the flexure of a long, rigid rod.

Strata. See *stratum*.

Stratification. The layered arrangement of sediments, sedimentary rocks, or extrusive igneous rocks.

Stratified drift. Drift that is both sorted and stratified.

Stratigraphic superposition (principle of). In a sequence of strata, not later overturned, the order in which they were deposited is from bottom to top.

Stratigraphy. The study of strata.

Stratovolcanoes (also called *composite volcanoes*). Volcanoes that emit both fragmental material and viscous lava, and that build up steep conical mounds.

Stratum (plural = *strata*). A tabular layer of sedimentary rock distinct from layers above and below it.

Streak. A thin layer of powdered mineral made by rubbing a specimen on a nonglazed porcelain plate.

Stream. A body of water that carries rock particles and dissolved substances and flows down a slope along a definite path.

Stream capture (also called *piracy*). The diversion of a stream by the headward growth of another stream.

Streamflow. The flow of surface water in a well-defined channel.

Stress. The magnitude and direction of a deforming force.

Striations. See *glacial striations*.

Strike. The compass direction of a horizontal line in an inclined plane.

Strike-slip fault. A fault on which displacement has been horizontal.

Structural geology. The branch of geology devoted to the study of rock deformation.

Structure (of minerals). See *crystal structure*.

Subatomic particles. The small particles that combine to form an atom—electrons, protons, and neutrons.

Subchron. See *magnetic subchron*.

Subduction zone (also called a *convergent margin*). The line along which a plate of lithosphere sinks down into the asthenosphere.

Submarine canyon. A steep-sided valley on the continental shelf or slope resembling a river-cut canyon on land.

Submergence. A rise of water level relative to the land so that areas formerly dry are inundated.

Subpolar glacier. A glacier in which surface temperature reaches 0°C in summer, but beneath several meters the temperature is below the pressure melting point.

Subsequent stream. A stream whose course has become adjusted so that it occupies belts of weak rock or other geologic structures.

Superposed stream. A stream that was let down, or superposed, from overlying strata onto buried bedrock having composition or structure unlike that of the covering strata.

Surf. Wave activity between the line of breakers and the shore.

Surface waves. Seismic waves that are guided by the Earth's surface and do not pass through the body of the Earth.

Surge. An unusually rapid movement of a glacier marked by dramatic changes in glacier flow and form.

Suspended load. Fine particles suspended in a stream.

Swash. The surge of water up a beach caused by waves moving against a coast.

S waves. Seismic body waves transmitted by an alternating series of sideways (shear) movements in a solid. *S* waves cause a change of shape and cannot be transmitted through liquids and gases.

Symmetrical fold. A fold in which both limbs dip equally away from the axial plane.

Syncline. A downfold with a troughlike form.

System. The primary unit in a time-stratigraphic sequence of rocks.

Talus. The apron of rock waste sloping outward from the cliff that supplies it.

Tangential component of gravity. The component of gravity that acts along and down a slope.

Tar (also called *asphalt*). An oil that is viscous and so thick it will not flow.

Tarn. A small, generally deep mountain lake occupying a cirque.

Tectonics. The study of movements and deformation of the crust on a large scale.

Temperate (warm) glacier. A glacier in which the ice is at the pressure melting point sometime during the year.

Temperature. The hotness, or degree of heat possessed by a body.

Tephra. A loose assemblage of pyroclasts.

Terminal moraine. An end moraine deposited at the front of a glacier.

Terminus. The outer, lower margin of a glacier.

Terrace. An abandoned floodplain formed when a stream flowed at a level above the level of its present channel and floodplain.

Terrane. A large piece of crust with a distinctive geological character.

Terrestrial planets. The innermost planets of the solar system (Mercury, Venus, Earth, and Mars) which have high densities and rocky compositions.

Terrigenous sediment. Sediment derived from sources on land.

Tethys. The name of a narrow sea separating Gondwanaland from Laurasia.

Texture. The overall appearance that a rock has because of the size, shape, and arrangement of its constituent particles.

Thalweg. A line connecting the deepest parts of a stream channel.

Thermal metamorphism. See *contact metamorphism.*

Thermal plume. A vertically rising mass of heated rock in the mantle.

Thermal spring. A natural spring that emits hot water.

Thermoremanent magnetism. Permanent magnetism that is a result of thermal cooling.

Thin section. A thin slice of rock glued to a glass slide and used for microscopic examination.

Thrust faults (also called *thrusts*). Low angle reverse faults with dips less than 45°.

Tidal bore. A large, turbulent wall-like wave of water caused by the meeting of two tides or by the rush of tide up a narrowing inlet, river, estuary, or bay.

Tidal bulge. A bulge in bodies of marine and fresh water, produced by the gravitational attraction of the Moon and Sun, that moves around the Earth as it rotates.

Tidal current. A current of water generated by the twice daily tidal bulges that pass around the Earth.

Tidal inlet. A coastal inlet through which water flows alternately landward with rising tide and seaward with falling tide.

Tide. The twice-daily rise and fall of the ocean surface resulting from the gravitational attraction of the Moon and Sun.

Till. A nonsorted sediment deposited directly from glacier ice.

Tillite. A nonsorted sedimentary rock of glacial origin.

Till matrix. The fine-grained sediment enclosing stones and rock fragments in till.

Tilt (of axis). The angle of the Earth's rotational axis with respect to the plane of the Earth's orbit.

Time line. A line of constant age in a stratigraphic section.

Time-stratigraphic unit. A unit representing all the rocks, and only those rocks, that formed during a specific interval of geologic time. Compare *rock-stratigraphic unit.*

Titius–Bode rule. The distance to each planet is approximately twice as far as the next inner one, measuring from the Sun.

Tombolo. A ridge of sand or gravel that connects an island to the mainland or to another island.

Topography. The relief and form of the land.

Topset layer. A layer of stream sediment that overlies the foreset layers in a delta.

Trade wind. A globe encircling belt of winds in the low latitudes. They blow from the northeast in the northern hemisphere, and the southeast in the southern hemisphere.

Transform. The junction point where one of the major deformation features—a midocean ridge, a seafloor trench, or a strike–slip fault—meets another.

Transform faults. The special class of strike–slip faults that link major structural features.

Transgression. A spreading of the sea over the land.

Transpiration. The process by which plants release water vapor from their leaves.

Transverse dune. A sand dune forming a wavelike ridge transverse to wind direction.

Trap. A reservoir rock plus a roof rock that serve to accumulate petroleum.

Trenches. Long, narrow, very deep and arcuate basins in the seafloor.

Triple junction. The place where three plate edges meet.

Tsunami. See *seismic sea wave.*

Turbidite. Sediment deposited by a turbidity current.

Turbidity current. A gravity-driven current consisting of a dilute mixture of sediment and water having a density greater than the surrounding water.

Type section (also called a *stratotype*). A section that displays the primary characteristics of a stratigraphic unit in a typical manner.

Typhoon. A term used in the western Pacific Ocean for a tropical cyclonic storm. See *hurricane.*

Unconformity. A substantial break or gap in a stratigraphic sequence that marks the absence of part of the rock record.

Unconformity-bounded sequence. A grouping of strata that is bounded at its base and top by unconformities of regional or interregional extent.

Uniform layer. A layer of sediment or sedimentary rock that consists of particles of about the same diameter.

Uniformitarianism. See *Principle of Uniformitarianism.*

Unsaturated zone (zone of aeration). The groundwater zone in which open spaces in regolith or bedrock are filled mainly with air.

Valley glacier. A glacier that flows from a cirque or cirques onto and along the floor of a valley.

Valley train. A body of outwash that partly fills a valley.

van der Waals bond. A weak electrostatic attraction that arises because certain ions and atoms are restored from a spherical shape.

Varve. A pair of sedimentary layers deposited during the seasonal cycle of a single year.

Ventifact. Any bedrock surface or stone that has been abraded and shaped by windblown sediment.

Vesicle. A small opening, in extrusive igneous rock, made by escaping gas originally held in solution under high pressure while the parent magma was underground.

Viscosity. The internal property of a substance that offers resistance to flow.

Volcanic ash. See *ash.*

Volcanic neck. The approximately cylindrical conduit of igneous rock forming the feeder pipe immediately below a volcanic vent.

Volcanic sediment (in the ocean). Sediment from submarine volcanoes, together with ash from oceanic and nonoceanic volcanic eruptions.

Volcanic shard. A particle of ash-sized, glassy tephra.

Volcano. The vent from which igneous matter, solid rock, debris, and gases are erupted.

Volcanogenic massive sulfide deposit. A mineral deposit formed by deposition of sulfide minerals from a submarine hot spring.

Wadi. A term used in the Middle East and northern Africa for a dry stream channel in a desert region.

Wall-rock alteration. Changes produced in the mineral assemblages of rocks lining the flow channel of a hydrothermal solution.

Water quality. The fitness of water for human use, as affected by physical, chemical, and biological factors.

Water table. The upper surface of the saturated zone of groundwater.

Wave. An oscillatory movement of water characterized by an alternate rise and fall of the water surface.

Wave base. The effective lower limit of wave motion, which is half of the wavelength.

Wave-cut bench. A bench or platform cut across bedrock by surf.

Wave-cut cliff. A coastal cliff cut by surf.

Wave-cut notch. A notch along the base of a sea cliff produced by wave erosion.

Wavelength. The distance between the crests of adjacent waves.

Wave refraction. The process by which the direction of a series of waves, moving in shallow water at an angle to the shoreline, is changed.

Waxes. A family of solid organic compounds that are one of the precursors of coal.

Weathering. The chemical alteration and mechanical breakdown of rock materials during exposure to air, moisture, and organic matter.

Weathering rind. See *rind*.

Welded tuff (also called *ignimbrite*). Pyroclastic rocks, the glassy fragments of which were plastic and so hot when deposited that they fused to form a glassy rock.

Well. An excavation in the ground designed to tap a supply of underground liquid, especially water or petroleum.

Westerlies. Globe encircling belts of winds centered at about 45° latitude in both hemispheres.

Xenoliths. Fragments of country rock still enclosed in a magmatic body when it solidifies.

Selected References

Chapter 1

Broecker, W. S., 1983, The ocean: Sci. American, v. 249, no. 3, p. 146–161.

Cloud, P., 1983, The biosphere: Sci. American, v. 249, no. 3, p. 176–189.

Garrels, R. M., Mackenzie, F. T., and Hunt, C., 1975, Chemical cycles and the global environment: Los Altos, Calif., William Kaufman.

Ingersall, A. P., 1983, The atmosphere: Sci. American, v. 249, no. 3, p. 162–174.

Scientific American, 1983, The ocean: San Francisco, W. H. Freeman.

Siever, R., 1983, The dynamic Earth: Sci. American, v. 249, no. 3, p. 46–55.

Essay

Kozlovsky, Ye. A., ed., 1987, The superdeep well of the Kola Peninsula: Berlin, Springer-Verlag, 558 p.

Kozlovsky, Ye. A., 1984, The world's deepest well: Sci. American, v. 251, no. 6, p. 98–104.

Chapter 2

Berry, L. G., Mason, B., and Dietrich, R. V., 1983, Mineralogy, 2nd ed.: San Francisco, W. H. Freeman.

Dietrich, R. V., 1969, Mineral tables—Hand specimen properties of 1500 minerals: New York, McGraw-Hill.

Dietrich, R. V., and Skinner, B. J., 1979, Rocks and rock minerals: New York, John Wiley.

Klein, C., and Hurlburt, C. S., Jr., 1985, Manual of mineralogy: New York, John Wiley.

Pough, F. H., 1976, A field guide to rocks and minerals, 4th ed.: Boston, Houghton Mifflin.

Walton, A. J., 1983, Three phases of matter, 2nd ed.: Oxford, Oxford University Press.

Essay

Skinner, B. J., 1976, A second Iron Age ahead?, American Sci., v. 64, p. 258–269.

Chapter 3

Blong, R. J., 1985, Volcanic hazards: A sourcebook on the effects of eruptions: Sydney, Academic Press.

Decker, R., and Decker, B., 1981, Volcanoes: San Francisco, W. H. Freeman.

Dietrich, R. V., and Skinner, B. J., 1979, Rocks and rock minerals: New York, John Wiley.

Eaton, G. P., Christiansen, R. L., Iver, H. M., Pitt, A. M., Mabey, D. R., Blank, H. R., Jr., Zietz, I., and Gettings, M. E., 1975, Magma beneath Yellowstone National Park: Science, v. 188, p. 787–796.

Ehlers, E. G., and Blatt, H., 1982, Petrology: Igneous, sedimentary and metamorphic: San Francisco, W. H. Freeman.

Hamilton, W., and Myers, W. B., 1967, The nature of batholiths: U.S. Geol. Survey Prof. Paper 554-C, p. C1–30.

Williams, H., and McBirney, A. R., 1979, Volcanology: San Francisco, Freeman, Cooper.

Essay

Simkin, T., and Fiske, R. S., 1983, Krakatau 1883. The volcanic eruption and its effects. Washington, D.C., Smithsonian Inst. Press, 464 p.

Stommel, H., and Stommel, E., 1983, Volcano weather. The story of 1816, the year without a summer. Newport, R.I., Seven Seas Press, 177 p.

Chapter 4

Blatt, H., Middleton, G. V., and Murray, R. C., 1980, Origin of sedimentary rocks, 2nd ed.: Englewood Cliffs, N.J., Prentice-Hall.

Damuth, J. E., and Kumar, N., 1975, Amazon cone: Morphology, sediments, age, and growth pattern: Geol. Soc. America Bull., v. 86, p. 863–878.

Dunbar, C. O., and Rodgers, J., 1957, Principles of stratigraphy: New York, John Wiley.

Kennett, J., 1982, Marine geology: Englewood Cliffs, N.J., Prentice-Hall.

LaPorte, L. F., 1979, Ancient environments, 2nd ed.: Englewood Cliffs, N.J., Prentice-Hall.

Reineck, H. H., and Singh, I. B., 1980, Depositional sedimentary environments, 2nd ed.: New York, Springer Verlag.

Essay

CLIMAP Project Members, 1981, Seasonal reconstructions of the Earth's surface at the last glacial maximum: Geological Society of America Map and Chart Series MC-360.

Chapter 5

Best, M. G., 1982, Igneous and metamorphic petrology: San Francisco, W. H. Freeman.

Dietrich, R. V., and Skinner, B. J., 1979, Rocks and rock minerals: New York, John Wiley.

Ehlers, E. G., and Blatt, H., 1982, Petrology. Igneous, sedimentary and metamorphic: San Francisco, W. H. Freeman.

Ernst, W. G., ed., 1975, Metamorphism and plate tectonic regimes: New York, Dowden, Hutchinson, and Ross.

Hyndman, D. W., 1985, Petrology of igneous and metamorphic rocks, 2nd ed.: New York, John Wiley.

Turner, F. J., 1981, Metamorphic petrology: Mineralogical, field and tectonic aspects, 2nd ed.: New York, McGraw-Hill.

Winkler, H. G. F., 1979, Petrogenesis of metamorphic rocks, 5th ed.: New York, Springer-Verlag.

Chapter 6

Dalrymple, G. B., and Lanphere, M. A., 1969, Potassium-argon dating. Principles, techniques and applications to geochronology: San Francisco, W. H. Freeman.

Dunbar, C. O., and Rodgers, J., 1957, Principles of stratigraphy: New York, John Wiley.

Eicher, D. C., 1976, Geologic time, 2nd ed.: Englewood Cliffs, N.J., Prentice-Hall.

Faure, G., 1986, Principles of isotope geology, 2nd ed.: New York, John Wiley.

Matthews, R. K., 1974, Dynamic stratigraphy: Englewood Cliffs, N.J., Prentice-Hall.

North American Commission on Stratigraphic Nomenclature, 1983, North American stratigraphic code: Amer. Assoc. Petroleum Geologists Bull., v. 67, p. 841–875.

Palmer, A. R., 1984, Decade of North American Geology geologic time scale: Geol. Soc. of America, Map and Chart Series MC-50.

Tarling, D. H., 1983, Paleomagnetism: London, Chapman and Hall.

Chapter 7

Birkeland, P. W., 1984, Soils and geomorphology: New York, Oxford University Press.

Carroll, Dorothy, 1970, Rock weathering: New York, Plenum Press.

Colman, S. M., and Dethier, D. P., 1986, Rates of chemical weathering of rocks and minerals: New York, Academic Press.

Gauri, K. L., 1978, The preservations of stone: Sci. American, v. 238, p. 126–136.

Hunt, C. B., 1972, Geology of soils. Their evolution, classification, and uses: San Francisco, W. H. Freeman.

Essay

Campbell, I.B., and Claridge, G.G.C., eds., 1987, Antarctica: Soils, weathering processes and environments. New York, Elsevier.

Chapter 8

Costa, J. E., and Wieczorek, G. F., 1987, Debris flows/avalanches; Process, recognition, and mitigations: Geol. Soc. America Reviews in Engineering Geology VII.

Crandell, D. R., 1971, Postglacial lahars from Mount Rainier volcano, Washington: U.S. Geol. Survey Prof. Paper 677.

Hsu, K. J., 1975, Catastrophic debris streams (sturzstroms) generated by rockfalls: Geol. Soc. America Bull., v. 86, p. 129–140.

Porter, S. C., and Orombelli, G., 1981, Alpine rockfall hazards: Am. Scientist, v. 69, p. 67–75.

Selby, M. J., 1982, Hillslope materials and processes: Oxford, Oxford University Press.

Voight, B., ed., 1978, Rockslides and avalanches, 1. Natural phenomena: New York, Elsevier.

Essay

Crandell, D.R., Miller, C.D., Glicken, H.X., Christiansen, R.L., and Newhall, C.G., 1984, Catastrophic debris-avalanche from ancestral Mount Shasta volcano, California: Geology, v. 12, p. 143–146.

Moore, J.G., 1964, Giant submarine landslides on the Hawaiian Ridge. U.S Geological Professional -Paper 501-D, p. 95–98.

Chapter 9

Dolan, R., and Goodell, H. G., 1986, Sinking cities: Am. Scientist, v. 74, p. 38–47.

Dunne, T., and Leopold, L. B., 1978, Water in environmental planning: San Francisco, W. H. Freeman.

Freeze, R. A., and Cherry, J. A., 1979, Groundwater: Englewood Cliffs, N.J., Prentice-Hall.

Heath, R. C., 1983, Basic groundwater hydrology: U.S. Geol. Survey Water-Supply Paper 2220.

Heath, R. C., 1984, Ground-water regions of the United States: U. S. Geol. Survey Water-Supply Paper 2242.

Jennings, J. N., 1983, Karst landforms: Am. Scientist, v. 71, p. 578–586.

Moore, G. W., and Sullivan, G. N., 1978, Speleology. The study of caves: Teaneck, N.J., Zephyrus Press.

Trudgill, S., 1985, Limestone geomorphology: White Plains, N.Y., Longman.

Chapter 10

Czaya, E., 1981, Rivers of the world: New York, Van Nostrand Reinhold.

Dolan, R., and Goodell, H. G., 1986, Sinking cities: Am. Scientist, v. 74, p. 38–47.

Dunne, T., and Leopold, L. B., 1978, Water in environmental planning: San Francisco, W. H. Freeman.

Freeze, R. A., and Cherry, J. A., 1979, Groundwater: Englewood Cliffs, N.J., Prentice-Hall.

Gregory, K. J., and Walling, D. E., 1973, Drainage basin form and process: London, Edward Arnold.

Leopold, L. B., Wolman, M. G., and Miller, J. P., 1964, Fluvial processes in geomorphology: San Francisco, W. H. Freeman.

Morisawa, M., 1986, Rivers; form and process: New York, John Wiley.

Schumm, S. A., ed., 1972, River morphology: Stroudsburg, Pa., Dowden, Hutchinson, & Ross.

Schumm, S. A., 1977, The fluvial system: New York, John Wiley.

Essay

Baker, V.R., and Nummendal, D., 1978, The Channeled Scabland. Washington, D.C., National Aeronautics and Space Administration.

Chapter 11

Brookfield, M. E., and Ahlbrandt, T. S., eds., 1983, Eolian sediments and processes: New York, Elsevier.

Greeley, R., and Iversen, J., 1985, Wind as a geological process: Cambridge, Cambridge University Press.

Hadley, R. F., 1967, Pediments and pediment-forming processes: Jour. Geol. Education, v. 15, p. 83–89.

Mabbutt, J. A., 1977, Desert landforms: Cambridge, M.I.T. Press.

McGinnies, W. G., Goldman, B. J., and Paylore, P., eds., 1968, Deserts of the world: Tucson, Univ. of Arizona Press. (Contains maps of all deserts.)

McKee, E. D., ed., 1979, A study of global sand seas: U.S. Geol. Survey Prof. Paper 1052.

Péwé, T. L., ed., 1981, Desert dust: origin, characteristics, and effect on man: Geol. Soc. America Spec. Paper 186, 303 p.

Pye, K., 1987, Aeolian dust and dust deposits: New York, Academic Press.

Sheridan, D., 1981, Desertification of the United States: Washington, D.C., Council on Environmental Quality.

Walker, A. S., 1982, Deserts of China: Am. Scientist, v. 70, p. 366–376.

Chapter 12

Agassiz, L., 1967, Studies on glaciers (Neuchatel, 1840), Translated and edited by A. V. Carozzi: New York, Hafner.

Chorlton, W., 1983, Ice ages: Alexandria, Va., Time-Life Books.

Flint, R. F., 1971, Glacial and Quaternary geology: New York, John Wiley.

Flint, R. F., 1974, Three theories in time: Quaternary Research, v. 4, p. 1–8.

Imbrie, J., and Imbrie, K. P., 1979, Ice ages: Solving the mystery: Short Hills, N.J., Enslow.

LaChapelle, E. R., and Post, A. S., 1971, Glacier ice: Seattle, University of Washington Press.

Prest, V. K., 1983, Canada's heritage of glacial features: Geological Survey of Canada Misc. Rept. 28, 119 p.

Sugden, D. E., and John S., 1976, Glaciers and landscape: London, Edward Arnold.

Swiss National Tourist Office, 1981, Switzerland and her glaciers: Berne, Kummerly and Frey.

Chapter 13

Bascomb, Willard, 1964, Waves and beaches. The dynamics of the ocean surface: Garden City, N.Y., Anchor Books, Doubleday.

Dolan, R. B., Godfrey, P. J., and Odum, W. E., 1973, Man's impact on the barrier islands of North Carolina: Amer. Sci., v. 61, p. 152–162.

Gross, M. G., 1986, Oceanography, 3rd ed.: Englewood Cliffs, N.J., Prentice-Hall.

Inman, D. L., 1954, Beach and nearshore processes along the southern California coast: California Division of Mines Bulletin 170, Chap. 5, p. 29–34.

Kennett, J. P., 1982, Marine geology: Englewood Cliffs, N.J., Prentice-Hall.

Shepard, F. P., and Wanless, H. R., 1971, Our changing coastlines: New York, McGraw-Hill.

Chapter 14

Board of Earth Sciences, National Academy of Sciences, 1983, The lithosphere: Washington, D.C., National Academy of Sciences.

Bolt, B. A., 1978, Earthquakes: A primer: San Francisco, W. H. Freeman.

McKenzie, D. P., 1983, The Earth's mantle: Sci. American, v. 249, no. 3, p. 114–129.

Press, F., 1975, Earthquake prediction: Sci. American, v. 232, no. 5, p. 14–23.

Walker, B., 1982, Planet Earth. Earthquake: New York, Time-Life Books.

Wyllie, P. J., 1975, The Earth's mantle: Sci. American, v. 232, no. 3, p. 50–63.

Essay

Atwater, B. F., 1987, Evidence for great Holocene earthquakes along the outer coast of Washington State: Science, v. 236, p. 942–944.

Monastersky, R., 1987, The Juan de Fuca Plate: A sticky situation: Science News, v. 132, p. 42–43.

Chapter 15

Billings, M. P., 1972, Structural geology, 3rd ed.: Englewood Cliffs, N.J., Prentice-Hall.

Burchfiel, B. C., 1983, The continental crust: Sci. American, v. 249, no. 3, p. 130–145.

Davis, G. H., 1984, Structural geology of rocks and regions: New York, John Wiley.

Park R. G., 1983, Foundations of structural geology: Glasgow, Blackie and Son Ltd.

Suppe, J., 1985, Principles of structural geology: Englewood Cliffs, N.J., Prentice-Hall.

Chapter 16

Anderson, D. L., 1971, The San Andreas Fault: Sci. American, v. 225, no. 5, p. 52–68.

Cox, A., and Hart, R. B., 1986, Plate tectonics; How it works: Palo Alto, Calif., Blackwell.

Macdonald, K. C., 1982, Mid-Ocean Ridges: Fine scale tectonic, volcanic and hydrothermal processes within the plate boundary zone: Ann. Rev. Earth and Planetary Sci., v. 10, p. 155–190.

Raymond, L. A., ed., 1984, Mélanges: Their nature, origin and significance: Geol. Soc. America, Special Paper 198.

Tarling, D. H., ed., 1978, Evolution of the Earth's crust: London and New York, Academic Press.

Wilson, J. T., ed., 1976, Continents adrift and continents aground: San Francisco, W. H. Freeman.

Wyllie, P. J., 1976, The way the Earth works: New York, John Wiley.

Chapter 17

Landscape Evolution

Chorley, R. J., Schumm, S. A., and Sugden, D. E., 1984, Geomorphology: London, Methuen.

Hart, M. G., 1986, Geomorphology: pure and applied: London, Allen and Unwin.

Hunt, C. B., 1974, Natural regions of the United States and Canada: San Francisco, W. H. Freeman and Co.

The Roof of the World

Liu Dong-sheng, ed., 1981, Geological and ecological studies of the Qinghai-Xizang Plateau: Beijing, Science Press (New York, Gordon and Breach), 2 vols.

Wager, L. R., 1937, The Arun River and the rise of the Himalaya: Geographical Journal, v. 89, p. 239–250.

Volcanoes in the Sea

Clague, D. A., and Dalrymple, G. B., 1987, The Hawaiian-Emperor volcanic chain. Part I. Geologic evolution, in Decker, R., T. L. Wright, and P. H. Stauffer, eds., Volcanism in Hawaii: U.S. Geol. Survey Prof. Paper 1350, p. 5–54.

Macdonald, G. A., Abbott, A. T., and Peterson, F. L., 1983, Volcanoes in the sea: the geology of Hawaii, 2nd ed: Honolulu, Univ. Hawaii Press.

Glaciated Landscapes of Northeastern North America

Bell, M., and Laine, E. P., 1985, Erosion of the Laurentide region of North America by glacial and glaciofluvial processes: Quaternary Research, v. 23, p. 154–174.

Sugden, D. E., 1978, Glacial erosion by the Laurentide Ice Sheet: Journal of Glaciology, v. 20, p. 367–391.

Chapter 18

Bates, R. L., 1960, Geology of the industrial minerals and rocks: New York, Harper & Row.

Brobst, D. A., and Pratt, W. P., eds., 1973, United States mineral resources: U.S. Geol. Survey Prof. Paper 820.

Darmstadter, J., Landsberg, H. H., and Morton, H. C., 1983, Energy today and tomorrow: Englewood Cliffs, N.J., Prentice-Hall.

Jensen, M. L., and Bateman, A. M., 1981, Economic mineral deposits, revised printing: New York, John Wiley.

Perry, H., 1983, Coal in the United States: A status report: Science, v. 222, p. 377–384.

Sawkins, F. J., 1984, Metal deposits in relation to plate tectonics: Berlin, Springer-Verlag.

Skinner, B. J., ed., 1981, Economic Geology Seventy-Fifth anniversary volume, 1905–1980: El Paso, Tex., The Economic Geology Publishing Co.

Skinner, B. J., 1986, Earth resources, 3rd ed.: Englewood Cliffs, N.J., Prentice-Hall.

Chapter 19

Carr, M. H., 1981, The surface of Mars: New Haven, Conn., Yale University Press.

Carr, M H., ed., 1984, The geology of the terrestrial planets: NASA SP-469, Washington, D.C. National Aeronautics and Space Administration.

Greeley, R., 1985, Planetary landscapes: London, Allen and Unwin.

Head, J. W., and Solomon, G. C., 1981, Tectonic evolution of the terrestrial planets: Science, vol. 213, p. 62–76.

Hunten, D. M., Colin, L., Donahue, T. M., and Moroz, V. I., eds., 1983, Venus: Tucson, University of Arizona Press.

Morrison, D., ed., 1982, Satellites of Jupiter: Tucson, University of Arizona Press.

Murray, B. M., Malin, C., and Greeley, R., 1981, Earth-like planets: New York, W. H. Freeman.

Scientific American, 1983, The Planets: New York, W. H. Freeman.

Photo Credits

Preface: John S. Shelton.

Part 1 Opener: S. C. Porter.

Chapter 1
Opener: Courtesy ESA. Figure 1.9: S. C. Porter. Figure 1.16: Landsat/USGS/John S. Shelton. Figure 1.19: K. R. Gill. Figure B1.1: Robert S. Andrews.

Chapter 2
Opener: William Sacco. Figure 2.11: William Sacco. Figure 2.12: William Sacco. Figure 2.13 and 2.14: William Sacco. Figure 2.15: William Sacco. Figure 2.16a: William Sacco. Figure 2.17: William Sacco. Figure 2.19: Brian J. Skinner.

Chapter 3
Opener: Brian J. Skinner. Figure 3.2: R. S. Fiske. Figure 3.13: Craig Johnson. Figure 3.15: Brian J. Skinner. Figure 3.17: William Sacco. Figure 3.18: Brian J. Skinner. Figure 3.19: S. C. Porter. Figure 3.20: R. V. Fisher. Figure 3.21: William Sacco. Figures 3.22 and 3.23: S. C. Porter. Figure 3.24: Bruce Marsh. Figure 3.25: S. C. Porter. Figure 3.26: Lyn Topinka/USGS/David A. Johnson Cascades Volcano Observatory. Figure 3.27: S. C. Porter. Figure 3.28a: Agnus/ WHOI. Figure 3.28b: Fred Grassle/WHOI. Figure 3.31: John S. Shelton. Figure 3.33: Bruce K. Goodwin.

Chapter 4
Opener: S. C. Porter. Figure 4.1: Josef Muench. Figures 4.2a and b: John S. Shelton. Figure 4.2c: S. C. Porter. Figure 4.2d: Brian J. Skinner. Figure 4.3: S. C. Porter. Figure 4.4: Richard J. Stewart. Figures 4.5, 4.6, 4.7, and 4.8: S. C. Porter. Figure 4.9: I. Aarseth. Figure 4.10: Betty Crowell/ Faraway Places. Figure 4.11a: S. C. Porter. Figure 4.11b: Joanne Bourgeois. Figure 4.12a: Josef Muench. Figure 4.12b: S. C. Porter. Figure 4.13: Tom Bean. Figure 4.14: S. C. Porter. Figure 4.15: Gary Ladd Photography. Figure 4.18: David H. Krinsley. Figure 4.19: Bob Krist/Leo de Wys. Figure 4.20: S. C. Porter. Figure 4.22: Deep Sea Drilling Project/Scripps Institute of Oceanography.

Chapter 5
Opener: Brian J. Skinner. Figures 5.2 and 5.3: Brian J. Skinner. Figure 5.5: Brenda Sirois. Figures 5.7a and c: Brian J.

Skinner. Figure 5.7b: E. R. Degginger/Earth Scenes. Figures 5.8 and 5.9: Brian J. Skinner. Figure 5.10: Craig Johnson. Figure 5.11: Brian J. Skinner. Figure 5.12: Simon Hanmer. Figure B5.1: Craig Johnson.

Chapter 6
Opener: S. C. Porter. Figure 6.1: John S. Shelton. Figure B6.1: Robert C. Bostrom/Geological Research Corp.

Part 2 Opener: S. C. Porter.

Chapter 7
Opener: F. C. Ugolini. Figure 7.1: William E. Ferguson. Figure 7.3: S. C. Porter. Figure 7.4: Maurice and Sally Landre/ Photo Researchers. Figures 7.5, 7.6, 7.7, and 7.8: S. C. Porter. Figure 7.9: Robert S. Anderson. Figure 7.11: Barrie Rokeach. Figure 7.14a: F. C. Ugolini. Figures 7.14b, 7.14c, 7.16, and B7.1: S. C. Porter.

Chapter 8
Opener: S. C. Porter. Figure 8.3: John S. Shelton. Figure 8.4: W. R. Hausen/USGS. Figures 8.5, 8.6, 8.7, and 8.9: S. C. Porter. Figure 8.11: G. R. Roberts. Figures 8.12 and 8.13: S. C. Porter. Figure 8.15: B. Mears. Figure 8.17: Chalmers M. Clapperton. Figure B8.1: C. D. Miller.

Chapter 9
Opener: S. C. Porter. Figure 9.16: S. C. Porter. Figure 9.17: Gary Ladd Photography. Figure 9.18: Josef Muench. Figure 9.19: David Hiser/Photographers Aspen. Figure 9.21: Sam C. Pierson, Jr./Photo Researchers. Figure 9.22: Alex Soto. Figure B9.1: Gary Millburn/Tom Stack and Associates.

Chapter 10
Opener: Courtesy NASA. Figure 10.1: Geopic/Earth Satellite Corp. Figure 10.3: Thomas Dunne. Figure 10.9: Courtesy NASA. Figure 10.13: S. C. Porter. Figure 10.19: John S. Shelton. Figure 10.20: Geopic/Earth Satellite Corp. Figure B10.1: Courtesy NASA.

Chapter 11
Opener: David Hiser/Photographers Aspen. Figure 11.2: Gary Ladd Photography. Figure 11.3: Gordon Wiltsie/Bruce Coleman. Figure 11.4: Peter Fronk/Click, Chicago. Figure

11.5: G. R. Roberts. Figure 11.6: S. C. Porter. Figures 11.7 and 11.8: John Adams. Figure 11.9: E. R. Degginger. Figure 11.10, 11.12, and 11.13: S. C. Porter. Figure 11.14: M. J. Grolier/USGS. Figure 11.16: Gary Ladd Photography. Figure 11.18: Galen Rowell/High and Wild Photography. Figure 11.19: Eric Cheney. Figure 11.21: Courtesy NASA. Figures 11.22 and 11.23: S. C. Porter. Figure 11.25: Victor Engelbert/Photo Researchers. Figures B11.1 and B11.2: S. C. Porter. Figure B11.3: Courtesy NASA.

Chapter 12
Opener: S. C. Porter. Figure 12.1a: S. C. Porter. Figures 12.1b and c: Austin S. Post/USGS. Figure 12.1d: Friedrich Röthlisberger. Figure 12.9: Austin S. Post/USGS. Figure 12.10: Josef Muench. Figures 12.11, 12.12, 12.13, 12.14, 12.15, 12.16, 12.17, and 12.18: S. C. Porter.

Chapter 13
Opener: G. R. Roberts. Figure 13.3: G. R. Roberts. Figures 13.6 and 13.8a: John S. Shelton. Figure 13.8b: Nicholas Devore III/Photographers Aspen. Figure 13.10: S. C. Porter. Figure 13.11: Ray Manley/Shostal Associates. Figure 13.12: G. R. Roberts. Figures 13.14 and 13.17: Earth Satellite Corp. Figure 13.18a: Matt Bradley/Tom Stack and Associates. Figures 13.20a and b: Nicholas Devore III/Bruce Coleman. Figure 13.20c: Michael Friedel/Woodfin Camp. Figure 13.22: Arthur L. Bloom. Figure 13.23: P. Dunwiddie. Figure 13.24: Photri. Figure B13.1: Courtesy Advanced Satellite Productions, Inc. and Earth Observation Satellite Company.

Part 3 Opener: David Muench.

Chapter 14
Opener: Stan Wayman/Life Magazine. Figure 14.2: Joe Caravetta/Gamma-Liaison. Figure 14.10: Asahi Shimbun. Figure 14.15: Brian J. Skinner. Figure 14.18: A. M. Dziewonski and D. L. Anderson, *American Scientist*, v. 72, pp. 483–494, 1984. Figure 14.24: Ken Ewing.

Chapter 15
Opener: G. R. Roberts. Figure 15.4: R. S. Fiske. Figure 15.5: John S. Shelton. Figure 15.7b: Brian J. Skinner. Figure 15.10a: Earth Satellite Corp. Figure 15.10b: Peter Arnold. Figure 15.12a: John S. Shelton. Figure 15.14a: Greg Davis. Figure 15.14b: William E. Ferguson. Figure 15.15: Barrie Rokeach. Figure 15.18c: B. Caulfield. Figure 15.27: John S. Shelton.

Chapter 16
Opener: G. R. Roberts. Figure 16.15: Warren Hamilton.

Chapter 17
Opener: S. C. Porter. Figure 17.1: David Hiser/Photographers Aspen. Figure 17.5: Courtesy NASA. Figure B17.1: Brian J. Skinner.

Part 4 Opener: Brian J. Skinner.

Chapter 18
Opener: William Sacco. Figure 18.3: Illinois Geological Survey. Figure 18.14: Dudley Foster/WHOI. Figure 18.15: Brian J. Skinner. Figure 18.16: John Allcock. Figures 18.19, 18.21: Brian J. Skinner. Figure 18.23: William Sacco. Figure 18.24a: J. Bahr. Figure 18.24b: H. Murray. Figure 18.26: Jeff Rotman/Peter Arnold. Figures 18.27 and B18.1: Brian J. Skinner.

Part 5 Opener: Art Wolfe/Aperture.

Chapter 19
Opener: Courtesy NASA. Figure 19.2: John S. Shelton. Figures 19.4 and 19.5: Courtesy NASA. Figure 19.7: U.S. Department of the Interior, Astrogeology Department. Figures 19.9 and 19.10: Courtesy NASA. Figure 19.11a: Brian J. Skinner. Figure 19.11b: John A. Wood. Figures 19.12, 19.14, 19.15, and 19.16: Courtesy NASA. Figures 19.17, 19.18, 19.19, 19.20, and 19.21: Courtesy Space Photography Laboratory, Arizona State University.

Index

Numbers of pages on which terms are defined are in **boldface italic.**
Asterisks indicate illustrations.
A indicates page is Appendices.